MOLECULAR CLUSTERS: A BRIDGE TO SOLID-STATE CHEMISTRY

Despite the fact that clusters can be viewed as solids at the nanoscale, molecular cluster chemistry and solid-state chemistry traditionally have been considered as separate topics. Such treatment makes it difficult to appreciate commonalities of structure and bonding. This book connects the electronic structure models of small clusters with conceptually analogous models of solid-state materials, thereby creating a bridge. The approach also establishes boundary conditions on the electronic structure of nanoclusters: species that lie in-between the two. Although the focus is on clusters, sufficient attention is paid to solid-state compounds at each stage of the development to establish the interrelationship between the two topics. Comprehensive coverage of cluster types by composition, size, and ligation is provided, as is a synopsis of selected research.

Written in an accessible style with numerous illustrations, exercises, problems and solutions to aid comprehension, this book is suitable for graduate students and researchers in inorganic chemistry, physical chemistry, materials science, and condensed matter physics.

THOMAS P. FEHLNER is Emeritus Grace Rupley Professor of Chemistry at the University of Notre Dame, Indiana. He received his Ph.D. in physical chemistry from the Johns Hopkins University, Maryland, in 1963. His current research interests include the systematic chemistry of metalloboranes and the application of mixed valence metal chemistry to molecular electronics.

JEAN-FRANÇOIS HALET is a Director of Research in the chemical sciences laboratory at the *Centre National de la Recherche Scientifique* and the *Université de Rennes 1*, France. He received his Ph.D. in physical chemistry in 1984 from the *Université Pierre-et-Marie Curie*, Paris. His research interests focus on understanding the chemical bond in transition metal inorganic chemistry using different quantum chemical computational tools. In particular, he attempts to analyze structural similarities between molecular and solid-state compounds.

JEAN-YVES SAILLARD is a Professor of Chemistry in the chemical sciences laboratory at the *Université de Rennes 1*, France, and in the *Institut Universitaire de France*. He received his Ph.D. in chemistry in 1974 from the *Université de Rennes 1*, France. His current research interests include the rationalization of structures, and the reactivity and physical properties of inorganic compounds, particularly organometallic complexes, clusters and solid-state compounds.

MOLECULAR CLUSTERS

A Bridge to Solid-State Chemistry

THOMAS P. FEHLNER
University of Notre Dame, Indiana

JEAN-FRANÇOIS HALET
JEAN-YVES SAILLARD
CNRS-Université de Rennes, France

CAMBRIDGE UNIVERSITY PRESS
Cambridge, New York, Melbourne, Madrid, Cape Town, Singapore, São Paulo

Cambridge University Press
The Edinburgh Building, Cambridge CB2 8RU, UK

Published in the United States of America by Cambridge University Press, New York

www.cambridge.org
Information on this title: www.cambridge.org/9780521852364

© T. P. Fehlner, J.-F. Halet, J.-Y. Saillard 2007

This publication is in copyright. Subject to statutory exception
and to the provisions of relevant collective licensing agreements,
no reproduction of any part may take place without
the written permission of Cambridge University Press.

First published 2007

Printed in the United Kingdom at the University Press, Cambridge

A catalog record for this publication is available from the British Library

ISBN-13 978-0-521-85236-4 hardback

Cambridge University Press has no responsibility for the persistence or accuracy of URLs for external or third-party internet websites referred to in this publication, and does not guarantee that any content on such websites is, or will remain, accurate or appropriate.

Contents

Preface		page ix
1	Introduction	1
	1.1 Molecular orbitals without the nasty bits	2
	1.2 Three-center bonds	18
	1.3 An orbital explanation of electron-counting rules	22
	Problems	29
	Additional reading	32
2	Main-group clusters: geometric and electronic structure	33
	2.1 Definition of a cluster	33
	2.2 Three-connect clusters	34
	2.3 Four-connect and higher clusters	39
	2.4 The Wade–Mingos electron-counting rule	42
	2.5 Closed-cluster composition and structure	43
	2.6 Open *nido*-clusters	45
	2.7 Open *arachno*-clusters	48
	2.8 The *closo*-, *nido*-, *arachno*-borane structural paradigm	51
	2.9 Localized bonds in clusters	53
	2.10 Clusters with nuclearity greater than 12	59
	2.11 Ligand-free clusters	64
	2.12 When the rules fail	69
	Problems	79
	Additional reading	81
3	Transition-metal clusters: geometric and electronic structure	85
	3.1 Three-connect clusters	85
	3.2 Small four-connect metal-carbonyl clusters and the $14n + 2$ rule	90
	3.3 Variations characteristic of metal clusters	96
	3.4 Naked clusters	120

3.5	High-nuclearity clusters with internal atoms	122
3.6	Nanoscale particles	129
	Problems	132
	Additional reading	135

4 Isolobal relationships between main-group and transition-metal fragments. Connections to organometallic chemistry ... 139

4.1	Isolobal main-group and transition-metal fragments	139
4.2	Metal variation with fixed ancillary-ligand set	148
4.3	Metal–ligand complex vs. heteronuclear cluster	150
4.4	p-Block–d-block metal complexes	152
4.5	Carborane analogs of cyclopentadienyl–metal complexes	155
	Problems	161
	Additional reading	163

5 Main-group–transition-metal clusters ... 165

5.1	Isolobal analogs of p-block and d-block clusters	165
5.2	Rule-breakers	178
	Problems	201
	Additional reading	203

6 Transition to the solid state ... 205

6.1	Cluster molecules with extended bonding networks	205
6.2	Outline of the electronic-structure solution in a one-dimensional world	210
6.3	Complex extended systems	236
6.4	From the bulk to surfaces and clusters	246
	Problems	253
	Additional reading	255

7 From molecules to extended solids ... 257

7.1	From a single atom to an infinite solid: the example of C	258
7.2	B clusters in solids: connections with molecular boranes	282
7.3	Molecular transition-metal complexes in solids	287
7.4	Molecular vs. solid-state condensed octahedral transition-metal chalcogenide clusters: rule-breakers again	291
7.5	Cubic clusters in solids	296
	Problems	299
	Additional reading	301

8 Inter-conversion of clusters and solid-state materials ... 303

8.1	Cluster precursors to new solid-state phases	303
8.2	Solid-state phases to molecular clusters	309
8.3	Clusters to materials	314
8.4	A final problem	319
8.5	Conclusion	321

Appendix: Fundamental concepts: a concise review	323
A1.1 Elements	324
A1.2 Atomic properties	324
A1.3 Homoatomic substances	330
A1.4 Heteroatomic substances	337
Problem Answers	349
Chapter 1	349
Chapter 2	352
Chapter 3	357
Chapter 4	360
Chapter 5	362
Chapter 6	363
Chapter 7	366
References	369
Index	371

Preface

Who, what, where, why, when and how – the elementary prescription for a news squib is also appropriate for a preface.

Who? The book is intended primarily as a text for advanced undergraduates and graduate students. It can also serve the needs of research workers in the wide area of nanochemistry, as molecular clusters and extended solid-state materials constitute the structural "bookends" of nanoparticles: species that are not large enough to be treated with solid-state concepts but too large to follow the simple rules of molecular clusters. Those interested in a wide-ranging introduction to models of electronic structure applicable to delocalized, three-dimensional systems will also find it useful.

What? This text circumscribes a non-traditional area of inorganic chemistry. The focus is on a class of compound that exhibits cluster bonding. Emphasis is on connections between the problems of small molecular clusters, where the vast majority of atoms are found at the surface, to large crystals, where most atoms are found in the bulk. A review of bonding in molecular compounds (Chapter 1) is followed by the fundamentals of cluster bonding in p-block clusters (Chapter 2) and transition-metal clusters (Chapter 3). After making connections with organometallic chemistry (Chapter 4), mixed p–d-block clusters are developed (Chapter 5). A bonding model for periodic extended structures (Chapter 6) is developed in the style of Chapter 1. Chapter 7 then illustrates some of the similarities and differences between the bonding of clusters and related solid-state structures. The finale (Chapter 8) abstracts a selection of recent research to illustrate real connections between clusters and solid-state systems.

Where? Time will tell where this text will fit in the curricula of relevant departments. Presently, there is no common course in chemistry that it could serve as a primary text. However, since the mid-1990s a first-year graduate course along the lines of the material contained within this text has been offered at Notre Dame.

Drafts of the present text have been used twice in Chem 616 "Solid-state and cluster chemistry" further developing the material herein.

Why? In inorganic texts solid-state chemistry appears ancillary to the main emphasis of molecular chemistry. The title of one first-year chemistry text proclaims chemistry a "molecular science." Clusters fare less well. Service on standard inorganic exam committees reveals many teachers of inorganic chemistry who are uncomfortable with both solid-state and cluster chemistries. The conceptual barrier involves the delocalized bonding networks required for an understanding of electronic structure. This book attempts to smooth the transition between simple localized bonding models and the delocalized ones by using clusters to bridge molecular and solid-state chemistries. From the localized two-center bonds of three-connect clusters to the band structure of metals, cluster bonding provides a unifying paradigm.

When? Both solid-state chemistry and main-group cluster chemistry can be considered mature areas. Transition-metal cluster chemistry is of more recent origin; however, it too has been well defined in a number of edited works. There are texts on solid-state chemistry and one on cluster chemistry but there is no text that exploits connections between the two using simple models. Simplified models are the tools of the working chemist but the power of the simple models within an area also creates barriers to inter-area understanding. The time is right to show that the molecular and solid-state boundary conditions on clusters reveals the exciting problems of structure and properties that remain to be discovered in the region lying between small clusters and bulk materials – nanoparticles.

How? The text is representative, not comprehensive, and we attempt to balance simplification and detail. Additional sources are gathered at the end of each chapter but this list is far from complete. Literature is cited when it is felt the reader might benefit from following the original arguments or when a more comprehensive monograph provides access to the details of a given topic or area. For non-chemists the Appendix contains an outline of the fundamental concepts of chemistry prerequisite to the body of the text. With the exception of Chapter 8, each chapter includes worked exercises and homework problems at the end of the chapter. A number of problems are drawn from the research literature to illustrate the approach advocated. They are challenging by design and a few of the solutions are not published.

Caveats. Those familiar with cluster chemistry will mark the absence of cluster synthesis, framework dynamics and reactivity. Considerable information exists and these topics for selected cluster types are well developed in cluster reviews and edited volumes. However, our focus on electronic structure is deliberate. We wished to compare and contrast geometric and electronic structure across the large sweep of element composition and cluster size up to and including bulk materials. To keep the book of manageable size relative to a typical one-semester advanced course yet still bridge the disparate areas encompassed in "cluster chemistry" we chose a focus consistent with our scope. The approach is unabashedly qualitative but we hope the reader finds the material an *hors d'oeuvre* leading to more satisfying *entrées* in the literature of the many broad topics touched upon.

Thank yous. First, we owe an intellectual debt to the masters of both cluster and solid-state chemistries from whom the models presented arose. Some are acknowledged by name in the text where appropriate. Many others, unacknowledged, created and described the magnificent bodies of chemistry, experimental and theoretical, which constitute the foundations of this work. That is the nature of science – most of us are ants piling our grains of chemistry so that those with longer sight can see even farther.

One of us (TF) held a Leverhulme Visiting Professorship at the University of Bath in the spring of 2004 during which the writing of this book was begun. Dr. Andrew S. Weller made this happen. Many thanks to Andy and his department for both a productive and pleasant sojourn in England. We are also grateful to the CNRS and the NSF for the support of our independent research as well as a joint project of cooperative research which facilitated our writing efforts. Our thanks go to Nancy Fehlner who read the entire manuscript in its final form as well as Dr. Mouna Ben Yahia who kindly performed some theoretical calculations to check qualitative (sometimes not) ideas we had in mind. Still, it is highly unlikely we have produced an error-free book – entropy rules – and the errors that remain are ours.

1
Introduction

A modern chemist has access to good computational methods that generate numerically useful information on molecules, e.g., energy, geometry and vibrational frequencies. But we also have a collection of models based on orbital ideas incorporating concepts of symmetry, overlap and electronegativity. In this text we focus on the latter as these ideas have been a huge aid in understanding the connections between stoichiometry, geometry and electronic structure. The connections can be as simple as an electron count yielding user-friendly "rules." Our problem here, the electronic structure of a cluster or a more extended structure of the type encountered in solid-state chemistry, requires the application of models beyond those reviewed in the Appendix. Models are like tools – they permit us to disassemble and assemble the electronic structure of molecules. For each problem we choose a model that will accomplish the task with minimum effort and maximum understanding. Just as one would not use a screwdriver to remove a hex nut, so too we cannot use highly localized models to usefully describe the electronic structures of many clusters and extended bonding systems. We must use a method that is capable of producing a sensible solution as well as one that is sufficiently versatile to treat both the bonding in small clusters and bulk materials.

The proven method we will use is one that generates solutions based on the orbitals and electrons that the atoms or molecular fragments bring to the problem. For molecules, it is the linear combination of atomic orbitals molecular orbital (LCAO-MO) method. Hence, as a prelude to subsequent chapters on clusters and extended structures, a qualitative review of the application of this model to simple molecules is presented. In all cases the intrinsically complex results are pruned to the essentials according to the guidance of several prize-winning chemists. In certain cases the ultimate simplification generates the familiar, easy-to-apply and handy electron-counting rules. We assume the reader has a strong background in the descriptive chemistry that is outlined in the Appendix. The Appendix or an inorganic text should be used as needed to refresh the memory of the chemical facts as well as the popular localized descriptions of the bonding of simple molecules. If more is needed, general texts of inorganic or organometallic chemistry should be consulted.

1.1 Molecular orbitals without the nasty bits

The Appendix includes a few examples of non-cluster systems where the intrinsic limitations of the two-center–two-electron bond are revealed. By and large, however, the model is a good one. Many cluster systems demand a more flexible model to explain even less complex aspects like stoichiometry and geometry. The model that chemists have adopted is that of molecular orbitals with a Hoffmann-style approach, i.e., an approach in which the essence of the problem is identified with a small subset of molecular orbitals describing the system. This conceptual, essentially qualitative, approach has become the language of modern experimental chemistry. In the following, some of the essential aspects of the model are described utilizing experimental results of valence-level photoelectron spectroscopy for selected empirical support. In essence, the ionization energies of molecules are used in the same way as the ionization energies of atoms are used to justify the H atom model for the electronic structure of atoms.

In Section A1.3 the united atom model for H_2 is described. As a consequence, molecules may be viewed as "atoms" that contain multiple nuclei at different positions in space. Molecular orbitals (MOs) are thus "atomic orbitals" (AOs) distorted by a complex "nucleus." These modified "atomic orbitals" can be correlated with the real atomic orbitals of the united atom as well as with linear combinations of the atomic orbitals of the separated atoms from which the molecule is constructed. Once one goes beyond simple diatomics, the united atom model rapidly loses its usefulness; however, the linear combination of atomic orbitals approach does not. It constitutes a productive approach to the generation of MOs. There are several good texts that present molecular orbital ideas for the experimental chemist, e.g., Albright, Burdett and Whangbo, and here a pragmatic approach to the utilization of MO models is presented. The examples and exercises given will produce sufficient familiarity that application of the approach to clusters and extended systems in successive chapters will produce understanding rather than confusion.

1.1.1 The H_2 model

Let's begin with H_2. As shown in Figure 1.1, the combination of two H 1s orbitals yields two molecular orbitals – one bonding and one antibonding. For an electron in the bonding combination, additional electron density is placed between the nuclei (more than would be present if two non-bonding H atoms were placed at the

1.1 Molecular orbitals without the nasty bits

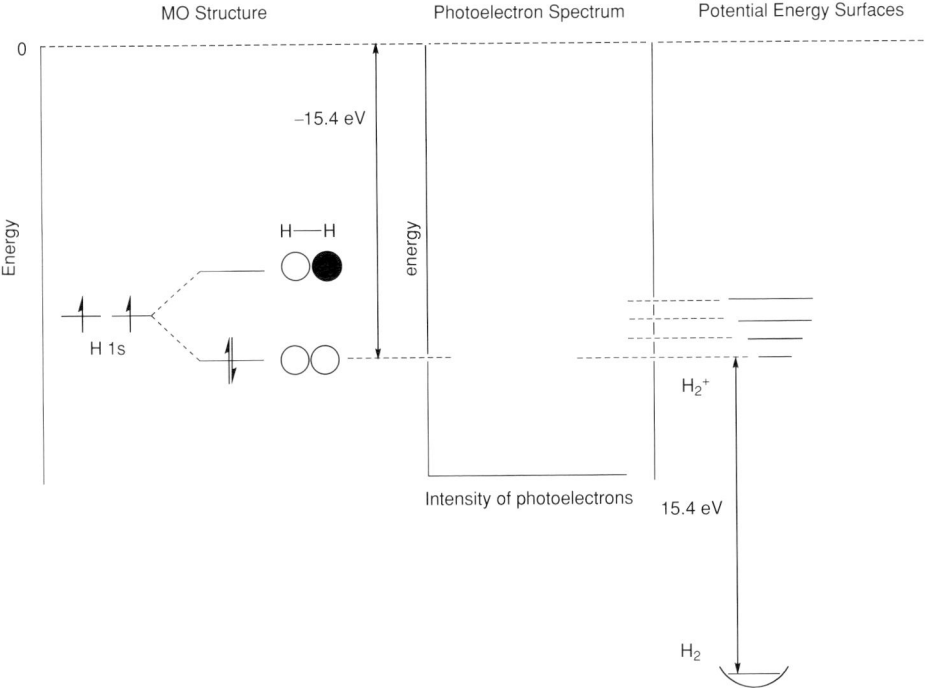

Figure 1.1

same positions). For an electron in the antibonding combination, electron density is removed from the region between the nuclei relative to two non-interacting H atoms separated by the same distance. In the former case, the nuclei are better shielded from each other and the electrons between the nuclei experience the attraction of two nuclei. This net attractive interaction is just balanced at the equilibrium internuclear distance of H_2 by electron–electron and nucleus–nucleus repulsions. The energy of the bonding MO is lower than that of the AOs for the non-interacting atoms. The opposite is true of the antibonding orbital. The ground state is represented by placing two electrons of opposite spin in the bonding MO. In this one-electron MO approach electrons are added after MO formation in the manner of the hydrogen atom model of heavier atoms.

Ionization of H_2 can be described as removing an electron from the bonding MO and Koopmans' theorem states that the ionization energy $IE = -\varepsilon_{MO}$. The MO model suggests that $IE(H_2)$ should be larger than $IE(H) = 13.6\,eV$. As shown by its photoelectron spectrum, $IE(H_2) = 15.4\,eV$. The photoelectron spectrum gives us additional information about the nature of the occupied molecular orbital from the fine structure observed in the photoelectron band. This fine structure corresponds to vibrational excitation of the molecular ion H_2^+ and reports on the role of the electron

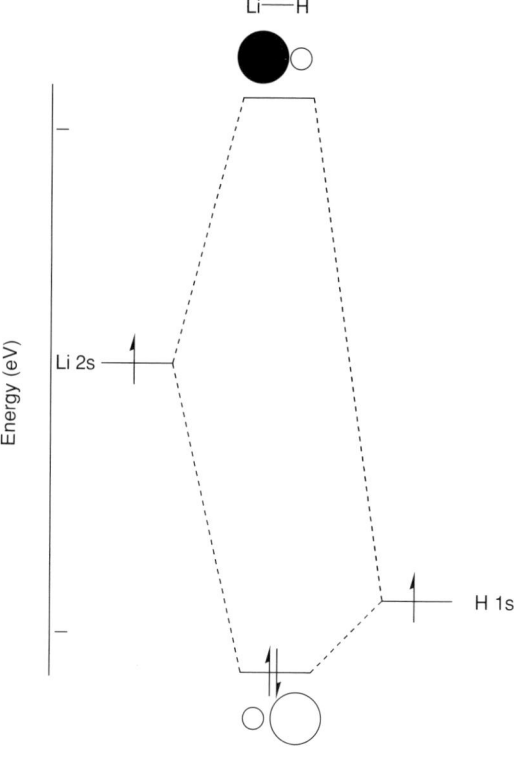

Figure 1.2

removed in the bonding of the molecule H_2. The H–H stretching frequency in the molecule is 4820 cm^{-1} whereas in the molecular ion it is 2260 cm^{-1}. Removing the electron generates a more loosely bound ion; therefore the electron removed was a bonding electron and the MO from which it was removed is H–H bonding. The fact that a long vibrational progression is observed in the ion confirms that the structure (H–H distance) is considerably different in the ion vs. the molecule. The qualitative agreement of experiment and model suggest that even though MOs have no more real existence than AOs, they serve as a powerful tool for discussing electronic structure in meaningful terms.

We can use the hypothetical molecule LiH to gain an idea of how electronegativity enters into the MO model. In Figure 1.2 the calculated MOs of LiH are illustrated. Note that there are still two MOs as we have not included the 2p functions of Li for simplicity. One is bonding and filled and one is antibonding and empty. However, neither is symmetrical relative to the amplitude of the MO at the two different atomic centers. The bonding MO contains a higher proportion of H character, the more electronegative atom, whereas electropositive Li has a higher amplitude in the empty antibonding MO. The distribution of electron density in the molecule is given by the square of the absolute magnitude of the wave function (empty orbitals do not contribute). Hence, the H end of the molecule will be more electron rich than the Li end and the molecule will have a dipole moment. How much? Mulliken devised a simple measure of charge distribution by assigning charge to a given atom center according to its AO contributions to filled MOs. Called a Mulliken population analysis, it provides a relative measure of charge distribution. For LiH the Mulliken charges are 0.46 on Li and −0.46 on H; however, quantitative agreement with a measured dipole moment cannot be expected.

Suppose we consider the excitation of an electron from the bonding to the antibonding MO. The net result is to transfer electron density from the H end to the Li end of LiH thereby reducing the strength of the Li–H interaction. The effects of differing electronegativities in polyatomic molecules on MO characters are less easily anticipated; however, the changes result from the same factors illustrated by LiH.

Exercise 1.1. Sketch out the MO energies and wavefunction for the molecular ion [HeH]$^+$. Do you expect the H atom to have positive or negative character relative to He? Does the MO picture agree with your intuition?

Answer. The He 1s function is at lower energy that the H 1s function; hence, the qualitative MO diagram is that of Figure 1.2 with He in the position of H and H in the position of Li. In the ground state, H shares less of the bond pair than He and, hence, more of the positive charge of the molecule.

1.1.2 Extension of the H$_2$ model to p-block elements

H$_2$ only requires 1s functions for an MO description. Next we have to consider atoms with p functions as well. So let's look at B$_2$. The results of a Fenske–Hall MO calculation on B$_2$ are shown in Figure 1.3 and Table 1.1. For clarity, the dashed correlation lines to each of the manifolds are only shown for one B atom each. Now, the MOs are represented by energies and linear combinations of the 2s and 2p functions of the two B atoms. The 2s and 2p functions are called the basis functions and the number of basis functions in any problem equals the number of MOs, i.e., here are four basis functions on each B atom so there are eight MOs. In the absence of any symmetry, each MO can contain a contribution from every basis function. But note that there are four MOs (counting from the lowest energy MO 3 and 5, each doubly degenerate) that contain only 2p$_x$ and 2p$_y$ functions (the z axis is the B–B axis). The reason is that functions with σ symmetry relative to the B–B axis (no change in sign on rotation about the B–B axis) are orthogonal to functions with π symmetry (one change in sign on 180° rotation about the B–B axis). As they do not mix, bonding MOs 3 and antibonding MOs 5 can be generated by a

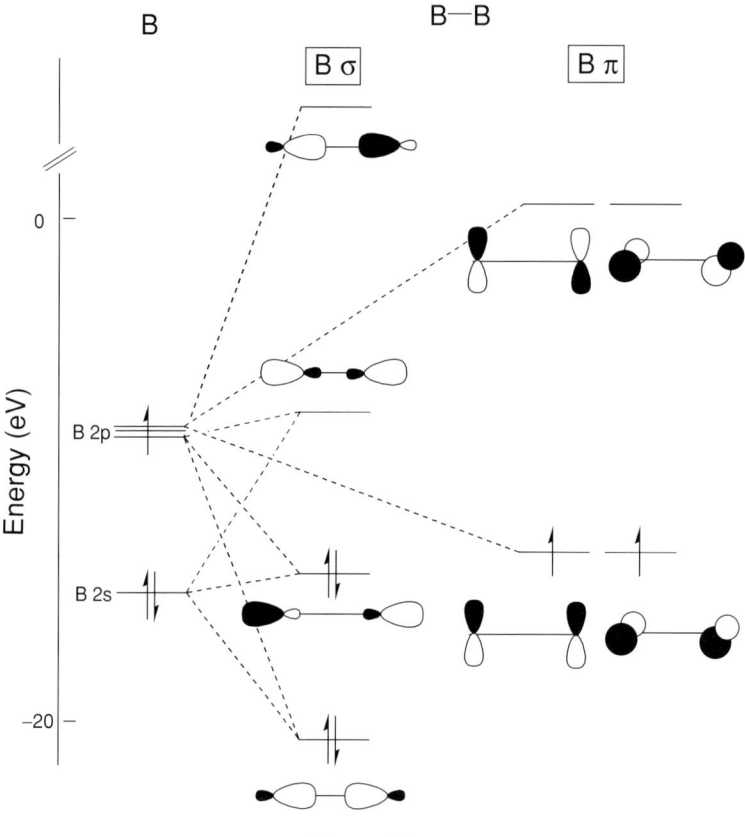

Figure 1.3

2×2 scheme just like the one used for H_2 above. Make yourself a little structure diagram with a coordinate system taking the molecular axis as the z axis. Then look at the table of eigenvectors (signs only) with AO %s for each eigenvector. Mark out the π MOs and draw MO pictures showing AO contributions to verify those in Figure 1.3. The eigenvectors give you the signs and the AO %s give the relative size of the AOs you draw. These are easy as each atom only contributes one AO.

The remaining four MOs, 1, 2, 4 and 6, formed from the 2s and $2p_z$ functions of σ symmetry require a 4×4 scheme. Take a closer look at these four orbitals and note in Table 1.1 that the lower ones have larger 2s character and the higher ones larger 2p character. Draw a picture of the lowest energy one. You should find it of predominantly 2s character and symmetric (no nodes). The 2s AO energy is lower than the 2p AO energy so the lowest energy orbital will be mainly of 2s character. Thus, the highest energy MO will have large $2p_z$ character. Check it the same way. It has a more complex 2s and 2p mixture so at the side add a 2s orbital to a 2p orbital and then subtract the two. What do you get? Yes, you get something that looks like

Table 1.1. *Eigenvalues and eigenvectors for* B_2 *from a Fenske–Hall calculation*

MO[a]	1	2	3	4
Energies (eV)	−21.6	−13.3	−12.1	−12.1
1B 2s (%)	+(40)	−(25)	(0)	(0)
1B $2p_x$ (%)	(0)	(0)	+(50)	(0)
1B $2p_y$ (%)	(0)	(0)	(0)	+(50)
1B $2p_z$ (%)	+(11)	+(25)	(0)	(0)
2B 2s (%)	+(40)	+(25)	(0)	(0)
2B $2p_x$ (%)	(0)	(0)	+(50)	(0)
2B $2p_y$ (%)	(0)	(0)	(0)	+(50)
2B $2p_z$ (%)	−(11)	+(25)	(0)	(0)
MO[a]	5	6	7	8
Energies	−7.1	0.7	0.7	80
1B 2s (%)	+(10)	(0)	(0)	−(25)
1B $2p_x$ (%)	(0)	+(50)	(0)	(0)
1B $2p_y$ (%)	(0)	(0)	+(50)	(0)
1B $2p_z$ (%)	−(40)	(0)	(0)	−(25)
2B 2s (%)	+(10)	(0)	(0)	+(25)
2B $2p_x$ (%)	(0)	−(50)	(0)	(0)
2B $2p_y$ (%)	(0)	(0)	−(50)	(0)
2B $2p_z$ (%)	+(40)	(0)	(0)	−(25)

[a] To simplify the table only the sign of the AO coefficients in the MOs are given along with the % characters in parentheses.

a hybridized orbital. But this happens automatically, when you turn the crank of the computer. It's not something you somehow should know. The highest energy MO is strongly B–B antibonding and has three nodes. Look at the remaining two MOs. They have one and two nodes, respectively, and the net bonding and antibonding characters are hard to judge from the drawings. Why? If the MO places electron density between the nuclei it has bonding character but if it places it outside it has antibonding character. Look at the σ MO with two nodes – this orbital places density both between the nuclei and outside. The photoelectron spectra discussed below show that this MO, when filled, is in fact nearly non-bonding in character. Note that in these rough drawings one only sketches out the major contributions plus the nodal and bonding/antibonding characters. One must pay attention to Table 1.1. Alternatively, plotting programs are available for precise MO drawings if desired.

In a one-electron model the electrons are added after the MOs are formed. Thus, the eight MOs of B_2 provide a qualitative description of any diatomic molecule with s and p valence functions only. Electrons are added using the same rules we

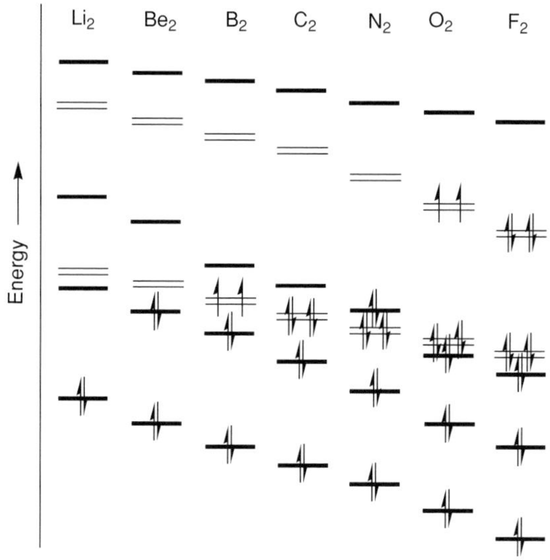

Figure 1.4

use for filling AOs of atoms. However, the placement of the σ and π manifolds relative to each other will depend on the relative 2s and 2p energies which, in turn, depend on the atom identities. For a given electron count, the measured multiplicity of the ground state provides information on the relative energy ordering. Figure 1.4 presents the accepted MO level diagrams of the first row diatomics and one sees, for example, that the paramagnetism of O_2, which is a problem for the two-center–two-electron bond model, can now be explained. The model also explains nicely why IE (O_2) < IE (N_2) even though the electronegativity of O > N.

The next step is to look at heterodiatomics, e.g., CO. One expects the same number and type of MOs as found with the homonuclear molecules and the number and symmetry types of basis functions are the same. However, the energies and compositions are distorted by the differing electronegativities of the two atoms just as they were for H_2 vs. LiH above. In Figure 1.5 the MO diagrams of N_2 and CO are compared as are the photoelectron spectra. The spectra clearly show that the highest occupied MOs (HOMOs) are nearly non-bonding. Both are sharp bands (little change in inter-nuclear distance on ionization) and the vibrational frequencies in the ion states are nearly the same as those in the molecules; 2191 vs. 2345 cm^{-1} for N_2 and 2200 vs. 2157 cm^{-1} for CO. Perhaps this is a problem for the reader as in the Appendix we describe these molecules as triply bonded and a triple bond is often represented by one filled σ-bonding orbital and two filled π-bonding orbitals. In the MO description the filled π-bonding MOs are obvious; however, the σ-bonding orbital is not. If it's not the highest lying filled σ MO where is it? In the MO model the σ bonding character is spread over all three filled σ orbitals!

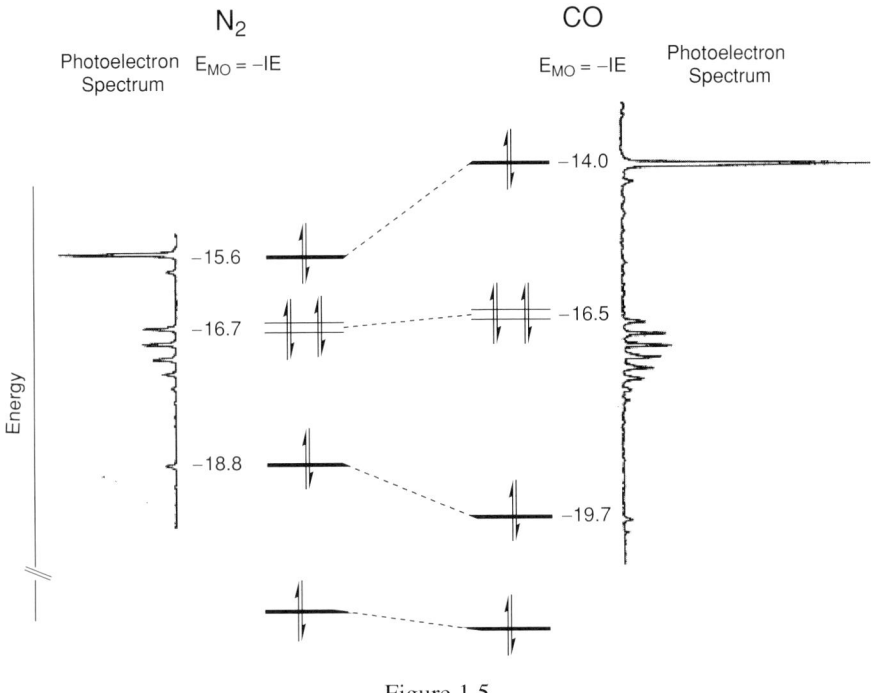

Figure 1.5

In contrast to the nice, neat two-center–two-electron bond model, it is not so easy to determine the overall bonding character from MO orbital drawings alone. We need another measure. This comes from the Mulliken overlap population which is a numerical indicator of bonding (positive) and antibonding (negative) character between a pair of atoms within a molecule. For N_2 in an approximate calculation the overall overlap populations are +0.68 for the three σ filled MOs and +0.54 each for the π MOs. If one considers each π interaction of bond order one then the overall bond order is clearly three.

1.1.3 Importance of frontier orbitals

One more elementary MO concept needs to be mentioned. Fukui shared in a Nobel award for his effective use of the frontier orbitals of a molecule (the highest occupied MO, HOMO, and lowest unoccupied MO, LUMO) to correlate and predict chemical behavior. Good Lewis bases are expected to have high lying HOMOs and good Lewis acids are expected to have low lying LUMOs. For CO the HOMO is a σ orbital, C–O non-bonding, with the highest amplitude on C, which is the more electropositive atom. This justifies carbon-bound CO when found as a ligand to a transition metal such as Fe (see Appendix): a fact that is counterintuitive based on a

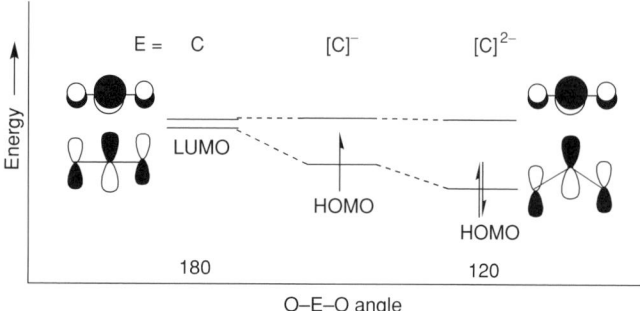

Figure 1.6

simple interpretation of the effect of electronegativities on charge distribution. The LUMO of CO has π symmetry and is CO antibonding but it, too, has its highest amplitude on C. To act as a Lewis acid when bound through the C atom, CO must interact with a metal center that has filled orbitals with π symmetry relative to the M–C axis. The later transition metals, such as Fe, possess the requisite filled orbtals. Thus, the primary CO donor interaction to the metal is buttressed by secondary π back donation to the CO ligand leading to a robust Fe–CO bond. Note that depopulation of the non-bonding σ MO of CO combined with population of the antibonding π MOs of CO leads to a net weakening of C–O bonding on coordination. In fact, the CO frequency decreases on binding to a transition metal Lewis acid, e.g., Fe, whereas it remains about the same if bound to a Lewis acid incapable of acting as a π acceptor, e.g., BH_3.

Walsh showed that the properties of the HOMO could be used to rationalize the shapes of polyatomic molecules. A good example is the O–E–O series of triatomic molecules, E = C, N and O. In Figure 1.6 the HOMO energy is plotted as a function of the O–E–O angle. It correlates with one component of the degenerate LUMO of CO_2 and decreases in energy because of the increasing O–O bonding interaction as the angle decreases. Consideration of the properties of a single MO neatly correlates with the observed O–E–O angles of 180°, 134° and 117°, respectively, for E = C, [C]$^-$ and [C]$^{2-}$ isoelectronic with the known series E = C, N and O. The importance of HOMO/LUMO properties provides a gratifying simplification of the MO approach.

1.1.4 Polyatomic molecules

An excursion into polyatomic molecules is next. An informative series from the point of view of two-center bonds is CH_4, NH_3, OH_2, FH, Ne. In Figure 1.7 a representation of the photoelectron spectroscopic bands (IEs) illustrate how the

1.1 Molecular orbitals without the nasty bits

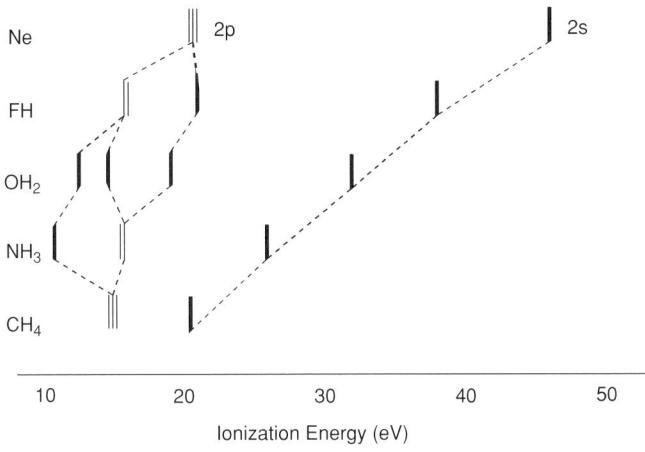

Figure 1.7

spherical symmetry of the filled AOs of Ne is broken for this series of isoelectronic and isoprotonic species. Moving protons from the nucleus into the valence region introduces structure into the former Ne 2p ionization feature. For example, consider FH as [F]$^-$ (isoelectronic with Ne but with a reduced nuclear charge) as perturbed by a proton on the z axis. The 2p ionization will no longer be a single feature as the $2p_z$ ionization will be stabilized relative to the $2p_x$ and $2p_y$ ionizations as seen in Figure 1.7. This correlation again harks back to the united atom model in which these H derivatives become pimply versions of Ne with modified "AOs" to match. All the trends in orbital and ionization energies for atoms learned in your first year of chemistry carry over nicely to these molecules. The connection becomes more obscure as the molecules become even more complex but it is still there.

The isoelectronic series in Figure 1.7 involve pretty small molecules yet the MOs of, e.g., H_2O, are considerably more complex than the simple Lewis dot structure. Hence, one anticipates that molecules of respectable size will be even more so. For example, beginning with atoms, the MO approach to P_3H_5 yields 17 MOs from the 17 basis functions of which 10 are filled – not something that one can dash off on the back of an envelope. Hence, we need ways of abstracting the essence of a problem from the large set of MOs describing a real system. That is, we don't need MO theory to discuss CH_4 at the level we are interested in here. Two ways have already been described: the emphasis on frontier orbitals as a guide to reactivity, e.g., Lewis acid–base interactions; and the HOMO as a guide to structure.

Another fruitful approach, employed to great effect by Hoffmann and members of his school of chemists, is to divide a molecule into fragments and examine the MO description of the fragment–fragment orbital interactions as the molecule is formed. If necessary, review in the Appendix how to easily generate a reasonable

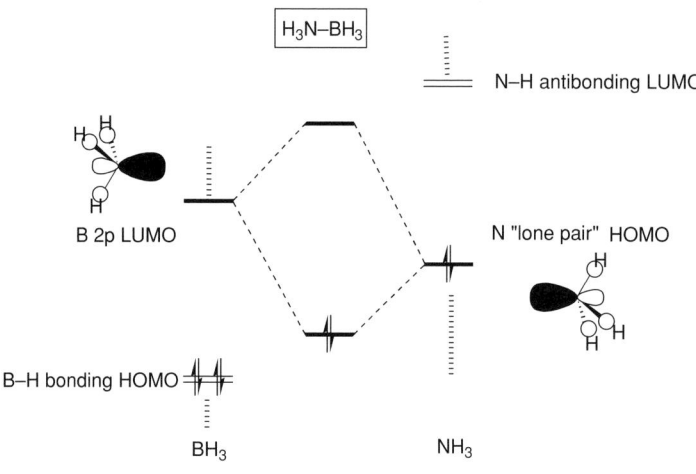

Figure 1.8

structure for P$_3$H$_5$ from that of PH$_3$ by replacing two H with two PH$_2$ groups. Fragment analysis applied to MO descriptions of complex electronic structures is an analogous procedure for simplifying the problem. In some cases, e.g., formation of a coordination compound, the fragments are obvious and, for the formation of a single bond type, no more difficult than the diatomic problem discussed above. In other cases, the choice of fragments requires considerable insight and/or work. Very helpful is the careful definition of the question one desires answered and the specification of the molecules which will provide the answer by inter-comparison of MO changes. For simple MO approaches a given MO is a rather crude entity and of little value in any absolute sense. However, the manner in which it varies as one changes, e.g., a substitutent, more often than not is of value in understanding what makes the electronic structure tick.

Let's apply the fragment approach to the simple Lewis donor–acceptor complex H$_3$BNH$_3$ which is isoelectronic with C$_2$H$_6$. Both have 14 MOs, 7 of which are filled. The obvious fragment analysis is to break H$_3$BNH$_3$ into BH$_3$ and NH$_3$. Borane is primarily an acceptor and ammonia is primarily a donor; hence, one need only consider the LUMO of the former and the HOMO of the latter to generate a description of H$_3$BNH$_3$ (Figure 1.8). This problem is no more difficult than that of LiH discussed above, i.e., we have reduced a 14 MO, seven-electron pair problem to a two MO, one-electron pair problem. With an MO program one can crank out the MO energies and compositions and then transform these AO basis results into a fragment orbital basis set generated by calculating the MOs of each fragment separately. Inspection of the results summarized in Figure 1.8 shows a diatomic-like model. We have not discovered anything new in this problem. It is simply a method of simplification which will be used in Section 1.2 to isolate and describe interactions that cannot be reduced to a diatomic-like description, e.g., multicenter two-electron bonds.

Exercise 1.2. In the discussion of Lewis acids and bases in the Appendix, the compound $[I_3]^-$ is analyzed as an adduct of the base I^- with the acid I_2. It probably is not clear how I_2, a diatomic that perfectly fits the two-center–two-electron bond model and the eight electron rule, can act as a Lewis acid. Show how a HOMO–LUMO analysis of the acid–base interaction rationalizes the interaction and predicts a linear structure.

Answer. The HOMO of I^- is a filled 5p AO whereas the LUMO of I_2 is the highest lying σ-antibonding MO shown in Figure 1.3. Best overlap of the donor and acceptor orbitals will be achieved in a linear structure. As $[I_3]^-$ is a homonuclear compound this HOMO–LUMO analysis cannot be pushed too far and we will defer presentation of the MOs of a species like $[I_3]^-$ until we develop a model for three-center bonds later in the chapter.

A word of caution here is in order. Like all approximate descriptions, the simplifications permitted depend on the question asked. No such model should be extrapolated outside of its range of applicability. Unfortunately, disregard of this caution has led to some famous chemical contretemps lasting literally years, e.g., the "non-classical" carbonium ion controversy. For some reason, despite living in the age of quantum mechanics where electronic structure is described in terms of electron distributions, chemists just love to argue about "where the electrons are." Like nervous parents worrying about their teenagers, they want to put electrons on the ligands rather than the metal, between two nuclei but not three, and so on. Perhaps some feel wandering electrons might get into trouble! This conceptual problem is not a joke, however, when one has to teach about cluster bonding and extended systems like metals to students brainwashed to believe that localized ideas of oxidation state, hybridization and two-center–two-electron bonds are some sort of fundamental, inviolate tenets of chemistry.

1.1.5 Coordination compounds

To continue this introduction to MO models, we look next at a coordination compound. Consider the 18-electron complex $Cr(CO)_6$ which has octahedral geometry (Figure 1.9). Here the metal acts as a Lewis acid but now one that can accommodate more than a single pair of electrons. How does this factor complicate the analysis? A chemically informative fragment analysis generates a Cr atom and six CO ligands.

14   Introduction

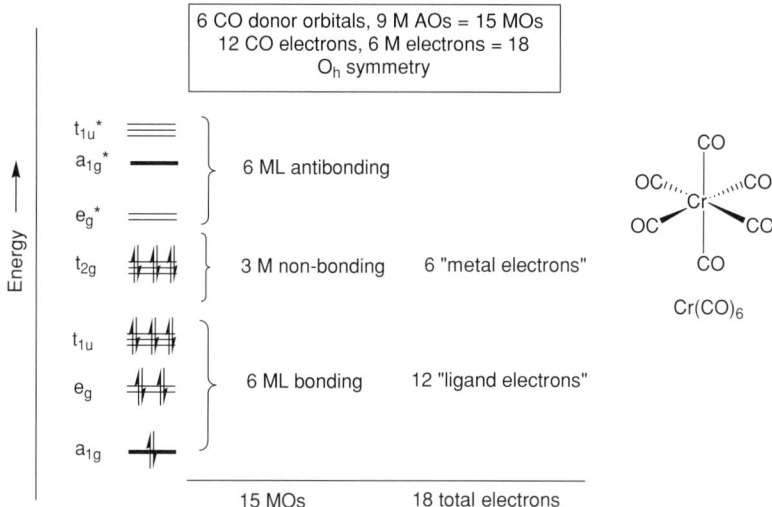

Figure 1.9

To produce an interaction diagram we need the frontier orbitals of Cr and CO. The former are 3d, 4s and 4p whereas the latter are 5σ and $2\pi^*$ and Cr possesses six valence electrons whereas each CO has two in the HOMO. For simplicity we will first consider only the CO donor orbitals and then add the acceptor orbital interactions later. For the first problem then we have nine metal functions and six ligand functions plus 18 electrons total. Thus, we must generate 15 MOs of which 9 will be filled. Is this a simplification? The full problem requires nine metal functions and $6 \times (4+4)$ ligand functions for a total of 57 MOs! Of course, in a real calculation, the fragment analysis is done by the computer on the MO solutions of the complete problem. Hence, the fragment analysis generates energies and MO compositions that are as valid as the original calculation.

We can use symmetry to simplify the conceptual problem (keep in mind that you don't need to know any symmetry to do the calculation on a computer but it does simplify sorting out the MOs that the computer spews out) but we need to go beyond σ and π and classify the orbitals according to their symmetry in the octahedral point group. Here we will just do it – many sources for learning how are available in standard inorganic texts. All we have to remember is that orbitals of like symmetry interact whereas those of unlike symmetry do not. Six low-energy filled MOs (a_{1g}, t_{1u}, e_g) result from the bonding interactions of the central metal AOs with the symmetry-adapted linear combinations of the ligand σ functions and six high-energy unfilled MOs result from their antibonding partners. There is no combination of ligand σ functions that has t_{2g} symmetry so the metal d set of t_{2g} symmetry (xy, xz, yz) retain their pure AO character and for Cr are filled.

This result is a MO description that could equally well be applied to a Werner complex, e.g., $[Co(NH_3)_6]^{3+}$, with 18 electrons. But the chloride salt of the latter is an orange solid whereas $Cr(CO)_6$ is a white solid. Why the difference in absorption properties? Absorption in the visible can be crudely associated with the magnitude of the HOMO–LUMO gap. Empirically it is found that, for the same geometry, the splitting increases with increasing metal oxidation number and nature of the ligands as reflected by position in the spectrochemical series. The first factor leads to a larger gap for Co(III) vs. Cr(0); hence the difference must be associated with ligand properties. The consensus is that NH_3 acts primarily as a Lewis base and the MO description at the top of Figure 1.9 is an adequate description. As mentioned more than once already, CO is both a Lewis base and a Lewis acid as the LUMO is of low enough energy and has the proper symmetry to interact with the filled t_{2g} metal orbitals on the metal center. Including this donor–acceptor interaction in our MO diagram (bottom of Figure 1.9) leads to a lowering of the filled t_{2g} orbitals thereby increasing the M–C bond strength as well as increasing the HOMO–LUMO gap for $Cr(CO)_6$. In the language of coordination chemistry, CO is a strong field ligand (high in the spectrochemical series) because it is a strong π acceptor.

Exercise 1.3. Consider a coordination compound formed from BH_3 and C_2H_4. From the HOMO–LUMO properties of each species predict the geometric structure of the Lewis acid–base adduct. Now predict the structure of a compound formed by replacing one CO ligand of $Fe(CO)_5$ with C_2H_4. Note the parallelism between the main group and transition metal examples. The second compound is a stable and isolatable compound, whereas the first is a transient intermediate in the hydroboration of ethylene to ethyl borane and has only been characterized as a transient intermediate in a fast-flow system by modulated mass spectrometry.

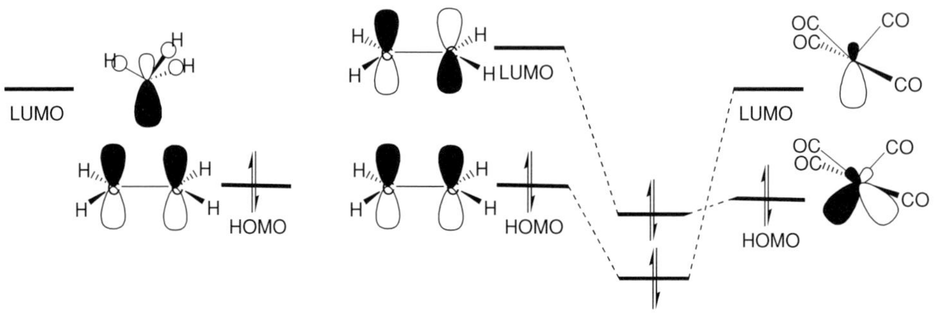

Exercise 1.3

Answer. The frontier orbitals of BH_3 and C_2H_4 are shown in the figure and the most favorable interaction between the σ acceptor orbital of BH_3 and the HOMO of C_2H_4 can only take place if BH_3 is placed on the symmetry axis bisecting the C–C bond and perpendicular to the plane of the C_2H_4 molecule. Do not be confused by the fact that it is the π bonding orbital of C_2H_4 that is the donor orbital. This is a consequence of treating complex molecules as if they were diatomics. The replacement of CO with C_2H_4 is also shown but now both the HOMO and LUMO of C_2H_4 participate in the same fashion as the HOMO and LUMO of CO. The interaction of the HOMO of C_2H_4 with the LUMO of the $Fe(CO)_4$ fragment is analogous to that of BH_3; however, the secondary interaction between the "t_{2g}-like" metal orbitals on the iron fragment and the LUMO of C_2H_4 distinguishes the metal system from the non-metal one. This two-part interaction constitutes the Dewar model and covers metal–olefin bonding situations ranging from a π complex (small participation of the LUMO and little pyramidalization at carbon) to a metallacyclopropane (large participation of the LUMO and considerable pyramidalization at carbon).

In the review of main-group diatomic molecules in the Appendix, we mention that C_2 is not expected to generate a quadruple bond. Now we are in a position to understand why. From the MO diagram for E_2 above it should be clear that in addition to the σ bonding interaction, only two π-type interactions are possible. We are also in a position to understand what would be required for a compound to possess a quadruple bond. It is worthwhile doing so as a final exercise. Cotton provided this analysis and the subsequent manipulation of this bonding feature in dinuclear complexes by changing metal type and/or oxidation state. We will analyze the bonding in $[Re_2Cl_8]^{2-}$ with the geometric structure shown in Figure 1.11. Key features of the structure are the short Re–Re bond distance (2.24 Å vs. 3.18 Å = sum of covalent radii) and the eclipsed conformation of the Cl ligands. Obviously, this complex is not a diatomic molecule but a fragment analysis that mimics a diatomic molecule can be used to explain these properties and show them to be characteristic of a M–M quadruple bond.

1.1 Molecular orbitals without the nasty bits

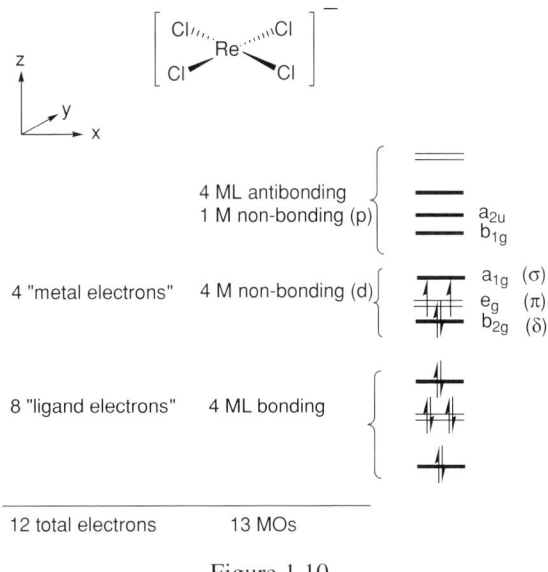

Figure 1.10

The appropriate fragment analysis is one in which we generate the Re–Re interaction from two D_{4h} [ReCl$_4$]$^-$ fragments shown in Figure 1.10. First, we form the frontier orbitals of [ReCl$_4$]$^-$ using an analysis similar to that employed for stable Cr(CO)$_6$. If we use a single donor orbital from each of the four Cl and the nine Re valence functions, 13 MOs will be produced (crudely: four M–L bonding, four M–L antibonding and five non-bonding M functions). The available electrons (four from four Cl, seven from Re and one from the negative charge) fill the four M–Cl bonding orbitals with the remaining four going into the M d functions, i.e., z^2 (a_{1g}), xz and yz (e_g), xy (b_{2g}) which have σ, π and δ symmetry relative to the Re–Re bond axis (z axis) of [Re$_2$Cl$_8$]$^{2-}$ (Figure 1.11). One metal p function (out of plane a_{2u}) is at high energy. Now bring the two square planar Re fragments together face to face in an eclipsed conformation as shown in Figure 1.11. Because functions of different symmetry do not mix, the σ, π and δ functions combine two by two (four H$_2$ problems, if you wish) to yield bonding and antibonding combinations. Note that for the δ functions this can only happen in the eclipsed structure – in the staggered structure the δ overlap is zero by symmetry. Eight electrons are available and they fill the four bonding combinations generating a formal quadruple bond and thereby rationalizing the short Re–Re distance and eclipsed conformation. The model is a simple one but it originates from an insightful fragment analysis of an ostensibly complex molecule. This is exactly what we must do in the upcoming chapters in order to understand cluster bonding as well as bonding in the solid state. The trick, and not an easy one to learn, is to simplify just the "right" amount! It's

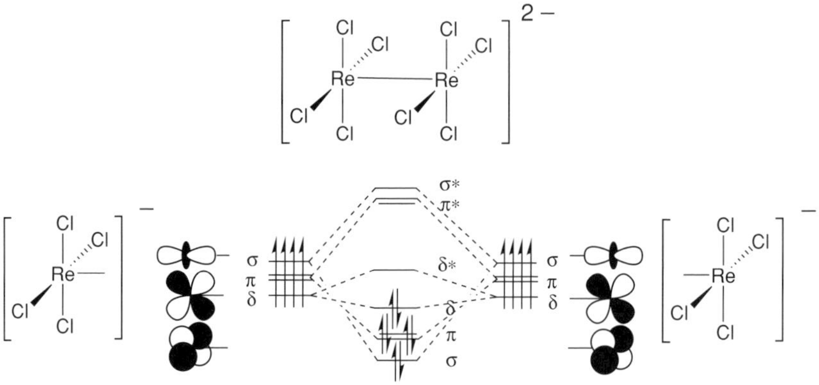

Figure 1.11

like the instruction found in old recipes – "cook until done." One learns by doing it, not by reading about it.

1.2 Three-center bonds

Boron chemists reading this text will know that the story of the boranes starts long before the cluster electron-counting rule, which will be developed in Chapter 2, was formulated. The place of the boron hydrides in the fabric of our understanding of chemical bonding has a long history. Those interested can find fascinating accounts in the literature of the struggles to accommodate molecules that just didn't seem to fit the perceived wisdom of the time. An essential part of this struggle was the development of the concept of the three-center–two-electron bond. It is a Nobel-quality concept that impacts on more than borane chemistry. But it was with the polyhedral borane structures that Lipscomb used it so effectively in generating the first useful descriptions of the electronic structures of the boranes. Application of the concept to clusters allows us to probe cluster electronic structure for features hidden by the simplicity of the electron counting rule. Hence, we go back now and look at some of this history.

1.2.1 Diborane

The story begins with the problem of the structure of diborane, B_2H_6, shown in Figure 1.12. Long a molecule of uncertain structure that caused much angst among the chemical community, the history of its development says much about the eagerness of chemists to apply models with gay abandon. All well and good: this is how one always begins. But when strange descriptions arise of shoe-horning the

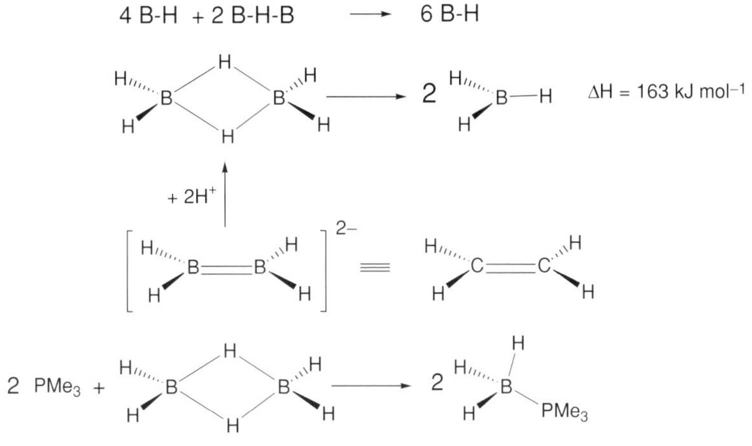

1.2 Three-center bonds

Figure 1.12

stoichiometry of B_2H_6 (eight nuclei, 12 valence electrons) into the structure of C_2H_6, it's time to review the model. Once the structure of diborane was settled experimentally, firm boundary conditions were established. As shown in Figure 1.12, each B atom is in a near tetrahedral environment of H atoms even though it does not have the geometric structure of ethane. Ignoring the bridges for a moment, it should be clear that each of the four non-bridging H atoms can be connected to the nearest B atom by a two-center–two-electron bond. Four equivalent B–H bonds utilize 8 of the 12 available electrons. Two B–H–B bridges are left but only four electrons or two per bridging interaction remain. Based on distances, the bridging H is bonded to both boron atoms. Hence, the bonding of diborane is said to be partitioned into four B–H two-center–two-electron bonds and two B–H–B three-center–two-electron bonds. The "electron-poor" borane monomer solves its electronic problem by dimerization and formation of two three-center bonds.

How good is a three-center–two-electron B–H–B bond? The chemistry and energetics tell the story. Each B atom in diborane is associated with eight electrons; hence, the eight-electron rule is obeyed. However, in the presence of Lewis bases such as PMe_3 it is symmetrically cleaved into two BH_3 base adducts as shown in Figure 1.12 ($[BH_4]^-$ isoelectronic with CH_4 for base = $[H]^-$). Shore showed that bases like NH_3 result in unsymmetrical cleavage into the salt $[BH_2(base)_2][BH_4]$ in which both cation and anion are eight-electron species. Clearly, two three-center bonds are not as energetically favorable as two base–B donor–acceptor bonds. The energetics of the cleavage of diborane into two monoboranes provides a different view. Cleavage requires $163\,kJ\,mol^{-1}$, which is about half the energy necessary to cleave a B–B single bond (Figure 1.12). Another way to express this result is that one three-center–two-electron bond is better than a two-center–two-electron B–H bond plus an empty orbital by about 84 kJ mol^{-1} – a chemically significant energy relative to kT at ordinary temperatures.

Exercise 1.4. Look up in a table of average bond energies the C–C single-bond energy. Compare it with the dissociation energy of diborane. A famous experiment in the history of organic chemistry is the detection of triphenyl methyl radicals on the attempted synthesis of hexaphenyl ethane in 1900 by Gomberg. The dissociation constant, K_{diss}, of hexaphenyl ethane at 25 °C measured in benzene is 2×10^{-4}. On this basis would you predict $(C_6H_5)_2BH$ to be monomeric or dimeric?

Answer. The C–C bond energy is 335 kJ mol^{-1} compared with 163 kJ mol^{-1} for B_2H_6. To get a ball-park answer assume the entropy of reaction is the same for the two dissociation processes and that the steric effect of two Ph groups on the bridging interaction in diborane is the same as three Ph on the C–C bond in ethane. Then as K_{diss} is proportional to $e^{-E(X)/RT}$ where X = B or C, at the same temperature $K_{\text{diss}}(B) = K_{\text{diss}}(C) \, e^{-E(B) - E(C)/RT} = 2 \times 10^{-4} \times 8 \times 10^{29} = 2 \times 10^{26}$. Monomeric $(C_6H_5)_2BH$, however, is reported to disproportionate into $(C_6H_5)_3B$ and $(C_6H_5)BH_2$ with the latter found as a dimeric species in solution.

One early and insightful model for diborane is the protonated double-bond model shown in Figure 1.12. Symmetrical protonation of the π bond of ethylene above and below the molecular plane leads to the structure of diborane. Replacement of each C in ethylene with [B]$^-$ leads to $[B_2H_4]^{2-}$. In fact $Li_2B_2R_4$, where R is a bulky substituent, has been structurally characterized and shown to possess B–B multiple-bond character.

We are talking boron hydrides here – compounds with BH bonds polarized with positive B and negative H. What is the nature of the bridging hydrogen atom? Is it protonic as the protonated double-bond model might suggest or is it still hydridic? The three-center bond provides a ready explanation. In the B–H–B bond two electrons are shared between three nuclei; hence, assuming equal electronegativities for simplicity, each has a 2/3 share and picks up a formal charge of +1/3 for the bond. As the bridging H forms no other bonds, its charge is +1/3. This formal charge is large enough to counter the electronegativity difference and the bridging hydrogens pick up a distinctly protonic character. The protonated double-bond model is more than a bedtime story! This counterintuitive feature of the three-center bridge bond lies behind a significant fraction of the borane chemistry developed over the last few decades in that the removal of a bridging H from a cluster as a proton generates a site of nucleophilic reactivity. This type of H bridge is found both in organometallic chemistry (C–H–M) as well as metal-cluster chemistry (M–H–M). As the model presented is a general one and independent of the identity of the bridged atoms, the fundamental properties associated with this bonding feature transfer as well.

1.2 Three-center bonds

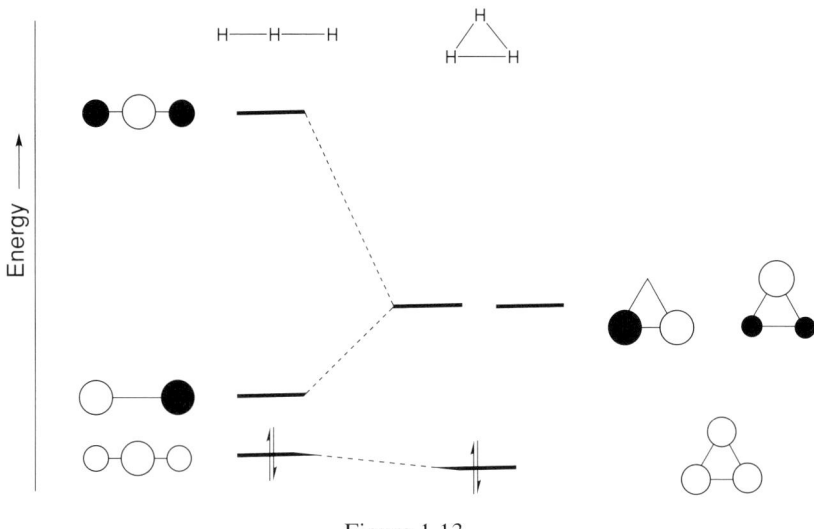

Figure 1.13

1.2.2 Trihydrogen cation: linear and ring structures

In terms of electronic structure, what are the characteristics of a three-center bond? A simple, but effective, model is that of $[H_3]^+$, a prominent ion in the mass spectrum of H_2 formed by an ion–molecule reaction. There are two limiting geometries for $[H_3]^+$, linear and triangular, and the results of a Fenske–Hall calculation are shown in Figure 1.13. By now you should be able to see how the MOs are generated from the three H 1s functions. Both linear and triangular $[H_3]^+$ are bonded by one filled MO distributed over three nuclei; hence, three-center–two-electron bond. These are the two types of bonds developed by Lipscomb to describe the bonding in polyhedral boranes which will be discussed in detail in the next chapter. Note that the linear (and bent) three-atom MO manifold consists of one bonding, one non-bonding and one antibonding MO. Hence, $[H_3]^-$ with four electrons should be a closed shell bonded species in the linear form. This is a model for a three-center–four-electron bond which will be explored briefly below. For yourselves you can show via an application of Walsh's approach that $[H_3]^+$ is predicted to be triangular whereas $[H_3]^-$ should be linear.

An understanding of $[H_3]^+$ permits the bent B–H–B three-center bond to be easily described as shown in Figure 1.14. The main difference is that two of the H 1s functions are replaced by hybrid B orbitals thereby generating bonding, non-bonding and antibonding MOs in the same manner.

Perhaps you will have noted that the three-center bond is a method for utilizing all the valence electrons and valence orbitals when the latter exceeds the former, e.g., diborane has 12 valence electrons and 14 valence orbitals. Each of the two

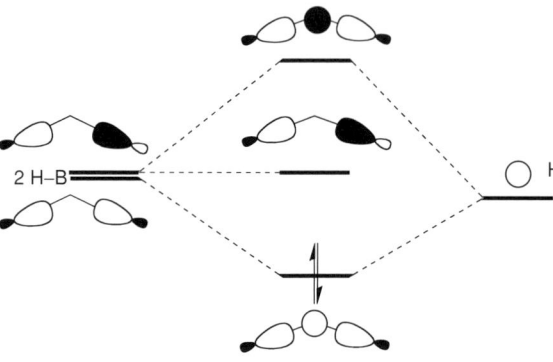

Figure 1.14

three-center bonds uses three orbitals and two electrons. It should then be clear that a three-center–four-electron bond can also be useful when there are insufficient valence orbitals to accommodate the valence electrons. The problem of $[I_3]^-$ is considered a Lewis acid–base adduct in the Appendix. It has 22 electrons and 12 orbitals. If each I atom uses three orbitals to accommodate three lone pairs one is left with four electrons and one orbital on each I atom. If we then use these three orbitals to form a three-center bond containing four electrons we not only use all the available orbitals and electrons but also there is no need to invoke an expanded octet at the central I atom. That is, the filled non-bonding MO has a node at this atom. The actual MO structure is more complex but this approach provides a reasonable alternative to the usual treatment.

1.3 An orbital explanation of electron-counting rules

There is a connection between an orbital description of electronic structure and the more elementary bonding discussions such as those reviewed in the Appendix. In this section we describe the connection of the 8- and 18-electron rules in order to provide a basis for understanding how the cluster electron-counting rules emerge from and are connected to molecular orbital descriptions of cluster bonding.

1.3.1 The electron-counting rules as consequences of the closed-shell principle

It is extremely important to keep in mind that any electron-counting rule is underlined by an orbital requirement which is called the closed-shell principle. This principle states that a stable molecule should have all its low energy MOs fully occupied and separated from the vacant high energy ones by a significant HOMO–LUMO gap (Figure 1.15). The larger this gap, the more stable the molecule. Indeed, the existence of this gap provides the molecule with Jahn–Teller stability (Jahn–Teller

1.3 An orbital explanation of electron-counting rules

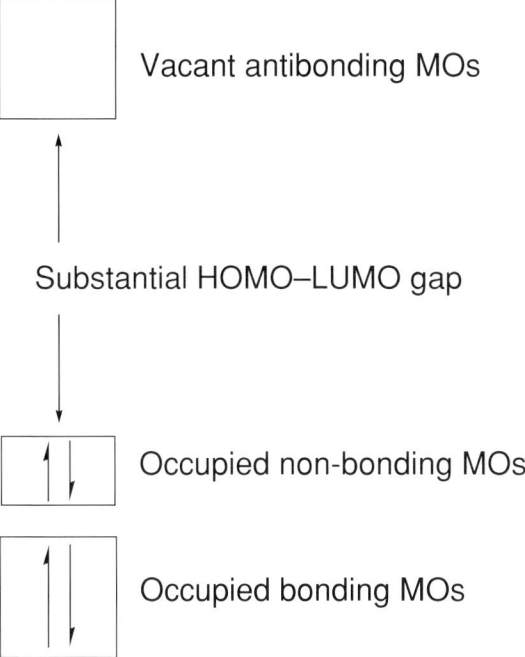

Figure 1.15

stability is nothing more than thermodynamic stability). A HOMO–LUMO degeneracy corresponds to the well-known first-order, i.e., real, Jahn–Teller instability. A small but non-zero HOMO/LUMO gap corresponds to the so-called second-order, i.e., pseudo, Jahn–Teller instability. For a molecule in some specified geometry, first- or second-order Jahn–Teller instability means that the molecule is unlikely to exist in the specified geometry and another geometry that minimizes the molecular energy will result from a structural modification which increases the HOMO–LUMO gap.

Three important points must be emphasized:

(i) In general, the occupied MOs of a molecule, which satisfy the closed-shell principle, are all bonding and non-bonding MOs whereas the empty MOs are all antibonding. Thus, in the most general case, the closed-shell principle can be reformulated as follows: a stable molecule should have all its bonding and non-bonding MOs fully occupied and separated by a significant HOMO–LUMO gap from the vacant antibonding MOs. There will be exceptions. We will see later that sometimes stability is achieved where some non-bonding orbitals are empty. In other cases, weakly antibonding orbitals can be occupied.

(ii) Consider now a stable molecule which satisfies the closed-shell principle as defined in (i). Let's assume that we can change its electron count by adding or removing an electron pair without any change in geometry. With its new electron count and its frozen geometry, the molecule is very likely to no longer satisfy the closed-shell principle. The first- or second-order Jahn–Teller instability which arises tells us that the molecular geometry has to change. If an electron pair is added, it must occupy an antibonding MO and to achieve greater stability, the molecule must somehow cancel the antibonding character of this new HOMO. Often this is accomplished by breaking a bond thereby changing the character of the HOMO from antibonding to non-bonding. It follows that adding electrons changes a molecular structure into a more open one. On the other hand, removal of an electron pair leads to a more compact structure. Thus, there is a close connection between electron count and geometric structure and conversely. But this oversimplifies real life. Think of isomers, for example, which have different structures but the same composition and electron count. Nevertheless, there are strong relationships between structure and valence-electron count in the whole area of chemistry, especially when one is dealing with stable compounds. These relationships are described by electron-counting rules such as the eight- and 18-electron rules. Each electron-counting rule applies to a particular family of compounds. None is universal but remember that all are based on the closed-shell principle. Application of such rules to molecules with small or no HOMO/LUMO gap is a dangerous business – thus, the third point.

(iii) Although a huge number of stable covalent compounds obey the closed-shell principle, there are many that do not. There are examples of closed-shell molecules with small HOMO/LUMO gaps and examples of paramagnetic open-shell molecules, e.g., O_2 and odd electron radicals, which are stable enough to be isolated. Such is the case, for example, with some transition-metal clusters whose electronic structure tends to look like that of bulk metals or of certain electron-conducting solid-state compounds (Chapter 3). For many of these molecular and extended compounds a particular structure is not associated with a single valence-electron count but with a range of allowed electron counts. This is a situation that will be the subject of further analysis later in the book.

1.3.2 The orbital explanation of the 8- and 18-electron rules

The 8-electron rule relies on the closed-shell principle and customarily assumes localized two-center–two-electron bonding between all atoms. Turned into orbital language, two-center–two-electron bonding means that each atom uses one valence AO (or a linear combination of AOs) per bonding electron pair. It follows that H, with one valence AO (1s), cannot participate in more than one localized bond and a main-group atom with four valence AOs (ns, np_x, np_y, np_z) cannot participate in more than four localized bonds (of course, less than four bonds is possible). Thus, two AOs, one on each center, combine to give a bonding and an antibonding MO with the former occupied by an electron pair. Consider a molecule of general formula AH_n, where A is a main-group atom. Assuming that all the A–H bonds are localized two-center–two-electron bonds means that n should be lower than or equal to four. Moreover,

1.3 An orbital explanation of electron-counting rules

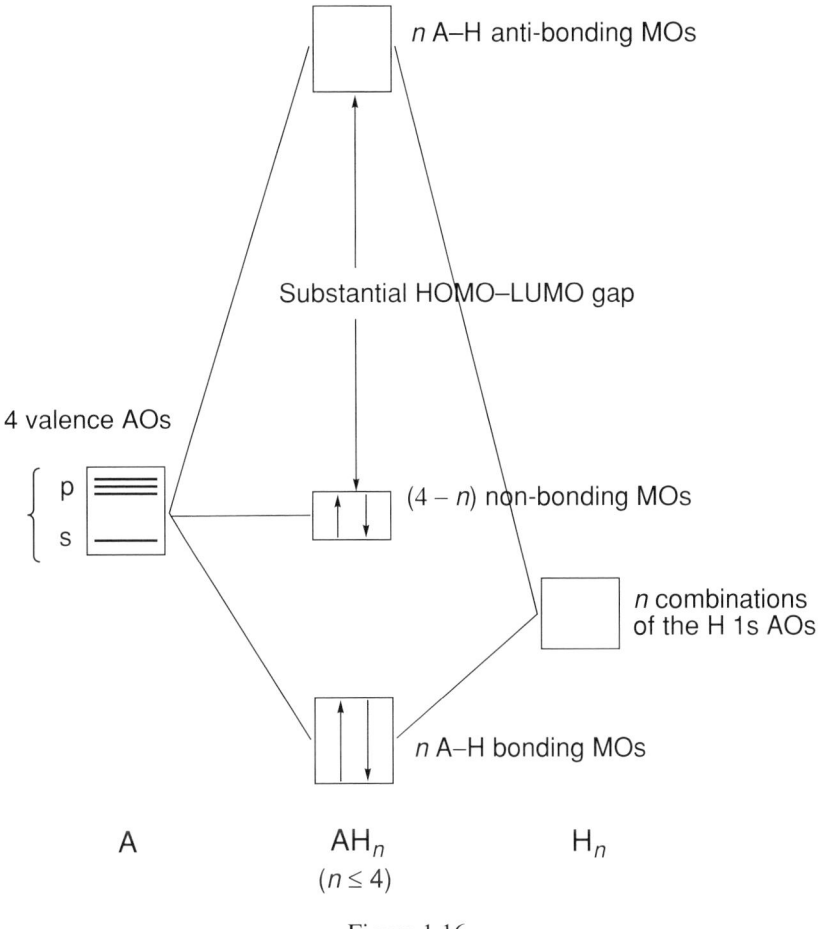

Figure 1.16

the number of bonding MOs should be equal to the number of antibonding MOs and equal to the number of A–H bonds, i.e., n. The simple MO diagram in Figure 1.16 easily follows. Now, to satisfy the closed-shell principle in its general form, all the bonding and non-bonding MOs have to be filled. This leads to the occupation of n bonding MOs + $(4-n)$ non-bonding MOs = 4 MOs for $0 \leq n \leq 4$. Thus, the combination of the closed-shell principle with a localized bonding mode tells us that the number of valence electrons lying in MOs in which A has significant participation should be equal to eight. The same demonstration can be made for the 18-electron rule which holds for many transition-metal compounds. For a transition metal having nine valence AOs just change four into nine in the diagram of AH_n in Figure 1.16.

Exercise 1.5. What is an electron-counting rule derived from application of the closed-shell principle to H or He? Why don't you find this rule in textbooks?

Answer. As both have a single valence function, $n = 1$ and the number of valence electrons lying in MOs in which a single H or He atom participate should be equal to two. A two-electron rule for two atoms seems unnecessary and counterproductive in the long run.

1.3.3 Limitations of the 8- and 18-electron rule: localized electron-deficient compounds

In Figure 1.16 we assumed that the closed-shell principle is satisfied when all the non-bonding MOs are occupied. In point (i) in Section 1.3.1 above, the reader was warned that in some cases the closed-shell principle can be satisfied without all the non-bonding MOs being occupied. This is actually the fact in main-group chemistry when non-bonding orbitals are pure np AOs of electropositive atoms such as B or Be. Indeed, the energy of the np shell of such atoms is naturally high in energy rendering these AOs less accessible to the electrons. Thus, a substantial energy gap can be created between non-bonding np AOs and the other non-bonding (if any) and bonding lower-lying orbitals. Viewed in this sense, the compound appears electron deficient with respect to the octet rule. This is the case with six-electron boron in trigonal planar BF_3 or BMe_3, for example. Although BH_3 is unstable with respect to dimerization, it provides a simplified model for further consideration. View it with the trigonal plane as the xy plane. By symmetry, this molecule has a non-bonding $2p_z$ AO which lies at too high energy to be occupied normally (the spectroscopic characterization of $[BH_3]^{2-}$ has been reported, however). In a diagram equivalent to Figure 1.16, one non-bonding AO would be vacant. Two electrons are missing with respect to the normal eight-electron requirement even though in this particular case there is a substantial HOMO/LUMO gap. Trigonal planar six-electron boron is not uncommon and the so-called sextet rule is often invoked for this element. Similarly, the linear BeH_2 molecule has two non-bonding $2p_\pi$ AOs which are empty thus leading to a count of four electrons.

One should note that the number and occupation of the bonding and antibonding MOs have not changed in the cases of BH_3 and BeH_2 as compared to the general case sketched in Figure 1.16. They are still equal to the number of B–H or Be–H bonds. Thus, we are still within a localized two-center–two-electron bonding mode and simple Lewis structures can be drawn for BH_3 and BeH_2 reflecting their electron deficiency relative to the eight-electron rule.

Finally, note that a very similar situation obtains for transition-metal complexes – the favored electron count of tetracoordinate square-planar complexes is $18 - 2 = 16$ and the favored electron count of dicoordinate linear complexes is $18 - 4 = 14$.

1.3.4 Limitations of the eight- and 18-electron rule: delocalized bonding

The development of these n electron rules, $n = 8, 18$, assumes a localized bonding mode. As simple Lewis structures cannot be drawn when the number of bonding contacts afforded by an atom with its neighbors is larger than its number of valence AOs, the rules will not apply in such a situation. Atoms in such a bonding situation have been called hypercoordinated and the number of bonding MOs either will not equal the number of antibonding MOs or are equal but lower than the number of bonding contacts. Hence, the points demonstrated above do not hold any more. Typical examples of hypercoordinate molecules are triangular $[H_3]^+$, B_2H_6 (see Section 1.2) and SF_6 (see Problem 2). This is also the case for many borane and heteroborane clusters which will be treated in detail in Chapter 2. For such compounds it is not possible to draw a Lewis structure which would obey the two- and eight-electron rules.

There are also compounds with main-group atoms which are not hypercoordinated, i.e., which have less than five bonds, but which appear to have more than eight electrons. XeF_2 and SF_4 are typical examples. Drawing simple Lewis structures of these molecules leads to the conclusion that the Xe and S are ten-electron atoms. With more than eight electrons in their surroundings, they have been called hypervalent. This is also the case for S in SF_6 (12 electrons). Thus hypercoordinated compounds can be hypervalent (SF_6) if electron rich or hypovalent ($[H_3]^+$ and B_2H_6) if electron poor. Always remember that the terms "electron rich" and "electron poor" are defined relative to the Lewis model for bonding. They lose relevance when multicentered bonding is introduced.

In earlier days it was believed that hypervalency was due to participation of the high-lying nd orbitals of the hypervalent atom in bonding. Today, it is clear from very accurate theoretical calculations that this participation is small and should be neglected in any qualitative description of the bonding in these molecules. The (unfortunately) still-popular idea of significant nd participation in the bonding in hypervalent molecules comes from the fact that it allows a very appealing (but wrong!) localized two-center–two-electron description of these molecules. One should realize that localized two-center–two-electron bonding is not a common feature of chemistry as a whole. As shown in Section 1.2, three-center (delocalized) bonding is the maximum simplification possible for molecules such as B_2H_6. There are many stable electron-poor and electron-rich compounds which do not obey the eight- or 18-electron rules if a localized two-center–two-electron bond model (Lewis model) is applied but none the less satisfy the closed-shell principle. The same is true of the electron-poor cluster compounds discussed in Chapter 2.

1.3.5 The 8- and 18-electron rule in the context of multicentered bonds

It is reasonable to ask at this point whether hypercoordinated molecules obey the eight- and 18-electron rules if multicentered rather than two-centered bonding is used. After all we touted three-center bonding in Section 1.2 as the answer to the problem posed by molecules such as B_2H_6. Consider the B atom in B_2H_6. As shown in Section 1.2, we need four two-center–two-electron bonds and two three-center–two-electron bonds to utilize all of the available valence orbitals (8 + 6 = 14) and electrons (6 + 6 = 12), i.e., each three-center bond "consumes" three orbitals and holds two electrons. The B atom participates in two two-center–two-electron B–H interactions and two three-center–two-electron B–H–B interactions or it is associated with eight electrons! The terminal H is obviously associated with two electrons and the bridging H, being involved in one three-center–two-electron bond, is also associated with two electrons. What about SF_6? If you succeed in solving Problem 2 (SH_6) you will find that one needs two two-center–two-electron bonds and two three-center–four-electron bonds to utilize all the available valence orbitals (4 + 6 = 10) and electrons (6 + 6 = 12), i.e., each three-center bond in this case "consumes" three orbitals and holds four electrons. Now in the latter case, two electrons are in the bonding orbital and two electrons are in the non-bonding orbital. Look carefully at Figures 1.13 and 1.14. Do you see that by symmetry the non-bonding orbital has no contribution from the central atom? Thus, the three-center–four-electron bond only places two, not four, electrons on the S atom of SH_6. The S is associated with only eight electrons and therefore obeys the eight-electron rule.

There is no contradiction here – the eight-electron rule only fails if one insists on applying a localized two-center–two-electron bond model where it is not valid. Basically this is the tail (bonding model) wagging the dog (electronic structure) and any such dog is bound to be hyper(valent)! Even for the electron-poor cluster compounds, construction of a bonding description with an appropriate set of two- and n-center–two-electron bonds will satisfy the eight-electron rule for the cluster atoms. Thus, it has been bandied about that there is no such thing as an electron-deficient (poor) compound, only theory-deficient chemists!

Exercise 1.6. Use three-center bond ideas to show that Xe in hypervalent linear XeF_2 obeys the eight-electron rule.

Answer. As a crude approximation, let the Xe atom in XeF_2 use one np AO to form one three-center–four-electron bond with one p orbital from each of the F atoms. The other nine ns and np AOs contain 18 of the 22 valence electrons as lone pairs. Figure 1.14 with the remaining four electrons occupying the bonding and non-bonding three-center orbitals applies. Thus, at the Xe center, we have three lone pairs (six electrons) and one three-center–four-electron bond (two electrons in the only orbital with Xe character) giving a total of eight electrons associated with the Xe atom.

1.3.6 Why count?

In the chapters that follow you will find numerous exercises in counting electrons for clusters – elaborations of the eight- and 18-electron rules for these complex structures. The same factors that cause the eight- and 18-electron rules to fail will similarly limit cluster-counting rules based upon them. Like these fundamental rules, even when satisfied, the cluster-counting rules yield no detailed information on electronic structure. Hence, the bolder student occasionally asks, "Why count?" by which he or she means "Of what real value are these counting exercises if little is learned about where the electrons really are?"

Let's address the question now as it equally applies to the 8- and 18-electron rules. Compounds that follow the rules are classified as "normal" and define the electronic "accounting" favored when no other factors are of overriding importance. The rules give the experimental chemist a simple method to rationalize and predict compound stoichiometry and connectivity. The rules permit logical categorization of the myriad of compounds via similarities in an electron count. Both facilitate more rapid development of a field such as cluster chemistry where both structure and composition can seem intimidating in the absence of an organizing principle.

But it is also important to appreciate the fact that all rules have a limited domain in which they are valid. Compounds that do not follow the rules become objects of interest often because they are associated with properties of value, e.g., the Lewis acidity of six-electron BF_3. But, as we will discover, very large clusters cannot follow the existing counting rules as they lie outside the domain of validity. Yet these large clusters, nanoparticles, must have a "drummer" to which they march. A shadowy outline of this presently unknown "drummer" appears in the context of the existing rules. That is, counting is a place to start!

Problems

1. (a) Using paper chemistry and the 18-electron rule, "make" three compounds for each of the metals Cr, Mn and Fe using one or more of the ligands, H, CO and η^5-C_5H_5. Restrict yourself to neutral, mono- and dinuclear complexes. Check some of them out in the *Dictionary of Organometallic Compounds* to see if they have been isolated and what their properties are. (b) Chances are you chose compounds with two-center–two-electron bonding only. Hence, try your hand at explaining the bonding in $[(CO)_5Cr-H-Cr(CO)_5]^-$, which has a linear Cr–H–Cr interaction.

Table 1.2. *Coordinates*

1 B	0.000 0	0.000 0	0.000 0
2 F	0.000 0	0.000 0	1.400 0

Table 1.3. *Mulliken atomic charges*

1 B	0.024
2 F	−0.024

Table 1.4. *Net orbital populations (diagonal) and overlap populations (off-diagonal)*

		1B 2s	1B 2px	1B 2py	1B 2pz	2F 2s	2F 2px	2F 2py	2F 2pz
1 B	2 s	1.465 96							
1 B	2 px	0.000 00	0.138 73						
1 B	2 py	0.000 00	0.000 00	0.138 72					
1 B	2 pz	0.000 00	0.000 00	0.000 00	0.778 98				
2 F	2 s	0.078 30	0.000 00	0.000 00	0.167 02	1.656 24			
2 F	2 px	0.000 00	0.218 71	0.000 00	0.000 00	0.000 00	1.642 57		
2 F	2 py	0.000 00	0.000 00	0.218 71	0.000 00	0.000 00	0.000 00	1.642 56	
2 F	2 pz	0.035 37	0.000 00	0.000 00	0.187 33	0.000 00	0.000 00	0.000 00	1.630 80

2. To test your understanding of the MO model for a typical octahedral coordination complex, construct an appropriate, qualitative MO diagram for O_h SH_6 (a model for known SF_6). Hint: first calculate the total number of MOs you should end up with from the number of available basis functions (AOs). Second, compare the valence AO functions of S with those of a transition metal (refer to Figure 1.9 and realize that, for a coordinate system with the H atoms on the x, y and z axes, the AO functions of the central atom and the symmetry-adapted linear combinations of ligand functions transform as: s, a_{1g}; p, t_{1u}; d_{xy} d_{xz} d_{yz}, t_{2g}; $d_{x^2-y^2}$ d_{z^2}, e_g in the O_h point group). Now count the number of filled MOs and the number of S–H bonding interactions.

3. Tables 1.2–1.6 contain the results of an approximate MO calculation (Fenske–Hall) on the BF molecule. From this output: (a) construct a MO diagram showing MO energy levels and qualitative AO compositions in MO drawings; (b) examine the HOMO and LUMO relative to Lewis acid/base behavior and compare it with CO. Would BF be suitable for coordination to, e.g., a Cr center?; (c) use the Mulliken charges to predict the direction of the dipole moment; (d) examine the Mulliken overlap populations and decide whether it is proper to describe the B–F bond as a single, double or triple bond.

Table 1.5. *Eigenvalues and eigenvectors*

		1–( )	2–( )	3–( )	4–( )	5–( )	6–( )	7–( )	8–( )
Energy =		−44.926 50	−22.329 90	−20.480 10	−20.480 10	−11.347 80	−4.707 40	−4.707 40	33.099 80
1 B	2 s	0.138 92	−0.312 11	0.000 00	0.000 00	0.785 03	0.000 00	0.000 00	−0.722 30
1 B	2 px	0.000 00	0.000 00	0.263 37	0.000 00	0.000 00	0.992 99	0.000 00	0.000 00
1 B	2 py	0.000 00	0.000 00	0.000 00	0.263 37	0.000 00	0.000 00	0.992 99	0.000 00
1 B	2 pz	0.153 09	−0.088 05	0.000 00	0.000 00	−0.598 58	0.000 00	0.000 00	−0.973 98
2 F	2 s	0.876 95	0.241 76	0.000 00	0.000 00	0.025 20	0.000 00	0.000 00	0.772 36
2 F	2 px	0.000 00	0.000 00	0.906 25	0.000 00	0.000 00	−0.483 84	0.000 00	0.000 00
2 F	2 py	0.000 00	0.000 00	−0.000 01	0.906 25	0.000 00	0.000 00	−0.483 84	0.000 00
2 F	2 pz	−0.011 08	0.851 75	0.000 00	0.000 00	0.299 66	0.000 00	0.000 00	−0.593 94

Table 1.6. *% Character of molecular orbitals*

		1( )	2( )	3( )	4( )	5( )	6( )	7( )	8( )
Energy =		−44.93	−22.33	−20.48	−20.48	−11.35	−4.71	−4.71	33.10
1B	2s	5.58	14.82	0.00	0.00	55.74	0.00	0.00	23.86
1B	2px	0.00	0.00	12.40	0.00	0.00	87.60	0.00	0.00
1B	2py	0.00	0.00	0.00	12.40	0.00	0.00	87.60	0.00
1B	2pz	8.10	1.24	0.00	0.00	38.47	0.00	0.00	52.19
2F	2s	86.24	2.70	0.00	0.00	.01	0.00	0.00	11.05
2F	2px	0.00	0.00	87.60	0.00	0.00	12.40	0.00	0.00
2F	2py	0.00	0.00	0.00	87.60	0.00	0.00	12.40	0.00
2F	2pz	0.09	81.23	0.00	0.00	5.79	0.00	0.00	12.89

In case you are wondering BF is a known species – an example of a high-temperature molecule.

4. There are substantial numbers of small molecular species that are important species at high temperatures that are not normally discussed in standard inorganic courses. For example, the vapor over solid NaCl contains diatomic NaCl molecules as well as NaCl dimers. Does the stoichiometry NaCl fit with the simple ideas of valence derived from molecules like $SiCl_4$? Explain. Construct a qualitative MO diagram with orbital drawings for the NaCl diatomic molecule and compare it with that for a homonuclear, isoelectronic diatomic molecule. Does this help in answering the first question?

5. (a) Draw the HOMO and LUMO of H_2. (b) Consider the interaction of H_2 with the Lewis acid BH_3 and predict the structure of the initial acid–base adduct. (c) Consider the replacement of one CO of $Cr(CO)_6$ with H_2 coordinated in the same manner as with BH_3. Can the LUMO of H_2 interact with any of the filled Cr functions? See (Kubas 2001) for a discussion of the importance of this model in explaining the coordination of dihydrogen.

6. In Figure A1.20 a complex of Cp_2Zr and 2,3-dimethyl-1,3-butadiene is shown and in the accompanying text it is stated that on coordination the outer formal C=C distances increase whereas that between the central carbon atoms (formal C–C single bond)

decreases. Show that these structural changes are consistent with depopulation of the butadiene HOMO and population of the butadiene LUMO on coordination to the metal fragment.

Additional reading

Section 1.1

Albright, T. A., Burdett, J. K. and Whangbo, M.-H. (1985). *Orbital Interactions in Chemistry*. New York: Wiley.

Hoffmann, R. (1981). *Science*, **211**, 995.

Cotton, F. A. and Walton, R. A. (1993). *Multiple Bonds Between Metal Atoms*. Oxford: Oxford University Press.

Section 1.2

Lipscomb, W. N. (1963). *Boron Hydrides*. New York: Benjamin.

Muetterties, E. L. (Ed.) (1975). *Boron Hydride Chemistry*. New York: Academic Press.

Moezzi, A., Olmstead, M. M. and Power, P. P. (1992). *J. Am. Chem. Soc.*, **114**, 2715.

2
Main-group clusters: geometric and electronic structure

Clusters – a form of matter with structure and properties lying somewhere between those of atoms and solid-state substances – impact a substantial fraction of chemistry, drawing the attention of both inorganic and physical chemists. The larger the cluster the stronger the connection with solid-state chemistry and the greater the ramifications for modern materials science in the area of nanochemistry. The term cluster is used to designate a three-dimensional assembly of atoms and cluster structures may be found in s-, p-, and d-block element chemistries. When composed of a single element, the cluster motif complements the chains and rings of molecular catenates and the chains, sheets and networks of solid substances. Clusters are found with external ligands as well as without. Cluster structure is the focus of this text and we intend to show that cluster electronic structure serves as a bridge between molecular compounds and non-molecular solid-state compounds. These connections will become more readily apparent as the structural properties of clusters are developed.

The story begins in this chapter with the clusters of simplest geometric and electronic structure. These are clusters of p-block elements with defined stoichiometry and structure in which the cluster surface-atom valences are "terminated" with ligands. The large number known provide the factual base from which clever people have derived models that connect atomic composition with structure. In turn, these p-block models provide a foundation on which to build an understanding of more complex clusters such as condensed clusters, bare clusters and transition-metal clusters. A more comprehensive account of the structural chemistry will be found in older books and reviews, a selection of which will be found in the reading list at the end of each chapter.

2.1 Definition of a cluster

Definitions are useful if they permit a body of information to be organized around a model. The more information organized the better the model. But a caution must be

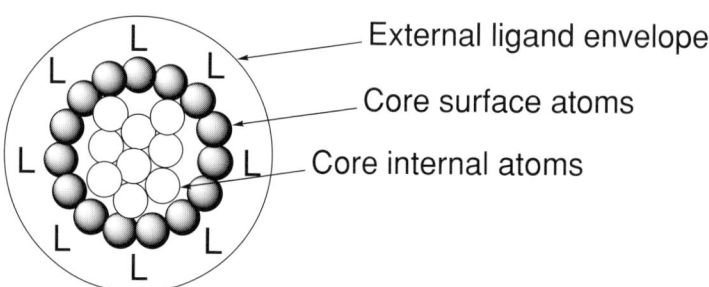

Figure 2.1

raised. The better the model the greater the temptation to give it authority beyond the system upon which it is based. Obviously, when new "square pegs" are forced into the "round holes" of an inadequate model, nothing is learned. But, unlike a child learning shapes, it is not always easy to see that the model has "round holes." Recognition is the key, and understanding the model the first step. Hence, although the models discussed below are firmly established, they are not by any means inclusive of all cluster chemistry and clusters yet to be discovered may require additional models.

Figure 2.1 illustrates schematically an idealized cluster which is spherical and surrounded by a ligand envelope. The surface atoms of the cluster are bound to the external ligands. If a surface atom does not have a ligand, it usually possesses a lone pair. As size and nuclearity increase, the cluster may contain internal atoms not directly bound to the ligands. A cluster possesses a geometric structure or shape which in many cases can be directly related to core stoichiometry and ligand number and type. Shape is an important property as it is directly measurable and reflects the connection between geometric and electronic structure. Similar to the fundamental bonding models of Chapter 1, successful cluster models permit stoichiometry and shape to be predicted. The entire cluster can carry a positive or negative charge and, if charged, will be associated with appropriate counterions. This chapter deals with covalently bonded (strong bonds relative to thermal energy) clusters of the p-block elements with and without external ligands. Most contain no internal atoms; however, clusters with internal atoms will be introduced towards the end of the chapter.

2.2 Three-connect clusters

The surface atoms of three-connect clusters have three nearest neighbors within bonding distance (Figure 2.2). Many clusters formed from p-block elements are three-connect clusters. No new bonding ideas outside of the localized two-center–two-electron bond and the eight-electron rule are required to understand most

2.2 Three-connect clusters

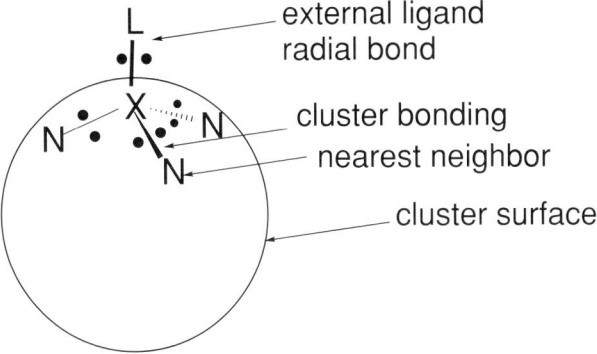

Figure 2.2

clusters of this type. Why? We know a p-block element has four valence functions and can form four localized electron-pair bonds pointed at the corners of a tetrahedron. For the chosen atom the spherical geometry of an ideal cluster illustrated in Figure 2.2 requires one pair to be directed outwards in a radial direction leaving three pairs to form bonds in the cluster surface. Thus, as long as the number of cluster atoms within covalent bonding distance is no larger than three, two-center bonds suffice. Note that the out-pointing orbital can accommodate a lone pair, the donor pair of a Lewis base, or a bonding pair to a one-electron ligand, e.g., CH_3.

2.2.1 Two-center bonding

For this reason cluster shapes defined by polyhedra with three or less vertices of connectivity can often be described by a set of two-center–two-electron bonds and are no more complicated than many simple molecules, e.g., CH_4. Thus, the trigonal prismatic and "butterfly" clusters shown in Figure 2.3 fit into this category whereas a square pyramidal shape with one vertex of connectivity four would not (see Figure 2.6). Because the bonding can be described with two-center bonds, these cluster types are sometimes called "electron precise." The number of cluster bonding electrons is precisely two times the number of edges of the polyhedron that defines its shape. As found with other simple molecules, the lines that define the shapes also express a localized version of the electronic structure. The frameworks in Figure 2.3 also illustrate connections between three-connect clusters and complex ring systems.

2.2.2 Electron-counting rules

In the common three-connect cluster shapes illustrated in Figure 2.3 the number of electrons that can be associated with the closed clusters is $5n$, where n is the

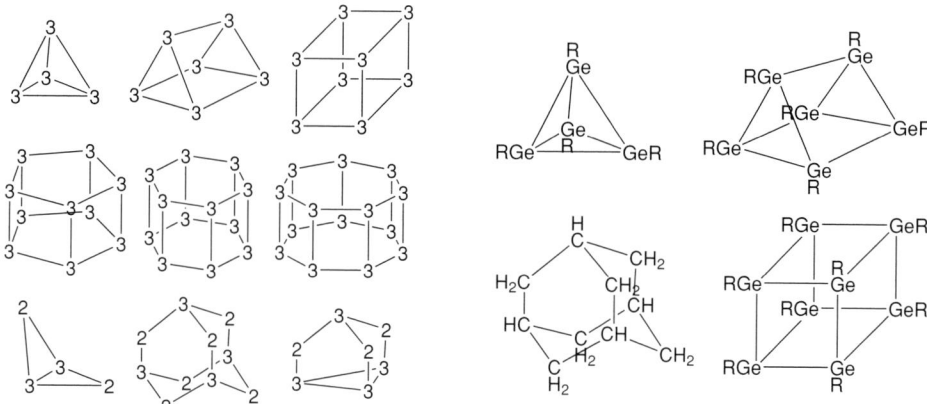

Figure 2.3

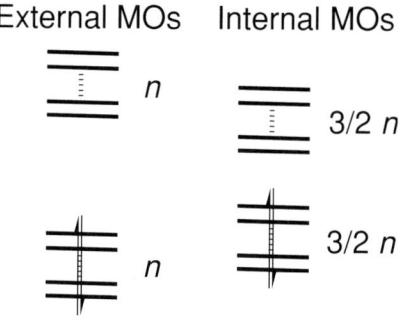

Figure 2.4

nuclearity of the cluster. The origin of this electron-counting rule is easily seen from a consideration of the number of filled and unfilled molecular orbitals (Figure 2.4). As each two-center–two-electron bond generates one bonding and one antibonding orbital, it follows that precisely half of the $3n$ skeletal orbitals are bonding and filled. The external cluster orbitals must also be filled to satisfy the eight-electron rule adding $2n$ additional electrons. This electron count, $5n$, is defined as the number of cluster valence electrons (cve).

Another way to count the electrons in clusters restricts the count to only those associated with cluster bonding: hence, skeletal-electron pairs (sep). For three-connect clusters sep $= 3/2n$. The assumption that orbitals associated with cluster ligand bonding can be considered independently of orbitals associated with cluster framework bonding results in an important simplification for more complex systems. However, it is a separation that is convenient rather than rigorous. We will see later that mixing between external and internal cluster orbitals can have interesting consequences.

Some readers will be mumbling that this electron-counting rule for electron-precise clusters is pretty useless. And they are perfectly correct! When two-center–two-electron bonds are sufficient, the bonding analysis is so simple that an electron-counting rule adds a frustrating layer of algebraic complexity and can inhibit understanding. However, the development of the $5n$ cluster valence-electron count (and its equivalent $3/2n$ skeletal-electron pair count) for three-connect clusters illustrates a point – it is possible to express the relationship of cluster shape to compound stoichiometry as a valence-electron sum. In systems in which the cluster bonding is more complex and where a higher-level treatment is necessary to obtain even a qualitatively correct description, the existence of an algebraic expression connecting valence-electron count and stoichiometry is desirable. It reveals relationships between known clusters (shape rationalization and correlations between compounds of disparate elemental composition) as well as predicting cluster shape from an electron count easily derived from cluster composition and charge.

2.2.3 Three-center bonding

R_4E_4, E = Al, Ga, In, also exhibit tetrahedral cluster shapes even though they possess only four sep and 16 cve. We still have 12 orbitals to utilize but only eight electrons contributed by four R–E fragments to fill them. We encountered a similar situation in Chapter 1 with B_2H_6 and three-center–two-electron bonding was introduced to solve the problem. A similar approach provides a solution to the R_4Ga_4 problem. Each R–Ga fragment contributes three orbitals and two electrons from which we can construct four three-center–two-electron Ga–Ga–Ga bonds lying in the four triangular faces of the cluster. Thus, as shown in the right-hand side of Figure 2.5, each Ga atom is associated with eight electrons: two from the R–Ga bond and two from each of the three-center Ga–Ga–Ga bonds it participates in. The same analysis can be used to describe B_4Cl_4.

2.2.4 MO models

Let us compare the MO descriptions of six and four sep R_4E_4 tetrahedral clusters. In Figure 2.5 the cluster MO energies for E = C and E = Ga are shown alongside the localized descriptions discussed above. The number of filled and unfilled MOs is equal for E = C: a consequence of the two-center–two-electron bond model. Conversely, the number of filled MOs is less than the number of empty MOs for E = Ga: a consequence of the three-center–two-electron bond model. Note also that an e symmetry pair of orbitals lies in between lower-energy filled orbitals and

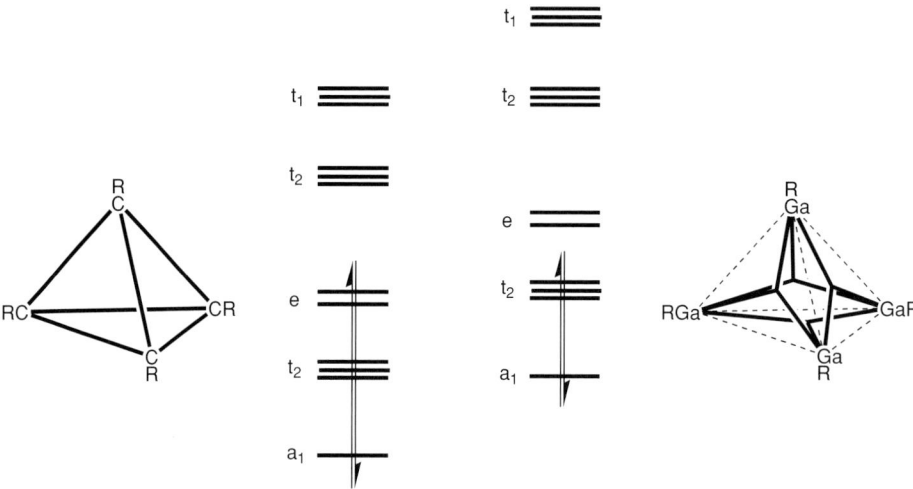

Figure 2.5

higher-energy unfilled orbitals. This is a situation we will encounter later in the text and is one which often leads to multiple electron counts for the same shape depending on the properties of the atoms that make up the cluster.

2.2.5 Model choice

Before going further we need to make a point concerning the usefulness of simple models vs. the validity of the electronic-structure model upon which they are based. Tetrahedral P_4 can be easily considered as being bonded by four two-center–two-electron bonds just like tetrahedral R_4C_4. The stability of tetrahedrane is reduced by its steric strain energy of about 502 kJ mol^{-1}. Hence its synthesis was an experimental achievement of note. In contrast, P_4 is a common form of elemental phosphorus and has an estimated strain of only 67 kJ mol^{-1} even though both molecules have 60° E–E–E angles. Clearly the electronic structures must be significantly different. On this basis it has been argued that although the localized two-center–two-electron bond model "works" for P_4, its electronic structure is better represented as a highly delocalized one.

This point may appear to be semantics to one interested primarily in the connection between stoichiometry and geometric structure; however, for more detailed considerations the model used matters. For example, the two-center–two-electron bond model with a lone pair on P suggests electrophilic attack at a single P atom, e.g., protonation leading to a P–H terminal bond. On the other hand, in the delocalized MO model the degenerate e symmetry HOMOs of P_4 are centered on the P–P edges. Frontier-orbital considerations, then, suggest electrophilic attack would

2.3 Four-connect and higher clusters

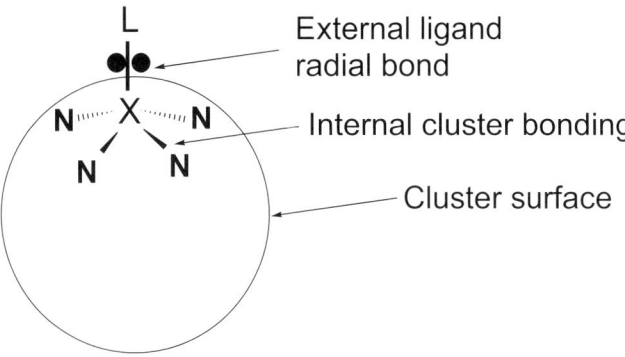

Figure 2.6

take place at a P–P edge. Indeed the structure of [HP$_4$]$^+$ in the gas phase exhibits a P–H–P edge bridging H. The lesson is that the success of an electron-counting rule does not imply deep understanding of the electronic structure, i.e., electron-density distribution over the nuclear framework.

2.3 Four-connect and higher clusters

Look at the ligated cluster atom X on the surface of the spherical cluster shown in Figure 2.6. Positioned within bonding distance of four other surface cluster atoms, it is said to reside on a four-connect cluster vertex. In common with the three-connect cluster, an outward-pointing two-center–two-electron bond to an external ligand or an outward-pointing lone pair is assumed. Thus, the 8-electron rule restricts to six the number of electrons associated with the remaining bonding connections to the four adjacent atoms. Hence, partitioning the valence electrons exclusively into two-center–two-electron bonds is not possible. Multicentered bonding is required. To do so a general MO model of closed clusters with at least one four-connect vertex is introduced now. Later, more localized bonding representations with two- and three-center–two-electron bonds will be discussed. Historically, the latter came first.

2.3.1 Deltahedra

The polyhedral shapes pertinent to this class of clusters are the deltahedra shown in Figure 2.7. Deltahedra are polyhedra containing exclusively triangular faces. The vertex connectivities are shown on the drawings. Those for $n = 4$ (not shown), 6 and 12 have uniform connectivities of three, four and five; hence, the cluster atom centers lie on a sphere. The others with less uniform connectivities are less spherical in shape if all edge lengths are equal. As it is a three-connect cluster, the tetrahedron

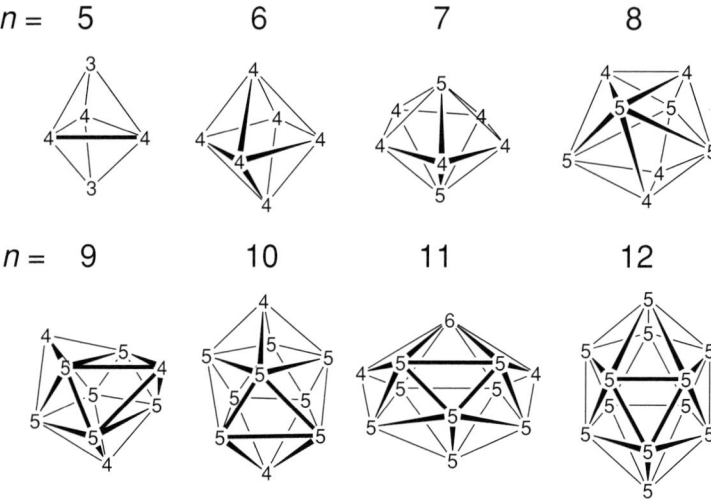

Figure 2.7

need not be considered a member of this group of clusters. Others do include it but we will not.

The trigonal bipyramid, $n = 5$, contains two three-connect and three four-connect vertices and the interested reader will find that its electronic structure has spawned considerable discussion in the literature concerning the nature of the atom–atom bonding in the equatorial triangle. Only one of the deltahedra in Figure 2.7 has a vertex of connectivity larger than five and that occurs for $n = 11$ where a single vertex has connectivity six. Those with $n = 5$, 7 and 10 are known as polar deltahedra and the cluster atoms lying on the highest-order rotational axes (the poles) possess equal connectivities, whereas the remaining atoms lie in an equatorial belt on vertices also of equal connectivities although differing from those of the poles.

2.3.2 Electronic requirements

Historically, the first general insight into the electronic structures of deltahedra came from Longuet-Higgins' theoretical considerations of boron compounds. He predicted the stoichiometry of boron hydrides produced conceptually from solid-state borides containing six-vertex octahedral clusters and from elemental B containing 12-vertex icosahedral clusters by rupture of inter-cluster bonding followed by terminating the dangling B bonds with H. The generalized result of these considerations for an n-vertex deltahedron is shown in Figure 2.8 in the same style as that of three-connect clusters in Figure 2.3. Consider specifically the 12-vertex icosahedron made up of BH fragments. There will be $5 \times 12 = 60$ MOs, $2 \times 12 = 24$ of which will be associated with the external B–H bonding. Recall that the

2.3 Four-connect and higher clusters

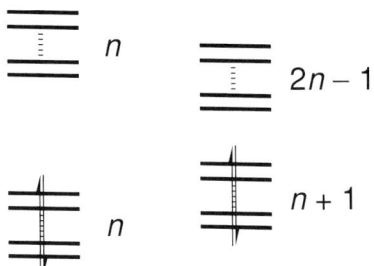

Figure 2.8

qualitative MO criterion for stability is a large HOMO–LUMO energy gap. $B_{12}H_{12}$ possesses 48 valence electrons; however, the large gap was found between the 25th and 26th MOs. Hence, the 25th MO was designated as HOMO and $[B_{12}H_{12}]^{2-}$ was predicted to be the stable stoichiometry and charge for an icosahedral molecular borane. Subtraction of the 12 external filled B–H MOs leaves 13 or $n+1$ skeletal-bonding MOs. Subsequent to this work $[B_{12}H_{12}]^{2-}$ was synthesized and shown to have a high barrier to decomposition dramatically confirming the stability of this structure and composition. Why dramatic? The contrast between the reactivity of B_2H_6 and $[B_{12}H_{12}]^{2-}$ is enormous.

If we compare Figures 2.4 and 2.8 we see that the number of filled and unfilled radial external cluster orbitals, n, is identical to that of a three-connect cluster of the same order. However, the remaining $3n$ orbitals associated with the skeletal bonding are no longer divided equally between filled and empty. In fact, the number of high-lying cluster core MOs is $n-2$ larger ($2n-1-n-1$) than the number of low-lying cluster core orbitals. As only the latter are filled, the number of sep is $n+1$ vs. $3/2n$ for three-connect clusters whereas the number of cve is $4n+2$ vs. $5n$ for three-connect clusters.

For the $n=6$ and 12 clusters with 12 and 20 bonding edges the cluster bonding is accomplished with 7 and 13 bonding pairs, respectively. Because the number of occupied cluster orbitals is less than the number generated by treating each cluster edge as a two-center–two-electron bond, these compounds have been called "electron-deficient" compounds. As noted in Chapter 1, some have taken strong exception to the use of the term; however, it does contain an element of truth in the sense that many borane clusters are prone to add Lewis bases thereby increasing the average number of electrons per B center.

Consider the example of $[B_{12}H_{12}]^{2-}$, $n=12$. There are 12 filled BH bonding orbitals and 12 empty BH antibonding orbitals. In addition, there are 13 filled cluster bonding orbitals and 23 unfilled cluster non-bonding and antibonding orbitals. Alternatively, if no external–internal cluster-orbital separation is made 50 cve fill the 25 bonding orbitals leaving 35 high-lying unfilled orbitals.

In order for each vertex to satisfy the eight-electron rule, it must be associated with six cluster bonding electrons to supplement the external B–H bonding pair. Hence, some complex sharing must be taking place. This is another way of saying the cluster valence electrons are delocalized throughout the cluster bonding network. An important corollary is that, in general, the lines drawn to define shapes of high connectivity clusters do NOT correspond to two-center–two-electron bonds. The idea of expressing both geometry and bonding with straight lines is so convenient for the chemistry of C and the elements to the right of C that the separation of the two constitutes a conceptual problem for many students and those not familiar with cluster bonding concepts.

The fact that the lines between cluster atoms do not in general represent two-center–two-electron bonds is reflected empirically in the distances between cluster atoms which are longer and vary more than they do for non-cluster compounds. For example, in tricapped trigonal prismatic $[B_9H_9]^{2-}$ the B–B edge distances range from 1.68 Å to 1.93 Å compared to twice the covalent radius of boron of 1.64 Å. Of course in the clusters of higher symmetry, the variation is much less, e.g., $[B_{12}H_{12}]^{2-}$, 1.76 Å to 1.78 Å.

2.4 The Wade–Mingos electron-counting rule

The conclusions presented above were formulated into a rule which is sufficiently general to be useful. The implications of the results were described nearly contemporaneously by Wade and Mingos with emphasis on main-group and transition-metal clusters, respectively. They saw that the principle, based on the series of deltahedra in Figure 2.7, also applied to open clusters. Both recognized that the idea forged a link between main-group and metal clusters. Once the initial concept was published in the early 1970s, ramifications rapidly followed, culminating in the definitive monograph of Mingos and Wales. However, developments leading up to the formulation of this rule took many years as the structural data were generated one piece at a time and the initial attempts to fit patterns were based on insufficient structural data. It wasn't until the geometric patterns were put together in the correct way by Williams that the puzzle was solved. The picture painted is a pretty one! Let's take a look.

Cast in terms of cve, the rule for a main group cluster is:

For a deltahedron of order n, *there are* $4n + 2$ *cve associated with cluster bonding.*

In terms of sep, the rule is:

For a deltahedron of order n, *there are* $n + 1$ *sep associated with cluster bonding.*

Some may protest a rule based on approximate calculations may not reflect anything "real" in nature. Others will counter with the argument that it works, that is, other rules and explanations of chemistry stand on spongy physical support but chemists still use them for convenience. Here, on the contrary, work by Stone subsequent to the publication and use of the cluster electron-counting rules provides strong theoretical support. Application of the theory of tensor surface harmonics to the analysis of the bonding problem presented by closed spherical clusters verified the $n + 1$ sep rule. The theory provides strong physical justification for the counting rule, and it has been used effectively in the discussion of selected cluster problems. Elegant though the theory be, it adds little more understanding to the beginning student. Hence, it will not be dealt with further in this text. In fact, for discussions of the detailed electronic structure of a real cluster going beyond electron counting, explicit calculations rather than idealized general theory should be used.

2.5 Closed-cluster composition and structure

Both versions of the rule are easy to use for closed clusters. There is no restriction to element type, external-ligand type or even number of occupied vertices of a deltahedron – only the number of sep associated with a deltahedral shape is fixed. However, in terms of sep count, one cannot stray too far from an average of two electrons per cluster fragment for closed-cluster shapes without having to deal with large charges. Hence, the B–H fragment is an ideal cluster fragment. Thus, it is no surprise that the canonical deltahedral shapes in Figure 2.7 are illustrated by the homologous series of *closo*-boranes $[B_nH_n]^{2-}$, $n = 6$–12. (*Closo* refers to a fully closed deltahedral cluster.)

In Figure 2.9 a variety of *closo*-heteroborane clusters are illustrated for $n = 10$ and 12. The extra numbers preceding the compound formula designate the positions of the heteroatoms. The accepted numbering scheme is shown by the numbered framework in the figure. Position one is taken to be an atom on the axis of highest symmetry (C_4 for $n = 10$) and numbering is clockwise around this axis proceeding layer by layer.

The deltahedron for $n = 10$, a bicapped square antiprism, exhibits two four-connect and eight five-connect vertices. Hence, for one heteroatom in a ten vertex *closo*-cluster we have 1- and 2-isomers and two heteroatoms in 1,10-, 1,6- 1,2-, 2,3-, 2,4-, 2,6- 2,8-isomers. Different placements generate different cluster stabilities. A rule of thumb is that the more electronegative element prefers the lower-connectivity vertex. Multiple heteroatoms more electronegative than B prefer non-adjacent positions as far apart as possible. Rearrangement to the most stable isomeric form need not be fast. In the case of icosahedral clusters, for example, the barrier to rearrangement is large and isomers can be isolated.

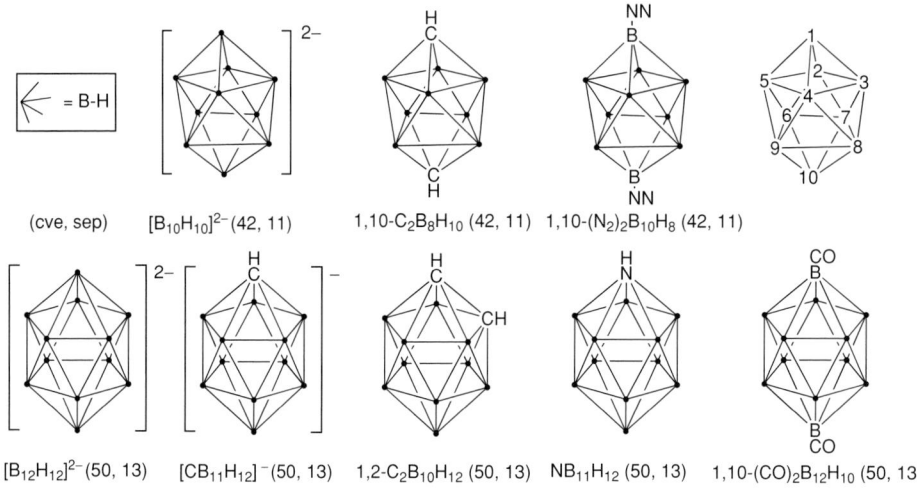

Figure 2.9

Exercise 2.1. Verify the cluster electron counts for the closed clusters in Figure 2.9. $NB_{11}H_{12}$ is an interesting one. Can you think of a reasonable composition that would correspond to a 12-vertex closed cluster containing one bare N atom and 11 B atoms?

Answer. $[B_{10}H_{10}]^{2-}$: The number of valence electrons for B is three, H ligands are one-electron ligands, and the cluster charge is 2. cve = $10 \times 3 + 10 \times 1 + 2 = 42 = 4n + 2$. Consistent with the discussion above, the B–H external cluster bond is taken to involve one electron from H and one from B; hence, the B–H fragment contributes two electrons to cluster bonding. sep = $(10 \times 2 + 2)/2 = 22/2 = 11 = n + 1$.

1,10-$C_2B_8H_{10}$: cve = $2 \times 4 + 8 \times 3 + 10 \times 1 = 42$; sep = $(2 \times 3 + 8 \times 2)/2 = 11$. The C–H fragment contributes three electrons to cluster bonding – C is one step to the right of B.

1,10-$B_{10}H_8(N_2)_2$: cve = $10 \times 3 + 8 \times 1 + 2 \times 2 = 42$. The end-on bound dinitrogen ligand is a two-electron ligand to the cluster. sep = $(8 \times 2 + 2 \times 3)/2 = 11$. The B–H are two-electron fragments but the two B–N≡N fragments are three-electron fragments as its two-electron donor ligand allows the B atom to use all three valence electrons in cluster bonding.

$[CB_{11}H_{12}]^-$: cve = $1 \times 4 + 11 \times 3 + 12 \times 1 + 1 = 50 = 4n + 2$. sep = $(1 \times 3 + 11 \times 2 + 1) = 26/2 = 13 = n + 1$.

$C_2B_{10}H_{12}$: cve = $2 \times 4 + 10 \times 3 + 12 \times 1 = 50 = 4n + 2$. sep = $(2 \times 3 + 10 \times 2) = 26/2 = 13 = n + 1$.

$NB_{11}H_{12}$: cve = 1 × 5 + 11 × 3 + 12 × 1 = 50 = 4n + 2. sep = (1 × 4 + 11 × 2) = 26/2 = 13 = n + 1.

$B_{12}H_{10}(CO)_2$: cve = 12 × 3 + 10 × 1 + 2 × 2 = 50 = 4n + 2. sep = (2 × 3 + 10 × 2) = 26/2 = 13 = n + 1.

A 12-vertex cluster with a bare N atom is $NB_{11}H_{11}$: cve = 1 × 5 + 11 × 3 + 11 × 1 = 49; thus, one more electron is needed giving $[NB_{11}H_{11}]^-$.

Rationalization of known compounds provides a level of usefulness that justifies the rule. But the rule also permits observed molecular stoichiometries of newly synthesized compounds to be translated into a cluster shape. For example, $[Al_{12}{}^tBu_{12}]^{2-}$ has cve = 50 or sep = 13 consistent with n = 12 and a deltahedral structure. The compound has been synthesized and an X-ray diffraction study reveals an icosahedral shape. The ability to suggest reasonable structures based on knowledge of a molecular formula generated by a technique like mass spectrometry accelerated the development of cluster chemistry simply because rapid spectroscopic methods can be more productively applied. Although efficient X-ray crystallographic structure determination reduces its importance for compounds that can be isolated in pure crystalline forms, transient intermediates detected in a reaction mixture can now be given reasonable structures.

2.6 Open *nido*-clusters

Open clusters add some complexity to the cluster structure problem. These clusters are considered to be incomplete spherical clusters. Those that can be completed by the addition of a single fragment are called *nido*-clusters (net-like).

2.6.1 nido-*Clusters without bridging hydrogens*

Consider the problem posed by the known molecular compound with the formula $C_4B_2H_6$. First do the electron count. The sep = 8 = n + 1: n = 7. This suggests a structure based on a pentagonal bipyramid. OK, but there are only six cluster fragments (four C–H and two B–H) so if this rule applies, one vertex must remain empty. Is this correct? The observed structure, shown in Figure 2.10, shows it is. As expected from the relative electronegativities and our rule of thumb, a B–H rather than C–H fragment is found in the five-connect vertex. Heteroatom positional placement is not the only way of generating isomers in this open cluster. Isomerism can also be generated from the choice of whether a five- or four-connectivity vertex of the bicapped pentagon is left vacant. The rule of thumb for borane-based clusters is that the higher connectivity vertex is left vacant.

Many texts specify the sep count for a *nido*-cluster as 2n + 4 where n is the number of cluster atoms rather than the order of the deltahedron upon which the

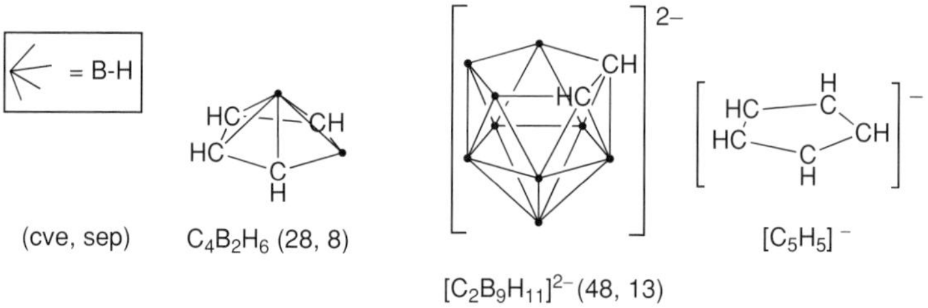

46   *Main-group clusters*

[Figure 2.10 showing: box with arrow = B–H (cve, sep); C₄B₂H₆ (28, 8) structure with HC, HC, CH, CH, C-H; [C₂B₉H₁₁]²⁻ (48, 13) icosahedral cage with HC, CH vertices; [C₅H₅]⁻ pentagonal structure]

Figure 2.10

structure is based. As one must go back to the parent deltahedron to generate all the possible open shapes anyway, this refinement will NOT be used here. Unless specifically noted otherwise, the parameter n will always be used to refer to the number of vertices of the parent deltahedron – with thoughtful application, one rule, the $n + 1$ rule, suffices for both closed and open clusters.

Why have we ignored the cve count. Naively applied, the cve = 28 = $4n + 2$: $n = 6.5$! This is not very helpful. What went wrong? If a vertex is vacant there is one missing radial external cluster orbital containing a pair of electrons. Hence, the cve for a cluster with one vacant vertex is (external cluster electrons) + (cluster bonding electrons) = $(2n - 2) + (2n + 2) = 28$ for $n = 7$. This is the same answer obtained by the sep method. Note that two electrons were subtracted from the $2n$ normally associated with a closed cluster of order n. In general, then, the cve count can be written as $4n + 2 - 2x$ where x is the number of vacant vertices. It is easy to see that $n = 7$ and $x = 1$ is a solution for cve = 28; however, the sep approach avoids the headaches of keeping track of those pesky external bonds. Hence, for open clusters the sep approach exhibits a greater ease of application. In mixed metal–main-group clusters, considered in Chapter 5, we will see that there are some advantages to the cve count simply because it does not force a division of the valence electrons into external and core cluster bonding electrons. The method used is a personal choice. Both are found in the literature.

Exercise 2.2. Justify the shape of $[C_2B_9H_{11}]^{2-}$ shown in Figure 2.10.

Answer. With two C–H fragments, nine B–H fragments and a charge of -2, the cluster has $(2 \times 3 + 9 \times 2 + 2)/2 = 13$ sep. Thus, the structure must be derived from an icosahedron (12-vertex deltahedron) but there are only 11 fragments. One vertex (all are equivalent in an icosahedron) is left vacant (a *nido*-structure). Show for yourself that the cve count of 48 gives the same result.

$[C_2B_9H_{11}]^{2-}$ is prepared from 1,2-$C_2B_{10}H_{12}$ (Figure 2.9) by vertex "decapitation" with base. The B–H vertex preferentially removed is that situated adjacent to two C–H fragments (the position 3-vertex as shown in Figure 2.9) and its removal generates

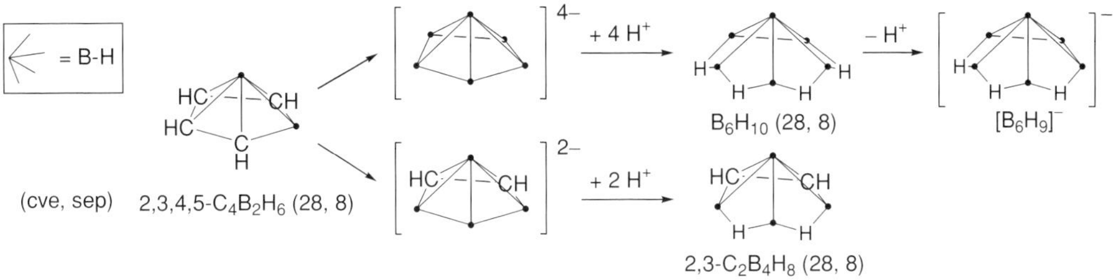

2.6 Open nido-clusters 47

Figure 2.11

an open face with the C atoms in adjacent positions. This open cluster $[C_2B_9H_{11}]^{2-}$ is an important one as it presents a pentagonal face with five orbitals containing six valence electrons to a binding partner. The latter point will not be obvious now but later in this chapter you will see how it comes about. If this point is just accepted for now, $[C_2B_9H_{11}]^{2-}$ constitutes an analog of $[C_5H_5]^-$, that ubiquitous ligand of organometallic chemistry that binds η^5- (all five C atoms bound to the metal center) to transition metal centers. Indeed, Hawthorne has developed the metal chemistry of $[C_2B_9H_{11}]^{2-}$ into a sub-area to rival that of metal-cyclopentadiene chemistry. More examples of this "ligand" chemistry will be described in Chapter 4.

2.6.2 nido-Clusters with bridging H atoms

In the examples considered thus far utilizing a sep count, it was possible to partition the molecular formula into E–L fragments, E = main-group atom, L = ligand, with nothing left over. Life would be boring if simple, so let's see what composition nature gives us for an all-B analog of $C_4B_2H_6$. $[B]^-$ is isoelectronic with C and substitution of all four C atoms with $[B]^-$ yields $[B_6H_6]^{4-}$. This anion has not been synthesized but B_6H_{10} and $[B_6H_9]^-$ have been and possess the structures shown in Figure 2.11. Both species exhibit pentagonal pyramidal, *nido*-cluster cores like that of $C_4B_2H_6$. They also exhibit B–H–B bridging H atoms on the pentagonal faces. This structural feature is characteristic of open boranes. The identical cve counts of $C_4B_2H_6$ and B_6H_{10} immediately tell the story – they have the same cluster shapes. In a comparison of known compound with unknown, the cve count rules. To do a sep count, on the other hand, we partition B_6H_{10} into six B–H fragments (six sep) and are forced to use the four "extra" H bridging atoms to provide the remaining two sep needed to make the required eight sep. Bridging H atoms, which we will refer to as *endo*-cluster H atoms, contribute one electron each to the sep count. Hence, careful partitioning of a molecular formula must precede a sep count.

Let's test our understanding by working a problem the other way around. Consider the molecular formula of the known compound $C_2B_4H_8$. We partition it into two C–H, four B–H and two *endo*-H giving sep $= (6 + 8 + 2)/2 = 8$. The structure should be based on an $n = 7$ deltahedron (pentagonal bipyramid) with one missing vertex. If we follow the rule of thumb and leave one connectivity-five vertex vacant, a pentagonal pyramidal core structure results for which a number of positional isomers are possible. The C atoms with higher electronegativity go to lower-connectivity vertices in basal positions. Electronegativity considerations suggest the C-apart isomer would be more stable. But note that the 2,4-isomer would force an *endo*-hydrogen to bridge a C–B edge which is not as favorable as bridging a B–B edge. The 2,3-isomer is the one observed (Figure 2.11). A cve count of 28 requires no partitioning into fragments (cve $= 4n + 2 - 2x = 28$; $x = 1$ gives $n = 7$). In this discussion we blithely assume that the observed isomer is the most stable one. This, of course, need not be true as a less stable isomer might be formed in the synthetic reaction and, if the barrier to rearrangement to a more stable arrangement is large relative to thermal energy, the less stable isomer will be isolated. Many of the smaller carboranes have been synthesized under forcing conditions so at least for the open systems, the product observed is most likely the most stable one.

2.7 Open *arachno*-clusters

Incomplete spherical clusters that require the addition of two fragments to be closed are called *arachno*-clusters (spider web-like). As the shape is based on a deltahedron of order n but with only $n - 2$ vertices occupied, the number of cluster bonding electrons per cluster fragment is larger than that for *closo*- and *nido*-clusters. It follows that a continuation of the *closo*-, *nido*-, *arachno*- progression eventually leads to "electron precise" rings or chains.

2.7.1 An arachno-*borane*

Consider the known compound B_5H_{11} shown in Figure 2.12. Partitioning gives five B–H and six *endo*-H for sep $= 16/2 = 8$; thus, $n = 7$ and the structure is based on a pentagonal bipyramid with two vacant vertices. The four possibilities are shown in Figure 2.12. Based on our rule of thumb (leave the highest connectivity vertex vacant), a planar five-membered ring would be chosen! Indeed eight sep $[C_5H_5]^-$, which is isoelectronic with B_5H_{11}, is planar. However, the observed structure for the borane is based on a pentagonal bipyramid with one five- and one four-connect vertex vacant. Why? Known structures show us that the nearest neighbor environment of a given B atom is approximately tetrahedral. Try to place eleven

2.7 Open arachno-clusters

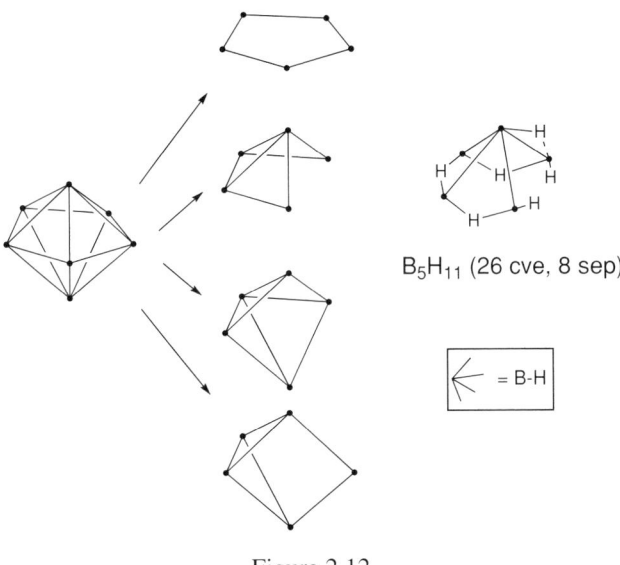

Figure 2.12

H atoms aound a five-membered planar ring. The other three structures are much better in this regard. And of the three, the one generated by removal of adjacent five- and four-connect vertices of the parent bicapped pentagon is the observed one. Note that each two-connect vertex contains a BH_2 group and that one H of each BH_2 must be considered an *endo*-H atom in order to get the right sep count. Once again no such assumption is necessary for the cve count as we have cve = $4n + 2 - 2x = 26$ for $x = 2$ and $n = 7$.

If B_5H_{11} were a new compound, ^{1}H and ^{11}B NMR spectroscopy would be used to distinguish between the possibilities generated by the counting rule. The importance of the rule is that it provides a limited set of satisfactory structures from which to choose.

2.7.2 Role of endo-H atoms

The discussion above suggests *endo*-H atoms are not just contributors to the cluster electron count. They have a role in determining which of the possible cluster core structures will be observed. Another characteristic of these skeletal H atoms is evident from NMR studies. Movement of *endo*-H atoms on a cluster framework can be quite facile relative to the timescales on which one does bench chemistry. Hence, in contrast to the movement of heteroatoms in many cluster frameworks, *endo*-H atoms are found in their most stable positions.

endo-H atoms are important to other stability considerations as well. Although $[B_5H_{10}]^-$ is known and even some dianions have been isolated by Shore, it is highly unlikely that $[B_5H_5]^{6-}$ will be synthesized in solution. On the other side of the coin, protonation of *closo*-borane anions, e.g., $[B_6H_6]^{2-}$ generates face-protonated $[B_6H_7]^-$, but all attempts to produce the neutral B_6H_8 have failed. Theoretical studies suggest the cluster framework bonding is perturbed on face protonation forcing localization in the vicinity of the proton. Do not ignore the fact the proton has non-trivial structural demands often hidden by the electron-counting rule.

Like unbridged B–B distances in *closo*-clusters, there is a range of B–B distances when H bridged. For boranes it is 1.71–1.86 Å. B–H distances in symmetrical bridges are about 1.35 Å; however, many bridging H atoms are unsymmetrically placed between the B atoms.

2.7.3 Ambiguities in open clusters

One implication of Figure 2.12 presages complications to come. The *arachno*-structure in Figure 2.12 generated by removal of two adjacent four-connect vertices is a distorted square pyramid – very much like a *nido*-structure! With the electron count of an *arachno*-cluster, but the apparent shape of a *nido*-cluster, you can see the potential for heated discussion. A clear implication is that cluster shape distortions can accomplish the same effect as qualitative shape changes. On the other hand, a cluster with a heterogeneous cage composition will not have an ideal cluster shape. You see the problem. Hence, a simple categorization by qualitative geometry with no attention to electronic structure can be misleading.

A final observation brings us back to the earlier discussion of three-connect clusters. Note that the electronic structures of the planar ring and the "butterfly" shape with a handle (removal of two non-adjacent four-connect vertices) in Figure 2.12 can also be described with two-center–two-electron bonds (think of the latter as C_5H_6). Considering how many valence isomers there are for a simple organic formula like C_5H_6, then it should come as no surprise that the same is true for clusters. The bridged butterfly structure has not been seen for pure boranes but we will encounter it in metal cluster chemistry. These electron-precise structures, sometimes called classical structures, are higher-energy alternatives for the boranes. If barriers for inter-conversion are high, they may well be generated by proper choice of synthetic route. Indeed in a few cases this has been accomplished both by judicious choice of B/C cluster content as well as external ligand choice (F for H).

Exercise 2.3. Work out possible structures for the compositions $[B_{10}H_{10}]^{6-}$ and $[B_{10}H_{10}]^{4-}$. Consider only structures where the vertex of highest connectivity plus an adjacent vertex are removed. Discuss any ambiguities discovered. *Nido*-$B_{10}H_{14}$ is a known compound.

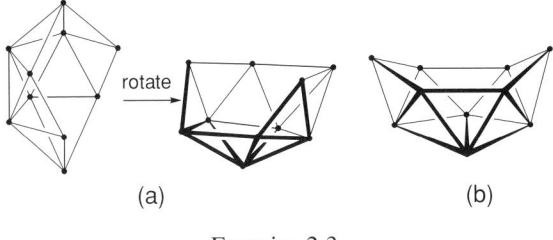

(a) (b)

Exercise 2.3

Answer. Use the sep method. $[B_{10}H_{10}]^{6-}$ has 13 sep and ten cluster fragments; hence the structure should be based on a deltahedron of order 12 with 2 adjacent vertices vacant. All the vertices of an icosahedron are identical and the shape produced is *arachno* as shown in (a) above. $[B_{10}H_{10}]^{4-}$ has 12 sep and ten cluster fragments; hence the structure should be based on a deltahedron of order 11 with 1 vertex vacant. Removing the connectivity-six vertex (see Figure 2.7 for the shape) generates (b). The two clusters possess the same qualitative shape! Electron count, not geometry, defines $B_{10}H_{14}$ as *nido*.

2.8 The *closo-*, *nido-*, *arachno-*borane structural paradigm

Figure 2.13 reflects the breakthrough structural correlation of Williams. For simplicity the B–H–B and BH_2 groups on the open faces of the *nido-* and *arachno-* frameworks are not shown. As already discussed these have a role in determining which of the possible open structures is most stable. In the form due to Rudolph, the chart may be found in many inorganic texts. It stands as a structural paradigm for stable main-group clusters containing external ligands. The *closo-*boranes, the embodiment of the shapes of the canonical deltahedra, are shown in the first column, the *nido-*boranes in the second, and the *arachno-*boranes in the third. The horizontal lines connecting the three structures of fixed value of n (n is the number of vertices in the parent deltahedron and NOT the number of atoms in the cluster) and identical sep and cve express and summarize the discussion of open clusters presented immediately above. Effectively, a B–H vertex is removed from a parent deltahedron and replaced with an electron pair. This is equivalent to removing a $[BH]^{2+}$ fragment and is often referred to as a *debor* process.

The diagonal lines connecting clusters of equal nuclearity reveal another important relationship – as the number of cluster electron pairs increases in a step-wise fashion the cluster opens up into a "net" and then a "web." As was pointed out in Chapter 1, this is a recurring motif – the more electron rich, the more open the structure. One can continue. Addition of three pairs of electrons to a *closo-*borane generates an even more open shape albeit a less well-accepted nomenclature. The

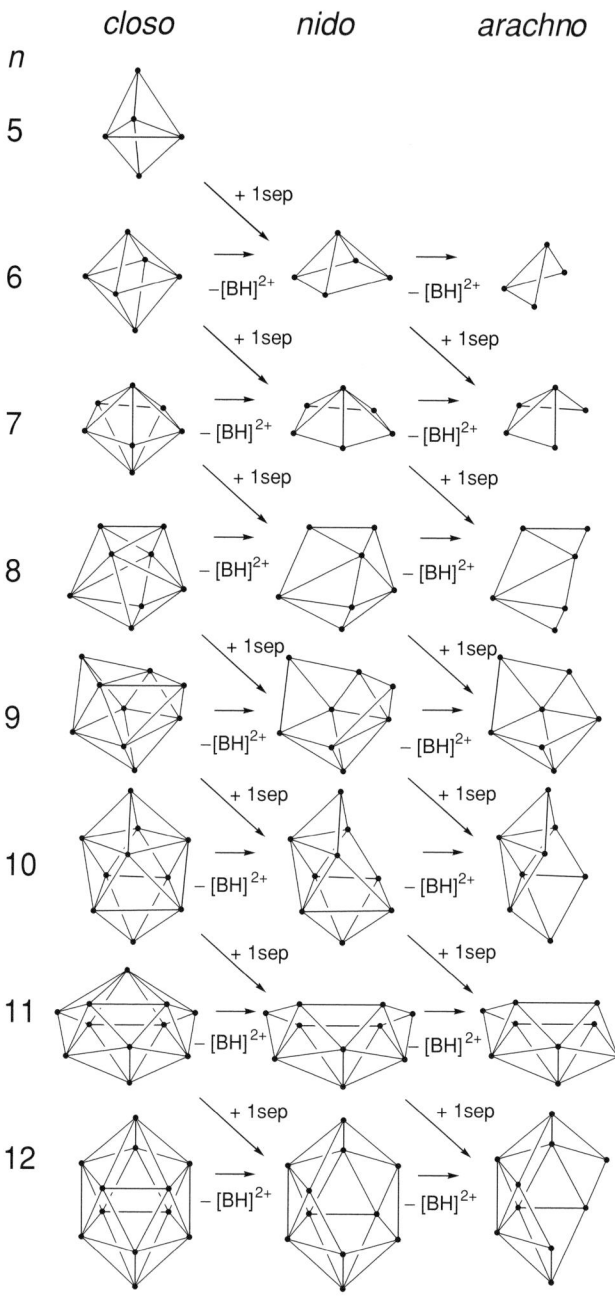

Figure 2.13

term *hypho*-boranes is often used to describe the relatively few examples known arising, e.g., from addition of Lewis bases to a framework. As we saw above, a B–CO fragment is a three-electron fragment like C–H. Hence, coordination of two PMe$_3$ ligands to two of the boron atoms of B$_5$H$_9$ yields *hypho*-B$_5$H$_9$(PMe$_3$)$_2$ with a very open structure. The trend is clear. As more electron pairs are added to a cluster the electronic situation approaches that of a hydrocarbon. Vertex connectivities decrease and eventually multicentered bonding is no longer required.

The correlations illustrated in Figure 2.13 should not be considered a fence excluding other possibilities. Recall the earlier discussions of the possibilities of *nido*-isomers for deltahedra with more than a single type of vertex connectivity. In the last section we saw that the geometric possibilities for *arachno*-clusters are even larger. Later we will see that heteronuclear clusters permit access to some of these structure varients. Factors, such as electronegativity, charge and number of *endo*-H atoms, can be important contributors to the net stability of a given shape. Variations from the borane structural paradigm contained in Figure 2.13 are expected and will be illustrated in succeeding chapters.

2.9 Localized bonds in clusters

The electronic structures of borane clusters were first successfully described using localized three-center– and two-center–two-electron bonds. These treatments have been replaced by the cluster electron-counting rule based on MO methods; hence, why bother with the three-center bond model in a book about clusters? Let's consider why there is value in a more localized approach.

The electron-counting rule abstracts from the MO results an essential aspect of the composition–structure problem important for understanding a fundamental aspect of the area – the relationship between structure and stoichiometry. In doing so the finer details of structure relevant to reaction properties and physical properties are ignored, e.g., the frontier-orbital characteristics. Counting electrons says little about properties or reactivities. In addition, the simplicity associated with localized bonding models is missing. For example, we have already seen that the H atom in a three-center B–H–B bond in a cluster has Brønsted acidic character. Deprotonation can generate a site of nucleophilic activity. Hence, when we see a compound with an H bridge, activation by deprotonation comes to mind. It is these kinds of simple considerations that aid chemical understanding and lead to new ideas. When do localized cluster model three-center bonds suffice? Alternatively, when should one restrict consideration to full MO methods even for qualitative considerations? Let's look at a few examples.

2.9.1 Ring-cap analysis of B_5H_9

Much has been written about three-center bond models in cluster structure analysis including the modern treatments of metallaboranes by King. Here the single example of B_5H_9 is sufficient to address our questions. For simplicity, we begin with $[B_5H_5]^{4-}$ which is made up of atoms possessing a total of 25 valence orbitals and 24 valence electrons. The resulting MOs are divided as shown in Figure 2.14(a): five external cluster bonding plus five antibonding; seven internal cluster bonding orbitals and eight antibonding. How do we get this result from that for the parent octahedron $[B_6H_6]^{2-}$? It has 30 valence orbitals and 26 valence electrons. In forming the *nido*-square pyramid a B–H fragment is lost (five valence orbitals and four electrons) and the negative charge is increased by two (adding two electrons). This results in the loss of two external B–H MOs (one filled and one unfilled) and three unfilled internal MOs – the number of framework bonding orbitals remains the same.

Now let us fragment the square pyramid into a planar $[B_4H_4]^{4-}$ ring and a capping B–H fragment (Figure 2.14(b)). The square $[B_4H_4]^{4-}$ fragment is isoelectronic with *cyclo*-butadiene. Hence its frontier orbitals are the four π MOs populated with four electrons. Both of these fragments are satisfactorily described with two-center–two-electron bonds so the multicenter bonding must be localized in the interaction between the ring- and capping-fragment frontier orbitals found within the box in the scheme. This should not be a surprise as it is the capping (apical) vertex which is of connectivity four and a difficulty for the two-center bond model. The bonding between the ring and the capping B–H, Figure 2.14(c), is derived from the σ and π symmetry-adapted linear combinations of the out-of-plane ring 2p orbitals. These form three bonding and antibonding MOs with the orbitals of corresponding symmetry on the B–H fragment whereas the δ ring combination remains nonbonding. The three bonding combinations are filled with the six available electrons to generate a five-center–six-electron ring–cap bond.

The final step is the addition of four protons (four empty orbitals) that stabilize four of the seven filled internal orbitals and generate four more empty orbitals (each B–H–B bond is a three-center–two-electron bond generating one filled and two unfilled MOs) giving the total of 29 MOs for the neutral compound (Figure 2.14(d)). We can also see now that it is the multicenter bonding, ring–cap and B–H–B, that generates the imbalance between filled internal cluster orbitals and unfilled internal cluster orbitals. It is this feature that distinguishes these clusters from most three-connect clusters.

It should be no surprise that in a full MO treatment, the HOMO of B_5H_9 is the degenerate ring–cap π MO set and the LUMO is the non-bonding ring δ orbital. For the binary boron hydrides, the frontier orbitals and, by implication, the reactivity

2.9 Localized bonds in clusters

(a)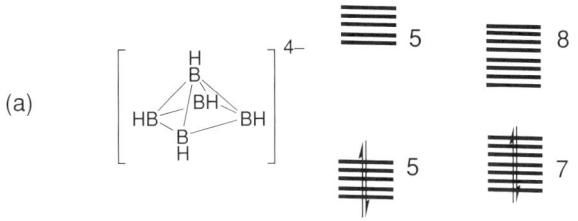

external MOs   Internal MOs

(b)

ring fragment — B–H fragment

(c)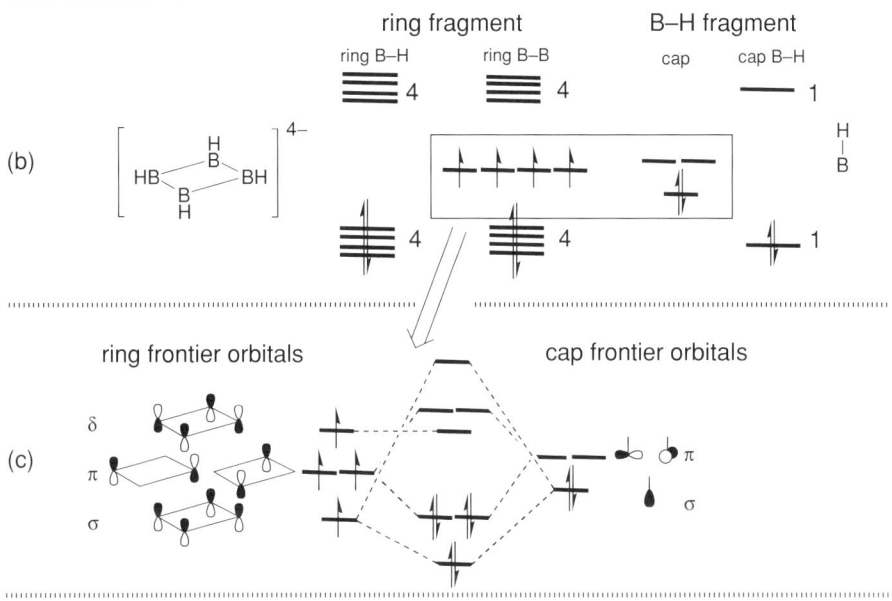

ring frontier orbitals   cap frontier orbitals

(d)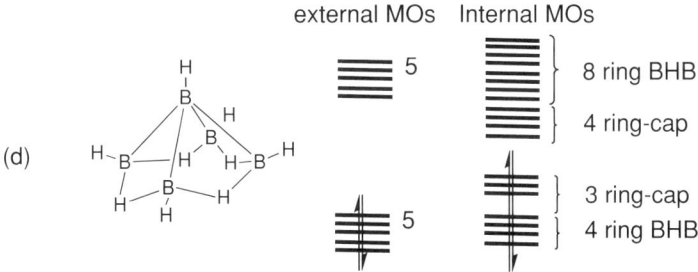

external MOs   Internal MOs

Figure 2.14

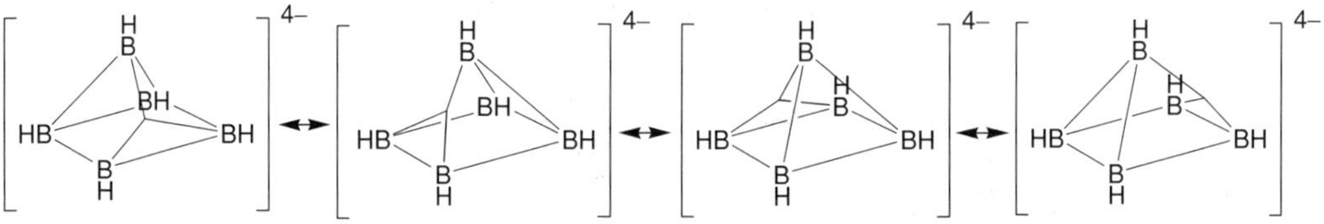

Figure 2.15

will be connected with the multicenter framework bonding. The central idea here, ring–cap analysis, was described very early by Hoffmann and Lipscomb and it has been used effectively recently by Jemmis to analyze a related set of cluster problems.

2.9.2 Cluster analog of the cyclopentadienyl ligand

Recall that in Section 2.6 we mentioned that the open cluster $[C_2B_9H_{11}]^{2-}$ presents five orbitals containing six electrons to a potential bonding partner. Now we can understand why. Consider the removal of the capping B–H vertex from B_5H_9 as $[BH_2]^{2+}$. What remains is a four-membered ring with a set of four out-of-plane orbitals containing six electrons. Now a ring-capping interaction is not restricted to a four-membered ring and works equally well for a five-membered ring (see Problem 9). Hence, the generation of $[C_2B_9H_{11}]^{2-}$ from $C_2B_{10}H_{12}$ by the removal of a $[BH_2]^{2+}$ fragment generates a five-membered open face possessing an out-pointing set of σ and π orbitals containing six electrons.

2.9.3 Three-center bond model for B_5H_9

A systematic treatment of B_5H_9 as well as the rest of the boranes by Lipscomb employed two- and three-center–two-electron bonds – no five-center bonds. Let's see how to generate a three-center description of the ring–cap interaction. Go back to $[B_5H_5]^{4-}$ in Figure 2.14. The ring–cap interaction shown in Figure 2.14(c) generates four empty orbitals over three filled ones so a description in terms of two- and three-center bonds requires two two-center and one three-center bond. The difficulty is how to place them on the framework without losing the C_{4v} symmetry. It is the same problem faced when trying to describe benzene with three double bonds and three single bonds yet retain D_{6h} symmetry. Resonance structures rear their ugly heads. Figure 2.15 illustrates a set of four resonance structures describing the ring–cap interaction with two B–B bonds and one "closed" B–B–B bond thereby retaining the C_{4v} symmetry of B_5H_9. In these resonance structures the lines now DO designate two-center– or three-center–two-electron bonds. Pleasingly each B center has near tetrahedral bond coordination (but not connectivity) and is associated with four electron pairs. Resonance structures that fail this test are less favored.

The triangular B–B–B three-center–two-electron bond represented in Figure 2.15 (two equal-energy empty orbitals over one filled – Figure 1.13) is related to the B–H–B bond. It was introduced by Lipscomb to treat the framework bonding problem. The numerical generalization of the accounting process we have just gone through was formalized in the *styx* rule – parameters that specify the numbers of two- (x B–H and y B–B) and three-center bonds (s B–H–B and t B–B–B) necessary to describe a borane of specified stoichiometry. As with the sep cluster count, the external B–H two-center–two-electron bonds are not considered in the *styx* count so B–H refers to "extra" terminal H as found in the BH_2 groups of B_5H_{11} for example (Figure 2.12). The *styx* rule suffers from the fact that more than a single *styx* configuration can be generated for a single composition and even though rationalization of structure is possible, prediction is difficult. The *styx* rule actually works better for open structures of low symmetry. A symmetrical *closo*-borane cluster requires more resonance structures (Exercise 2.5). But it is exactly these closed borane structures that are least flexible in terms of available shapes. They are the canonical structures from which the open ones are derived. Hence, the *styx* rule never revealed the underlying geometric and electronic correlations summarized in the cluster electron-counting rule. As already described, the deltahedral geometries of the *closo*-frameworks plus MO analyses pointed to the heart of the problem.

Exercise 2.4. Use the localized-bond model to rationalize the dipole moment exhibited by B_5H_9 (1.7 D).

Answer. Recall that in a three-center bond, the two electrons are now shared by three atomic centers. For simplicity ignore the differences in the electronegativities of B and H and assume they are shared equally, i.e., all the terminal H atoms will bear zero charge. The unique apical B shares three two-center and one three-center bonds; hence, the charge is $+3 - (3 \times 1 + 1 \times 2/3) = -2/3$. The charges on the four basal B atoms in any one bonding representation (Figure 2.15) is averaged by resonance among the four; hence the charge is $+3 - (2 \times 1 + 2 \times 2/3 + 1 \times 1 + 3 \times 2/3)/2 = -1/6$. Finally the charge on each bridging H atom is $+1 - 2/3 = +1/3$. The total charge on the molecule is zero ($-2/3 - 4/6 + 4/3$). The dipole moment lies along the C_4 symmetry axis with the negative end pointing towards the apical B atom.

Exercise 2.5. Develop a three-center bond description of one resonance structure for $[B_6H_6]^{2-}$ and show that each B atom satisfies the eight-electron rule (review Section 1.3.5).

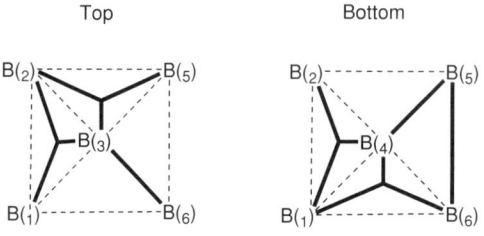

Exercise 2.5

Answer. There are a total of 30 orbitals and 26 electrons to be utilized in bonding. The external B–H bonds utilize 12 orbitals and 12 electrons leaving 18 orbitals and 14 electrons for cluster bonding. Four three-center B–B–B and three two-center B–B bonds utilize $12 + 6 = 18$ orbitals and $8 + 6 = 14$ electrons. They may be placed on the framework as shown in the diagram above where the top and the bottom of the octahedron are shown separately. By counting you can find that B(1) and B(2) are associated with three three-center–two-electron bonds (and a B–H bond), B(3) and B(4) with two three-center and one two-center bonds (and a B–H), and B(5) and B(6) with one three-center and two two-center bonds (and a B–H), i.e., eight electrons around each B. Notice that one would need to draw a considerable number of resonance structures to give all the boron atoms the same electronic environment.

2.9.4 Elemental B

Now we are in a position to show that the structure of the simplest allotrope of elemental B can be considered as a covalently bonded solid related to diamond (review Section A1.3 in the Appendix). The two geometric structures (Figures A1.9 and A1.11) look very different and if one's bonding repertoire is limited to the two-center–two-electron bond, it is hard to reconcile the available valence electrons with the number of bonding interactions based on geometry. Yours should not be so limited by this point. So let's count electrons. Each icosahedral B_{12} cluster requires 13 sep or 26 electrons. Each icosahedral cluster unit is bound externally by a total of six two-center–two-electron bonds to the layers above and below its layer. It is also bound by six three-center–two-electron bonds within its layer. The former requires six electrons (six two-electron bonds shared by two B atoms each) and the latter four electrons (six two-electron bonds shared by three B atoms each). For the 12 B atoms $26 + 6 + 4 = 36$ electrons are required or exactly the three per B atom available. All this covalent bond model requires are two-center, three-center and cluster network bond models to make it work. With strong cluster bonding and strong inter-cluster bonding it is no surprise then that elemental B is a high melting-point solid (2300 °C) like diamond.

2.10 Clusters with nuclearity greater than 12

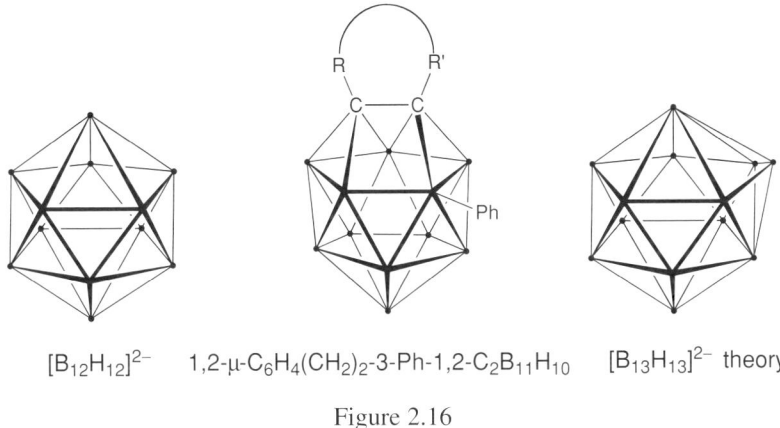

[B₁₂H₁₂]²⁻   1,2-μ-C₆H₄(CH₂)₂-3-Ph-1,2-C₂B₁₁H₁₀   [B₁₃H₁₃]²⁻ theory

Figure 2.16

2.10 Clusters with nuclearity greater than 12

The set of deltahedra used to describe the geometries of the boranes and closely related compounds stop at the icosahedron with nuclearity 12. But even in the early days of the structural characterization of the boranes higher-nuclearity clusters were known, e.g., $B_{18}H_{22}$. However, these clusters can be described as linked or fused deltahedra with nuclearities 12 or lower. For many years no single main-group cluster with $n > 12$ was found and a barrier to higher-nuclearity main-group clusters was thought to exist.

2.10.1 Supra-icosahedral clusters

For many years researchers looked beyond the icosahedron wondering about the possibilities of single p-block clusters with greater than 12 atoms in the cluster framework. Calculations suggested a combination of the high stability of the 12-vertex icosahedral cluster combined with poor stability of the 13-vertex deltahedron that requires a vertex of connectivity 6 creates an energy barrier to the higher nuclearities. But in 2003 a 13-vertex carborane (Figure 2.16) was generated by Welch in a masterful cluster-expansion reaction sequence. The framework structure found is shown in Figure 2.16 where it is compared with icosahedral $[B_{12}H_{12}]^{2-}$ and the lowest energy structure of a 13-vertex *closo*-borane from calculations. The carborane has trapezoidal faces, i.e., it does not have a deltahedral shape. The calculations show that the observed structure of the carborane is favored over the borane structure type by 29 kJ mol⁻¹. However, if the two CR fragments are replaced with [BH]⁻ fragments, the fully triangulated structure is more stable by 16 kJ mol⁻¹. The key point is that energy differences are small and the established preference of C for lower-over higher-connectivity vertices is expressed in the observed geometric structure. Clearly, if Figure 2.13 is to be extended to nuclearity 13, it is the calculated deltahedral shape appropriate for the borane that should be added. But we must also

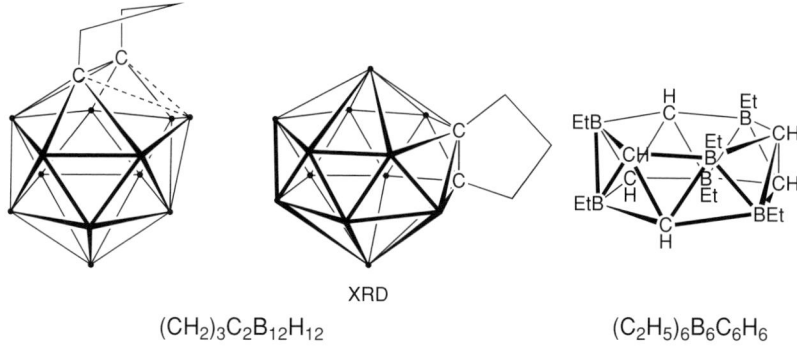

Figure 2.17

expect that non-deltahedral shapes are likely on replacement of BH fragments with heteroatom fragments.

In 2005, Xie showed that it is differing reducing properties of the reductively opened C_2B_{10} framework isomers, C-apart vs. C-adjacent, rather than framework energetics of BH addition that inhibit cluster expansion beyond the icosahedral *closo*-framework. In an elegant synthetic accomplishment, he showed that, with an external ring holding the C_2 cluster unit in the C-adjacent position, not only was the 13-vertex cluster accessible but also two isomeric forms of a 14-vertex cluster (Figure 2.17). The latter two species do not inter-convert thermally but can be inter-converted under reducing conditions. Further, the more symmetrical isomer has been crystallographically characterized and it exhibits a true deltahedral shape, a bicapped hexagonal antiprism, as predicted for $[B_{14}H_{14}]^{2-}$ by calculations. Interestingly the known *arachno*-hexacarborane with 15 sep shown at the right is based on the same shape with the two six-connect vertices unoccupied.

2.10.2 Linked and fused clusters

Examples of linked and fused clusters are illustrated in Figure 2.18. The development of the transition-metal cluster fusion reaction by Grimes generated a variety of fused boranes as well as carboranes. The electronic structural problem posed by these compounds ranges from trivial to complex. The isomers of the dimer of B_5H_9 shown in Figure 2.18(a) lie in the former category. They are named 1,1′-, 2,2′- and 1,2′-*conjuncto*-$[B_5H_8]_2$. One B–H terminal bond of each framework is replaced by a B–B two-center–two-electron bond. The sep and cve counts of each cluster are unaffected as each individual cluster acts as a one-electron ligand to the other.

The problem presented by the fused cluster $B_{20}H_{16}$ with 76 cve is a more difficult one to analyze. Look at $B_{20}H_{16}$ and note the four atoms held in common. Take it apart at the points of fusion with each fragment retaining the four common

2.10 Clusters with nuclearity greater than 12

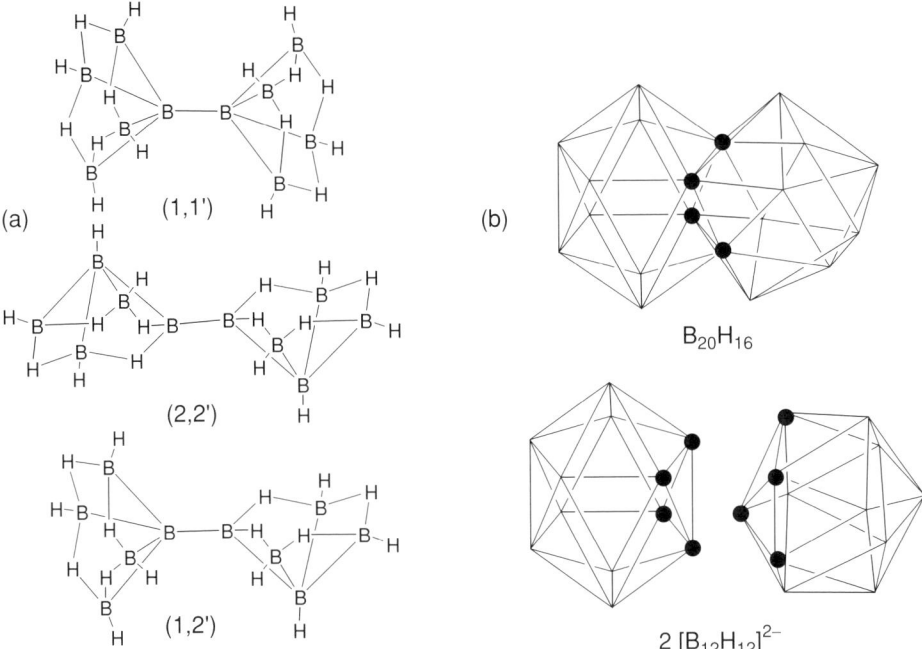

Figure 2.18

atoms. Two icosahedra are generated each of which requires 50 cve. By sharing vertices, somehow the electronic requirements of each icosahedron are satisfied. Jemmis provided a simple analysis, the *mno* rule, which is useful even though it shields the user from the intimate electronic details. In developing his rule Jemmis recognized similarities between fused clusters and fused aromatic systems, e.g., benzene vs. naphthalene in which two carbon atoms are shared in the π system. Mingos and Wales have analyzed the problem in detail for transition-metal clusters and the interested reader is referred to their book in the reading list at the end of this chapter. Likewise Burdett has treated some of the difficult cases of fused main-group clusters.

In the *mno* rule, m is the number of clusters, n the number of vertices in the fused cluster and o the number of single-vertex shared atoms. Unfortunately n does not have the same meaning we have given it above where n defines the number of vertices in the deltahedron upon which the cluster structure is based, thus for open clusters parameter p must be added, where p is the total number of missing vertices. The sum of m, n and o (and p if there are missing vertices) is the total number of cluster bonding electron pairs needed for a stable, condensed fused cluster. Applying this rule to $B_{20}H_{16}$ we have, from the molecular formula, a total count of 22 pairs (16 B–H and 4 B giving $(16 \times 2 + 4 \times 3)/2 = 22$). The *mno*

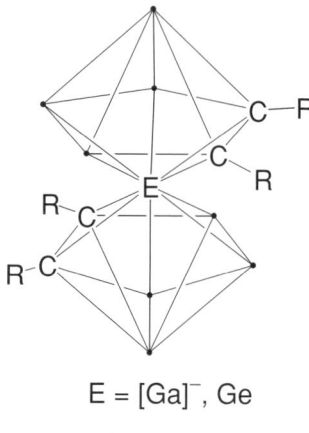

E = [Ga]⁻, Ge

Figure 2.19

rule gives: $m = 2$, $n = 20$, $o = 0$ for $m + n + o = 22$. Note that all three electrons from the shared B atoms are contributed to the cluster electron count. We will find a similar situation with interstitial atoms in metal clusters in the next chapter. The facile nature of the rule makes it a simple job to work out the stoichiometry of four-vertex sharing dimers like $B_{20}H_{16}$ as well as face, edge and single-vertex sharing. The analysis has also been used to suggest trimeric species presently unknown and the possibility of cationic as well as anionic clusters, e.g., $[B_{28}H_{24}]^{2+}$, a four-vertex sharing trimer of three fused icosahedra.

If there is only one atom shared between two clusters (an unknown situation for B) then $o = 1$. Thus, $[(R_2C_2B_4H_4)_2E]$, $E = Ga^-$, Ge, (Figure 2.19) exhibits a structure consisting of two pentagonal pyramids fused at one apical vertex occupied by the atom E. There are 16 pairs in the isoelectronic clusters (2 $R_2C_2B_4H_4$ fragments = 14 pairs + E = 2 pairs) and $m + n + o = 2 + 13 + 1 = 16$. The need for parameter o is discussed in the Jemmis article in the reading list.

Fused metal clusters are much more abundant than fused main-group clusters and Mingos developed an effective approach to their description (Section 3.3.3). His rule is that in fusing two clusters, the fragment formally eliminated must obey normal valence rules. Thus we have:

$$B_{20}H_{16}(\text{cve} = 76) \to 2[B_{12}H_{12}]^{2-}(\text{cve} = 100) - [B_4H_8]^{4-}(\text{cve} = 24)$$

Note that the common fragment appears to be a 22 cve butterfly fragment (like B_4H_{10}) rather than a ring isoelectronic with C_4H_8. We will encounter this ambiguity for metal clusters as well when cluster fusion involves four common atoms.

The fused deltahedra need not be connected at four vertices, nor closed, nor of the same size. Thus, the number of fused compounds possible is large and the number of "macropolyhedral" clusters, as they are sometimes described, is growing.

2.10 Clusters with nuclearity greater than 12

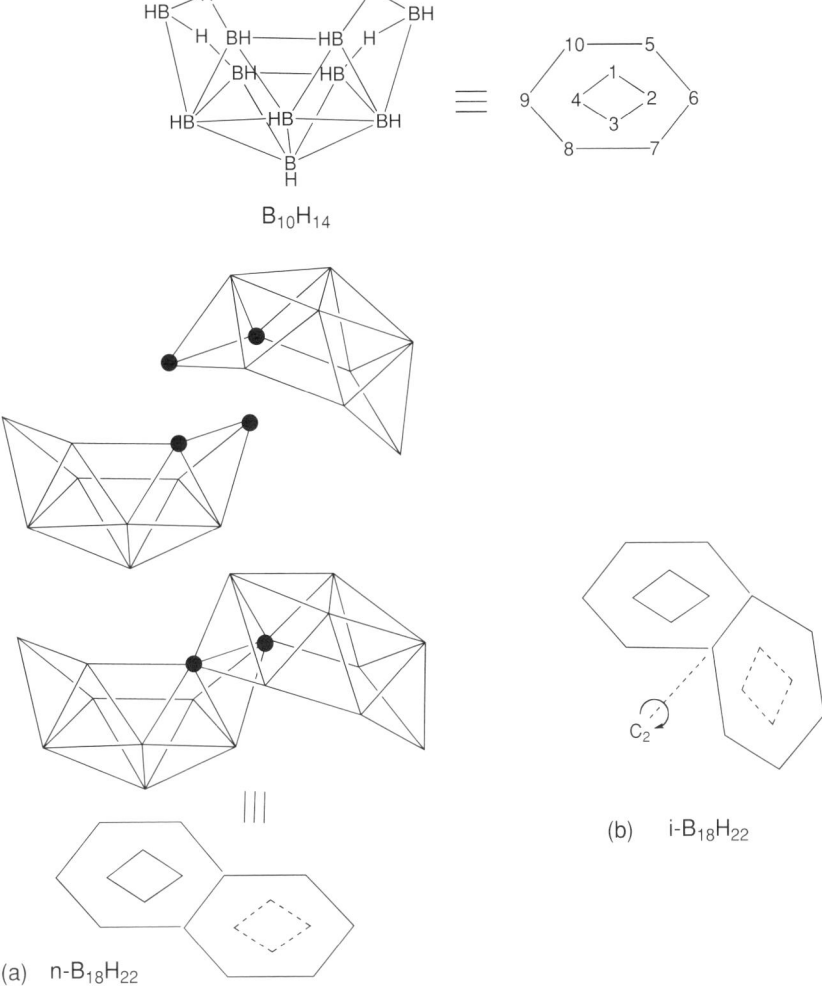

Figure 2.20

Considering the fact that the number of polyaromatic compounds known is large, this result should be no surprise. Of course neither the *mno* rule nor the cluster-fusion rule by itself gives any insight on the synthesis of the compounds. But they do suggest many such fused systems should be possible to make and that is added justification for seeking their syntheses.

Exercise 2.6. In Figure 2.20(a) the structure of one of the two known isomers of $B_{18}H_{22}$ is shown in stick form as well as its "retro-construction" from two $B_{10}H_{14}$ clusters. Schematic representations of this isomer as well as the other known isomer are shown in Figure 2.20 (a and b). Justify the compositions as well as the isomerism.

Answer. The *mno* rule gives $2 + 18 = 20$: sep $= 18$ B–H $+ 4$ H $= (36 + 4)/2 = 20$. Application of the cluster-fusion analysis is also straightforward. Inspection shows that $B_{18}H_{22}$ can be thought of as derived from two *nido*-$B_{10}H_{14}$ clusters as shown in Figure 2.20(a). Hence, $B_{18}H_{22}$ (cve $= 76$) $\rightarrow 2$ $B_{10}H_{14}$ (cve $= 88$) $- B_2H_6$ (cve $= 12$). The isomer shown in detail is centrosymmetric ("normal") and results from fusion of the 5,6-edge of one $B_{10}H_{14}$ with the 6,7-edge of the other (see numbering diagram). The non-centrosymmetric isomer ("iso") results from fusion of two 6,7-edges of two $B_{10}H_{14}$ as shown in Figure 2.20(b). When the geometry is known the electron count is easy to rationalize. If not known, the spectroscopic data must be used to limit the possibilities.

2.11 Ligand-free clusters

Hypothetical $[C_6]^{2-}$ with O_h symmetry should exhibit the same cluster bonding orbitals as $[B_6H_6]^{2-}$ and approximate MO calculations (Figure 2.21) show that this is true. There are significant differences. Specifically, in going to $[B_6H_6]^{2-}$ the highest six occupied MOs of $[C_6]^{2-}$ (t_{1u}, a_{1g}, e_g) are stabilized in that largely non-bonding orbitals become B–H bonding MOs. A smaller HOMO–LUMO gap and lower ionization energy is found for $[C_6]^{2-}$ even though C has a higher electronegativity than B. As the validity of electron-counting rules in general depend on large HOMO–LUMO gaps, bare clusters will have a greater likelihood of adopting cluster shapes that do not follow the cluster electron-counting rule or exhibiting cluster electron counts less than that specified for the observed shape.

Let us look at bare clusters from a different point of view. What would a naked $[B_6]^{2-}$ octahedral cluster be like? It has 20 cve, 6 short of the requirement. This number of electrons is sufficient to serve as the required seven sep and provide three external lone pairs; however, three of the out-pointing external cluster orbitals would be empty and a structural rearrangement would be required to create a significant gap between MO 10 and the t_{1u} set 11–13 (left side of Figure 2.21). Hence, in the same way that BH_3 is only found as base adducts, so too octahedral $[B_6]^{2-}$ would be expected to be found coordinated to bases.

2.11.1 Oxidative coupling

How strong is the tendency to fill an empty external cluster orbital? The linked cluster pair shown in Figure 2.22 is instructive. Oxidation of $[B_{10}H_{10}]^{2-}$ results in the loss of two electrons and a proton (effectively H^- is lost). The shape remains the same so the 11 sep associated with cluster bonding are unaffected. Hence, an external outward-pointing empty orbital is generated. Dimerization to form $[B_{20}H_{18}]^{2-}$ arises by formation of two three-center–two-electron bonds between the clusters.

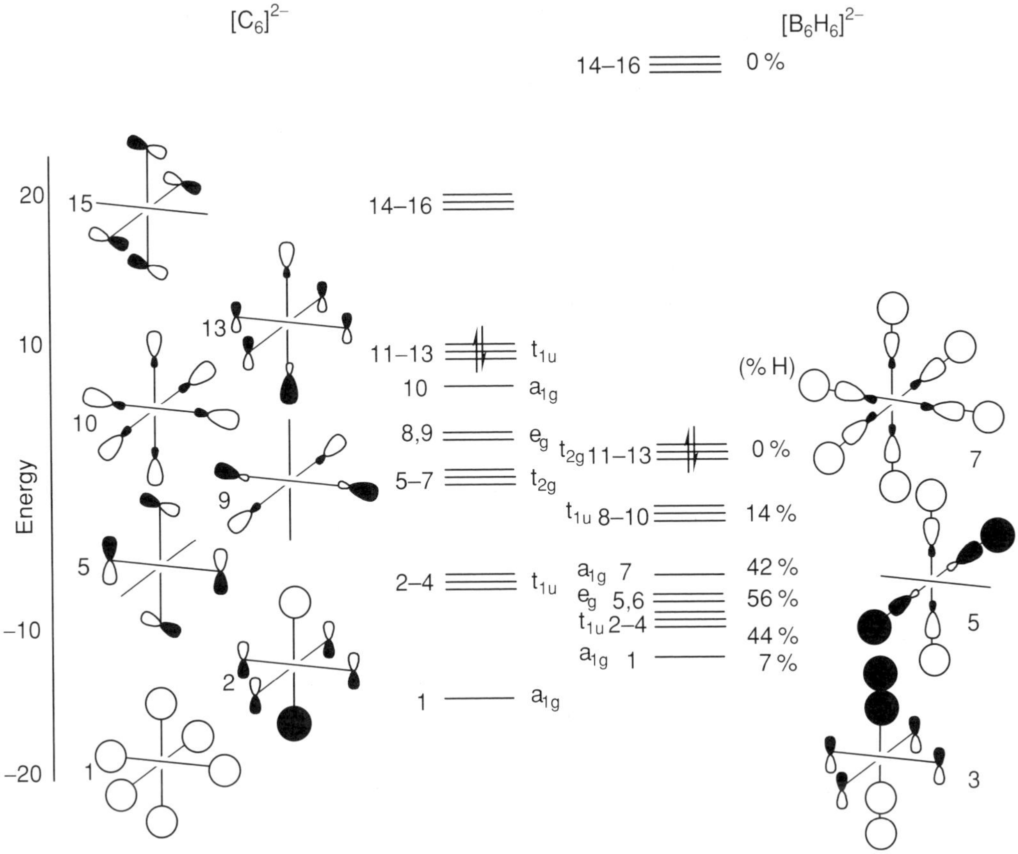

2.11 Ligand-free clusters

$[C_6]^{2-}$ $[B_6H_6]^{2-}$

Figure 2.21

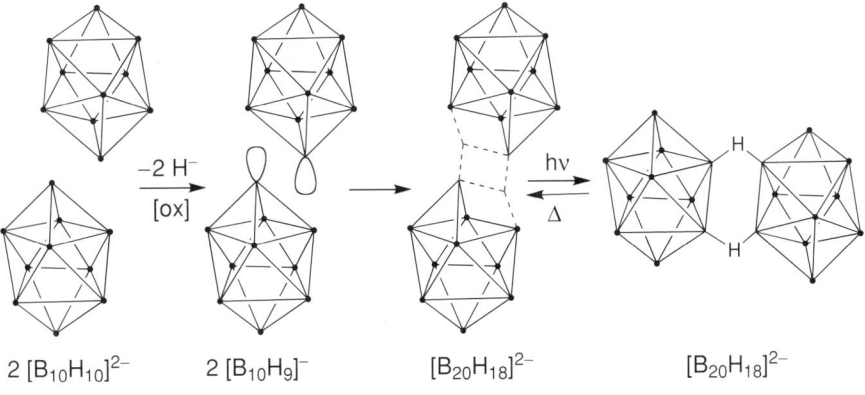

Figure 2.22

This linked cluster is similar to that formed from two [B_5H_8] radicals (Figure 2.18) except three-center instead of two-center bonding joins the clusters. This dimerization also parallels that of diborane in which two B–H–B three-center interactions are formed from two B–H bonds and two empty orbitals. In fact photoexcitation leads to the generation of an isomeric species in which the cluster linkage now consists precisely of two three-center B–H–B bonds rather than B–B–B bonds. Conversion back to the more stable B–B–B linked form takes place on heating. The message from this experiment is that external vacant orbitals on a cluster need to be satisfied in some fashion. Thus, alternative structures become energetically competitive.

2.11.2 Zintl clusters

Bare clusters bearing negative charges are known and these species are derived from the pioneering work of Zintl. Solid-state polar intermetallics, known as Zintl phases, when dissolved in liquid NH_3, provide solutions with properties consistent with the existence of multiply charged cluster ions. These species are known as Zintl ions. Once structurally characterized, it was clear that the Zintl phases did not contain the cluster species postulated to exist in solution; hence, cluster formation took place on dissolution. In addition, clusters, e.g., tetrahedral, found in the solid state were not found in solution. Eventually, with the availability of ligands for alkali metal such as the crown ethers, isolation and structural characterization was achieved. Recently, known Zintl cluster anions have been found in Zintl phases thereby forging a link between solution and solid-state structural chemistry. This connection is amplified in Chapter 8. Here we consider the alkali metal–p-block element compounds as p-block element anions in a "sea" of alkali-metal cations with or without the presence of amines or crown ethers. Keep in mind that the compounds are made from the elements and retain the high sensitivity to moisture that the alkali metals exhibit. In the case of solid-state systems we will ignore for the time being the sometimes non-trivial task of sorting the contents of the unit cell into clusters and other less interesting bits. It's not the atom positions that are a problem as these are defined precisely by the X-ray diffraction experiment, but rather the assignment of charges. That is, if some of the alkali-metal atoms retain their valence electrons, the compound will exhibit metallic character and the assignment of cluster charge can be difficult.

Let us focus on Ge and begin with three-connect clusters. A compound containing [Ge_4]$^{4-}$ tetrahedra has been prepared from Na and Ge (Figure 2.23). [Ge]$^-$ is isoelectronic with As. In the sep count it is a three-electron cluster fragment like C–H. So the cluster is analogous to P_4 with six two-center Ge–Ge bonds and four external lone pairs. One can see the power of the Zintl idea: ME is equivalent to E′ where E′ is a p-block element one column to the right.

2.11 Ligand-free clusters

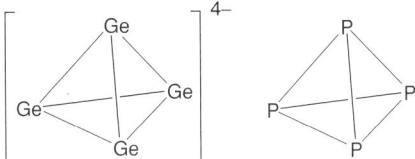

Figure 2.23

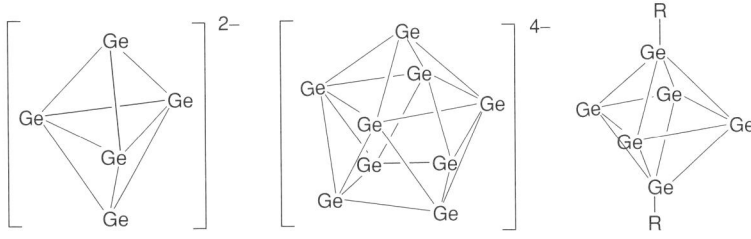

Figure 2.24

Things become more interesting when larger clusters are addressed. Consider the clusters shown in Figure 2.24. $[Ge_5]^{2-}$ with six sep has the deltahedral shape expected from the borane paradigm. Neutral Ge atoms are now behaving like B–H fragments. The phase $Na_{12}Ge_{17}$ contains a 2:1 ratio of $[Ge_4]^{4-}$ and $[Ge_9]^{4-}$ clusters. The former are tetrahedral whereas the latter are often represented as capped square antiprisms (Figure 2.24). That is, in conformance with 11 sep the structure is based on a deltahedron of order 10 (Figure 2.13) with one four-connect vertex vacant. Direct connections with ligated clusters arise when sufficient ligands are added to alleviate the need for an overall cluster charge. Ge_6R_2, where R is a bulky ligand, is neutral and exhibits an octahedral cluster shape in accord with its sep count of seven (Figure 2.24).

Exercise 2.7. A Zintl ion with the formula $[TlSn_9]^{3-}$ has been synthesized and characterized. Suggest a structure.

Answer. A count of 42 cve or 11 sep yields $n = 10$ so a bicapped square antiprismatic *closo*-cluster shape is predicted and was found in a solid-state structure determination.

2.11.3 Localized bonding analyses

Mixed species such as Ge_6R_2 can generate interesting bonding puzzles that are often effectively described by localized models. The key to solving the puzzles is to utilize all the valence electrons and orbitals. Take the structure of $[Ge_{10}(Si^tBu_3)_6I]^+$ shown in Figure 2.25(a). Analyze the problem. There are 47 valence AOs and

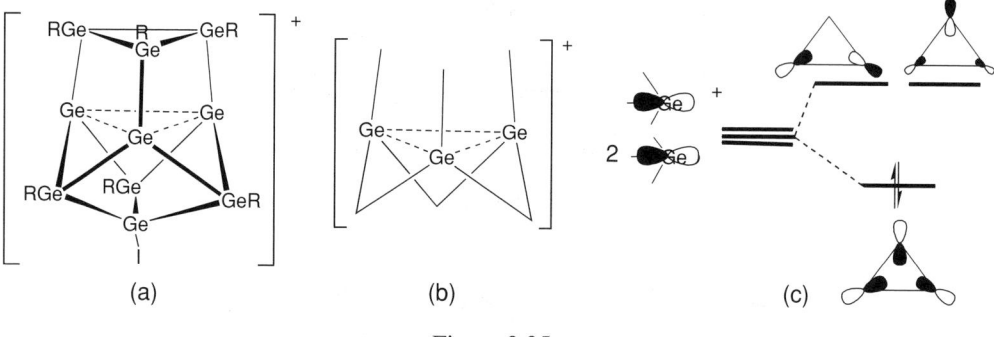

Figure 2.25

46 electrons, all of which must be used. Viewed as a three-connect cluster (ignore the long Ge–Ge distances indicated by dashed lines), $5n = 50$ electrons would be required. Clearly, some delocalization is necessary. R–Ge is a three-orbital–three-electron fragment and you should now be able to recognize that the triangular Ge_3R_3 fragment on the top and the irregular Ge_4R_3R' fragment on the bottom can be viewed as three-connect networks (electron precise) that require 44 AOs and 44 electrons. Thus, the essence of the problem can be abstracted as a triangle of three $(RGe)_3Ge$ fragments one of which carries a single positive charge (Figure 2.25(b)). As shown in Figure 2.25(c) the three radial orbitals of the three X_3Ge fragments combine to form two antibonding MOs over one bonding MO. The latter contains the two available electrons leading to a net three-center bonding interaction. Recall triangular $[H_3]^+$. Consistent with the model, the Ge–Ge distances in this triangular fragment are longer than a Ge–Ge single bond but somewhat shorter than expected if non-bonding. This set of three orbitals is expected to be the HOMO/LUMO set and good calculations verify this point.

The three-center Ge–Ge–Ge bond is similar to the B–B–B three-center bond used in a localized model of borane clusters. In fact, we can view this Ge cluster as a hybrid of an electron-precise, three-connect cluster and an electron-poor borane cluster. A difference is that the filled orbital also has lone-pair character. As pointed out in the comparison of $[C_6]^{2-}$ and $[B_6H_6]^{2-}$, the presence of high-lying lone-pair orbitals enhances mixing with cluster bonding orbitals. One implication of the result is that one cannot assume a bare atom at a cluster vertex has an external lone pair removed from the cluster bonding network. As emphasized in the introduction of Section 2.11, the electron-counting rules developed for ligated clusters must be applied to clusters containing bare atoms with caution. We will return to this point in considering larger clusters with ligand-free atoms below.

A facility with the use of multicenter bond models is very helpful in thinking about problems of this type. Insight into new structure types arises more rapidly if one is able to break down the problem in an approximate fashion with partially localized bonding models. Let us return now to the electron-counting rules and examine some systems where the connection between stoichiometry and structure is less precise than one might desire.

Exercise 2.8. The structure of $Ga_{10}R_6$ is shown below. Consider the cluster as made up of edge-fused octahedra and compare the observed cve with the calculated one. Now apply the *mno* rule to this fused cluster system and calculate the number of cluster electrons required from each of the bare Ga atoms in order to satisfy the rule.

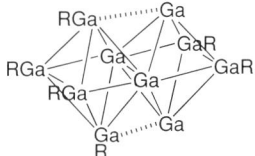

Exercise 2.8

Answer. For two octahedral clusters fused on an edge, the cve count is $(2 \times 26 - 14) = 38$ whereas the observed count is $(10\, Ga + 6\, R) = 30 + 6 = 36$. Thus, we cannot assume non-cluster bonding lone pairs on the bare Ga atoms. With the *mno* rule, $m = 2$, $n = 10$ and $o = 0$ giving $m + n = 12$ sep. Each of the two Ga atoms shared between the clusters contributes all three valence electrons. Hence, we have 6 RGa + 2 Ga(shared) + 2 Ga(unshared) = $(12 + 6 + 2x)/2 = 12$ sep, where x is the contribution of the unshared cluster Ga atoms. Clearly $x = 3$ in this cluster, which suggests there are no formal lone pairs on these two Ga vertices. Indeed, the structure shows the Ga–Ga distances between the apical RGa and Ga centers (broken lines in the drawing) are about 0.2 Å shorter than the other Ga–Ga distances. Electron counting identifies the cluster bonding problem but does not solve it. We will have more to say about this cluster type below.

2.12 When the rules fail

There are some cluster shapes in which the energy difference between two geometric shapes for the same composition is relatively small. We have already encountered a 13-vertex closed cluster where two shapes differ little in energy and factors other than electron counting determine the shape observed (Figure 2.16). Two other sources of non-conformity are pathological features of a particular shape and the difficulties of accommodating high charge. Very large clusters with internal atoms do not follow the simple rules and these are considered separately below and later

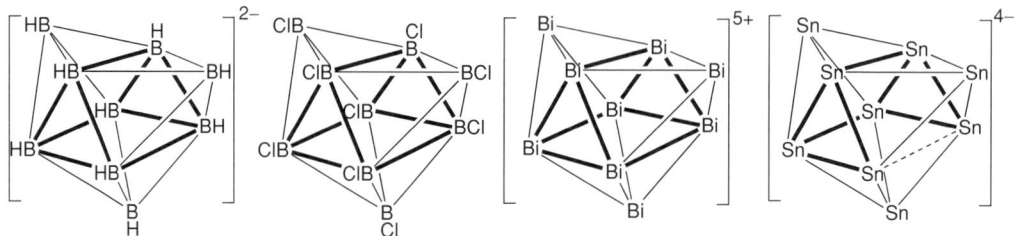

Figure 2.26

in the text. Examples illustrate the problems and a more comprehensive discussion will be found in Mingos and Wales.

2.12.1 Three-fold axes

Recall the situation encountered for the tetrahedral cluster earlier in the chapter where the e set lies in the gap between the t_2 cluster bonding and antibonding sets (Figure 2.5). Two counts, four and six sep, are observed. The presence of a C_3 axis in other cluster shapes leads to a similar situation. For example, in the *closo*-nine-vertex deltahedron (tricapped trigonal prism) there is no clear-cut separation between the energies of the $n+1$ cluster bonding orbitals and the $2n-1$ antibonding orbitals. In fact, this cluster shape possesses three high-lying cluster orbitals as potential HOMOs and examples of this cluster shape are known for sep counts of n, $n+1$ and $n+2$, i.e., 9, 10 and 11 sep (Figure 2.26). $[B_9H_9]^{2-}$ exhibits a tricapped trigonal prismatic shape and contains 10 sep in accord with the "normal" counting rule. However, both B_9Cl_9 and Ga_9R_9 exhibit the same shape but only possess nine sep. Likewise $[Bi_9]^{5+}$ exhibits a tricapped trigonal prismatic shape but contains 11 sep. The fact that six sep $[Bi_5]^{3+}$ exhibits the expected trigonal bipyramidal shape suggests that the varied counts for the nine-atom cluster is characteristic of the shape rather than the atom type. The structural distortions observed in the $[Bi_9]^{5+}$ framework relative to that of $[B_9H_9]^{2-}$ correlate nicely with the Bi–Bi bonding and antibonding character of the $n+1$ and $n+2$ frontier cluster orbitals. Cluster distortion, rather than cluster opening, accompanies the addition of one sep.

A nine-fragment Zintl system with 11 sep reinforces this point. The structure of $[Sn_9]^{4-}$ with 11 sep has a short diagonal on the lower square face (dashed line, Figure 2.26). Note that opening this edge, which connects two five-connect vertices of the tricapped trigonal prism, generates a capped square antiprism. So is it a distorted *closo*-cluster with $n+2$ sep or a distorted *nido*-cluster with $n+2$ sep? Good calculations or geometric parameters are needed to go beyond an electron-counting description. That is, debate of the question *nido* vs. *closo* is not a productive one. We are only talking rules here, not commandments!

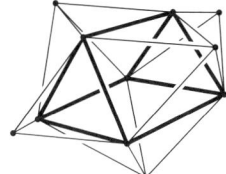 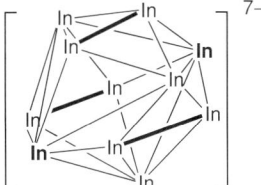

Figure 2.27

2.12.2 High charge

Cluster shapes with no obvious connection to the borane paradigm are observed. For example, the $[In_{11}]^{7-}$ cluster is observed in the metallic phase K_8In_{11}. The cluster is formulated as $[In_{11}]^{7-}$ rather than $[In_{11}]^{8-}$ as the material is metallic and one of the eight K atoms retains its electron which is "delocalized." The observed shape (shown at the right in Figure 2.27) can be derived from the pentacapped trigonal prism shown to the left by six diamond–square–diamond, dsd, rearrangements. In a dsd rearrangement a cross–diamond connection is broken to produce a square and then the alternative cross–diamond connection formed to regenerate a diamond. The net result is a change in the relative connectivities of the four corners but no change in overall connectivity. In this case, the six edges of the two triangular faces of the tricapped trigonal prism are broken (bold lines of the left-hand structure missing in the right-hand structure). As a result the two In atoms capping these faces (these two indium atoms are shown in bold in the right-hand structure) go from three-connect to six-connect. To retain In–In bonding, compression of the cluster accompanies the rearrangement. We will learn in Section 3.3.2 that capped clusters only require the sep count of the uncapped central cluster, i.e., in this case only ten sep are required for a bicapped tricapped trigonal prism. $[In_{11}]^{7-}$ has nine sep and we saw in the very last section this count is one possibility for a tricapped trigonal prism. So why the different structure? Notice that the vertex connectivities of the left-hand structure are: two three-connect, three four-connect, and six six-connect (total = 54); whereas those for the right-hand structure are: six four-connect and five six-connect, i.e., the observed structure is more spherical. As vertex connectivity is related to cluster charge distribution the more spherical shape will be favored.

2.12.3 Boron wheels

A spectacular example of the unusual shapes possible for bare clusters is the recent development of naked B "wheels" and other planar shapes generated by calculational chemistry and supported by quantitative fits to experimental photoelectron spectroscopic data. Two examples, $[B_8]^{2-}$ and $[B_9]^-$, are shown in Figure 2.28

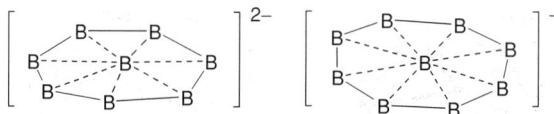

Figure 2.28

where it is seen that planar rings of seven or eight B atoms surround a central B atom to generate a wheel-like shape – quite the antithesis of the closed deltahedral clusters $[B_nH_n]^{2-}$! The lines that create the spokes are bonding interactions but the reader will now appreciate that they are not two-center–two-electron bonds. In fact the stability is described as arising from the combined effects of π aromaticity (as found in benzene) and σ aromaticity (derived from delocalization of the σ bonding network which constitutes another way of describing multicenter bonding).

Exercise 2.9. The cluster $[Al_{14}R_6I_6]^{2-}$, with two bare Al atoms, exhibits the unusual shape shown below and has been called a "nano-wheel." In light of the discussion above of rule-breakers, consider its electronic structure at the level of electron counting and suggest a reasonable analysis.

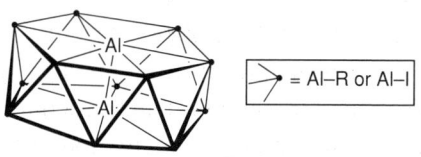

Exercise 2.9

Answer. $[Al_{14}R_6I_6]^{2-}$ possesses $(14 \times 3 + 12 \times 1 + 2) = 56$ cve. However, considered as a borane analog we expect cve $= 4n + 2 = 58$ as found for $R_2C_2B_{12}H_{12}$ (Figure 2.17). Note that the shape of $[Al_{14}R_6I_6]^{2-}$ is compressed relative to a spherical cluster so that there can be a cross-cage Al–Al interaction. Hence, if the two bare Al atoms utilize one electron each to form a cross-cage Al–Al bond it leaves two per Al atom for cluster bonding. In fact, the Al–Al distance is 2.73 Å which falls in the range observed for these Al cluster types and, thus, is consistent with a bonding interaction. Hence, in terms of sep count we have (12 AlR + 2 Al + 2(−)) = (24 + 4 + 2)/2 = 15 sep appropriate for a 14-vertex *closo*-cluster – the bare Al atoms do not have lone pairs. A similar mechanism is adopted by some ReB clusters we will encounter in Section 5.2.2. See also the discussion below in Section 2.12.5 on the electronic behavior of a bare atom E nearly coplanar with ER fragments. Contrary to the statement contained in the original publication, the counting rules do provide a rationale of observed composition and shape.

2.12 When the rules fail

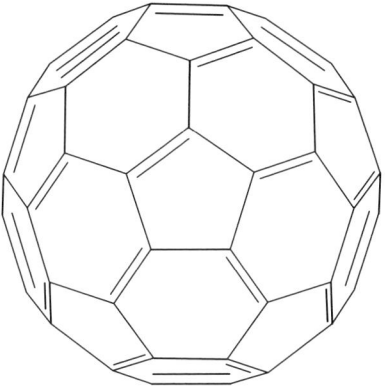

Figure 2.29

2.12.4 Buckyball

And what about C_{60} and its larger siblings? How are these attention-getting bare clusters related to our discussion? Inspect its structure shown in Figure 2.29. It is a naked three-connect cluster made up of five- and six-membered rings, i.e., it is not deltahedral. If we construct a two-center–two-electron bonding network using the three C sp^2 orbitals lying in the surface of the cluster we are left with one out-pointing p orbital per C atom containing one electron. This is a similar situation to a polyaromatic albeit we are not dealing with a planar species. On the other hand, the size of C_{60} means the curvature at any atom is less than that for an icosahedron, for example. It is not too much of a stretch to conclude that C_{60} contains an aromatic system related to that in graphite. That is, C_{60} does not have external lone pairs but rather an external delocalized aromatic system, i.e., one external electron per C atom. Like the bare Al atoms in $[Al_{14}R_6I_6]^{2-}$ (Exercise 2.9) only one electron is not used in skeletal bonding. C_{60} is to the prismanes (Figure 2.3) what polyaromatics are to aliphatic compounds.

The reactivity of C_{60} with metal fragments appears to correlate more closely with that of dienes than aromatics, however. This suggests that localization to form, e.g., organometallic complexes does not involve large destabilization energies due to loss of aromaticity. We will come back to the interesting properties of C_{60} in Chapter 7.

2.12.5 Elementoid clusters

Mixed ligated/naked-atom clusters, which Schnöckel has depicted as elementoid or metalloid, are clusters with internal as well as surface atoms. They illustrate the generalized cluster shown in Figure 2.1 and follow neither the borane paradigm nor

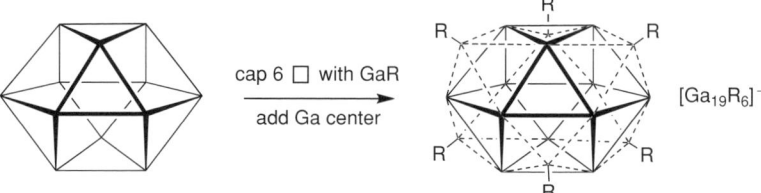

Figure 2.30

a localized bond treatment. For the large ones the structures of the central cores have connections with the solid-state element structures – hence, the term elementoid. They form a bridge between this chapter and those on solid-state systems (Chapters 6 and 7).

We begin with mid-sized clusters with a single internal atom (called "interstitial" for metal clusters, Section 3.3.1) that are deltahedral clusters but exhibit irregular surfaces. For example, $[Ga_{19}R_6]^-$ with 64 cve has the shape shown in Figure 2.30. It can be imagined to be constructed by capping each of the six rectangular faces of a 12-vertex cubooctahedron with a GaR fragment and adding a centering Ga atom. Approached as a borane we expect $4n + 2 = 74$ cve to be required. Treated as a capped metal cluster (see Section 2.10.2 and the more extended discussion in Sections 3.3.2 and 3.3.3 in the next chapter) we have $4n + 2 = 50$ ($n = 12$) plus 12 for the 6 caps for a predicted cve count of 62. The observed cve of 64 is bracketed to be sure but can we do better? There is a smell of the $[Ge_{10}R_7]^+$ problem analyzed in Section 2.11.3, where we found the triangle of bare Ga atoms involved in a three-center–two-electron bond that incorporated lone-pair character. The result of Exercise 2.8 is also pertinent. Perhaps the same factors are important in $[Ga_{19}R_6]^-$ and we need to break it down into more localized building blocks. How?

A way was suggested by Schleyer in a discussion of a set of calculated clusters whimsically described as a "sea urchin" family of boranes and carboranes. In essence he suggests that an organic polyhedrane atomic framework with triangular, rectangular, pentagonal and hexagonal faces, capped on all faces larger than triangular with ER fragments, yields a stable cluster for a specific electron count. From the compositions and charges observed to be stable minima by theoretical calculations, the number of cluster bonding electrons (cve less the pairs associated with the R groups) was reproduced by twice the number of triangles in the original polyhedrane skeleton plus six times the number of larger faces capped. The rationale is simple: as we saw from the $[Ge_{10}R_7]^+$ problem the number of electrons associated with bonding the Ga_3 triangle is two. Further, as we saw from the ring-cap model of Section 2.9.1 and Problem 8 the number of electrons associated with apical–basal bonding in capping a rectangular or pentagonal ring is six. The total number of

2.12 *When the rules fail* 75

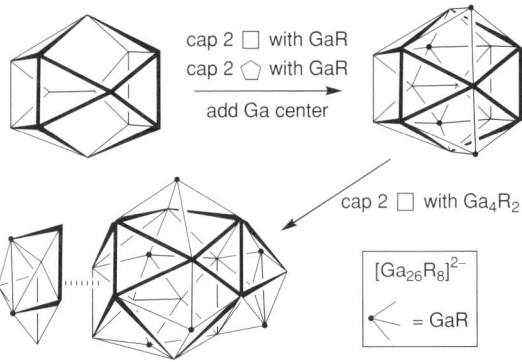

Figure 2.31

electrons associated with the cluster (cve) is then the sum associated with all the polygons that make up the framework plus those that serve to bond the external R groups. Excluding the external R group bonds, Schleyer called the $6m + 2n$ sum a new skeletal electron-counting rule!

To our knowledge this approach has not been applied to experimentally determined cluster compositions/shapes but let's try it on $[Ga_{19}R_6]^-$. As shown in Figure 2.30 we have eight triangles and six capped rectangles plus six R groups giving us: 8×2 (triangles) $+ 6 \times 6$ (capped faces) $+ 6 \times 2$ (R groups) $= 64$ exactly as observed. Now there is a difference between this count and that for the Schleyer molecules as here the polyhedrane is made up of Ga atoms not C atoms. For example, in Schleyer's approach C atoms were replaced with BH to generate borane cluster structures. Hence, our count for $[Ga_{19}R_6]^-$ effectively assumes each Ga has no defined lone pair. This should no longer be a surprise for E_mR_n clusters where $m > n$.

This approach also works for $[Ga_{26}R_8]^{2-}$ (88 cve) which has a more complex structure built from a centered 13-vertex polyhedron (8 triangles, 4 rectangles, 2 pentagons shown in Figure 2.31). Two of the rectangles and the two pentagons are capped with single GaR groups. The other two rectangles are fused to bicapped trigonal prismatic fragments – more complexity! But the latter can be treated the same way, i.e., the trigonal prism has three rectangles and two triangles – cap two rectangles (one is reserved for fusion to the 13-vertex polyhedron) giving a count of $2 \times 2 + 2 \times 6 = 16$. The total count is then: 8×2 (triangles) $+ 4 \times 6$ (capped faces) $+ 2 \times 16$ (complex caps) $+ 8 \times 2$ (R groups) $= 88$. It remains to be seen if this approach has generality. The greatest difficulty for more complex examples is in determining the uncapped polyhedron as well as the various appendages.

Despite these difficulties, the success of this simple analysis suggests that complex cluster species can be built up utilizing partially delocalized (σ aromatic) building blocks. The approach of Section 2.9, that probably seemed extraneous to you, now becomes a useful tool. Perhaps also you can see a connection with the completely delocalized external orbital system of C_{60}. A corollary is that these clusters, like C_{60}, will not, in general, follow the rules established for the near spherical borane clusters. Note particularly that in this method all of the bare Ga atoms contribute three valence electrons to the cluster count. There are no formal lone pairs. To test your understanding, try Problem 12 on Sn_8R_4. It is another example where an assumption of lone pairs on bare atoms leads one astray.

The obvious question arises: when does a bare atom exhibit a lone pair and when doesn't it? We have already seen that many bare clusters behave as expected with lone pairs, whereas it is the clusters with a mix of E and ER fragments that sometimes do not. An important paper by Burdett and Canadell suggests a rationale. To summarize their argument briefly, they suggest that as the radius of curvature of a cluster increases, i.e., cluster nuclearity increases, the occupation of cluster lone-pair orbitals becomes less likely. The origin of this effect is an antibonding (repulsive) interaction between the E lone pair orbitals and ER bonding orbitals leading to destabilization of the former. Thus, for Ge_6R_2 (Figure 2.24) the low radius of curvature leads to filled lone-pair orbitals whereas for the larger $[Ga_{19}R_6]^-$ the bare Ga centers act effectively as trigonal planar six-electron fragments. The almost flat top and bottom of $[Al_{14}R_6I_6]^{2-}$ (Exercise 2.9) is also consistent with no lone pairs on the two Al atoms.

Now we come to the even larger clusters such as $[Al_{69}R_{18}]^{3-}$ (Figure 2.32) with many internal atoms. The geometry of $[Al_{69}R_{18}]^{3-}$ may be described as an Al atom (indicated with an open circle containing a "c" in the drawing) centered polyhedron of 12 nearest neighbor Al atoms surrounded by a polyhedron of 38 next-nearest neighbor Al atoms and completed by a still larger and more open polyhedron of 18 AlR fragments. To emphasize the layers, Figure 2.32 only defines the polyhedron of the nearest neighbors of the central Al atom (after all the lines between atoms are imaginary). The 12-vertex polyhedron approximates a bicapped pentagonal prism with 5-fold symmetry – a symmetry only found in so-called quasi-crystalline materials (one cannot tile a floor with pentagons). The distinct layered structure sometimes described as "onion-like" or "stuffed" has its own nomenclature, i.e., $Al@Al_{12}@Al_{38}@(AlR)_{18}$ for $[Al_{69}R_{18}]^{3-}$.

Note that the cluster surface layer is not fully formed and that Al atoms of the second layer are exposed. In this sense the outer shells are reminiscent of $[Ga_{19}R_6]^-$ discussed above. Hence, one expects that the exposed atoms in the Al_{38} shell will be engaged in complex multicentered bonding and the presence of "lone pairs" cannot be assumed (see discussion immediately above). The structural parameters of $[Al_{69}R_{18}]^{3-}$ are revealing. The coordination numbers of the Al atoms decrease in going from the center to the surface of the cluster. Correspondingly, the Al–Al distances decrease from the center out suggesting an increase in bond localization in

2.12 When the rules fail

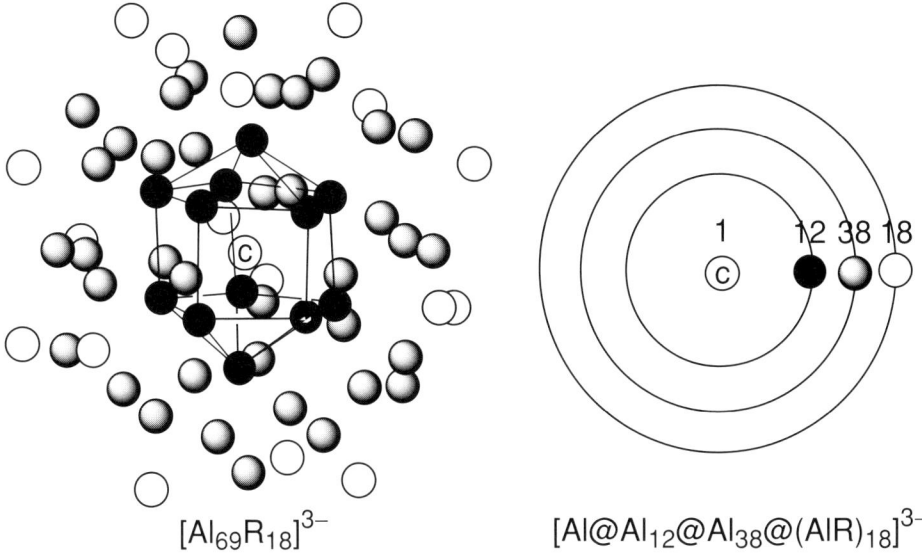

Figure 2.32

the same direction. The internal packing is sensitive to cluster size as well as cluster surface changes, i.e., $[Al_{77}R_{20}]^{2-}$ has an $Al@Al_{12}@Al_{44}@(AlR)_{20}$ layer structure but the geometry of the Al_{13} core is a distorted, centered icosahedron. All these observations are consistent with viewing these clusters as hybrids lying between molecular main-group clusters of the types already described and elemental crystallites. The connection between these metalloid clusters and the bulk pure elements is clear – they are nanoparticles of the bulk element preserved from further condensation by the ligands bound to the external surface of the particle. These ligands are not innocent but perturb the internal structure. Although only two examples are discussed here, other similarly large clusters of Ga establish the generality of the observations. We will revisit these interesting molecules in Chapters 3 and 6 as these nanoparticles truly constitute a bridge between small molecular derivatives of an element and the bulk element itself.

Exercise 2.10. The content of this chapter shows us that no single counting rule serves for all main-group clusters. This exercise illustrates how a reasoned analysis of a problem at the level of counting can tease out the core of the problem. The cubic Al clusters including "carbaalanes" have been cited as examples that "do not fit." Consider the three shown below and provide an analysis in the spirit of this chapter.

Answer. $SiAl_8(AlCp^*)_6$ is a capped cube of Al atoms centered with Si. With a mix of bare and ligated atoms, a cve count is the best approach (no assumption

78   *Main-group clusters*

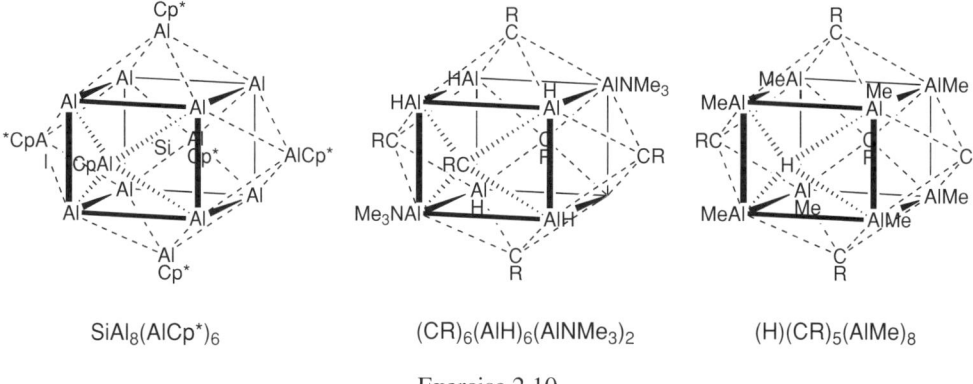

Exercise 2.10

concerning the existence of an Al lone pair is required). Observed cve = Si + 14 Al + 6 Cp* = 4 + 42 + 30 = 76. Three models are suggested by the structure: a *closo*-deltahedral borane-like cluster, a capped polyhedrane and a capped three-connect cube. A variation is that the external bond to the capping Al atoms involves six electrons not the usual two. Hence, for the first model we have 4(14) + 2 + 6 × 4 = 82 cve. For the second, we have six capped squares plus the electrons in the Al–Cp* bonding giving 6 × 6 + 6 × 6 = 72 cve. Finally, for a three-connect cube with Al–Cp* caps we have 5 × 8 + 6 × 6 = 76 cve and a match. It seems there are lone pairs on the Al atoms presumably because the radius of curvature is small.

$(CR)_6(AlH)_6(AlNMe_3)_2$ is also a capped cube but now the Al atoms that make up the cube are ligated and the capping atoms are the more electronegative C atoms. We have 8 Al + 6 H + 2 NR$_3$ + 6 C + 6 R = 24 + 6 + 4 + 24 + 6 = 64 cve. On your own apply the first two models used above for SiAl$_8$(AlCp*)$_6$ (58 and 48 cve). Here we go directly to the model that worked – a capped three-connect cube. We have 5 × 8 + 6 × 2 = 52! Three strikes and you're out. But look, we have 12 "extra" electrons to break Al–Al bonds incorporated in the model applied. This suggests the compound is bonded together mainly through the electronegative C caps. In essence, there are six CAl$_4$ five-center–six-electron bonds plus the 6 + 8 external bonding pairs (= 64 cve). Indeed, good calculations show that there are no Al–Al bonds and that the proper description of this cluster is a 14-vertex shape with 12 quadrilateral faces. Beware of associating all lines that show geometry with bonds.

$(H)(CR)_5(AlMe)_8$ is very similar to the second example in that one CR cap is replaced with an H atom. We have: H + 5 C + 5 R + 8 Al + 8 Me = 1 + 20 + 5 + 24 + 8 = 58 cve. With 5 × 6 = 30 electrons associated with the five CR caps plus 2(8 + 5) for the external bonding pairs on Al and C we have a total of 56 cve. Hence, the H cap can be reasonably considered a five-center–two-electron bond, i.e., H has no p functions to form the other two bonding interactions (review Figure 2.14.).

Problems

1. The structure of B_4H_{10} is shown below at the left. Adjacent to it is one structure proposed earlier which is a dimer of diborane. Generate and compare isoelectronic hydrocarbon analogs of the two valence isomers.

Problem 2.1

2. The compounds $[Sn_nR_n]$, $n = 6$, 8 and 10, where R is a bulky one-electron ligand, are known. Postulate a cluster structure for each.
3. Use the electron-counting rule to propose a structure for each of the following closed cluster compounds. Indicate which will possess a non-zero electric dipole moment: 1,2-$C_2B_3H_5$, 1,6-$C_2B_4H_6$, 2-Cl-1,6-$C_2B_4H_5$, 2,4-$C_2B_5H_7$, 1,2-Me_2-1,2-$Si_2B_{10}H_{10}$, $SB_{11}H_{11}$ and $[AsB_{11}H_{11}]^-$.
4. (a) Use the electron-counting rule to generate an open or closed structure for the following known cluster compounds: 2,3-$C_2B_4H_8$, 1,2-$C_2B_4H_6$, 1,5-$C_2B_3H_5$, B_4H_{10}, $C_3B_3H_7$. (b) Design a neutral open cluster that contains Si, P, H and B that exhibits a pentagonal pyramidal shape.
5. The structure of B_6H_{12} (Gaines and Schaeffer, 1964) was derived from the NMR data. The ^{11}B NMR data show the presence of three types of B in the ratio of 1:1:1. B–H coupling suggests one is a BH_2 group (dd) and the other two are BH groups (d). The ^{1}H spectrum shows four types of BH terminal protons and two types of BHB bridging protons in the ratio of 1:1:1:1:1:1. Suggest a structure. Analyze the structure and determine how many three-center bonds are required to describe the electronic structure with a localized model.
6. As part of a larger research project, closed clusters are proposed as corner units in the construction of molecular cluster rings utilizing Lewis acid–base interactions to make the connections. A cluster building block containing both a Lewis-basic site and a Lewis-acidic site (protected with a weak base such as THF) in the proper geometrical relationship for the ring size is required. Use the electron-counting rules and the geometrical relationships between cluster vertices to design the stoichiometry and shape of a cluster that would be a suitable building block for a square.
7. A bicapped hexagon is a deltahedral shape with the same total connectivity as a dodecahedron. The latter is shown in Figure 2.7. (a) Write down the set of vertex connectivities for each shape and compare. (b) If the B–B bonding distance is taken to be 1.9 Å and constant, is a bicapped hexagonal shape possible for $[B_8H_8]^{2-}$? (c) Even given a range of allowed B–B bonding distances, which of the two shapes is most spherical?

8. Work out a localized bonding model for B_6H_{10} using the ring–cap model described. Hint: planar $[B_5H_5]^{4-}$ is analogous to planar $[C_5H_5]^-$ with two fewer electrons.
9. The *nido*-$[2,3\text{-}C_2B_4H_6]^{2-}$ anion formed by the double deprotonation of $2,3\text{-}C_2B_4H_8$ has been shown to be an analog of the pentahapto $[C_5H_5]^-$ ligand (Grimes, 1992). Use the *debor* concept to justify this analogy.
10. The structure of $[Ga_9R_6]^-$ is shown below (Kehrwald *et al.*, 2001). It contains three bare Ga centers which are problematical for a simple cluster electron-counting approach. Apply an electron-counting rule and describe any unusual aspects of the cluster's electronic structure.

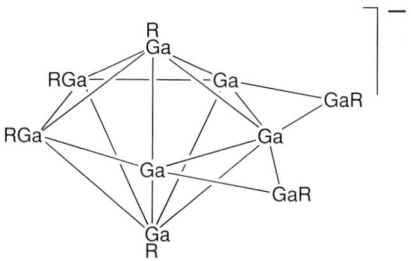

Problem 2.10

11. Both $As_7(SiMe_3)_3$ and $[As_7]^{3-}$ are known and exhibit the same cluster structure (Schmettow and Schnering, 1977). The observed structures have three types of As atoms in the ratio of 1:3:3. Suggest possible cluster structures.
12. The compound Sn_8R_4, where R is a bulky, one-electron external ligand, exhibits the cubane structure shown below (Eichler and Power, 2001). The dashed lines correspond to Sn–Sn distances of 3.1 Å (sum of Sn covalent and van der Waal's radii are 2.8 and 4.3 Å, respectively) whereas the other Sn–Sn distances lie between 2.9 and 3.0 Å. (a) Calculate the total number of valence electrons and valence orbitals available from which to construct a bonding model. How many molecular orbitals will be formed? How many will be filled? Is a localized two-center bond model possible or is some multicenter bonding required? (b) Keeping in mind the fundamental idea of using all valence electrons and orbitals, develop a bonding model for the interaction between the two Sn_2 fragments. Hint: review the analysis of $[Ge_{10}R_7]^+$ (Figure 2.25) and that of $Ga_{10}R_6$ (Exercise 2.7).

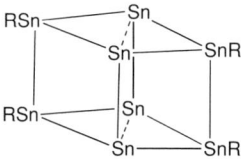

Problem 2.12

Additional reading 81

13. The compound $[(\eta^5\text{-Me}_5\text{C}_5)\text{Si}]^+$ has been described in a communication as "a stable derivative of HSi^+" (Jutzi *et al.*, 2004). Show that it is better described as a *nido*-cluster.

14. The cluster $[\text{Ga}_{13}\text{R}_6]^-$ has the structure shown below (see citation in Problem 10). It may be described as an incomplete cube (bold lines) of seven Ga atoms (solid circles) capped on the three complete faces by GaR fragments (open circles) and with a Ga_3R_3 fragment occupying the missing corner of the cube. Discuss its electronic structure at the level of electron counting.

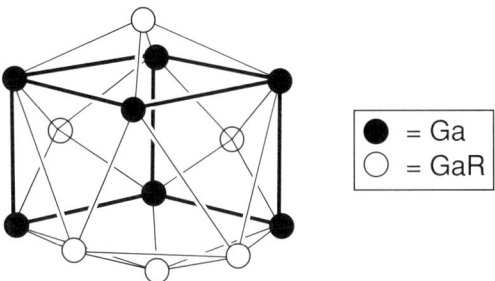

Problem 2.14

15. In a recent communication, the cluster $[\text{Sn}_{15}\text{R}_6]$ has been shown to have the structure below (Brynda *et al.*, 2006). Described as a metalloid cluster and an advance in the quest for a "bottoms up" synthesis of Sn nanoparticles, the electronic structure of this compound in terms of electron counting was not addressed. Do so.

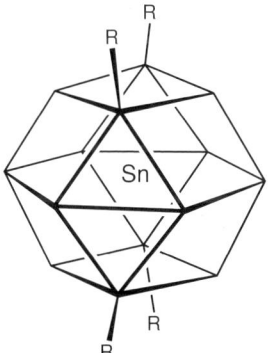

Problem 2.15

Additional reading

Lipscomb, W. N. (1963). *Boron Hydrides*. New York: Benjamin.
Wade, K. (1971). *Electron Deficient Compounds*. London: Nelson.

Section 2.1

Driess, M. and Nöth, H. (Eds.) (2004). *Molecular Clusters of the Main Group Elements*. Weinheim: Wiley-VCH.

Abel, E., Stone, F. G. A. and Wilkinson, G. (Eds.) (1995). *Comprehensive Organometallic Chemistry II*. Oxford: Pergamon.

Section 2.3

Longuet-Higgins, H. C. and Roberts, M. D. (1955). *Proc. R. Soc.* London, **A230**, 110.

Section 2.4

Wade, K. (1972). *Inorg. Nucl. Chem. Lett.*, **8**, 559.

Mingos, D. M. P. (1972). *Nature (London) Phys. Sci.*, **236**, 99.

Mingos, D. M. P. and Wales, D. J. (1990). *Introduction to Cluster Chemistry*. New York: Prentice Hall.

Mingos, D. M. P. (Ed.) (1997). *Structural and Electronic Paradigms in Cluster Chemistry*. Berlin: Springer.

Stone, A. J. (1980). *Molec. Phys.* **41**, 1339.

Section 2.6

Hawthorne, M. F. (1975). *J. Organomet. Chem.*, **100**, 97.

Section 2.8

Williams, R. E. (1970). *Prog. Boron Chem.*, **2**, 37.

Rudolph, R. W. (1976). *Acc. Chem. Res.*, **9**, 446.

Section 2.9

King, R. B. (1999). *Inorg. Chem.*, **38**, 5151.

Hoffmann, R. and Lipscomb, W. N. (1962). *J. Chem. Phys.*, **36**, 2179.

Section 2.10

Burke, A., Ellis, C., Giles, B. T., Hodson, B. E., Macgregor, S. A., Rosair, G. M. and Welch, A. J. (2003). *Angew. Chem. Int. Ed.*, **42**, 225.

Deng, L., Chan, H.-S. and Xie, Z. (2005). *Angew. Chem. Int. Ed.*, **44**, 2128.

Jemmis, E. D., Balakrishnarajan, M. M. and Pancharatna, P. D. (2001). *J. Am. Chem. Soc.*, **123**, 4313.

Grimes, R. N. (1982). *Pure and Appl. Chem.*, **54**, 43.

Mingos, D. M. P. (1983). *Chem. Commun.*, 706.

Section 2.11

Schnepf, A. (2004). *Angew. Chem. Int. Ed.*, **43**, 664.
Kehrwald, M., Köstler, W., Rodig, A., Linti, G., Blank, T. and Wiberg, N. (2001). *Organometallics*, **20**, 860.

Section 2.12

Zhai, H.-J., Zlexandrova, A. N., Birch, K. A., Boldyrev, A. I. and Wang, L.-S. (2003). *Angew. Chem. Int. Ed.*, **42**, 6004.
Wang, Z.-X. and Schleyer P. vR. (2003). *J. Am. Chem. Soc.*, **125**, 10484.
Burdett, J. K. and Canadell, E. (1991). *Inorg. Chem.*, **30**, 1991.
Schnepf, A. and Schnöckel, H. (2002). *Angew. Chem. Int. Ed.*, **41**, 3532.
Schnöckel, H. (2005). *Dalton Trans.*, 3131.
Uhl, W., Breher, F., Grunenberg, J., Lützen, A. and Saak, W. (2000). *Organometallics*, **19**, 4536.

3
Transition-metal clusters: geometric and electronic structure

To a large extent, we expect the cluster bonding principles established for main-group clusters in Chapter 2 to carry over to transition-metal clusters. However, the AO basis sets for building MOs differ for transition metals which means that the expression of the cluster bonding principles in geometric and electronic structure will also differ. That is, the observed cluster compositions and shapes differ and these differences can be associated with the participation of the metal d functions in cluster bonding. The d functions are the "wild cards" that make transition-metal chemistry so interestingly different from main-group chemistry. In writing Chapter 3 we have assumed that the reader has a basic understanding of the principles of cluster bonding as expressed by p-block clusters (Chapter 2). Emphasis here is placed on the varied expression of d-block metal character within a cluster context. A number of monographs on metal clusters are suggested at the end of this chapter as additional reading for those interested in pursuing a topic in more depth.

3.1 Three-connect clusters

Main-group clusters that exhibit three-connect shapes can often be described using localized two-center bonds. What is the situation for metal clusters?

3.1.1 Localized two-center bonds

Two-center–two-electron bonding and the eight-electron rule adequately rationalize three-connect clusters like P_4; hence, we expect the 18-electron rule to suffice for three-connect transition-metal clusters like tetrahedral $Ir_4(CO)_{12}$ (Figure 3.1) with 12 terminal carbonyl ligands. Indeed it does. Each of the four $Ir(CO)_3$ fragments is a 15-electron fragment (9 from Ir and 6 from 3 CO ligands) analogous to a five-electron P atom in terms of bonding requirements, i.e., three frontier orbitals with three electrons. Four such fragments can form six two-center–two-electron edge

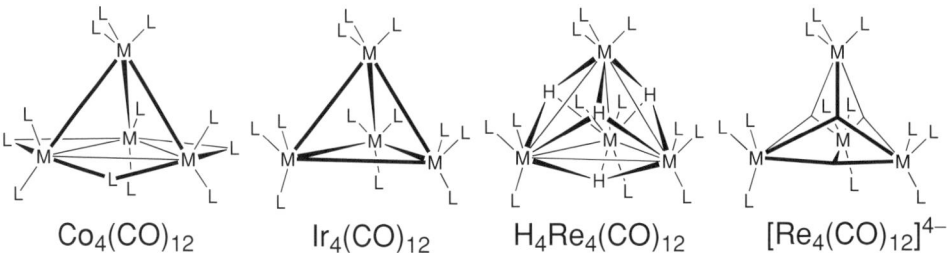

Figure 3.1

bonds. Note that the cobalt analog, $Co_4(CO)_{12}$, possesses a structure with three bridging CO ligands. This feature certainly affects the nature of the Co–Co bonding but, in terms of counting electrons for core structure predictions, the differences caused by bridging vs. terminal CO ligands can be ignored (see the discussion of $[CpFe(CO)_2]_2$ in the Appendix). Hence, in discussions of structure based on counting electrons the disposition of carbonyl ligands on the cluster framework is ignored and in some cases will not even be shown.

3.1.2 Localized three-center bonds

There are two significant differences between the structures of tetrahedral $H_4Re_4(CO)_{12}$ and $Ir_4(CO)_{12}$. First, the former has four face-bridging H atoms and the three CO ligands on each metal center eclipse the three adjacent tetrahedral edges rather than being staggered and lying over the three faces. The Re cluster has four fewer valence electrons and was originally described with metal bond orders greater than one. Indeed the Re–Re distances are about 0.1 Å shorter than that in $Re_2(CO)_{10}$ with a single bond. However, the fact that the three valence orbitals of each $Re(CO)_3$ fragment point to the adjacent faces rather than edges as well as the presence of three triply bridging H atoms suggests a more acceptable explanation. In the manner of a borane cluster with bridging H atoms, we can simplify by removing the four triply bridging H atoms as protons to generate $[Re_4(CO)_{12}]^{4-}$. A $[Re(CO)_3]^-$ fragment is a 14-electron fragment – one electron less than $Ir(CO)_3$. If the $Ir(CO)_3$ fragment behaves as a three-orbital–three-electron fragment, a $[Re(CO)_3]^-$ fragment should behave as a three-orbital–two-electron fragment. Multicenter bonding is required to generate the Re–Re bonding network and utilize all orbitals and electrons. Four three-center Re–Re–Re face bonds utilize all the valence electrons as well as valence orbitals. Each Re center shares in three three-center–two-electron bonds and satisfies the 18-electron rule. This localized bonding model is shown at the far right in Figure 3.1. The short Re–Re distances in the protonated cluster are explained by the necessity of good overlap with the triply bridging protons. Can you see that the bonding is analogous to that found for R$_4$Ga$_4$ discussed in Section 2.2.3?

Exercise 3.1. The cluster [Cp*$_4$Rh$_4$H$_4$]$^{2+}$, where Cp* is the five-electron η^5-C$_5$H$_5$ ligand, exhibits the same structure as H$_4$Re$_4$(CO)$_{12}$ shown in Figure 3.1. Propose a localized cluster bonding model based on two- and three-center bonds.

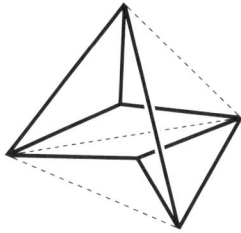

Exercise 3.1

Answer. Simplify the problem by considering [Cp*$_4$Rh$_4$]$^{2-}$ with four 14-electron Cp*Rh fragments analogous to the [Re(CO)$_3$]$^-$ fragments. Hence, we have 12 orbitals and ten electrons (don't forget the charge of 2 −) to utilize in cluster bonding. This is accomplished by forming two three-center–two-electron bonds (six orbitals and four electrons) and three two-center–two-electron bonds (six orbitals and six electrons). They may be placed on the tetrahedral framework as shown above. As you might expect, the measured Rh–Rh distances are not equal – two are short (2.66 Å) and four are long (2.83 Å).

3.1.3 Skeletal electron pair (sep) count

Above we separated the cluster framework bonding from the external bonding so we now look at the metal-cluster skeletal electron counts. Doing so highlights an ambiguity arising from the metal d functions and electrons which is an important feature of metal clusters and, thus, must be understood. Recall that for a main-group electron-precise cluster the sep count was $3/2\, n$ where n is the order of the three-connect cluster, e.g., six sep for $n = 4$, P$_4$. What is the sep for Ir$_4$(CO)$_{12}$? Now think! There is the same number of Ir–Ir two-center–two-electron bonds, but what do we do with the remaining three filled orbitals on the metal? Clearly, if a five-electron P fragment (or CH) indeed is analogous to a 15-electron Ir(CO)$_3$ fragment, we have to ignore them as far as cluster bonding is concerned, i.e., they are cluster non-bonding. This set of three filled orbitals is often viewed as a vestige of the t$_{2g}$ metal set in a classical octahedral metal coordination complex. Isolobal fragments, as they have been called, will be investigated more thoroughly in Chapter 4.

The designation of a set of metal valence electrons as cluster non-bonding has significant consequences. Recall that in order to do a sep count for a main-group cluster the external cluster bonding network is considered separately from the internal cluster bonding network. In a totally pragmatic sense, this separation causes little or no problem in a majority of cluster systems treated. Likewise, treating the t_{2g} metal set as cluster non-bonding works in many cases. But later in the chapter we will encounter a significant number of cluster systems where it doesn't.

The division of the metal valence electrons into cluster bonding and non-bonding electrons is not universally accepted. Woolley argues that it is not possible to separate a set of metal non-bonding d orbitals from the remaining s, p and d AOs. He contends that there are major differences between the electronic structures of metal and main-group clusters and any success in correlating structures and electron counts in the manner described in this chapter is fortuitous. We concede the possibility, but the assumption behind sep counts is so useful in describing and thinking about mixed main-group–transition-metal clusters of the later metals (Chapter 5) that there is ample justification for using it. Further, the whole question is one lacking an effective empirical test; hence, Woolley's position has been of little consequence to date.

3.1.4 Cluster valence electron (cve) count

An alternative approach which avoids d orbital/electron separation is that of cluster valence electron counts. Here the observed geometries along with models of the electronic structures define characteristic cve counts for two- and three-connect clusters (Table 3.1). We have seen that for a main group cluster of order n the counts are $6n$ and $5n$, respectively. Hence, it follows from the eight-eighteen-electron rules that for analogous transition metal clusters the counts are $16n$ and $15n$. For example, for C_3H_6 cve $= 18$, whereas for $Os_3(CO)_{12}$ cve $= 48$; 6×3 vs. 16×3. Likewise, the cve $= 20$ for P_4 and 60 for $Ir_4(CO)_{12}$; 5×4 vs. 15×4. In Figure 3.2 cve counts are shown for other geometries. Verify a couple of them and find main-group analogs in Chapter 2. The larger three-connect clusters have ligands bridging rectangular or square faces and can equally well be considered as mixed main-group–transition-metal clusters of higher nuclearity (Chapter 5).

3.1.5 Shape change with cluster count

For three-connect (and lower) main-group clusters, we concluded in Chapter 2 that a cve count is not particularly useful. However, metal cluster structures are not always easily analyzed in terms of metal connectivity. Bond distance criteria for the existence of a M–M bond are not as precise as those for an E–E bond.

3.1 Three-connect clusters

Table 3.1. *Cluster valence electron counts for representative metal-cluster geometries*

Geometry	No. atoms	cve count
Monomer	1	18
Dimer	2	34
Trimer	3	48
Tetrahedron	4	60
Butterfly	4	62
Square Plane	4	64
Trigonal Bypyramid	5	72
Square Pyramid	5	74
Bicapped Tetrahedron	6	84
Octahedron	6	86
Pentagonal Pyramid	6	88
Trigonal Prism	6	90
Triangular Dodecahedron	8	112
Square Antiprism	8	114
Bicapped Trigonal Prism	8	114
Cube	8	120
Icosahedron	12	170
Cube Octahedron	12	170
Truncated Hexagonal Bipyramid	12	170

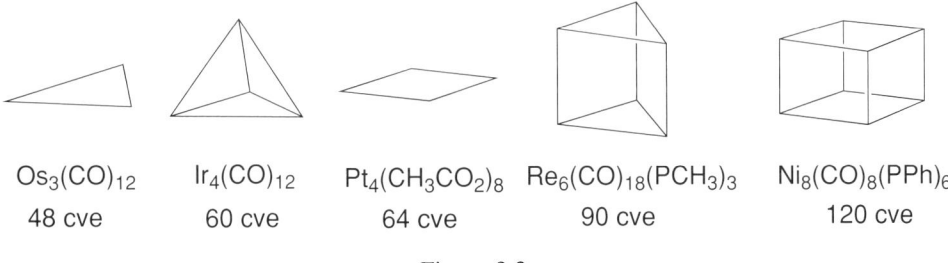

$Os_3(CO)_{12}$ $Ir_4(CO)_{12}$ $Pt_4(CH_3CO_2)_8$ $Re_6(CO)_{18}(PCH_3)_3$ $Ni_8(CO)_8(PPh)_6$
48 cve   60 cve   64 cve   90 cve   120 cve

Figure 3.2

Hence, the pioneering papers of Lauher contain cve counts as a function of cluster nuclearity. Table 3.1 in effect is the metal cluster analog of Figure 2.13. We can use these results to illustrate a difference between main-group and metal clusters. Thus, octahedral (four-connect vertices, cve = 86, derived below) and trigonal prismatic (three-connect vertices, cve = 90) clusters (Figure 3.3) are found in Table 3.1. With four more electrons, a cve = 90 count should correspond to a six-atom "*arachno-*" cluster. The "*arachno-*" borane cluster has the shape shown to the left in Figure 3.3 whereas the three-connect trigonal prismatic cluster is the common shape for the metal system. We must be prepared for the fact that different shapes are possible in

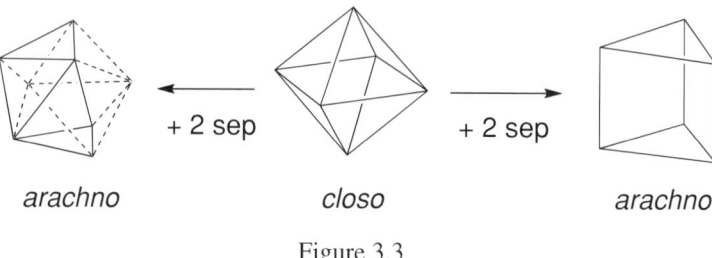

Figure 3.3

metal-cluster systems. But more about this aspect of the problem after we have dealt with four-connect clusters below.

In summary, three-connect clusters require no electron-counting rule to simplify analysis of the bonding but one is available and it can be useful for comparative purposes. The cve count includes all the metal d electrons without specifying to what extent they are involved in cluster bonding. On the other hand, the sep count requires explicit assignment of the d electrons in terms of their possible roles in cluster electronic structure. Appreciation of ambiguities in the role of the metal d electrons in cluster bonding is important because confusion on this point can lead to misunderstanding.

Exercise 3.2. Consider a trigonal prismatic cluster formed from six $Co(CO)_3$ fragments. Do the metal centers obey the 18-electron rule. Does the cve count for the cluster follow Lauher's prediction? Although no such compound exists, is it a reasonable target molecule? As this is a three-connect cluster, in principle, other three-connect clusters could be constructed with 15-electron $Co(CO)_3$ fragments. Use Figure 2.3 to design another three-connect metal cluster that might constitute a synthetic target.

Answer. The $Co(CO)_3$ fragment is a 15-electron fragment analogous to a five-electron main-group fragment, e.g. CH. If it forms three two-center–two-electron bonds it obeys the 18-electron rule (6 electrons from the 3 CO ligands, 9 electrons from the metal itself and 3 electrons from the three two-center–two electron bonds to the three nearest-neighbor metal atoms). The cve count is $6\,Co + 18\,CO = 54 + 36 = 90$ which agrees with the value in Table 3.1. Any of the three-connect structures shown in Figure 2.3 are, in principle, viable at this elementary level of structure.

3.2 Small four-connect metal-carbonyl clusters and the $14n + 2$ rule

Four-connect vertices and the limited valence functions of a main-group fragment demanded a delocalized bond model. Here we explore the same situation but with metal fragments where there is no similar orbital restriction. Each metal has nine

3.2 Small four-connect clusters and the 14n + 2 rule

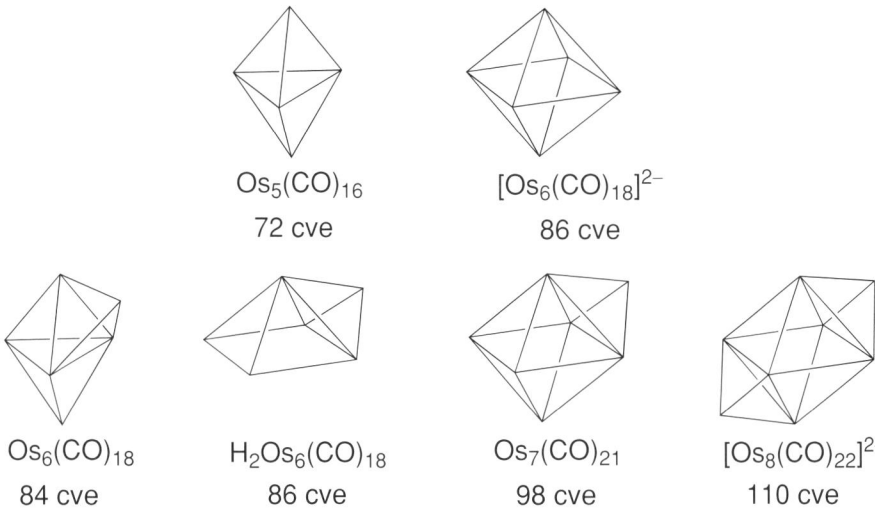

Figure 3.4

rather than four valence functions and coordination numbers greater than four are easily accommodated.

3.2.1 Geometry

Many main-group and metal clusters exhibit the same nuclearities but there are differences in the preferred geometries for a given nuclearity. Main-group element clusters with nuclearities from four to 12 display single cluster structures but, with rare exception, clusters of higher nuclearity require cluster fusion of some type. Single metal clusters are found with five and six cluster atoms (Figure 3.4), but there are no examples of single, metal deltahedral clusters of orders 7–12. Higher nuclearities either involve face-capping fragments, cluster fusion or the presence of internal cluster atoms. The most common geometry exhibited for nuclearities between six and ten is a face-capped deltahedron of nuclearity six or lower. The absence of icosahedral clusters for metals and the absence of capped clusters for boranes reflect metal vs. main-group bonding properties. A satisfactory description of the bonding of these systems should justify these characteristic differences.

3.2.2 Electronic structure: a localized approach

We begin with an obvious question. Given that a metal atom can accommodate larger coordination numbers than main-group atoms, why cannot clusters containing vertex connectivities greater than three be accommodated with two-center–two-electron bonding plus application of the 18-electron rule? Consider an octahedral

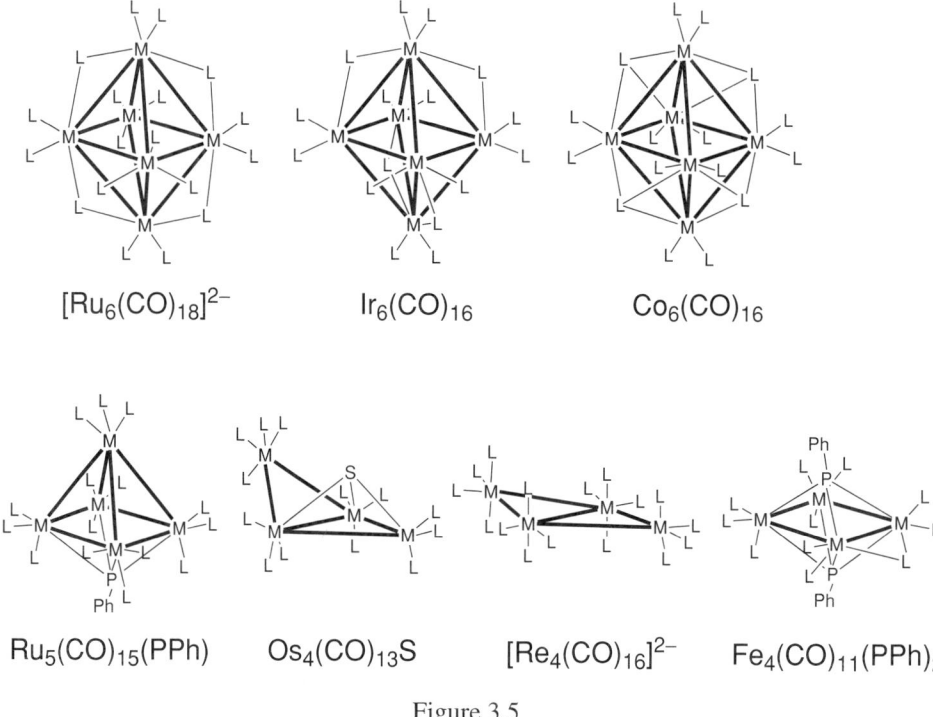

Figure 3.5

cluster formed from six 14-electron Ru(CO)$_3$ fragments. Of the nine orbitals, we use three Ru orbitals for the six CO electrons and four orbitals plus four electrons to form four nearest-neighbor Ru–Ru two-center–two-electron interactions. This leaves two filled Ru orbitals unused. Hence, [Ru$_6$(CO)$_{18}$] obeys the 18-electron rule and has a cve count of 84 (6 × 14). So, it is possible to generate a satisfactory bonding model without multicenter bonding. But models don't dictate chemistry. What does Nature tell us? Nature says that [Ru$_6$(CO)$_{18}$]$^{2-}$ (Figure 3.5) is isolated as an anion WITH a negative charge of −2. It has 86 cve (6 × 8 + 18 × 2 + 2 = 86) not 84. More importantly Os$_6$(CO)$_{18}$ (Figure 3.4) with 84 cve does NOT exhibit an octahedral structure as predicted by the 18-electron rule and two-center–two-electron bonding. Although a localized model does not work for these late metal-carbonyl clusters, we will encounter examples of different cluster types later in the chapter where it does.

3.2.3 Electronic structure: a delocalized approach

What model does provide a valence electron/composition connection for [Ru$_6$(CO)$_{18}$]$^{2-}$? Let's try this. Limit the metal to three orbitals for ligand binding and three orbitals for cluster bonding. Thus, we force it to act like a six rather

3.2 Small four-connect clusters and the 14n + 2 rule   93

than seven-coordinate metal center in a cluster environment. In effect we make 14-electron Ru(CO)$_3$ behave like four-electron BH in the manner of Section 3.1 and reserve three metal orbitals for a cluster non-bonding t$_{2g}$ metal set. This is tantamount to regenerating the main-group octahedral cluster problem in that the B–H fragment is strictly limited to three orbitals for cluster bonding. We know the solution. Octahedral [B$_6$H$_6$]$^{2-}$ possesses seven cluster bonding orbitals and six external BH bonding orbitals. So how can we partition the electrons suitably in the case of the metal? Each Ru(CO)$_3$ fragment has three filled Ru–CO bonding orbitals and three filled non-bonding t$_{2g}$-type orbitals. Six fragments contribute six pairs and the 2– charge one pair to cluster bonding. This metal fragment behavior is consistent with that found for three-connect metal clusters.

An obvious question is why metal-carbonyl clusters do not utilize more d orbitals in cluster bonding. Later in the chapter we will see that early metal clusters with π-donor ligands do adopt a localized model whereas these later metal clusters with π-acceptor ligands do not (Section 3.3.5). In the meantime, we can take pleasure in the similarities between the group 8/9 transition-metal and borane clusters. Even though the metal has sufficient orbitals to handle seven coordination, it still hews to a main-group party line. To review, the borane solution to the problem of accommodating a mono-ligated, four-coordinate, tetrahedral fragment to a four-connect vertex is mimicked in the transition-metal system as a response to the problem of accommodating a tri-ligated, six-coordinate, octahedral fragment to a four-connect vertex. Remember, in the case of the metal this solution is not required by the geometry alone. Although we can observe atom nearest-neighbor geometry, atom valence (orbitals utilized) must be inferred with the aid of calculations.

3.2.4 cve counts: closed clusters

With these preliminaries out of the way we are now prepared to take a look at the counting rules for the large class of metal clusters formed from group 8 and 9 metals with good acceptor ligands, e.g., CO, for which the bonding–non-bonding separation just discussed holds. A transition-metal deltahedral cluster of order n has a cve count of $14n + 2$ as per the discussion immediately above. This is pleasing as the cve count is just the main-group cluster rule ($4n + 2$) plus $10n$. The qualitative MO block diagram for a closed deltahedral transition-metal cluster made up of n M(CO)$_3$ fragments is shown in Figure 3.6. Compare it with the one for a p-block cluster made up of EH fragments (Figure 2.8). The total number of orbitals in each case is $5n$ and $12n$. In the main-group cluster there are $2n + 1$ filled and $3n - 1$ empty orbitals whereas in the transition-metal cluster there are $7n + 1$ filled and $5n - 1$ empty orbitals. If one removes the external cluster bonding orbitals in both cases, we have $n + 1$ filled, $2n - 1$ empty vs. $4n + 1$ filled, $2n - 1$ empty. Notice that

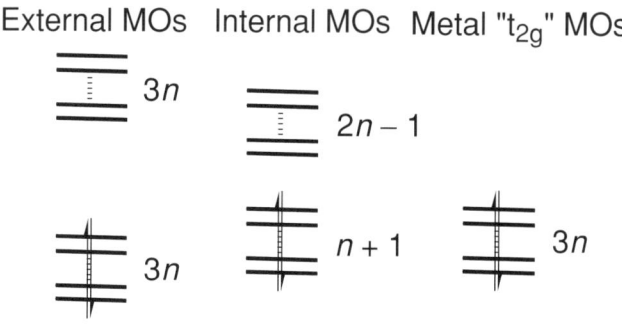

Figure 3.6

the number of empty orbitals is the same for the main-group and transition-metal clusters. Indeed, Mingos made this significant point early on. It is only when $3n$ metal orbitals are designated non-participants in cluster bonding that the numbers of filled cluster bonding orbitals become the same in p- and d-block clusters of the same shape.

3.2.5 cve counts: open clusters

For open metal clusters we encounter the same complication with the cve count as with main-group clusters – one vertex is empty and the metal non-bonding d electrons as well as ligand electrons must be subtracted from the $14n + 2$ count. So for a *nido*-cluster the count is $14n + 2 - 12$ and for an *arachno*-cluster the count is $14n + 2 - 24$ where n is the dimension of the closed deltahedron upon which the cluster geometry is based. Thus, for $Ru_5(CO)_{15}(PPh)$, Figure 3.5, with 74 cve (5 Ru + 15 CO + 1 PPh = 40 + 30 + 4) and a square pyramidal structure, we obtain $14(6) + 2 - 12 = 74$ cve from the rule. Later in Chapter 5 we will see that the PPh ligand can also be considered as a cluster fragment and that $Ru_5(CO)_{15}PPh$ is, alternatively, a *closo*-octahedral mixed main-group–transition-metal cluster. In fact there are no known square-pyramidal metal clusters without multisite bridging atoms or interstitial atoms (see below). This constitutes another significant difference between main-group and transition-metal systems.

For the compound $Os_4(CO)_{13}S$ (Figure 3.5) with 62 cve (4 Os + 13 CO + 1 S = 32 + 26 + 4) and an *arachno*-butterfly structure, we obtain $14(6) + 2 - 24 = 62$ from the rule. A source of possible confusion are clusters like $[Re_4(CO)_{16}]^{2-}$ (Figure 3.5) with 62 cve but its structure has a planar metal core rather than the butterfly structure expected. This is attributed to a very small energy change involved in flattening a butterfly structure and other factors, e.g., steric factors, becoming dominant. That is, we have now moved to group 7 metals and the number of CO ligands required to make the electron count is four per metal occupying considerable space above and

3.2 Small four-connect clusters and the 14n + 2 rule

below the two wings of the butterfly. Indeed, twisting of the "wing-tip" Re(CO)$_4$ fragments is observed in the solid state and attributed to ligand–ligand repulsions. Hence, it may well be that the 180° dihedral angle rather than the expected dihedral angle of an *arachno*-cluster (108° is typical) is caused by external-ligand steric effects. External-ligand steric requirements in metal-cluster chemistry affect both structure and accessible stoichiometries. These factors will be dealt with more generally below.

A possible source of confusion for the initiate is the practice of transition-metal chemists to sometimes denote a transition-metal cluster such as Ru$_4$(CO)$_{12}$(C$_2$Ph$_2$) as a 60 cve M$_4$ butterfly cluster with a four-electron alkyne ligand rather than a 66 cve octahedral E$_2$M$_4$ main-group–transition-metal cluster with delocalized bonding that fits the counting rule. If interested in more information on the metal complex vs. main-group–metal cluster problem right now, look ahead to Chapter 4 where you will find the Ru(CO)$_3$ fragment isolobal to BH and Ru$_4$(CO)$_{12}$(C$_2$Ph$_2$) analogous to 1,2-C$_2$B$_4$H$_6$.

3.2.6 Cluster isomers

With the open clusters one finds situations where more than a single count exists for the same shape. For example, the simple square shape can be derived from an octahedron by removing two non-adjacent vertices. Thus, the cve count is $14n + 2 - 24 = 62$, e.g., Fe$_4$(CO)$_{11}$(PPh)$_2$ (Figure 3.5) where μ_4-PPh is considered a four-electron donor. But a square is also possible with four two-center–two-electron bonds formed from four 16-electron metal fragments giving a cve count of 64, e.g., Fe$_4$(CO)$_{12}$(PPh)$_2$ which counts like Fe$_4$(CO)$_{16}$ = 4{Fe(CO)$_4$}, i.e., a ring with two-connect metal centers. See Chapter 5, Section 5.2 for a more complete analysis of this cluster type.

3.2.7 sep counts

The cve count is straightforward but what about the sep count? The sep count, you remember, requires a separation of the external cluster ligand bonding or lone-pair orbitals from the framework orbitals. For metal clusters we must, of necessity, also deal with the "extra" metal d orbitals. For these group 8/9 clusters we now see that three filled metal d orbitals can be designated cluster non-bonding such that the 14-electron Ru(CO)$_3$ behaves like the four-electron BH fragment as far as cluster bonding is concerned. Hence, both fragments are three-orbital–two-electron fragments for cluster bonding purposes and the sep count is 7 for both [B$_6$H$_6$]$^{2-}$ and [Ru$_6$(CO)$_{18}$]$^{2-}$. The same process can be used to generate *nido*- and *arachno*-cluster counts, e.g., sep = 7 for both five-atom square pyramid and four-atom "butterfly" clusters. Once one defines the metal fragment contribution then the cluster count follows the main-group cluster paradigm.

As with main-group clusters either type of count can be used to rationalize existing structures and to suggest structures based on composition data. There are advantages and disadvantages to both. The cve count is more easily obtained as one need only define ligand counts as one would in a mononuclear coordination compound. The cve count for open clusters suffers from having to deal with the empty vertices, but, on the other hand, there are fewer open metal clusters and those that do exist can often be advantageously described as mixed main-group–transition-metal clusters (Chapter 5). The sep count is easier to correlate with cluster geometry ($n + 1$ pairs for a deltahedron of order n) as ligand contributions are removed. This advantage is balanced by the need to determine the effective number of orbitals and electrons contributed by the metal fragment to cluster bonding. We will see in Chapter 5 that in mixed main-group–transition-metal clusters this need can cause problems.

Exercise 3.3. The compound $Os_5(CO)_{16}$ has the structure shown in Figure 3.4. Justify the cluster shape using both the cve count and the sep count.

Answer. The cve count is 5 Os + 16 CO = 40 + 32 = 72 = $14n + 2$ for a *closo*-cluster; thus, $n = 5$ as observed. The cluster can be divided into five $Os(CO)_3$ fragments and one extra CO. The former are 14-electron fragments and the fragment is equivalent to a four-electron BH fragment, i.e., it is a three-orbital–two-electron cluster fragment. The sep count, then, is 5 $Os(CO)_3$ fragments + 1 CO = 1/2 (5 × 2 + 2) = 6 which is appropriate for the trigonal bipyramidal shape observed.

3.3 Variations characteristic of metal clusters

Although the principles of cluster bonding developed for main-group clusters carry over to transition-metal clusters of the group 8/9 metals with carbonyl ligands, we fully expect transition metals to exhibit variations on this cluster bonding theme as well as novel behavior not seen in main-group systems. In this section we introduce those aspects of cluster chemistry characteristic of transition-metal clusters.

3.3.1 Interstitial atoms

In Figure 3.7, a selection of metal clusters containing interstitial atoms is shown. Examples with interstitial H atoms as well as transition-metal atoms are also known. Addition of an interstitial metal atom is the first step towards extended metal structures. The term "interstitial" derives from its use in solid-state chemistry where atoms are found in the interstices of metal lattices, e.g., the tetrahedral or octahedral

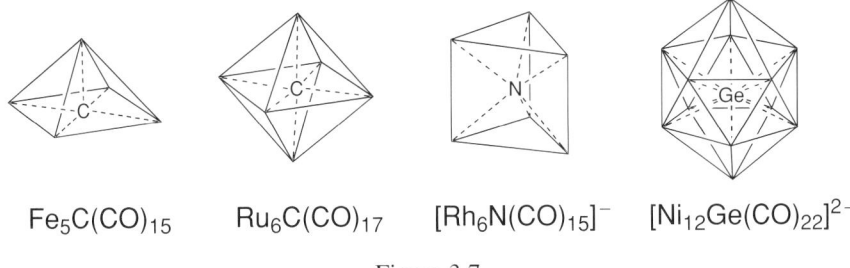

Figure 3.7

holes of a close-packed lattice. These atoms have extraordinary effects on properties. For example, the diffusion of H_2 into the metal walls of high-pressure H_2 cylinders causes metal embrittlement and can lead to eventual catastrophic failure. "Interstitial" in the case of metal clusters refers to atoms inside the metal-cluster framework that serve to stabilize the clusters. The icosahedral cluster $[Ni_{12}Ge(CO)_{22}]^{2-}$ provides an example of the stabilization of a single cage system with a nuclearity greater than six by the presence of an interstitial atom.

Incorporation of interstitial atoms into the electron-counting rule is as simple as their effect on electronic structure and properties is profound. Take the first example $Fe_5C(CO)_{15}$ (Figure 3.7). This is one precedent-setting structure of many originating in the laboratory of Dahl who is one of the pioneers of metal-cluster structure. It has the 74 cve required for a *nido*-square pyramidal geometry if all of the valence electrons of the interstitial C atom are included (5 Fe + 15 CO + C = 40 + 30 + 4). Likewise, we only obtain seven pairs by including four electrons of the C atom (5 Fe(CO)$_3$ + C = 5 × 2 + 4). $Fe_5C(CO)_{15}$ is thus a metal analog of B_5H_9 (Chapter 2) in terms of counting electrons where the interstitial C atom of the metal cluster serves the same role as the four B–H–B bridging H atoms of the borane.

Similarities in the electron counts between clusters do not equate to similarities in electronic structure. This is easily appreciated in the case of a *closo*-octahedral cluster in terms of the block MO diagram in Figure 3.8 for $[Ru_6(CO)_{18}]^{2-}$ and $Ru_6C(CO)_{17}$ (structures in Figures 3.5 and 3.7). For simplicity we assume ideal O_h symmetry and show only the cluster bonding orbitals, i.e., the sep approach is taken. The seven cluster bonding orbitals of the empty octahedral cluster have t_{2g}, t_{1u} and a_{1g} symmetries whereas the AOs of the interstitial atom have t_{1u} (2p) and a_{1g} (2s) symmetries. Hence, on placing the interstitial atom into the ruthenium cluster, four of the cluster bonding orbitals are significantly stabilized relative to the t_{2g} set.

A cluster with an interstitial atom requires fewer external ligands. The total number of external ligands are limited by the space available on a cluster surface. Hence, although $[Fe_6C(CO)_{16}]^{2-}$ is known, $[Fe_6(CO)_{18}]^{2-}$ is not. On the other hand, the larger surface of the Ru analog permits both $[Ru_6C(CO)_{17}]$ and $[Ru_6(CO)_{18}]^{2-}$ to be

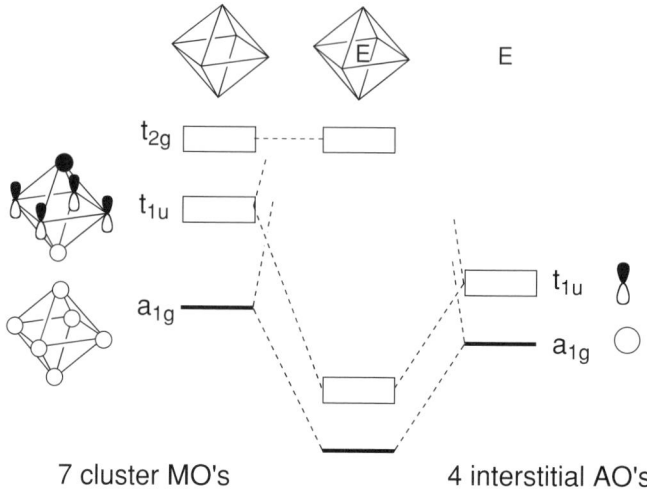

Figure 3.8

isolated. A more detailed consideration of external-ligand steric effects will be found in Section 3.3.4. The increase in stability imparted by an interstitial atom is also easy to appreciate. Octahedral $[Fe_6(CO)_{18}]^{2-}$, in the structure of $[Ru_6(CO)_{18}]^{2-}$, would be held together by relatively weak carbonyl bridged and unbridged Fe–Fe interactions whereas the known $[Fe_6C(CO)_{16}]^{2-}$ cluster has similar interactions, but supplemented by the six internal Fe–C interactions.

Perhaps you are thinking that the metal atoms in $[Fe_6C(CO)_{16}]^{2-}$ are now using more than three orbitals for cluster bonding. How else can they interact effectively with the interstitial atom? But notice that the interaction described in Figure 3.8 is no different in principle than the addition of four H atoms to a square pyramidal B_5H_5 cluster or the addition of 4 H^+ to $[B_5H_5]^{4-}$ as in the one electron MO method the electrons are added last. For the iron cluster, then, the equivalent model would be the insertion of a C^{4+} ion into the center of a $[Fe_6(CO)_{16}]^{6-}$ cluster. No additional metal orbitals are needed.

Exercise 3.4. Predict the cluster geometries for the following, using both cve and sep counts: $[Co_6N(CO)_{15}]^-$, $[Fe_4RhC(CO)_{14}]^-$, $[Fe_4C(CO)_{12}]^{2-}$, $[Ru_5N(CO)_{14}]^-$, $[Rh_6C(CO)_{13}]^{2-}$ and $[Rh_2Fe_4B(CO)_{16}]^-$.

Answer. cve count. $[Co_6N(CO)_{15}]^-$: 6 Co + N + 15 CO + (−) = 90 (trigonal prismatic, interstitial N, 15n, n = 6); $[Fe_4RhC(CO)_{14}]^-$: 4 Fe + 1 Rh + C + 14 CO + (−) = 74 (*nido*-square pyramid, interstitial C, 14n +2 − 12, n = 6); $[Fe_4C(CO)_{12}]^{2-}$: 4 Fe + C + 12 CO + 2 (−) = 62 (*arachno*-butterfly, interstitial C, 14n + 2 − 24, n = 6); $[Ru_5N(CO)_{14}]^-$: 5 Ru + N + 14 CO + (−) = 74 (*nido*-square pyramid, interstitial N, 14n +2 −12, n = 6); $[Rh_6C(CO)_{13}]^{2-}$: 6 Rh + C +

13 CO + 2 (−) = 86 (*closo*-octahedral, interstitial C, $14n + 2$, $n = 6$): [Rh$_2$Fe$_4$B(CO)$_{16}$]$^-$: 2 Rh + 4 Fe + B + 16 CO + (−) = 86 (*closo*-octahedral, interstitial B, $14n + 2$, $n = 6$).

sep count. [Co$_6$N(CO)$_{15}$]$^-$: N$^-$ equivalent to three CO so counts like six Co(CO)$_3$ = ½(6 × 3) = 9 sep consistent with dodecahedron with two vacant vertices or a three-connect polyhedron of order six, i.e., trigonal prism; [Fe$_4$RhC(CO)$_{14}$]$^-$: C is equivalent to two CO so counts like 4 Fe(CO)$_3$ + 1 Rh(CO)$_3$ + CO + (−) = ½(4 × 2 + 3 + 2 + 1) = 7 sep consistent with deltahedron of order six with one vertex unoccupied (*nido*-square pyramid); [Fe$_4$C(CO)$_{12}$]$^{2-}$: 4 Fe(CO)$_3$ + C + 2 (−) = ½(4 × 2 + 4 + 2) = 7 sep consistent with $n = 6$ and two vertices unoccupied (*arachno*-butterfly, interstitial C, note that if non-adjacent vertices are unoccupied a square shape is predicted but is unlikely with an interstitial atom); [Ru$_5$N(CO)$_{14}$]$^-$: N$^-$ is equivalent to three CO so counts like 5 Ru(CO)$_3$ + 2 CO = ½(5 × 2 + 4) = 7 sep (*nido*-square pyramid, interstitial N); [Rh$_6$C(CO)$_{13}$]$^{2-}$: C is equivalent to two CO so counts like 4 Rh(CO)$_3$ + 2 Rh(CO)$_2$ = ½(4 × 3 + 2 × 1) = 7 sep, $n = 6$, all vertices occupied (*closo*-octahedral, interstitial C); [Rh$_2$Fe$_4$B(CO)$_{16}$]$^-$: B$^-$ is equivalent to two CO so counts like 2 Rh(CO)$_3$ + 4 Fe(CO)$_3$ = ½(2 × 3 + 4 × 2) = 7 sep, $n = 6$, all vertices occupied (*closo*-octahedral, interstitial B).

Life is not always simple in cluster chemistry. Consider Os$_5$S(CO)$_{15}$ with a square pyramidal metal-cluster atom geometry. The S atom is centered in the open square face albeit not coplanar with the four metal atoms much like the structure of Fe$_5$C(CO)$_{15}$ in Figure 3.7. If S behaves as an interstitial atom, we obtain a cve count of 76 which is two higher than the expected number in Table 3.1. If S behaves as a four-electron ligand bridging the four-metal square face, we obtain a cve count of 74 appropriate for its geometry. Hence, the latter interpretation is correct, but such electronic information is not obvious from geometry alone. In Chapter 5 we will find that this cluster can also be considered as a M$_5$E mixed metal–main-group cluster with a distorted octahedral geometry for which a cve count of 76 is appropriate (for those of you who can't wait, replace one 18-electron rule M of a cve 86 M$_6$ cluster with one eight-electron rule E to generate 86 − 10 = 76 cve). Counting electrons must be done intelligently.

3.3.2 Face-capping

A second structural feature common to metal clusters, but very rare with main-group clusters, is face-capping. Examples of face-capped clusters are shown in Figure 3.4. Look at a couple. You should be able to see that each can be considered to be formed from a primary cluster by capping a triangular face. Mingos showed that the number of skeletal bonding electron pairs associated with a capped cluster is the same as the number associated with the primary cluster. Thus, a capped octahedral cluster like Os$_7$(CO)$_{21}$ has seven sep, i.e., 7 Os(CO)$_3$ = ½(7 × 2) = 7, which is the same

100   *Transition-metal clusters*

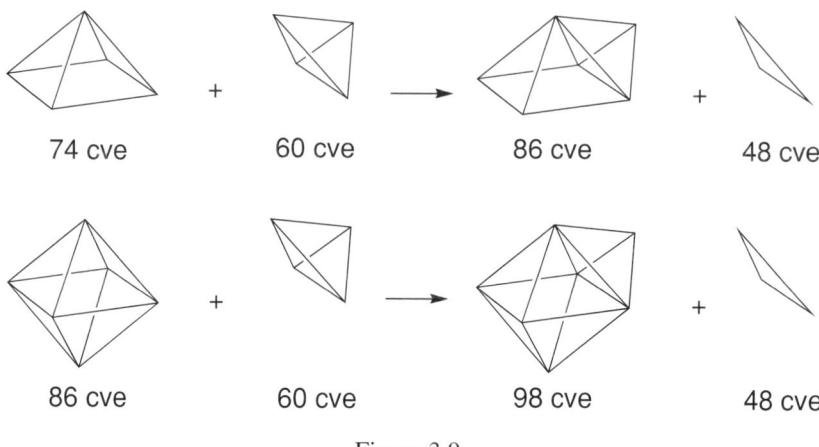

Figure 3.9

as that of an uncapped octahedral cluster such as $[Os_6(CO)_{18}]^{2-}$. The reasoning used to justify this result is similar to that used in Chapter 2 in the discussion of fused clusters. As shown in Figure 3.9, cluster capping can be considered to be a variation of cluster fusion. This particular fusion scheme, as well as others, will be more fully treated below.

The cve count, which includes the external ligands, hides the similarity in cluster bonding counts that exists for capped and uncapped clusters. The simple octahedral cluster has $14n + 2 = 86$ cve whereas the capped octahedral cluster possesses 98 cve, i.e., $14 \times 6 + 2 + 12$. The "correction" is numerically the same but in the opposite sense to that necessary when a single vertex of a closed cluster is left vacant, i.e., in the case of capping one adds the non-bonding and ligand electrons associated with the added vertex. It follows that one adds 12 electrons to the cve count of the primary cluster for each cap present.

Exercise 3.5. We have already shown that for late transition-metal carbonyl clusters, the metal fragments utilize only three orbitals in cluster bonding. Thus, we can use a main-group analog to explore the origins of the capping rule thereby minimizing the complexity of the MO picture. By referring to the analysis of $[B_5H_5]^{4-}$ given in Figure 2.14, show that capping one triangular face with a $[BH]^{2+}$ fragment does not increase the number of cluster bonding orbitals, i.e., seven sep are required for square pyramidal $[B_5H_5]^{4-}$ and seven sep are required for capped square pyramidal $[B_6H_6]^{2-}$.

Answer. As shown in the diagram below, the highest three filled MOs of $[B_5H_5]^{4-}$ can interact with the three frontier orbitals of a capping $[BH]^{2+}$ fragment. As the three are the highest filled orbitals, the bonding interaction will result in three stabilized, filled MOs and three destabilized, empty MOs thereby leaving the cluster electron count unchanged. A rare example of a capped main-group cluster is

3.3 Variations characteristic of metal clusters

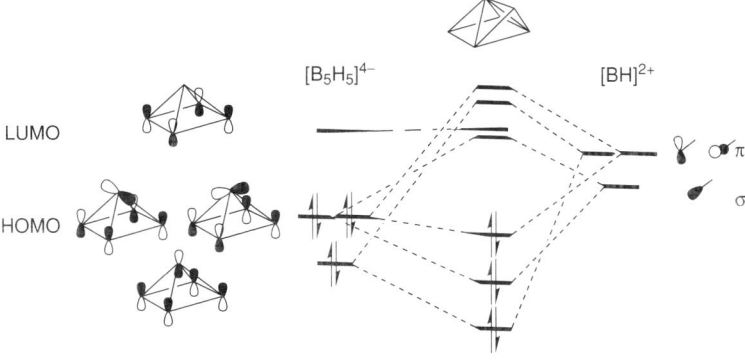

Exercise 3.5

In_8R_6 which consists of a In_6R_4 octahedron (bare In atoms trans) capped by two RIn fragments in a transoid geometry. Formally the RIn fragments contribute two electrons each and the In contribute one if a lone pair is assumed, yielding sep = 7.

For the beautiful tetracapped octahedral Os cluster, $[Os_{10}C(CO)_{24}]^{2-}$ with an interstitial C atom in the octahedral core, shown in Figure 3.10, the predicted cve count is $14(6) + 2 + 4(12) = 134$, which agrees with that of the observed stoichiometry. It's a little bit harder to count the sep but give it a try. Each tetrahedral cap consists of an $Os(CO)_3$ fragment and the other six fragments are $Os(CO)_2$ so we have $(4 \times 2 + 6 \times 0 + 4 + 2)/2 = 7$ appropriate for an octahedron. If you look ahead in Chapter 6 (Exercise 6.1), you will find that this trigonal bipyramidal ten-atom core can be excised from a cubic close-packed metal lattice (ABC layers). $[Os_{10}C(CO)_{24}]^{2-}$ can be considered a nano-sized metal particle stabilized by the ligands in the same manner as Ni atoms are stabilized when removed from Ni metal by CO as $Ni(CO)_4$ in the Mond process.

Capping permits a type of cluster isomerism not seen in main-group systems. Thus, for example, a six-atom capped square pyramidal cluster will have seven sep or 86 cve which are exactly the same counts observed for an octahedral cluster. If you have to deal with a metal-cluster system with this electron count, both structure possibilities must be considered. An example of a capped square pyramid, $H_2Os_6(CO)_{18}$, is shown in Figure 3.11. $H_2Os_6(CO)_{18}$ is prepared by protonation of octahedral $[Os_6(CO)_{18}]^{2-}$. Clearly the energies of these two shapes must be similar as it is the specific requirements of the bridging H atoms and/or lack of counterions that tips the balance in favor of the capped form. This example emphasizes another difference between metal and main-group systems. The weaker M–M vs. E–E bonding results in smaller energy differences between structural isomers and lower barriers for inter-conversions between them.

Different ways of fragmenting a capped cluster cause little problem. For example, $Os_6(CO)_{18}$ (cve = 84, sep = 6) possesses the cluster geometry shown in Figure 3.4.

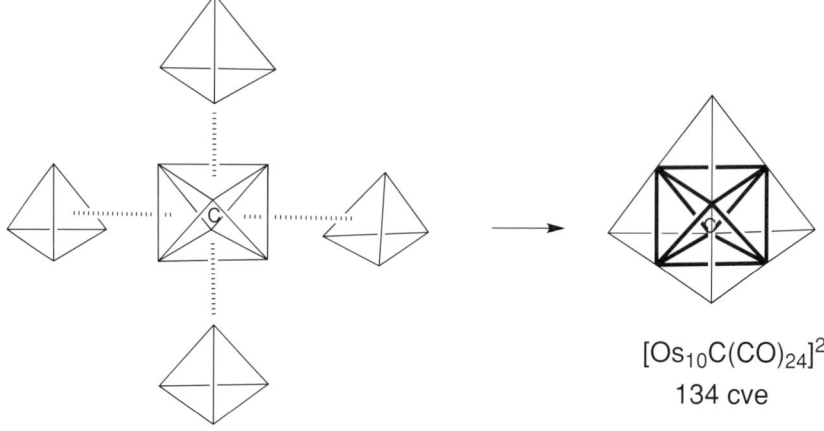

[Os₁₀C(CO)₂₄]²⁻
134 cve

Figure 3.10

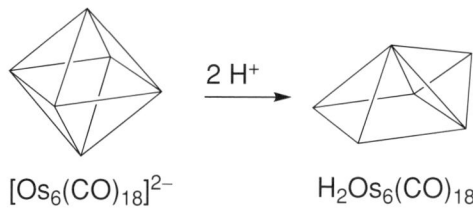

Figure 3.11

It may equally well be considered as a monocapped trigonal bypyramidal cluster (cve = 14(5) + 2 + 12 = 84, sep = 5 + 1 = 6) or a bicapped tetrahedral cluster (cve = 15(4) + 24 = 84, sep = 6). The latter description seems to be more popular.

Exercise 3.6. A cluster is found to have the molecular formula $[Os_8(CO)_{22}]^{2-}$. Generate a set of likely cluster geometries.

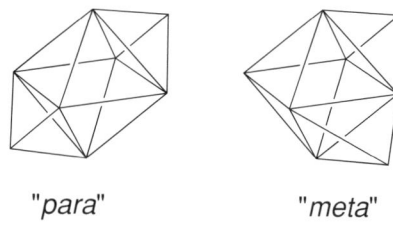

"para"   "meta"

Exercise 3.6

Answer. Let's stick with cve counts for this problem. The new cluster has a cve = 110. A single closed cluster of order eight would require a count of 14(8) + 2 =

114 and an open cluster based on a deltahedron of $n = 9$ would require $14(9) + 2 - 12 = 116$. A three-connect cluster would require $15(8) = 120$. So we try capping. A capped closed cluster of order seven requires a count of $14(7) + 2 + 12 = 112$ and a bicapped closed cluster of order six requires a count of $14(6) + 2 + 24 = 110$. Bingo! Now consider the isomeric possibilities. In the scheme above two known structural possibilities, sometimes labeled "*meta*" and "*para*", are shown. The one observed for this particular compound is the *para*-isomer. Much of this beautiful Os chemistry came from the Cambridge laboratory of Lewis and Johnson.

3.3.3 Cluster fusion

As already shown in Figure 3.9, $H_2Os_6(CO)_{18}$ can be generated by fusing a tetrahedron to a square pyramid. The counting analysis of Mingos shows that the cve count of the fused cluster is equal to the sum of the two fused clusters minus the count associated with the common fragment, which in this case is a metal triangle ($3 \times 16 = 48$ cve). So we have cve = cve (square pyramid) + cve (tetrahedron) $- 48 = 74 + 60 - 48 = 86$. How many ways can one skin a cat, you are probably wondering? But the beauty of the idea is illustrated in Figure 3.12. The common fragment eliminated is not limited to a triangle and can be a single vertex (-18), an edge (-34), a triangular face (-48 or 50), or even a butterfly or square face (-62 or 64). Examples of all are known in metal-cluster chemistry. The fact that two counts are possible for elimination of triangular or square faces reflects the fact that more than a single count is possible for these two-connect rings. The examples shown and broken down in Figure 3.12 illustrate a selection of known condensed clusters. Many more exist. Cluster fusion provides a mechanism for the construction of large, complex clusters. For example, we will have occasion to use the 90 cve raft cluster formed from four metal triangles later in discussing a 59-metal cluster.

Exercise 3.7. Show that the cve count associated with a bicapped tetrahedron sharing an edge with another tetrahedron (eight atoms in the final cluster) is 110.

60 cve   3 x 60 cve $[HOs_8(CO)_{22}]^-$ 34 cve   2 x 48 cve
                        110 cve

Exercise 3.7

Answer. As shown in the scheme below the structure can be generated from four tetrahedra by eliminating two triangles and one edge ($4 \times 60 - 2 \times 48 - 34 = 110$). In fact, this is the geometry observed on protonating a bicapped octahedral dianionic cluster (Exercise 3.6) to produce $[HOs_8(CO)_{22}]^-$. In metal-cluster chemistry, protonation is not a small structural perturbation.

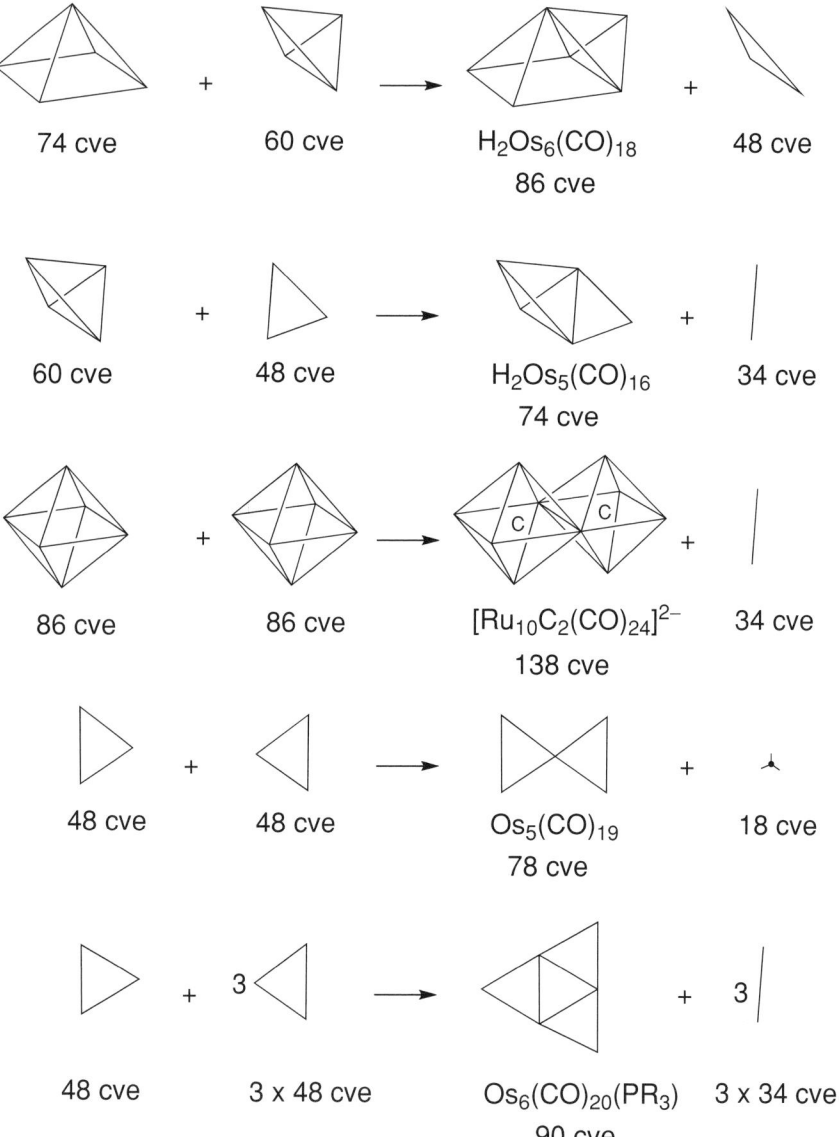

Figure 3.12

3.3 Variations characteristic of metal clusters

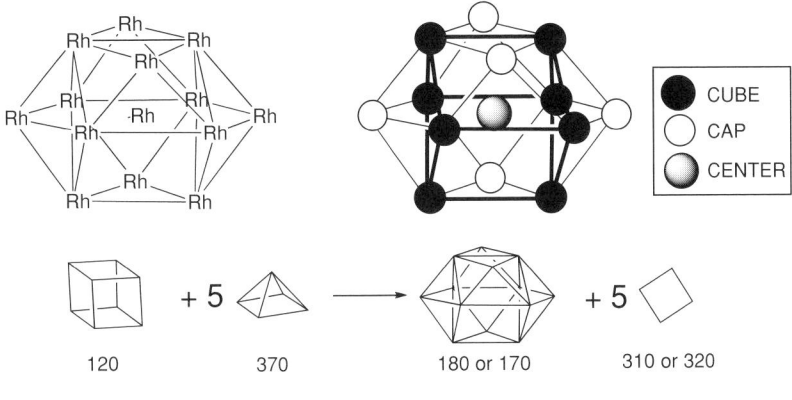

Exercise 3.8

Exercise 3.8. Justify the composition and structure (shown above) of the compound $[Rh_{14}(CO)_{26}]^{2-}$.

Answer. The first step is to analyze the geometry and break the cluster down into smaller clusters from which it may be reassembled using the cluster fusion principle of Mingos. As shown in the second drawing above, the cluster can be viewed as a pentacapped Rh-centered cube: hence, it may be assembled from a cube (120 cve, Table 3.1) and five square pyramids (74 cve). As shown above, the common fragment is a square and consultation of Table 3.1 indicates 64 cve. However, recall that a square can also be obtained from an octahedron by removing two non-adjacent vertices: hence, it has 62 cve, which is the same as for a butterfly (removal of two adjacent vertices from an octahedron). So there are two possibilities; 64 and 62 cve adding another variable to the mix. As shown below this leads to a prediction of cve = 170 or 180. The actual count is 180 (14 Rh + 26 CO + 2− = 126 + 52 + 2 = 180).

3.3.4 Ligand steric effects

A majority of main-group clusters exhibit single, one-electron external ligands, e.g., H, although clusters containing two-electron donors as well as naked clusters were discussed in Sections 2.5 and 2.11. The metal atoms in metal clusters often are coordinated by three two-electron ligands like CO or a polyhapto ligand like $[C_5H_5]^-$. But the possibilities for ligand variation and, therefore, metal and cluster properties are enormous. The electronic characters of transition metals change significantly in moving above and below groups 8/9 and the ancillary metal ligands play an important role in metal-fragment properties. In addition, the space occupied by the set of metal ligands is large and steric considerations play a role in determining accessible stoichiometries.

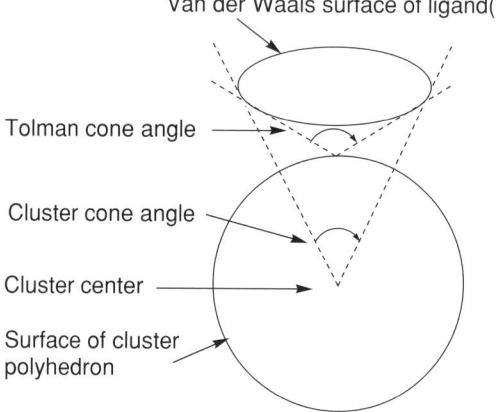

Figure 3.13

Although steric effects were appreciated earlier, it was Mingos who gave a systematic analysis of the situation and pointed out instances where steric effects appear to control observed geometry. Basically the cone angle argument, developed by Tolman to describe the steric bulk of phosphines, was adapted to describe the steric bulk of a cluster fragment. The essential ideas are sketched out in Figure 3.13. In this case the vertex of the cone containing the metal atom and its external ligands is centered at the cluster core. Hence, the cluster cone angle is more acute than an equivalent Tolman cone angle which would be centered at the metal atom itself. There is also a dependence on the nuclearity of the cluster.

The cluster cone angles for a $M(CO)_3$ fragment in four-atom tetrahedral, six-atom octahedral and twelve-atom icosahedral cluster geometries serve to illustrate the idea. For a M–M distance typical of a first-row transition metal, the respective cluster cone angles are 114°, 108° and 96°, whereas for a second-row metal with a longer M–M distance the same angles are 108°, 102° and 90°. As expected the space available for ligands is larger for second- and third-row metals. The cluster cone angles can be compared with ideal cone angles of 109°, 90° and 64° for the three geometries to estimate whether a steric problem exists for a given geometry. For example, one immediately sees that icosahedral $[Fe(CO)_3]_{12}$ is not going to be feasible.

An idea of the maximum number of ligands that a metal-cluster geometry can tolerate on steric grounds can be estimated from the ideal cone angle of the cluster shape times the number of metal atoms divided by the M(CO) cone angle. For the four-atom tetrahedron made of first-row metal atoms, this angle is 36° yielding 12.2 CO ligands. This is the limiting stoichiometry found for, e.g., $Co_4(CO)_{12}$ (Figure 3.1). Corresponding numbers for the octahedron and icosahedron are 15.8 and 25.3 for first-row metals. For second-row metals the numbers are 13.2, 17.2 and 27.8 in the three geometries.

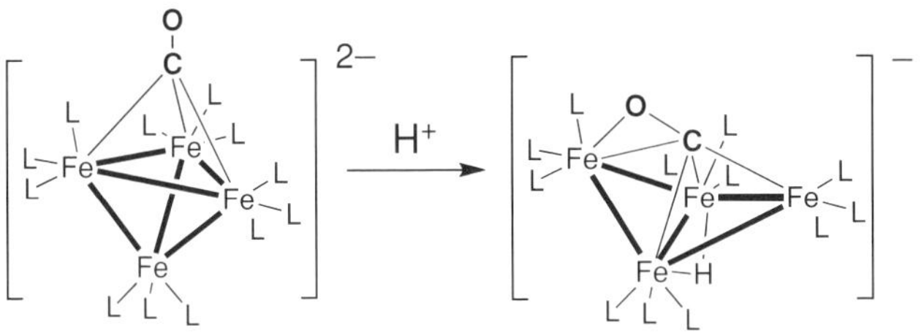

3.3 Variations characteristic of metal clusters 107

Figure 3.14

Some of the structural complexities presented earlier have their origin in ligand steric effects. For example, $Co_4(CO)_{12}$ possesses three bridging CO ligands whereas $Ir_4(CO)_{12}$ (Figure 3.1), with more room, exhibits 12 terminal CO ligands. $[Fe_4(CO)_{13}]^{2-}$ with a tetrahedral core and one triply bridging CO has no extra space for ligands. Hence, the addition of a relatively small ligand, H, formed by monoprotonation generates $[Fe_4H(CO)_{13}]^-$ with an open butterfly structure and a CO ligand curiously coordinated between the wing-tip metal atoms (Figure 3.14). The same type of steric factor may well lie behind the rearrangement of $[Os_6(CO)_{18}]^{2-}$ to $H_2Os_6(CO)_{18}$ by protonation (Exercise 3.7). Of course, it is always difficult to distinguish steric from electronic factors in driving structural change; however, the significant point of this section is that steric effects in metal clusters are real and need be considered for certain problems.

3.3.5 Metal effects: early transition metals with donor ligands

Metal and ligand variation can produce changes in electron count for a given structure. This fact is illustrated by a class of group 5/6 metal clusters bearing halogen or alkoxide ligands. They are of considerable significance in themselves but also can be seen in solid-state systems containing similar cluster units (Chapter 7).

In Figure 3.15 the structures of octahedral $[Mo_6Cl_8L_6]^{4+}$ and $[Ta_6Cl_{12}L_6]^{2+}$, where L represents a neutral two-electron terminal ligand, are illustrated. The former has eight face-bridging Cl ligands, whereas the latter has 12 edge-bridging Cl ligands. The two clusters illustrate two classes of compounds with electron-rich donor ligands. In order to count the cve for each, recall that these halide ligands act as four-(edge-bridged) and six-electron donors (face-bridged) when considered as anionic ligands. Thus we have a cve of 84 and 76 for the Mo and Ta clusters, respectively, neither of which is the value 86 found in Table 3.1.

But recall in Section 3.2 that if each metal fragment uses four valence functions and four electrons for M–M bonding, the octahedron can be bonded with 12 two-center–two-electron bonds. For the example considered, $[Ru_6(CO)_{18}]$ with 84 cve is the predicted composition. But it is the dianion $[Ru_6(CO)_{18}]^{2-}$ with 86 cve

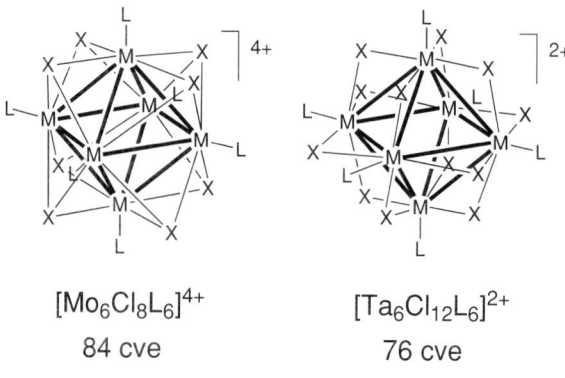

Figure 3.15

analogous to $[B_6H_6]^{2-}$ that is observed. From the experimental stoichiometry we are forced to conclude that the $Ru(CO)_3$ fragment utilizes only three orbitals and two electrons in cluster bonding and mimics a main-group fragment, i.e., a fourth metal orbital is not available for bonding. Below we will show that it is the combination of a later transition metal with acceptor ligands that leads to the observed behavior. But for now, 84 cve $[Mo_6Cl_8L_6]^{4+}$ is an example of a cluster that would fit a localized cluster bonding model. Does it? Let's see.

Consider each Mo center of $[Mo_6Cl_8L_6]^{4+}$ to be a square pyramidal $MLCl_4$ fragment. Of the nine metal orbitals, five are used for the Cl^- and L ligands leaving four available for framework bonding. The Mo_6^{12+} core contains 24 electrons. Hence, there are sufficient metal orbitals and electrons to form 12 two-center–two-electron Mo–Mo edge bonds. Can we approach $[Ta_6Cl_{12}]^{2+}$ the same way? Again we can consider the cluster as made up of $MLCl_4$ fragments; however, now with edge-bridging Cl^- ligands. Now we have a Ta^{14+} core with 16 electrons. With four metal orbitals per fragment, we have more orbitals than electrons. The solution? Eight three-center–two-electron face bonds perfectly utilize all the orbitals and electrons. The metal centers in these two clusters appear to utilize four valence orbitals in cluster bonding. Quite a difference from the metal-carbonyl clusters discussed earlier. These two metal clusters are analogous to tetrahedral clusters with edge vs. face bonding discussed in Section 3.1

So how can we justify using more metal d orbitals for earlier transition-metal clusters? An important aspect of the problem concerns the nature of the ligands. In a group 8/9 metal carbonyl the empty π^* orbital of the CO ligand serves an important role in delocalization of metal d electrons away from the metal center. In the process, selected metal d orbitals are stabilized by the π-acceptor ligands and are less likely to participate in cluster bonding. In the case of a π-donor ligand like Cl^-, interaction of the ligand with the metal d functions leads to a destabilization. It is

3.3 Variations characteristic of metal clusters

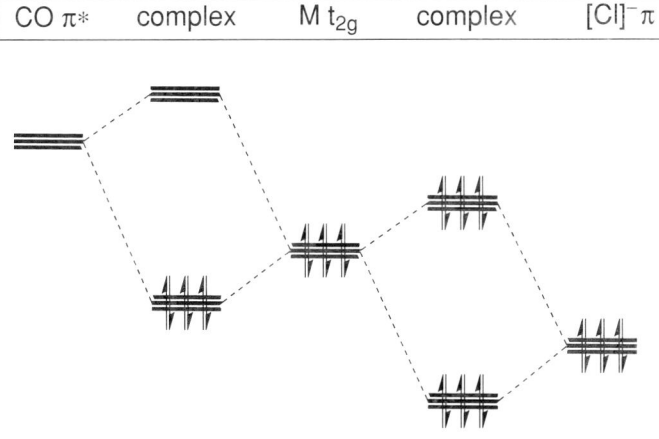

Figure 3.16

the same electronic mechanism used to explain why CO is high in the spectrochemical series whereas halogens are low, relative to a ligand like NH_3 (Appendix). Thus, both the π-donor ligand set as well as the lower nuclear charge of the earlier metals push the d set up into the valence energy region. The MO block diagram in Figure 3.16 illustrates the point. The net result is more filled metal orbitals in the valence region. A situation is created not unlike that for naked main-group clusters where some of the high-lying external lone-pair orbitals are able to mix more effectively with skeletal orbitals leading to unusual (relative to the borane paradigm) compositions and shapes.

Of course an MO approach is also possible just as it is for the tetrahedral main-group cluster. Consider the example of metal-alkoxide clusters, $[M_6(\mu_3\text{-L})_8(L)_6]^{n-}$. As illustrated in Figure 3.17 for M = Co, L = CO, $n = 4$, whereas for M = Mo, L = OR, $n = 2$, i.e., cve counts of 86 for M = Co and 84 for M = Mo if the face-bridging OR ligands are considered five-electron donors. Both clusters possess 16 π-acceptor or donor orbitals that stabilize or destabilize an equal number of metal-based orbitals, respectively. However, in O_h symmetry there is one metal orbital that is neither stabilized nor destabilized as there is no ligand combination of the proper symmetry to interact with it. Its energy relative to the other filled orbitals differs for π-acceptors or donors leading to its filling in the case of the late metal-carbonyl cluster and its remaining empty in the case of the earlier metal-alkoxide cluster.

The versatility of transition-metal–ligand combinations in stabilizing localized, partially localized and delocalized cluster bonding is evident in the three examples discussed to this point. But we still have no analog of a late metal-carbonyl cluster that follows the main-group counting paradigm for an octahedral shape. Centered zirconium chloride clusters provide an example of an octahedral cluster with seven

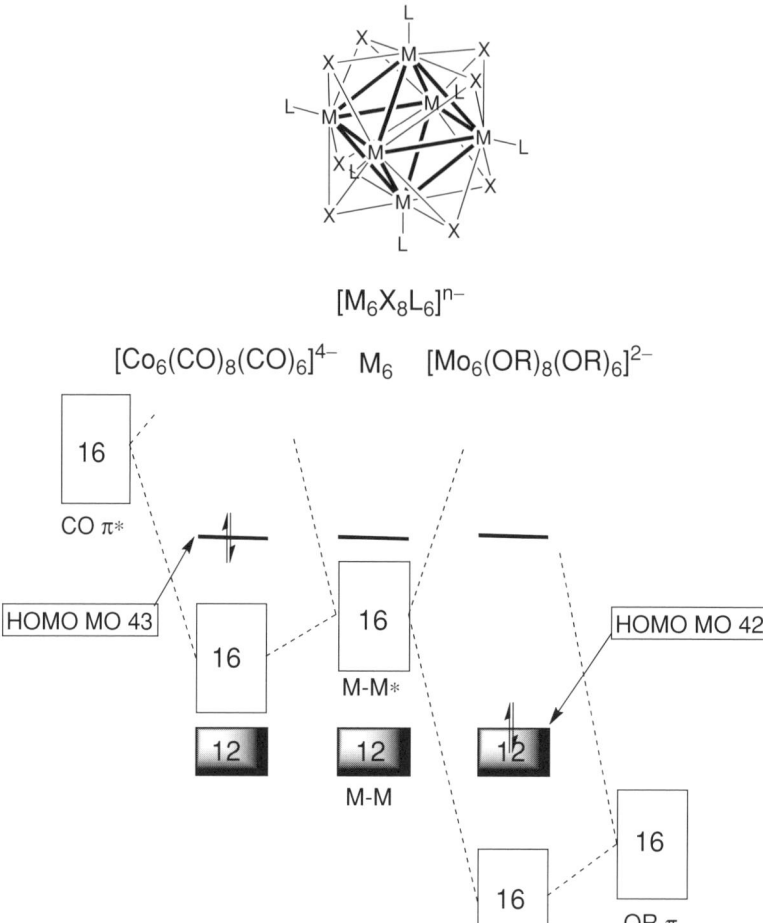

Figure 3.17

sep albeit not 86 cve. Take the cluster $[Zr_6Cl_{18}B]^{5-}$ which has the structure shown in Figure 3.18. It is of the structure type of the Ta cluster with which we began Section 3.3.5 but contains 74 rather than 76 cve. However, if we consider the interstitial atom as B^{3+}, then we have an octahedral $[Zr_6]^{10+}$ core with 14 electrons. But why not 86 cve? Recall that the cve count of $14n + 2$ includes $12n$ electrons associated with external bonding. If this assumption is not correct then the rule breaks down. Note that we can obtain the correct count here from $2n + 2$ for the core bonding $+ 10n$ for external bonding; however, the value of doing so is limited.

In this cluster type, the external terminal Cl^- ligands can be used to bridge adjacent clusters, i.e., they are shared between clusters. Hence, the M:Cl ratio can be increased from the 1:3 ratio found in $Rb_5Zr_6Cl_{18}B$. If the electronic requirements

3.3 Variations characteristic of metal clusters

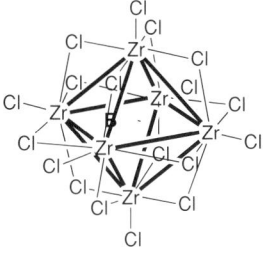

$[Zr_6Cl_{18}B]^{5-}$

Figure 3.18

of the octahedral building block are invariant to whether or not the external cluster ligands bridge, then the stoichiometry will be determined by the cluster core requirement of 14 electrons. This is most easily expressed in terms of the $[Zr_6Cl_{12}X]^{m+}$ core as $6 \times 4 - 12 + v_x - m = 12 + v_x - m = 14$, where v_x is the valence electron count of the main-group interstitial atom. As an example, consider the series of known compounds at fixed Zr:Cl ratio of 6:15: $K_3Zr_6Cl_{15}Be$, $CsKZr_6Cl_{15}B$, $KZr_6Cl_{15}C$, $Zr_6Cl_{15}N$, all of which possess a $[Zr_6Cl_{12}X]^{m+}$ core where $m = 0, 1, 2$ and 3, respectively. So we have for, e.g., N, $12 + 5 - 3 = 14$. Note that the role of the interstitial atom appears to be the same as found in group 8/9 transition-metal clusters (Figures 3.7 and 3.8).

Interestingly, members of this series of compounds are found containing transition-metal interstitial atoms as well. We can consider the effects of the

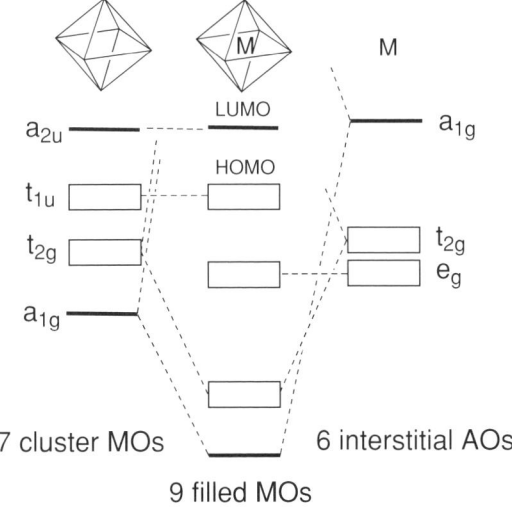

Figure 3.19

metal d functions of an interstitial atom in the manner of Figure 3.8, as shown in Figure 3.19, where for simplicity only the 3d and 4s valence functions are shown. The t_{2g} and a_{1g} metal functions find symmetry matches with four of the cluster bonding orbitals; however, there is no symmetry match for the metal e_g pair. As they are low enough in energy to be filled, the cluster electron count becomes $14 + 4 = 18$. Hence, we have $12 + v_x - m = 18$ for the relationship between identity of the interstitial atom and the charge on the $[Zr_6Cl_{12}X]^{m+}$ core. This permits us to explain three additional known compositions for 6:15 M:Cl ratio compounds: $Li_2Zr_6Cl_{15}Mn$, $LiZr_6Cl_{15}Fe$, $Zr_6Cl_{15}Co$. In the manner of the main-group example we have $m = 1, 2, 3$, respectively, and for, e.g., Co, $12 + 9 - 3 = 18$. The amazing result is that in some fashion B, C and N are behaving analogously to Mn, Fe and Co. Without this model of cluster electronic structure, a compositional mapping of this type would not be obvious.

Exercise 3.9. The compound $Ba_2Zr_6Cl_{17}X$ possesses a structure based on an octahedral cluster $Zr_6Cl_{12}X$ core with four terminal Cl ligands and *trans*-bridging Cl ligands such that linear chains are formed in the solid state (see drawing below). Assuming its composition is dominated by the electronic requirements of the cluster unit, suggest two possible identities for the interstitial atom X.

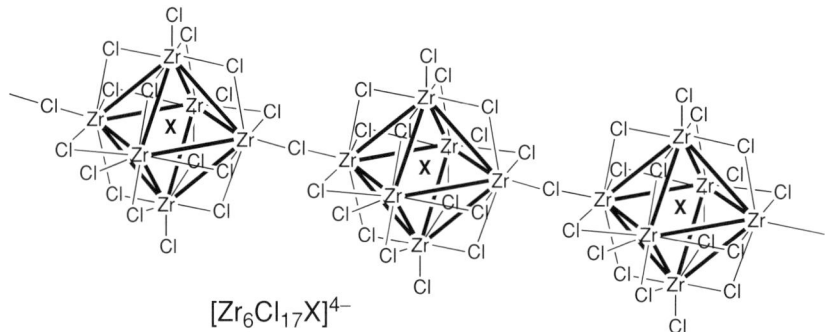

$[Zr_6Cl_{17}X]^{4-}$

Exercise 3.9

Answer. The $[Zr_6Cl_{12}X]^{m+}$ core charge is $-4 - (-5) = +1$. If X is a main-group atom then we have $12 + v_x - 1 = 14$. $v_x = 3 =$ B. If X is a transition metal the same sum $= 18$ and $v_x = 7 =$ Mn. The utilization of these cluster building blocks to generate nets and three-dimensional networks should be evident.

It should be clear that extension of these models for octahedral clusters to all early metal clusters with π-donor ligands is not possible. However, when one deals with a single metal and shape then the cluster electron count followed can be a useful tool. For example, it provides a ready rationalization of stoichiometry as well as

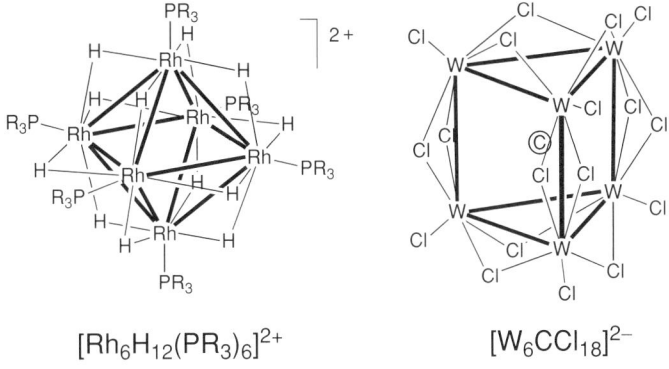

Figure 3.20

prediction of target stoichiometries. This information is particularly useful when dealing with compounds synthesized by high-temperature methods.

Recently, evidence has been produced for the existence of clusters with electronic structures lying between the earlier transition-metal systems with π-donating ligands and the later transition-metal systems with π-accepting ligands. The precedent setting compound, $[(^iPr_3P)_6Rh_6H_{12}]^{2+}$, shown at the left in Figure 3.20, exhibits an octahedral metal-cluster core with 12 edge-bridging H ligands and six terminal phosphines. It has an unambiguous cve count of 76, ten less than the count expected for an octahedral metal cluster of late transition metals. Its geometric structure, including the ligands, is the same as that of 76 cve $[Ta_6Cl_{12}L_6]^{2+}$, shown in Figure 3.15. The Rh cluster takes up two molecules of H_2 reversibly suggesting electronic unsaturation of the cluster bonding.

This Rh cluster provides an interesting counterpoint to our discussion of the Ta cluster at the beginning of this section. To review, the $[Ta_6]^{14+}$ cluster core provides 16 electrons, the 12 bridging Cl^- ligands provide 48 electrons and the six L ligands another 12 leading to 76 cve. In the Rh cluster, the $[Rh_6]^{14+}$ core provides 40 electrons, the 12 bridging H^- ligands 24 electrons and the six phosphines 12 electrons for a total of 76 cve. This would imply that 24 cve electrons provided by the bridging Cl ligands in the Ta cluster are supplied by the metal atoms in the Rh cluster, i.e., by moving from a group 5 metal to a group 9 metal, four additional electrons are contributed for each of the six metal atoms. So it's an interesting question then. What exactly is the role of these 24 ligand-derived or metal-derived electrons? Electron counting defines the problem but does not provide the answer – good calculations are required.

Another example of a new type of cluster that fits in this section is the C-centered, trigonal prismatic cluster $[W_6CCl_{18}]^{2-}$ (Figure 3.20) with 84 cve. High-quality calculation on the diamagnetic cluster generates multiple closely spaced levels in the frontier-orbital region. Consistent with these results the electrochemistry shows a total of five redox states: two on oxidation and two on reduction of the dianion (cve = 82, 83, 84, 85, 86). Chemical oxidation and reduction allow the monoanion and trianion to be isolated and structurally characterized. The structural distortions observed on one-electron oxidation and reduction correlate nicely with the bonding properties of the frontier orbitals which lose or gain an electron. Clearly this is a case where the electron-counting rule approach will not be useful as such rules are based on the existence of a large HOMO–LUMO gap for a characteristic electron population (Chapter 1). However, such compounds should not be shunned as their properties may well be of considerable value in selected applications. The downside of variability in electron count is balanced by an upside of rich redox behavior.

Exercise 3.10. The octahedral W cluster shown below has the formula $W_6S_8(PR_3)_6$ with the eight S atoms capping faces and the six phosphine ligands (represented by L) bound terminally to the metal atoms. Analyze the nature of the M–M bonding of the cluster core in the manner used for $[Mo_6Cl_8L_6]^{4+}$ and $[Ta_6Cl_{12}L_6]^{2+}$ and propose a bonding model using two-center and three-center bonds.

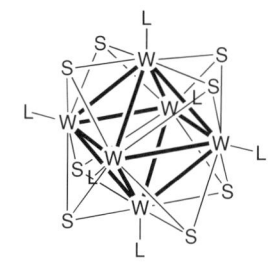

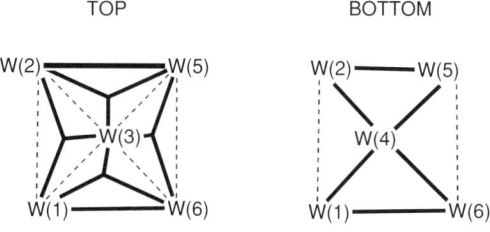

Exercise 3.10

Answer. With a cve count of 80, $W_6S_8(PR_3)_6$ fits neither the model proposed for 84 cve $[Mo_6Cl_8L_6]^{4+}$ nor that for 76 cve $[Ta_6Cl_{12}L_6]^{2+}$. Considering the capping S atoms as $[S]^{2-}$ ligands we have a $[W_6]^{16+}$ core with 20 valence electrons. Assuming that, like $[Mo_6Cl_8L_6]^{4+}$ and $[Ta_6Cl_{12}L_6]^{2+}$, four orbitals are used per metal atom for cluster bonding, we have 24 AOs to utilize. Hence, four three-center–two-electron

3.3 Variations characteristic of metal clusters

bonds (using 12 AOs and eight electrons) and six two-center–two-electron bonds (using 12 AOs and 12 electrons) satisfy these conditions. The final problem is to determine how to place them on the octahedral surface such that each metal center is associated with four bonding connections. Refer back to Exercise 2.5 where the analogous problem was dealt with for $[B_6H_6]^{2-}$ and you will see that the following placement of two- and three-center bonds meets the requirements. Clearly this is only one of several resonance structures.

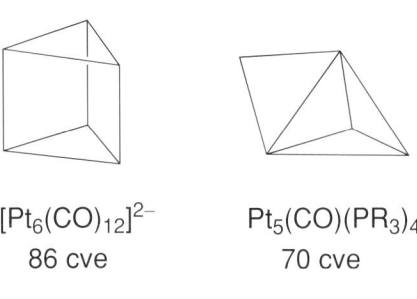

$[Pt_6(CO)_{12}]^{2-}$ $Pt_5(CO)(PR_3)_4$
86 cve 70 cve

Figure 3.21

3.3.6 Late transition metals

Group 10 metal clusters with carbonyl and phosphine ligands sometimes fail to follow the counting paradigm, but the reason for this failure differs from those discussed up to now. The situation is illustrated by the two Pt carbonyl clusters shown in Figure 3.21 where the cve counts are four less than those obtained for group 8/9 metal clusters for the same cluster shapes. What is the origin of these lower electron counts?

In coordination chemistry you learned that d^8 metal centers stabilize four-coordinate 16-electron square planar metal complexes – a coordination geometry favored by neither number of M–L bonds nor steric factors. In moving from Ti to Ni the separation between the 3d and 4p orbital energies increases. Hence, for Ni, and particularly for its heavier congeners, not all of the nine metal valence functions need be energetically accessible. Indeed, in a square planar complex, one of the p orbitals, the out of plane p_z orbital, is unused in primary bonding. Relative to a typical six-coordinate, octahedral complex, a square planar complex has an additional empty metal orbital in the metal–ligand antibonding orbital energy regime (Figure 1.10). Although unused in primary coordination, remember that the orbital has an important role in, e.g., facilitating higher coordination number intermediates in associative ligand substitution reactions.

How is this electronic feature of the metal atom incorporated into cluster bonding models? We need a little more cluster bonding theory to answer this question as well as to provide a basis for consideration of the gold clusters that follow. The MOs of a spherical cluster can be divided into radial and tangential orbitals loosely analogous to the σ and π systems in benzene. For our work horse model cluster, $[B_6H_6]^{2-}$, the a_{1g}, t_{1u} and e_g symmetry orbitals are radial (analogous to s, p_x, p_y, p_z, $d_{x^2-y^2}$ and d_{z^2}) whereas the t_{2g} orbitals are tangential (Figure 2.21). The HOMO set consists of tangential MOs that lie in the surface of the cluster sphere. In the case of a Pt cluster, not only are there $2n - 1$ inaccessible cluster antibonding orbitals as found for boranes and group 8/9 clusters, but also a small number (often two) tangential cluster bonding orbitals lying at high energy and empty. However, this conclusion is not sufficiently firm to justify a rule and approximate MO calculations are required to rationalize the electronic structure of new cluster systems of the late metals.

Exercise 3.11. Late transition-metal clusters can exhibit unusual properties. For example, trigonal bipyramidal 1,5-$\{Re(CO)_3\}_2Pt_3(P^tBu_3)_3$ is found to add three moles of dihydrogen at room temperature to yield 1,5-$\{Re(CO)_3\}_2Pt_3(P^tBu_3)_3$ (μ-H)$_6$ where the H atoms bridge the six Pt–Re edges (Adams and Captain, 2005). Use an argument based on cluster electron count to rationalize facile H addition.

Exercise 3.11

Answer. 1,5-$\{Re(CO)_3\}_2Pt_3(P^tBu_3)_3$ has a cve = 62 (2Re + 6 CO + 3 Pt + 3 PR$_3$ = 14 + 12 + 30 + 6 = 62 whereas a "normal" trigonal bipyramid exhibits 72. Even taking into account the two to four lower cve count of many late metal clusters, this mixed metal cluster is distinctly electronically unsaturated. Addition of six H atoms to the framework adds six electrons for a cve count of 68, four less that the "normal" count but acceptable for a Pt cluster. This hydrogenation reaction is formal reduction of an electronically unsaturated cluster by the addition of three moles of H_2.

3.3 *Variations characteristic of metal clusters* 117

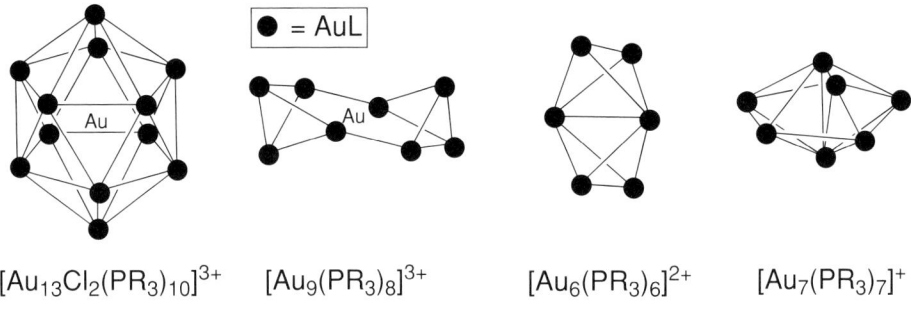

[Au$_{13}$Cl$_2$(PR$_3$)$_{10}$]$^{3+}$ [Au$_9$(PR$_3$)$_8$]$^{3+}$ [Au$_6$(PR$_3$)$_6$]$^{2+}$ [Au$_7$(PR$_3$)$_7$]$^+$

Figure 3.22

If some of the important tangential cluster orbitals, those that generate bonding within the surface of the cluster, are pushed to higher energy and emptied in moving to group 10 metals, can one empty even more cluster orbitals by moving to group 11? The answer is yes, as the farther one moves to the right in a transition metal series, the greater the energy gap between the $(n-1)$d, ns functions and np functions. In fact, Au clusters appear to be dominated by radial bonding as the d functions are now of low energy and contracted and the p functions of high energy and diffuse. Hence, M–M bonding takes place largely through the s functions. A consequence is that Au clusters have much in common with alkali-metal clusters. Let's take a brief look at some representative systems.

In Figure 3.22 a selection of Au clusters with or without an interstitial gold atom is shown each with its cve count. Mingos and Wales divide those with interstitial atoms into two groups in terms of gross structure – pseudospherical and toroidal (ring of atoms around the central atom), whereas the empty clusters are considered pseudospherical, prolate (elongated sphere) or oblate (squashed sphere) with only the latter two shown at the right in Figure 3.22. The building blocks are AuL fragments which are well known empirically to be isolobal replacements of, e.g., H atoms in M–H–M bridges. That is, structurally they appear to utilize a single valence orbital.

If we take the point of view that bonding in these clusters is essentially radial due to the limited valence set available to the metal atom, then the pseudospherical centered clusters can be considered as a central, bare gold atom using some or all its valence functions to bind n_s AuL fragments as "ligands." Provided all the valence functions of the central gold atom are used, this supra-coordination compound will exhibit an electron count of $18 + 12n_s$, i.e., 18 for the central atom plus the cluster non-bonding Au–L fragment electrons for n_s metal fragments. Thus, icosahedral [Au$_{13}$Cl$_2$(PR$_3$)$_{10}$]$^{3+}$ (Figure 3.22), has a calculated cve count of 162 ($18 + 12 \times 12$) vs. observed of 162 (13 Au + 2 Cl + 10 L − 3 = 13 × 11 + 2 + 20 −3). It is noteworthy that this analysis was used to predict the composition and structure

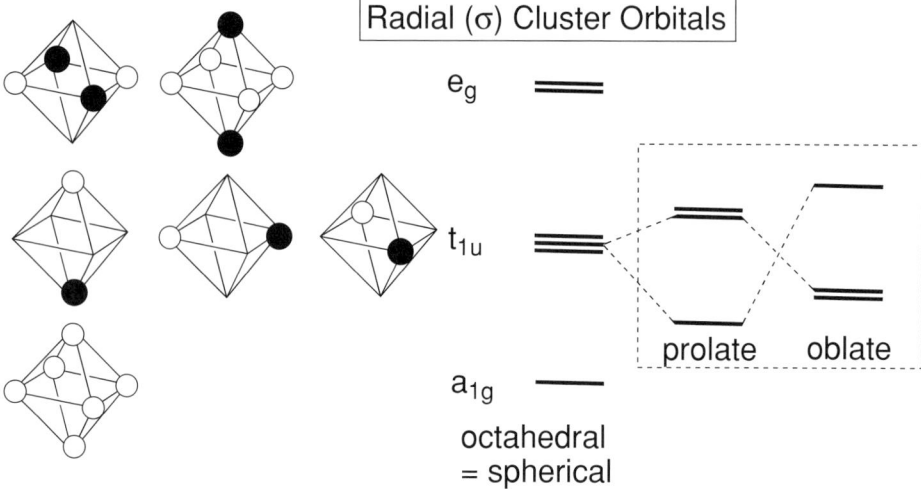

Figure 3.23

of this icosahedral gold cluster in advance of its synthesis. Note carefully that the icosahedral shape, and the lines associated with it, do not denote any kind of bonding connection with icosahedral boranes or heteroboranes in which radial as well as tangential MOs are filled (13 in all). In essence, then, this analysis suggests that the electronic structure of icosahedral $[Au_{13}Cl_2(PR_3)_{10}]^{3+}$ makes it a closer relative to a mononuclear metal complex than a cluster – it is a coordination compound of Au with coordination number 12! In terms of the theme of this text, observe that the central Au atom resides in an environment approaching that of an atom in bulk Au metal.

Moving to the toroidal Au clusters we find a ring-like array of AuL fragments surrounding a central Au atom with the axial positions vacant. These are to the pseudospherical clusters above as square planar complexes are to octahedral complexes. One p orbital of the centering Au atom, the one in the axial direction, remains unused and empty. Hence, if radial bonding alone suffices, the clusters count as $16 + 12n_s$, i.e., $[Au_9(PR_3)_8]^{3+}$, $16 + 12 \times 8 = 112$ vs. $9\,Au + 8\,L - 3 = 9 \times 11 + 16 - 3 = 112$. As already illustrated by the Pt clusters, this is an electronic option for late metal clusters and one that has compositional and structural consequences.

The counts of the non-centered prolate and oblate clusters, relative to spherical clusters, reflect shape-dependent electronic structure in a manner related to that of the toroidal clusters. The splitting of the degeneracy of the t_{1u} radial orbitals of the octahedral spherical cluster differs in the prolate and oblate geometries. As shown in Figure 3.23, stretching out the cluster reduces one cross-cluster antibonding axial interaction and increases two, whereas flattening the cluster decreases two and increases one. Hence, we have two over one splitting vs. one over two. If the

3.3 Variations characteristic of metal clusters

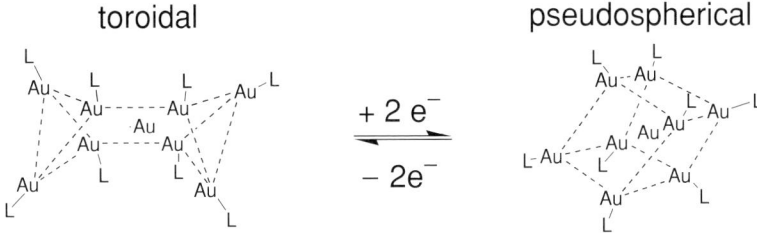

Figure 3.24

splitting is large enough, the prolate cluster will accommodate two cluster bonding electrons less than the oblate. To verify this, separate out the external ligand and metal cluster non-bonding electrons. Thus, $[Au_6(PR_3)_6]^{2+}$, which has 76 cve, has four cluster bonding electrons ($76 - 6 \times 12 = 4$) so that the stabilized orbital from the t_{1u} set and the lower lying a_{1g} (not shown in Figure 3.23) orbitals are filled, whereas oblate $[Au_7(PR_3)_7]^+$ with cve 90 has six cluster bonding electrons so that two stabilized orbitals from the t_{1u} set and the a_{1g} orbitals are filled. Note that the e_g (like $d_{x^2-y^2}$ and d_{z^2}) are not used and presumably lie at higher energy.

Although examples are limited, it is clear that the accommodation of composition and structure for a stable late group-11 cluster species marches to a different drummer than that of p-block clusters and late metal-carbonyl clusters. But there is a connecting theme in that geometry does change with electron count. A fine example is the demonstration that the addition of a pair of electrons to a toroidal cluster generates a pseudospherical cluster in a reversible process, e.g., the reduction of $[Au_9(PR_3)_8]^{3+}$ to $[Au_9(PR_3)_8]^+$ generates the cluster shape change shown in Figure 3.24.

Information on related alkali-metal clusters comes mainly from composition and abundance data from laser evaporation experiments. Geometric information comes from quantitative calculations, which give energies and structures corresponding to minima. A question sometimes raised is whether or not the complex potential energy surface has been sufficiently well explored to uncover the minimum of lowest energy (global minimum). Like the clusters formed from AuL fragments, alkali-metal atoms utilize a single valence orbital. Hence, clusters formed from these building blocks should exhibit similar cluster tectonics. The published calculations suggest as much.

To summarize this long section on metal effects, we can state: metal clusters can mimic main-group clusters (late-metal clusters with acceptor ligands); metal clusters with four-connect or higher vertices can be described with localized bond models (early-metal clusters with donor ligands); and metal clusters can have reduced (Pt clusters) or no tangential bonding (Au clusters). The characteristically more varied bonding modes available to the transition metals vs. main-group elements produce a rich structural chemistry.

Exercise 3.12. The Au cluster, $[Au_6C(PR_3)_6]^{2+}$ exhibits an octahedral metal-cluster structure and contains an interstitial C atom. Justify composition and cluster geometry.

Answer. The observed cve count is $6\text{ Au} + \text{C} + 6\text{ L} - 2 = 66 + 4 + 12 - 2 = 80$. If the AuL fragments are considered "ligands" on a C atom that obeys the eight-electron rule, we have a predicted cve count of $8 + 12\, n_s = 8 + 72 = 80$. Note that the $14n + 2$ rule gives 86 cve.

3.4 Naked clusters

In Section 2.11 naked and partially ligated main-group clusters were examined. Our development of fully ligated metal clusters up to this point with frequent comparison to the behavior of fully ligated main-group clusters suggests that naked main-group clusters will also provide the baseline behavior for the metal systems. This is important because the high reactivity of transition-metal clusters and the absence of naked transition metal Zintl ions make structural information difficult to obtain. These species are often "isolated" in, e.g., gas-phase beams or low-temperature matrices. Often the primary observation is mass and composition of ions and geometric structure is deduced from models or quantitative calculations. However, we have seen that transition-metal clusters have structural options not available to the main-group systems. Hence, idealized models based on non-transition-metal systems, e.g., rare-gas clusters, have little relevance. It has been stated that naked transition-metal clusters adopt shapes that maximize coordination numbers of surface atoms leading to spherical shapes. Rare-gas clusters, in contrast, maximize the number of interatomic distances optimal for pairwise atom–atom interactions. The bottom line is that naked metal-cluster structures need not correspond to those of fully ligated metal clusters, fragments of crystalline bulk metal, or rare-gas clusters.

So where can we get information on the problem of structure and properties? In the case of naked main-group clusters, modern quantum chemical treatments provide a means of generating structures at global energy minima and accessing energetics. In principle, the same is true of metal systems but, except for the very light metals, the size of the computing problem is much larger. There are indirect experimental methods that provide useful information. One measureable property, related to structure, is reactivity. Thus, cluster ions in a beam can be reacted with another gas, e.g., CO, and the addition products as well as fragmentation products can be identified by mass measurements. A single measurement is of little use but an advantage of the beam techniques is that one can often study a series of clusters

3.4 Naked clusters

of the same type going from small to large in single atom steps. In some cases, not all cluster sizes are observed suggesting that certain cluster atom numbers have special stability ("magic numbers"). Clusters without experimentally determined structures should not be dismissed out of hand. Recall that the first report of C_{60} was based on the dominant intensity of the $[C_{60}]^+$ ion in laser evaporated C vapor. Reactivity as a function of size contains significant information.

A systematic treatment of existing research on naked clusters, including naked-atom chemistry, is contained in a recent monograph by Klabunde. Sufficient for our purposes is a single study in which Ni_n^+, $n = 2$–13 were mass selected and reacted with CO gas *in situ*. The maximum number of CO ligands bound as a function of n was measured. In addition, the mass of fragmentation products gave maximum CO attachment numbers for Ni clusters that degraded under the reaction conditions. For the first set of data the total cve count of the $Ni_n(CO)_{max}$ clusters was compared to *closo*-cve counts ($14n + 2$). For $n = 4$, 6 and 7 observed and calculated counts agreed. For $n = 5$, a *nido*-cve count was observed whereas for $n = 13$ a count appropriate for a centered icosahedral cluster was found. For $n = 8$–12, however, the count was 2, 4, 6, 8 and 10 electrons below that appropriate for a single closed-cluster framework, i.e., 1, 2, 3, 4 and 5 CO ligands were missing.

The authors attributed the discrepancy to an unspecified structural difference (perhaps "metal crystallites" rather than molecular clusters). However, as we have seen, ligated metal clusters do not adopt the structure of *closo*-boranes for $n > 6$ largely because alternatives are available, e.g., face-capping. A subsequent interpretation of the data (Exercise 3.13 below) generates the cluster sequence shown in Figure 3.25. From nuclearity six and seven, the standard deltahedral shapes are observed followed by sequential face-capping until reaching nuclearity 12. Addition of the last Ni atom generates a centered icosahedral shape. Corroborative support for this interpretation comes from a similar analysis of the cluster-fragmentation results wherein the limiting $Ni_n(CO)_{max}$ stoichiometries can be accommodated by a series of open capped clusters systematically related to the closed series shown in Figure 3.25. Although the data and its interpretation is satisfying, it does not necessarily provide definition of the cluster shapes of the naked clusters. Nor does it prove that a given cluster size exhibits a single shape.

Exercise 3.13. The limiting stoichiometries observed for Ni_n^+ clusters after reaction with CO are: Ni_6CO_{13}, Ni_7CO_{15}, Ni_8CO_{16}, Ni_9CO_{17}, $Ni_{10}CO_{18}$, $Ni_{11}CO_{19}$, $Ni_{12}CO_{20}$ and $Ni_{13}CO_{20}$. Show that the cve counts are compatible with the geometric cluster shapes shown in Figure 3.25 (Mingos and Wales, 1990).

Answer. The cve counts for the cluster stoichiometries are: 86, 100, 112, 124, 136, 148, 160 and 170. For a *closo*-structure the cve count is $14n + 2$ where n is the cluster nuclearity. For a capped *closo*-structure the cve count is $14n + 2 + 12m$

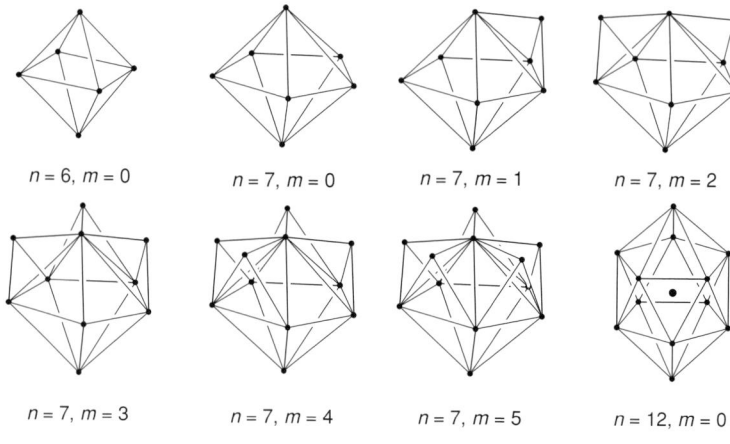

n = 6, m = 0   n = 7, m = 0   n = 7, m = 1   n = 7, m = 2

n = 7, m = 3   n = 7, m = 4   n = 7, m = 5   n = 12, m = 0

Figure 3.25

where n is the nuclearity of the *closo*-structure and m is the number of caps. The Ni_6 and Ni_7 compositions match six- and seven-vertex *closo*-cluster counts. The Ni_8–Ni_{12} do not, as the incremental increase in cve count is 12 rather than 14. This is perfectly consistent with capping five faces of a pentagonal pyramid (the Ni_7 cluster shape) in a step-wise fashion as shown in Figure 3.25. For the Ni_{13} cluster, the observed cve count is $14 \times 12 + 2$, exactly that expected for a 12-vertex icosahedron. Note from the discussion in Section 3.3.1 that a centered icosahedron will have the cve count of an icosahedron as the interstitial atom does not create any additional accessible MOs. It simply stabilizes a set of the cluster orbitals for this cluster shape.

3.5 High-nuclearity clusters with internal metal atoms

It was inevitable that we would arrive at transition-metal clusters in the size regime of the giant Al cluster $[Al_{69}R_{18}]^{3-}$ of Chapter 2. It is a problem we poked and prodded a bit from the perspective of deltahedral main-group clusters. We got a feel for the problem, but generated no connection between observed composition and structure. At this point in our development of cluster structure, you should be able to see why. These clusters lie somewhere in between small metal clusters (all surface and significant HOMO–LUMO gap) and metal crystallites (negligible surface and HOMO–LUMO gap less than kT). The HOMO–LUMO gap gets smaller as cluster nuclearity increases, which translates into more geometries available for a given composition. Indeed we have found that metal clusters exhibit more variations than main-group clusters. Eventually we must reach a size where a simple rule fails to encompass the possibilities. Does this mean we must give up attempts to understand high-nuclearity systems?

3.5 High-nuclearity clusters with internal metal atoms

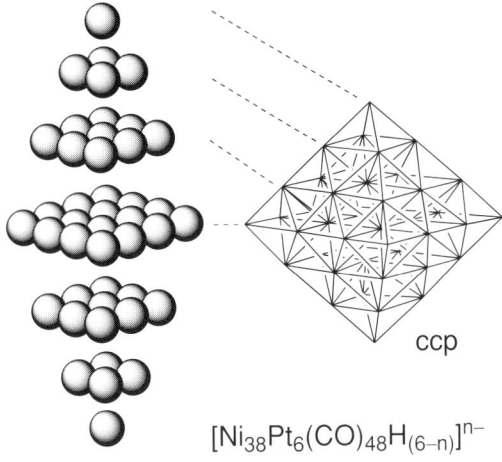

Figure 3.26

Not so. Using the tools already developed, large transition-metal cluster systems can be analyzed. Useful generalizations, if not predictability, are forthcoming. Beginning with the work of Chini and Dahl and their associates, many examples of large transition-metal clusters are available to us. Several are illustrated in Figures 3.26–3.29. Most can be considered as close-packed metal cores with ligands bound to the surface atoms. For example, consider $[Ni_{38}Pt_6(CO)_{48}H_{(6-n)}]^{n-}$ illustrated in Figure 3.26. The cluster is "exploded" to the left. Can you see that it has a ccp core structure?

The earlier discussions have given us more than one way to view a metal cluster. The concept of an interstitial atom and its role in cluster bonding is particularly important. For example, the possibility of larger interstitial fragments including polyhedra comes to mind. Further, we know that cluster electronic structure ranges from that analogous to deltahedral boranes with full radial and tangential (surface) bonding modes, to that with only radial bonding modes (Au clusters). In addition, the principles governing cluster fusion (vertex-, edge- and face-sharing) provide a systematic method for constructing large metal-cluster shapes including close-packed structures.

In this section we will attempt to dissect selected high-nuclearity transition-metal clusters to see what combination of these observed bonding principles explain composition and structure. We won't come up with any universal rule. As already stated, in going from small metal clusters to large metal clusters to bulk metal the HOMO–LUMO gap disappears. Hence, at some size counting rules will lose their usefulness. The analysis can still have value by emphasizing important differences between larger clusters vs. small clusters on the one hand, and metal crystallites on the other. In the process you should end up with a better understanding of the

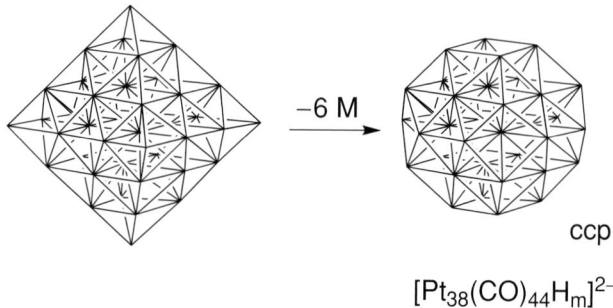

Figure 3.27

forms in which large metal clusters assemble. A more exhaustive development can be found in the book by Mingos and Wales.

3.5.1 Central cluster with radial metal "ligands"

There are two limiting cases. The first, considered here, is analogous to the situation found with centered Au clusters where radial interactions account for cluster bonding. The outer shell of metal fragments acts as a ligand envelope to the inner atom and the cve count is $18 + 12n_s$, where n_s is the number of metal atoms in the outer shell. In the same way that we moved from mononuclear complexes to dimers, trimers and clusters, so too we can envision a cluster made up of a centering polyhedron surrounded by a shell of metal fragment "ligands." In the manner of the Au cluster paradigm, it follows that the electronic requirements of the central polyhedron are determined by the $14n + 2$ rule and those of the outer shell by radial bonding only. Hence, the cve count is $\Delta + 12n_s$, where Δ is the cluster count associated with the central polyhedron.

Let's analyze a situation where this model obtains. In Figure 3.27 the cluster geometry of $[Pt_{38}(CO)_{44}H_m]^{2-}$ is shown as derived from the structure of $[Ni_{38}Pt_6(CO)_{48}H_{(6-n)}]^{n-}$ by clipping the six corner metal atoms off. In the case of this cluster the value of m is experimentally undetermined. Remember, the lines only express geometry not bonds. The cve count is 38 Pt + 44 CO + m H + 2 = 380 + 88 + m + 2 = 470 + m. The cve count associated with the central octahedron of metal atoms is 86 ($14n + 2$) and the radially bonded metal-carbonyl fragments will add $12n_s$ where n_s is 32. So we have a predicted cve = $86 + 12 \times 32 = 470$. It's a match for $m = 0$, i.e., no H atoms in the structure.

Exercise 3.14. Treat the bonding in the $[Ni_{38}Pt_6(CO)_{48}H_{(6-n)}]^{n-}$ with the structure shown in Figure 3.26 as a centered radial- and tangential-bonded cluster with radial bonding alone to the outer shell of metal fragments and compare observed and calculated cve counts.

3.5 High-nuclearity clusters with internal metal atoms

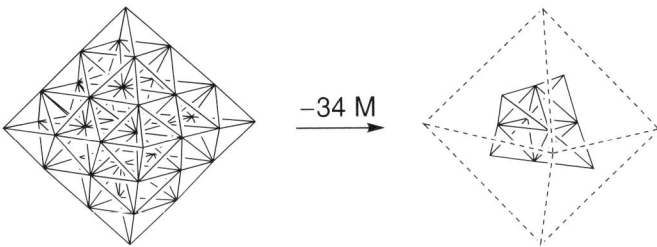

Exercise 3.14

Answer. The observed cve = $44 \times 10 + 48 \times 2 + 6 = 542$ for $n = 0$. The central polyhedron is an octahedron (86) and it is surrounded by 38 metal fragments (12×38) to give a calculated cve = $86 + 456 = 542$. Notice from the drawing above that if you strip off 34 metal atoms you can generate the structure of $[Os_{10}C(CO)_{24}]^{2-}$, a tetracapped octahedron (Figure 3.10). This suggests that the entire structure can be built up by capping an octahedron with metal fragments. As each cap on the central octahedron requires 12 additional electrons, the total cve count is again 542. The metal caps serve to replace electrons supplied by the external ligands they replace thereby reinforcing the analysis of the large cluster.

3.5.2 Fully bonded outer cluster with interstitial cluster

The second limiting case is modeled after the electronic situation that obtains for a group 8/9 metal-carbonyl cluster with an interstitial main-group atom. In this situation both radial and tangential bonding are fully utilized in the outer cluster and the electron count is determined solely by the number and geometry of the outer cluster atoms. For electron counting the interstitial atom contributes its valence electrons to cluster bonding but is not counted as, e.g., occupying a vertex. A high nuclearity cluster may be considered in which the electronic requirements of the outer cluster shell determine the electron count and the inner cluster functions solely to provide electrons to the cluster count.

Let's analyze an example. The structure of $Rh_{13}(CO)_{24}H_5$ with a cve count of 170 ($12 \times 9 + 24 \times 2 + 9 + 5$) is shown in Figure 3.28. It may be described as a truncated hexagonal bipyramid with a single interstitial Rh atom. With a centering metal atom, this shape corresponds to the fundamental coordination unit of a hcp metal lattice. A centered cube octahedron has the corresponding relationship to a ccp metal lattice. The cluster shell is a four-connect polyhedron and the cluster building blocks are group 9 metal-carbonyl fragments. Hence, a good guess is that the structure is governed by the $14n + 2$ rule ($14 \times 12 + 2 = 170$ cve) even though it has square faces. Indeed, consultation of Table 3.1 will show that Lauher calculated 170 cve for this shape, which he rationalized in his article. Fine, but suppose we

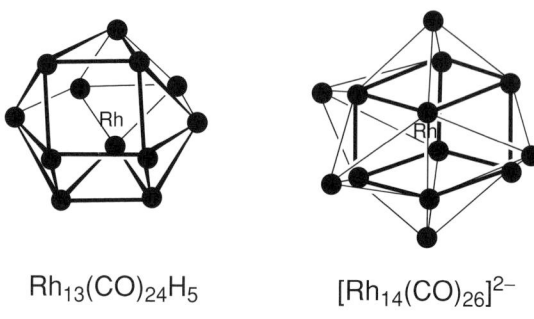

Figure 3.28

analyzed it under the first limiting electronic structure model? The central moiety is a single metal atom so the cve count predicted is $18 + 12 \times 12 = 162$ cve which is too low.

Have we forgotten sep counts? Well, no, for $Rh_{13}(CO)_{24}H_5$ the number of cluster bonding orbitals is $n + 1$ (cluster bonding electrons is $14n + 2 - 12n$ assuming six cluster non-bonding and six external ligand electrons per metal) which gives sep = 13. The geometry, however, is not icosahedral as the borane paradigm might suggest. Various factors, e.g., CO ligand packing, could be invoked to justify the stabilization of a non-deltahedral shape. Always keep in mind that there is a manifold of structures available for a given cluster nuclearity and, when energy differences between them are small, as is often the case for metal clusters with relatively weak M–M bonding, cluster bonding alone is not necessarily the determining factor.

3.5.3 Application of the cluster fusion principle

If these are the two limiting cases then you are probably gritting your teeth knowing that an example of a cluster lying in between is coming along. Let's deal with it. Consider the structure of $[Rh_{14}(CO)_{26}]^{2-}$ shown in Figure 3.28 with 180 cve and a centering Rh atom. Treat it first with the two limiting models. In the model driven by core valence requirements, we calculate $18 + 12 \times 13 = 174$. In the model driven by outer-shell valence requirements we calculate (assuming $14n + 2$ rule) $14 \times 13 + 2 = 184$. Equivalent ways of "making the cve count" are: (a) a model with three tangential cluster bonding orbitals filled; or (b) a model with two cluster bonding orbitals empty. But wake up! We satisfactorily treated this cluster in Exercise 3.8 as a cluster formed by fusing square pyramidal clusters on a centered cube. The lesson is that we must examine these complex clusters from several perspectives before deciding they are "strange."

These principles have been used to good effect by Dahl and coworkers in their work on very large metal clusters. In Figure 3.29, the complex geometric structure

3.5 High-nuclearity clusters with internal metal atoms

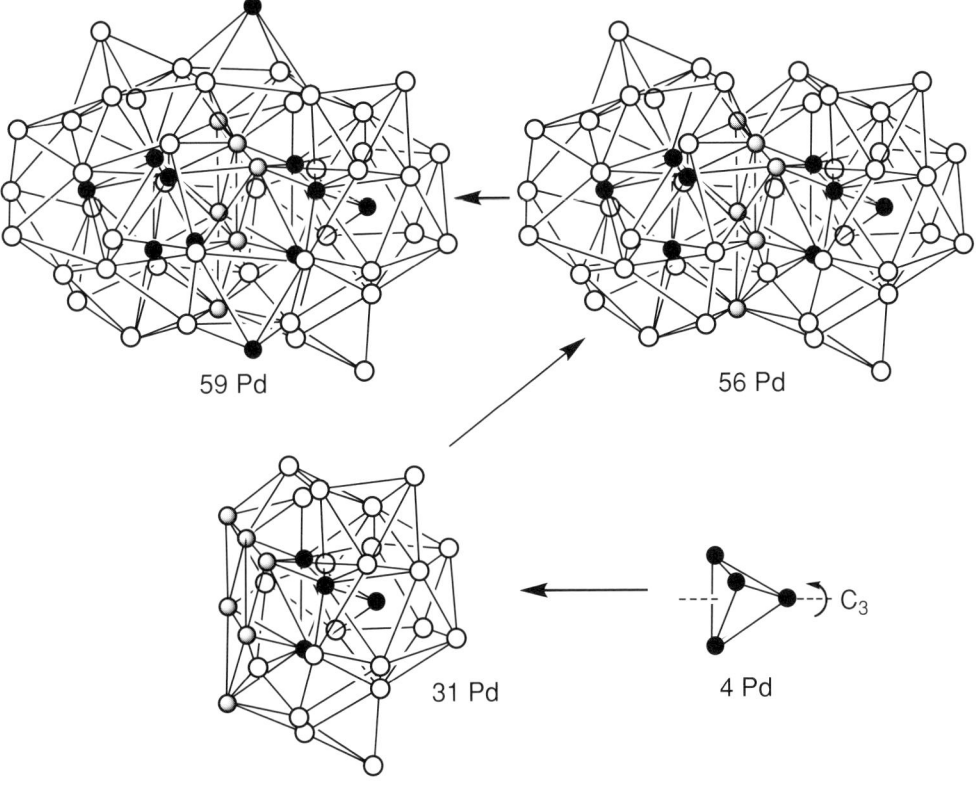

Figure 3.29

of the metal core of $Pd_{59}(CO)_{32}(PMe_3)_{21}$ is shown. At first sight it seems a structure one merely views in awe at the power of synthesis and crystallography. Even more so when its D_3 symmetry (C_3 axis passing through the two "apical" atoms of the two internal tetrahedra with dark spheres) and 11 internal atoms are fully appreciated. However, it can be, and was, successfully analyzed using two of the cluster principles we have discussed: (a) the large cluster model with a central polyhedral core with radial bonding alone to an outer shell of metals (cve count of $\Delta + 12n_s$); and (b) cluster fusion where the electron count of the common fragment is subtracted from that of the two clusters fused. In the "retrosynthesis" of the cluster in Figure 3.29 you can see that the cluster is considered a fused dimer of two 31-atom clusters containing tetrahedral cluster cores ($60 + 12 \times 27 = 384$ cve). The fusion is complex. It consists of a shared six-atom triangular raft (90 cve, see Figure 3.12) between the two 31-atom clusters yielding a 56-atom dimer. The remaining three atoms are the apical atoms of three additional square pyramidal clusters (74 cve) in which both basal pairs of atoms are shared (-2×34), one with each 31-atom cluster, thereby bridging the two fused clusters. This gives a cluster count of

$2 \times 384 - 90 + 3 \times 74 - 3 \times 68 = 696$ vs. the measured composition 59 Pd + 53 L = $59 \times 10 + 53 \times 2 = 696$. The presence or absence of H atoms in large clusters is always a difficult problem experimentally and the negative proton NMR experiments found support in the successful composition/electron count match. It is success stories such as this one that make the efforts devoted to model development worthwhile. But a cautionary reminder – an electron-counting match using a symmetry-based concept should not be construed as understanding the electronic charge distribution on the nuclear framework. As noted in Section 3.1.3 in the consideration of the view of Woolley, the important details of the electronic structure remain hidden and more rigorous approaches are required to reveal them.

3.5.4 Elementoid clusters

Does the discussion of large metal clusters help with our theme problem of Chapter 2 – the large Al cluster $[Al_{69}R_{18}]^{3-}$ with 228 cve? The metal clusters show how large numbers of internal atoms can be handled and just as we moved from main-group clusters to metal clusters so too we can do the reverse. Let's see how close we come to the electron count. Recall that the large Al cluster structure was described as $[Al@Al_{12}@Al_{38}@(AlR)_{18}]^{3-}$. One limiting model is to consider the elementoid cluster as a centered 12-vertex deltahedron (50 cve) surrounded by a 38-metal shell with radial bonding (38×2) surrounded in turn by an 18-metal shell with radial bonding (18×2) to give a total of 162. Obviously, we cannot neglect tangential bonding in the Al_{38} layer. To include full tangential bonding we view the cluster as a centered 12-vertex deltahedron within a 38-vertex polyhedron ($4n + 2$ in each case to give $50 + 154 = 204$ cve). Considering the low density of the 18 AlR fragments on the surface of the 38-vertex polyhedron, it is reasonable to assume they are attached exclusively by radial bonding (2×18) to give a total of 240 cve required for this model. A little high, but recall the discussion of Section 2.12.5 and $[Ga_{19}R_6]^-$. The above counts are based on assuming the eight-electron rule is followed for all Al atoms. If adjacent Al and AlR repulsion empties six of the lone-pair orbitals we also achieve the observed count.

But all this neglects the fact that the HOMO–LUMO gap may be, and as we will see in Chapter 6 is, small. One of the criteria for a valid electron-counting rule is no longer valid. In addition, we will see that the cubic clusters discussed in Section 5.2.5 exhibit highly variable electron counts suggestive of incipient metal-like properties. By all these measures, $[Al_{69}R_{18}]^{3-}$ should not be viewed in the same way as a small cluster. It has moved well along the bridge connecting clusters to metal crystallites. The electron-counting rules established in Chapters 2 and 3 for smaller species constitute a firm foundation for one end of our bridge but their utility decreases as cluster size increases.

Exercise 3.15. Using metal cluster ideas, try various electron counts on $[Al_{77}R_{20}]^{2-}$ with an $Al@Al_{12}@Al_{44}@(AlR)_{20}$ shell structure (consult Chapter 2).

Answer. The cve count for $[Al_{77}R_{20}]^{2-}$ is 253 so it is an odd electron cluster. Considering the inner two clusters with full radial and tangential bonding (the smaller one with an interstitial Al atom) yields a count of $50 + 178 = 228$. If the outer AlR fragments are attached by radial bonding alone the total count is $228 + 20 \times 2 = 268$ which is 15 more than observed and perhaps 14 more than a probable even electron species. Hence, one would have to assume seven tangential orbitals empty.

3.6 Nanoscale particles

The studies on high-nuclearity main-group and transition-metal clusters, plus the correlation of electron count with composition, have clear implications for nanochemistry. Thus, for example, we do not expect nanoparticles of Au to follow the same bonding paradigm as main-group or transition-metal nanoparticles. For the latter two, substantial tangential bonding in the internal layers is expected. External ligand effects, e.g., the surfactants used to isolate Au clusters, will be substantial as will mutual nanoparticle–substrate modification on binding in, e.g., catalysis of a reaction. The insight provided by our discussion of isolated and fully characterized high-nuclearity clusters should be of value in understanding nanoparticles.

Thus, to conclude this chapter, we examine one example of a nanoparticle system. The choice is an M_{55} ligand-stabilized colloid. This ligand-stabilized baremetal cluster is an unusual nanoparticle system but considerable evidence exists to establish it as monodisperse. Although this cluster type has never been structurally characterized by a classical single-crystal X-ray diffraction structure study, other less direct methods have established its cluster structure to the satisfaction of all but the purist demanding quantitative distances and angles. What makes it an appropriate model study for this text is the fact that a single-sized species can be discussed. Hence, the systematic spectroscopic and chemical studies apply to one cluster size. The results of, e.g., external ligands in tuning properties for a specific application can be directly compared with metal-cluster systems discussed above.

The $M_{55}L_{12}Cl_x$ nanoparticles, where M = Rh, Ru, Pt, Au, L = PR_3, AsR_3, $x = 6$ (Figure 3.30), are formed by the reduction of the appropriate metal salt with diborane in organic solvents. The structure shown in Figure 3.30 is a composite result derived from high-resolution transmission electron microscopy (HRTEM), scanning tunneling microscopy (STM) and extended X-ray absorption fine structure (EXAFS) on Pt and Au systems. The first method reveals metal atoms with face-centered cubic packing with size and shape consistent with the "two-shell" form

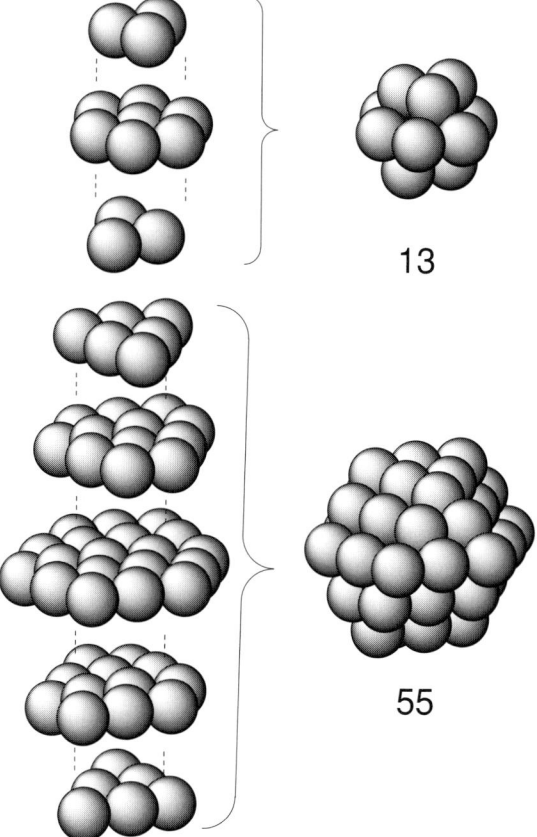

Figure 3.30

shown in the scheme. This consists of a cuboctahedral outer shell centered with a 13-atom close-packed core. The EXAFS on a Au system confirms the cuboctahedral structure and shows the Au–Au distance to be slightly less than that in bulk Au. Why M_{55}? Extra stability is attributed to the set of close-packed full shell clusters with $10n^2 + 2$ atoms per shell where n is the number of the shell, e.g., $1 + 12 + 42 = 55$. The STM experiments generate information on the external ligand envelope which can be modeled by calculations. The phosphine and arsine ligands coordinate to the 12 corners whereas the Cl ligands cover parts of the faces. In the $Au_{55}(PR_3)_{12}Cl_6$ cluster the six Cl ligands are thought to occupy the six square faces.

Exercise 3.16. Calculate the total number of atoms and the % surface atoms in full-shell clusters of the type represented by two-shell $Au_{55}(PR_3)_{12}Cl_6$ for 3, 4 and 5 shells.

Answer. From the $10n^2 + 2$ formula, the series beginning with the centering atom is $1 + 12 + 42 + 92 + 162 + 252 + \ldots$ yielding 147 (63 %), 309 (52 %) and 561 (45 %).

Other spectroscopy experiments are consistent with this picture and add detail. ^{197}Au Mössbauer spectroscopy of $Au_{55}(PR_3)_{12}Cl_6$ shows four types of atoms: 13 inner atoms, 24 uncoordinated surface atoms, 12 atoms coordinated to PR_3 and 6 coordinated to Cl. The phosphine ligands on this and the clusters of other metals are shown to be fluxional (mobile) in NMR experiments. These ligands are also mobile in a dissociative sense which has permitted effective ligand exchange reactions. $Au_{55}(PR_3)_{12}Cl_6$ is soluble in CH_2Cl_2. If the phosphines are replaced by a monosulfonated triphenyl phosphine $(Ph_2PC_6H_4SO_3Na)$ a water-soluble cluster is generated. Several other ligand-exchange reactions have been carried out to generate clusters with the same metal core but of different solubilities.

The electronic properties of these clusters and the magnitude of the perturbation caused by the external ligands have been probed. A low-temperature, ultrahigh-vacuum STM experiment on the $Au_{55}(PR_3)_{12}Cl_6$ cluster is representative. Discrete energy levels with an average spacing of 170 mV are observed and attributed to the Au_{55} core. The nature of the energy-level quantization and evidence for charge quantization (Coulomb blockade) are used to argue for a cluster model consisting of a metallic core extending slightly beyond the first close-packed shell of metal atoms. The exclusion of the outer shell is attributed to covalent interactions with the external ligands particularly electron depletion caused by the Cl ligands. Qualitatively this is in accord with earlier discussions in this chapter. One important point from this study is that this metal cluster exhibits different electronic transport properties than a completely non-metallic cluster in that there appears to be a non-zero density of states close to the Fermi level rather than a perfect gap (see Chapter 6 for a description of density of states and Fermi level as well as other characteristics of the electronic structure of metals). Other experiments with like objectives reveal additional aspects of the change in a measured property during the transition from small to large clusters.

The lability of the surface ligands leads to aggregation and, ultimately, metal precipitation, so the reaction properties of $Au_{55}(PR_3)_{12}Cl_6$ are limited to reactions involving the ligand envelope, e.g., ligand exchange mentioned above. This problem has been turned on its head by utilizing $Au_{55}(PR_3)_{12}Cl_6$ as a source of bare Au_{13} clusters that undergo controlled aggregation to $[(Au_{13})_{13}]_n$ metals that in turn convert into bulk metal between 400 and 500 °C. The scenario described, supported by HRTEM experiments, is one in which voltage applied to a metal electrode immersed in a solution of $Au_{55}(PR_3)_{12}Cl_6$ leads to removal of the outer layer of 42 Au atoms plus the external ligands. The bare Au_{13} cores form $(Au_{13})_{13}$ superclusters in a

132   *Transition-metal clusters*

stepwise fashion and lead, in turn, to $[(Au_{13})_{13}]_n$ particles. Results for Rh, Ru, Pt and Co analogs support a similar mechanism.

The hypothesis, or principle, of full-shell clusters based on close-packed metal arrays, and well illustrated by the $Au_{55}(PR_3)_{12}Cl_6$ work, has led to the study of larger cluster systems along the lines already sketched out for the M_{55} system. One example suffices for our purposes. The four-shell Pt_{309} cluster stabilized by 1,10-phenanthroline ligands (phen*), which in turn are substituted in the 4,7-positions by p-$C_6H_4SO_3Na$ moieties, has the formula $Pt_{309}phen*_{36}O_{30\pm10}$. The surface is partially covered with O_2 molecules. The clusters are irradiated by thermal neutrons thereby generating Pt_{309} clusters containing randomly distributed ^{197}Pt nuclei that can be observed in a Mössbauer experiment. The three different types of surface atoms are distinguished from the core atoms. Both the isomer shift and quadrupole splitting of the latter are the same as ^{197}Au in bulk metallic platinum. Consistent with the conclusion on the M_{55} clusters, it appears that the surface-ligand perturbation is mainly confined to the 162 surface atoms, and the 147 core atoms behave in a bulk-like manner at least in so far as measured by the 6s electron density.

This discussion brings us to the large area of nanoscale materials which goes beyond the scope of this text. However, the student using this book can now take up the chemical side of this area with a better understanding of the chemical principles that underpin the larger and more complex assemblies of nanochemistry. One point is clear. It is the inter-play between cluster surface atoms (with or without attached ligands) and internal cluster atoms that give rise to the unique properties of nanoparticles. We will return to this point from the perspective of extended structures in Chapter 6. Two books that provide an entry to nanochemistry are included on the reading list.

Problems

1. Justify the observed compositions and shapes for the clusters shown in the following figure. Discuss the problem presented by any that do not obey the counting rules.

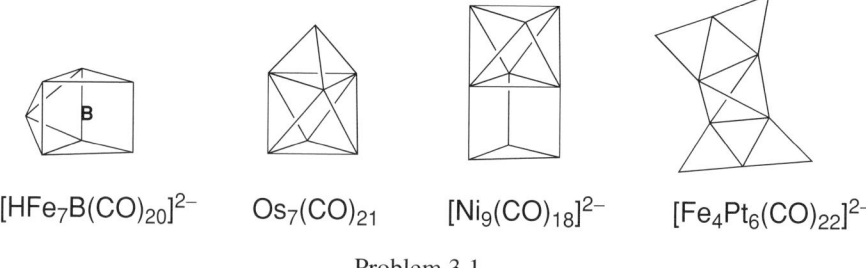

[HFe$_7$B(CO)$_{20}$]$^{2-}$ Os$_7$(CO)$_{21}$ [Ni$_9$(CO)$_{18}$]$^{2-}$ [Fe$_4$Pt$_6$(CO)$_{22}$]$^{2-}$

Problem 3.1

2. How many different cluster structures can you devise for the molecular formula $Os_6(CO)_{21}$? You should be able generate more than a single shape. Four shapes are known for six-atom metal clusters with the same electron count as this cluster.

3. Suggest a reasonable cluster core structure for $[Re_8C(CO)_{24}]^{2-}$ that exhibits C_{2v} symmetry with four types of metal positions. Suggest an isomer that has the same cluster building blocks but only two types of metal positions.

4. Both of the cluster shapes shown below for six-atom metal clusters have cve counts of 90. Construct a concise explanation suitable for a chemist with a basic knowledge of bonding (eight- and 18-electron rules, two-center–two-electron bonds) but unfamiliar with cluster bonding ideas.

Problem 3.4

5. Discuss the following in terms of cluster bonding theory. The equilibrium shown below has been established (Fumagalli et al., 1989). Both compounds have trigonal bipyramidal shapes with metal triangles of the same size. However, $[PtIr_4(CO)_{14}]^{2-}$ with the Pt atom in an equatorial position is significantly elongated in an axial direction relative to $[PtIr_4(CO)_{12}]^{2-}$ with the Pt atom in an apical position (the M–M distances in the equatorial triangle are 2.7 Å in both compounds, whereas the M–M apical–equatorial distances average 3.1 and 2.8 Å in the elongated vs. not elongated cluster).

$$[PtIr_4(CO)_{14}]^{2-} = [PtIr_4(CO)_{12}]^{2-} + 2CO$$

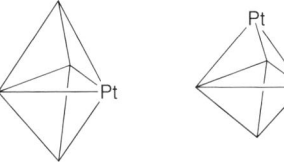

Problem 3.5

6. Count the electrons in $[Pt_{24}(CO)_{30}]^{2-}$ and justify its observed structure which is shown below. Hint: note that it can be derived from the 44-M cluster discussed in the text (Figure 3.26).

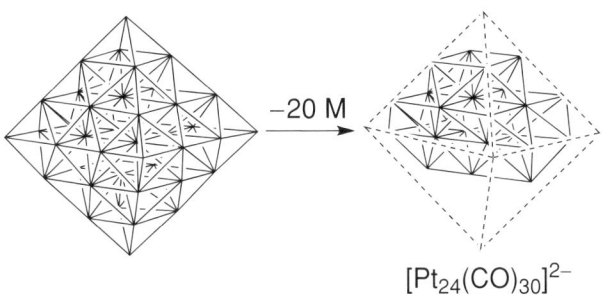

Problem 3.6

134   Transition-metal clusters

7. Consider the putative synthesis of a capped octahedral metal cluster shown in the diagram below. Postulate a reasonable structure for each of the intermediate species **1** and **2** in the reaction scheme given the structures of the reactants and product. Briefly justify your structures using established ideas of cluster bonding.

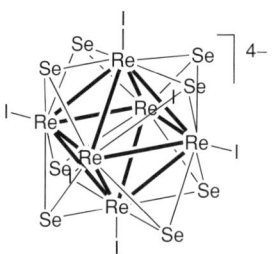

Problem 3.7

8. The cluster $[Re_6Se_8I_6]^{4-}$ has the face-capped octahedral structure shown below. Count the cluster electrons and comment on the geometry–count relationship.

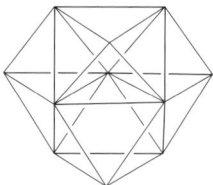

Problem 3.8

9. Treat the metal cluster $[Ru_{11}H(CO)_{27}]^{2-}$, with known metal core geometry shown below: (a) as a cluster that obeys the counting rules for main-group clusters; (b) as a fused cluster structure; and (c) as a radial- and surface-bonded core cluster with external cluster fragments bound by radial bonds only. Which provides a satisfactory "explanation" at the level of electron counting?

Problem 3.9

10. As shown below, the photolysis of the cluster $HPtOs_3(CO)_{10}(dppm)\{Si(OMe)_3\}$ (**1**) (dppm = bis-diphenylphosphinomethane) leads to a positional isomer (**2**) which can be

converted back to the original isomer on mild heating. Heating of **1** does not produce **2**. In the drawing, the lines correspond to CO ligands. The dihedral angles of the butterfly structures **1** and **2** are 176° and 172°, respectively, i.e., the metal atoms nearly lie in a single plane. The authors of this work (Adams *et al.*, 1993) propose that the isomerization proceeds via a tetrahedral cluster intermediate or excited state also shown below. Discuss the observations and the proposed mechanism in light of cluster bonding principles for transition-metal systems. In other words, how would you write up this work as a communication in terms of its relevance to cluster bonding?

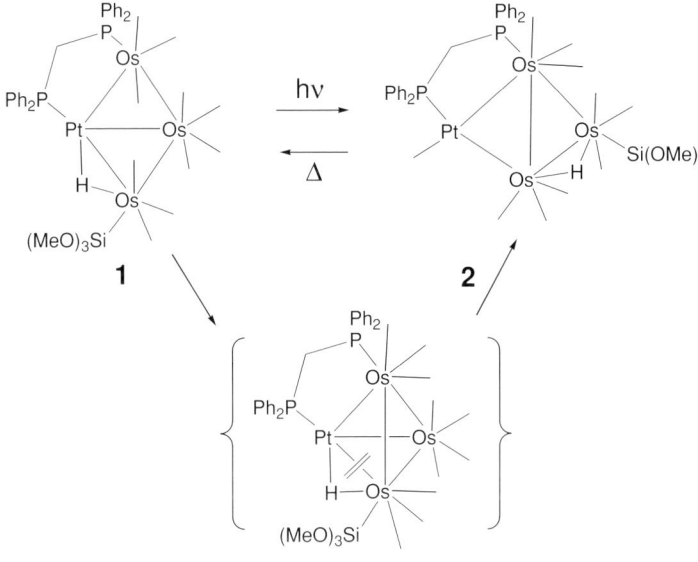

Problem 3.10

Additional reading

Section 3.1

Shriver, D. F., Kaesz, H. D. and Adams, R. D. (Eds.) (1990). *The Chemistry of Metal Cluster Complexes*. New York: VCH.

Mingos, D. M. P. and Wales, D. J. (1990). *Introduction to Cluster Chemistry*. New York: Prentice Hall.

Braunstein, P., Oro, L. A. and Raithby, P. R. (Eds.) (1999). *Metal Clusters in Chemistry*, Weinheim: Wiley-VCH.

Section 3.1.3

Woolley, R. G. (1985). *Inorg. Chem.*, **24**, 3519, 3525.

Section 3.1.5

Lauher, J. W. (1978). *J. Am. Chem. Soc.*, **100**, 5305.
Lauher, J. W. (1979). *J. Am. Chem. Soc.*, **101**, 2604.

Section 3.3

Lewis, J. and Johnson, B. F. G. (1982). *Pure and Appl. Chem.*, **54**, 97.
Johnson, B. F. G. and Lewis, J. (1981). *Adv. Inorg. Chem. Radiochem.*, **24**, 225.

Section 3.3.2

Mingos, D. M. P. and Forsyth, M. I. (1977). *J. Chem. Soc., Dalton Trans.*, 160.

Section 3.3.4

Mingos, D. M. P. (1982). *Inorg. Chem.*, **21**, 464.

Section 3.3.5

Chisholm, M. H. (Ed.) (1995). *Early Transition Metal Clusters with π-Donor Ligands.* New York: VCH.
Cotton, F. A. and Haas, T. E. (1964). *Inorg. Chem.*, **3**, 10.
Wade, K. (1976). *Adv. Inorg. Chem. and Radiochem.*, **18**, 1.
Chisholm, M. C., Clark, D. L., Hampden-Smith, M. J. and Hoffman, D. H. (1989). *Angew. Chem. Int. Ed. Engl.*, **28**, 432.
Hughbanks, T. (1989). *Prog. Solid St. Chem.*, **19**, 329.
Welch, E. J., Crawford, N. R. M., Bergman, R. G. and Long, J. R. (2003). *J. Am. Chem. Soc.*, **125**, 11464.
Ingleson, M. J., Mahon, M. F., Raithby, P. R. and Weller, A. S. (2004). *J. Am. Chem. Soc.*, **126**, 4784.

Section 3.3.6

Chini, P. (1980). *J. Organomet. Chem.*, **200**, 37.

Section 3.4

Klabunde, K. J. (Ed.) (2001). *Nanoscale Materials in Chemistry.* New York: John Wiley.
Stave, M. S. and DePristo, A. E. (1992). *J. Chem. Phys.*, **97**, 3386.
Fayet, M., McGlinchey, J. J. and Wöste, L. H. (1987). *J. Am. Chem. Soc.*, **109**, 1733.

Section 3.5

Mingos, D. M. P. and Johnston, R. L. (1987). *Structure and Bonding*, **68**, 29.

Section 3.5.3

Tran, N. T., Kawano, M., Powell, D. R. and Dahl, L. F. (1998). *J. Am. Chem. Soc.*, **120**, 10986.

Section 3.5.4

Schnepf, A. and Schnöckel, H. (2002). *Angew. Chem. Int. Ed.*, **41**, 3532.

Section 3.6

Schmid, G. (1992). *Chem Rev*, **92**, 1709.
Zhang, H., Schmid, G. and Hartmann, U. (2003). *Nano Letters*, **3**, 305.
Klabunde, K. J. (1994). *Free Atoms, Clusters and Nanoscale Particles*. New York: Academic Press.
Schmid, G. (2004). *Nanoparticles. From Theory to Application*. Weinheim: Wiley-VCH.
Ozin, G. A. and Arsenault, A. C. (2005). *Nanochemistry. A Chemical Approach to Nanomaterials*. Cambridge: Royal Society of Chemistry.

4

Isolobal relationships between main-group and transition-metal fragments. Connections to organometallic chemistry

The structural chemistry of main-group and transition-metal clusters has been set forth in the last two chapters. What more can be said about molecular clusters? Quite a bit, in fact. Although broad similarities between p-block and d-block cluster chemistries exist, we have illustrated important differences in structural preferences. The intriguing question, then, is what happens if p-block and d-block elements compete in a single cluster environment? Will the preferences of one element type dominate the other or will the merging of metal and main-group fragments generate possibilities not accessible to main-group or transition-metal systems alone. Perhaps clusters with novel hybrid properties will result.

But there is another perspective to mixed clusters. The transition-metal chemist sees the main-group fragment as a complex "ligand" through which structure and chemistry at the metal centers is perturbed. A p-block chemist may rather view metals as tools to systematically vary the structure and reactivity of a coordinated main-group moiety. Neither the cluster perspective nor the metal-complex view is wrong: one chooses a perspective optimal for the problem at hand. In this chapter we explore mixed p-block/d-block compounds as metal–ligand complexes with an emphasis on connections to organometallic chemistry. In Chapter 5 the focus will be the complementary cluster view.

4.1 Isolobal main-group and transition-metal fragments

In the first three chapters, instances were noted where the number, symmetry characteristics and occupation numbers of the frontier orbitals of a transition-metal fragment were similar to those of a main-group fragment. Such fragments are said to be isolobal to emphasize similar bonding capabilities. Since its enunciation by Hoffmann and Mingos, the concept has been used effectively for the analysis of both organometallic and cluster problems. Let's explore the idea in a more systematic

140   *Isolobal relationships between fragments*

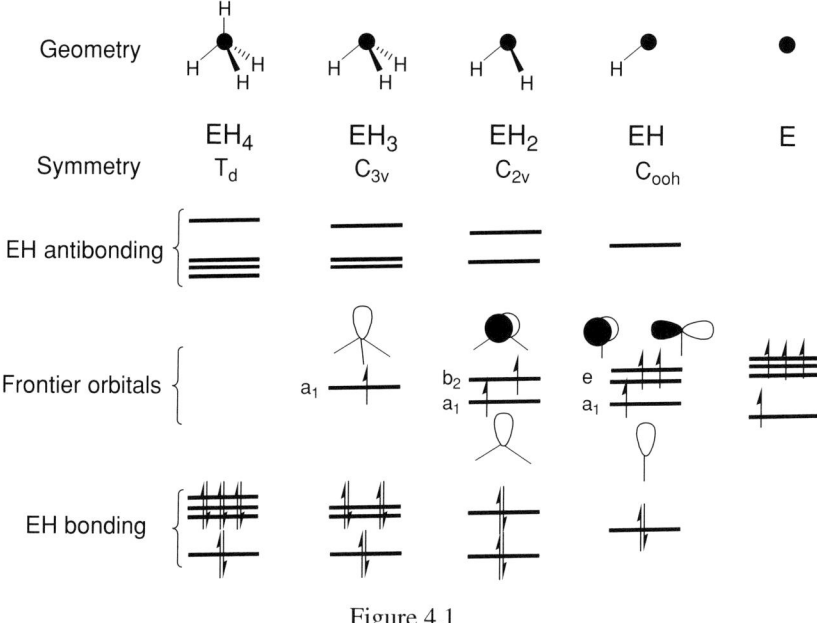

Figure 4.1

fashion to illustrate both the strengths and limitations of this p-block/d-block connection.

4.1.1 Fragment frontier orbitals

An effective way of presenting the isolobal concept is to begin with a main-group or transition-metal molecule that obeys the eight- or 18-electron rule and systematically strip off the ligands. Variations are possible: obvious ones, e.g., molecules not obeying the 8- or 18-electron rule, and subtle ones, e.g., π donor vs. acceptor ancillary ligands on a metal fragment. We have seen that one-electron ligands, e.g., H, are predominant on main-group fragments, whereas multielectron ligands, e.g., two-electron CO or five-electron η^5-C_5H_5, are common for transition-metal fragments.

Consider Figure 4.1 which illustrates the evolution of frontier-orbital sets beginning with a main-group EH_4 molecule where, for simplicity, E is a group-14 element. The geometry of eight-electron EH_4 is tetrahedral and fragments can be generated by sequential removal of H without structural relaxation. In each step, one orbital and one electron are removed and an EH bonding/antibonding MO pair collapses to a single orbital in the HOMO–LUMO gap and is populated by one electron. The orbitals generated constitute the frontier set. For any fragment, the frontier orbital populations can be varied by a change in the identity of the E atom. For E = group 13 the population decreases by one, whereas moving to group 15 increases it by one. Relative orbital energies (and sizes) also change as the atomic number changes.

4.1 Isolobal fragments

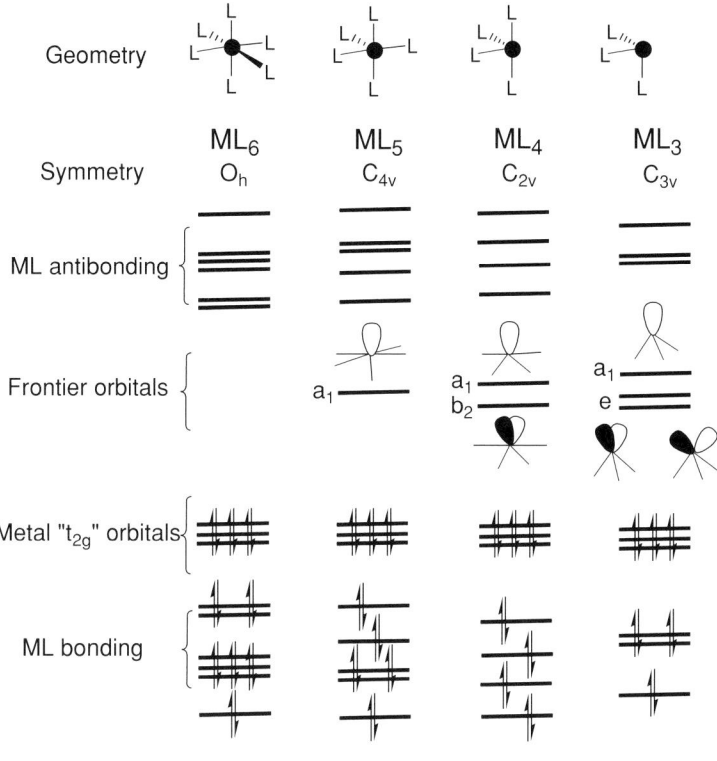

Figure 4.2

Take the EH$_2$ fragment, for example. It contains 3, 2, 1 electrons in the two frontier orbitals for E = N, C, B, respectively, with energies that increase in the order N < C < B. We assume this partitioning of the valence electrons and orbitals remains valid in the construction of molecules or clusters from these fragments.

Now consider application to a cluster. The main-group cluster, our structural paradigm [B$_6$H$_6$]$^{2-}$, can be generated from six three-orbital–two-electron B–H fragments which provide 18 frontier orbitals containing 12 electrons plus 2 from the negative charge. This results in 7 cluster bonding MOs ($n + 1$) containing 7 sep and 11 empty anti- or non-bonding MOs ($2n - 1$). The six (n) external B–H bonding MOs and six (n) antibonding MOs have been removed from the problem. As you know well, this is an approximation. The fragment analysis of this cluster is straight forward and differs little from our earlier treatment. The isolobal principle permits us to substitute one to six BH fragments with isolobal fragments. We will explore below some of the consequences.

In order to explore whether isolobal main-group and transition-metal fragments generate a simple description of mixed main-group and transition-element compounds, we need to generate the frontier-orbital sets of transition-metal fragments. In Figure 4.2 an octahedral complex with two-electron neutral ligands and a group-6

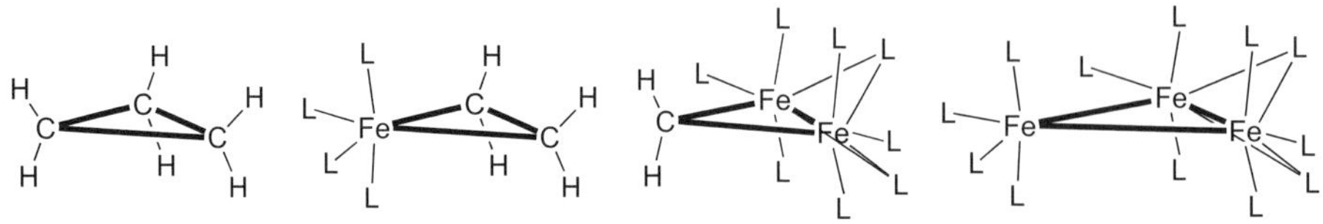

C$_3$H$_6$ C$_2$H$_4$Fe(CO)$_4$ CH$_2$Fe$_2$(CO)$_8$ Fe$_3$(CO)$_{12}$

Figure 4.3

metal (d^6) is sequentially stripped of three ligands. In each step, one orbital and two electrons are removed and an ML bonding/antibonding MO pair collapses to a single, empty orbital in the HOMO–LUMO gap. As with the main-group fragments, these orbitals constitute the frontier set.

There is a difference, however. The "t$_{2g}$" orbitals remain uncomfortably close in energy to the orbitals generated by the ligand removal process. Later transition metals with π-acceptor ligands produce a larger energy gap between the frontier set and the "t$_{2g}$" set. Earlier metals with π donors reduce the gap (Section 3.35). In the former case (group-8 and 9 metals with CO ligands) the frontier orbitals of EH$_n$ fragments, $n = 3, 2, 1$, are isolobal with ML$_m$ fragments, $m = 5, 4, 3$. Thus, a 16-electron C$_{2v}$ Fe(CO)$_4$ fragment is isolobal with a six-electron C$_{2v}$ CH$_2$ fragment. Hence, Fe$_3$(CO)$_{12}$ is a metal analog of cyclopropane C$_3$H$_6$ in the sense that both are constructed of three two-orbital–two-electron fragments. Cute, no? But the best part is that now one can mix and match these fragments thereby generating (μ-CH$_2$)Fe$_2$(CO)$_8$ and (η^2-C$_2$H$_4$)Fe(CO)$_4$ all of which are known compounds (Figure 4.3). Without an isolobal analysis, the relationship between the four compounds would not be obvious.

Exercise 4.1. Fragment ethane into two CH$_3$ radicals. Construct a neutral M(CO)$_m$ fragment isolobal with CH$_3$. Give the structures and compositions of all compounds analogous to ethane that can be formed with these two isolobal main-group and transition-metal fragments.

Answer. The CO ligand is a good π acceptor so it is likely that the t$_{2g}$ set will not be involved in the principal bonding interactions. Hence we need a ML$_5$ fragment (Figure 4.2) with a single frontier orbital containing 6 + 1 electrons, i.e., a group-7 metal. Mn is one possibility thereby generating Mn$_2$(CO)$_{10}$ and (CH$_3$)Mn(CO)$_5$ as the two metal compounds isolobal with C$_2$H$_6$.

4.1.2 Caveats

The effectiveness of a fragment analysis depends on the extent to which the perturbation of the fragment electronic structures is restricted to the frontier orbital set. To behave in the manner of a main-group fragment, an isolobal metal fragment must present to a bonding partner the same number of orbitals containing the same number of electrons as its main-group model does, e.g., to be isolobal with BH, three orbitals and two electrons only must be used. However, these frontier orbitals are not without contributions from the fragment ligands. Even with XH, the σ-symmetry frontier orbital contains H character. For the analogous ML_3 fragment, all three frontier orbitals contain M–L antibonding character. Considering the fact that metal–ligand interactions depend on geometry, the frontier and "t_{2g}" sets of a ML_m fragment will be sensitive to geometric changes on combining with another fragment. This point is explored further below. Just keep in mind that the behavior of a given metal fragment is more flexible than that of a main-group fragment. Thus, we saw in Chapter 3 that octahedral clusters utilizing three (cve = 86) or four (cve = 84) frontier orbitals per fragment are known.

These cautions are necessary because the isolobal idea works so well in many cases that one can easily be seduced into giving it a greater generality than justified. The message is that fragment analyses must be done thoughtfully. The saving feature is that classes of metal clusters do follow a $15n$ or $14n + 2$ cluster valence electron count analogous to the $5n$ or $4n + 2$ count for main-group clusters. It follows that similar isolobal fragment analyses should be possible leading to the generation of all the mixed main-group–transition-metal fragment possibilities. Thus, the isolobal concept is of value in providing connections between mixed main-group–transition-element clusters and their main-group and transition-metal parents provided it is used with understanding.

4.1.3 Applications

We are now prepared to apply the isolobal concept to simple p-block/d-block clusters. We begin with three-connect clusters amenable to localized bonding models. Consider P_4 to be constructed from four P fragments (Figure 4.4). We use four three-orbital–three-electron fragments, i.e., an E fragment in Figure 4.1 where E is a group-15 element and the lone-pair orbital is external to the cluster. From Figure 4.2, only the ML_3 fragment provides three orbitals if the t_{2g} set is at low energy. Hence, a metal with $6 + 3 = 9$ valence electrons, e.g., Co from group 9, is appropriate. A metal analog of P_4 then is $Co_4(CO)_{12}$ (Figure 3.1). Consequently, we predict mixed P_nM_{4-n} clusters to be tetrahedral with compositions $P_n\{Co(CO)_3\}_{4-n}$. Of these, $PCo_3(CO)_9$ is known. The others, however, constitute reasonable synthetic targets. For a given framework containing 2, 3, 4, ..., n fragments, the isolobal principle leads to ready prediction of 1, 2, 3, ..., $n - 1$ mixed compounds in terms of composition and structure. Hence, for large clusters the number of possible mixed-group–transition-metal clusters becomes

144   *Isolobal relationships between fragments*

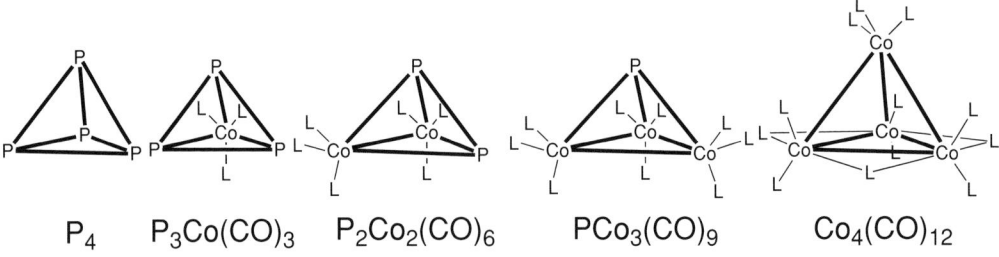

Figure 4.4

impressive. If one permits more than two different fragment types to be used, then the number of possible compounds becomes huge. Most importantly, we need no new bonding concepts in addition to those already discussed in Chapters 2 and 3.

Exercise 4.2. What hydrocarbon cluster is analogous to $Co_4(CO)_{12}$? Write the main-group and metal isolobal fragments. Write out the compositions and draw the structures of all the mixed Co/C clusters possible between these two homonuclear Co and C clusters.

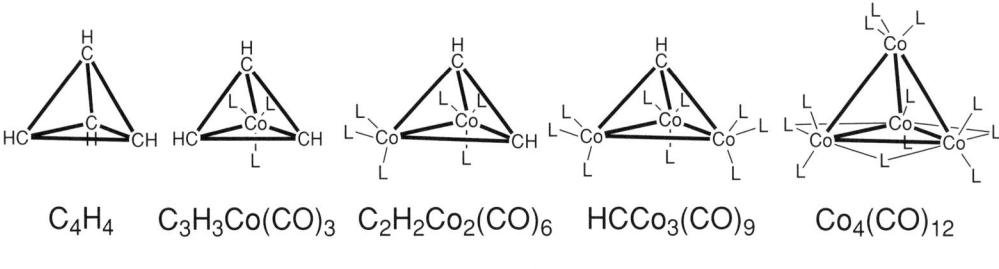

Exercise 4.2

Answer. $Co_4(CO)_{12}$ is made up of three $Co(CO)_3$ fragments which are three-orbital–three-electron fragments. From Figure 4.1 we require an EH fragment with three electrons populating the three frontier orbitals, i.e., E = C, Si, etc. fit our requirements. Thus, CH is isolobal with $Co(CO)_3$ and tetrahedrane, C_4H_4, is analogous to $Co_4(CO)_{12}$. The mixed main-group–transition-metal clusters that can be generated are: $(CH)_n\{Co(CO)_3\}_{4-n}$ which are shown above. As exemplified by the work of Seyferth and Nicholas, the chemistries of $RCCo_3(CO)_9$ and $(C_2R_2)Co_2(CO)_6$ are extensive ones.

4.1 Isolobal fragments

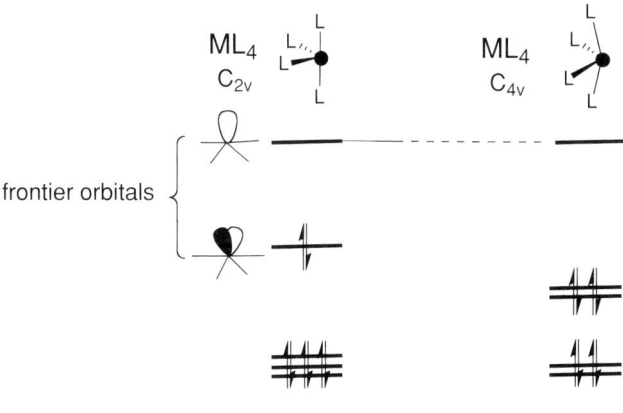

Figure 4.5

4.1.4 Variation of fragment ligands

The ancillary ligand variations possible for a main-group fragment are relatively modest compared to the possibilities for a transition-metal fragment. Thus, for B there are BR, R = H, CH_3, etc., and BL, L = a Lewis base such as CO or PR_3. BR is a three-orbital–two-electron fragment whereas BL is a three-orbital–three-electron fragment like CH. For a metal fragment, not only can the population of the frontier orbitals change but the number can change as well. For example, distortion of a group-8 C_{2v} ML_4 fragment into a C_{4v} fragment results in a stabilization of one frontier orbital leaving a single empty frontier orbital fragment (Figure 4.5). An $Fe(CO)_4$ fragment, then, can act either as a fragment isolobal with CH_2 (Figure 4.3) or BH_3 by simply adjusting geometry of the four ancillary ligands. So we would describe the metal fragments differently in $[(CO)_4Fe(\eta^2\text{-}C_2H_4)]$ (see also Exercise 4.3.) vs. $[(CO)_4FeH]^-$. The latter is analogous to $[BH_4]^-$. It follows that the steric requirements of the ancillary ligands can affect the electronic capabilities of a given metal fragment.

Another way of appreciating the dependence of frontier-orbital character on ancillary-ligand geometry is to repeat the exercise of generating ML_m fragments but begin with a square planar ML_4 complex instead. There are now sets of four ML strongly bonding and antibonding orbitals plus six M-based orbitals (five d and one p), shown in Figure 4.6. In a d^8 metal complex the lower four M orbitals are populated. Removing two ligands one by one in the manner done previously generates fragments that have one and two frontier orbitals. At this rather crude level, a C_{4v} ML_5 fragment, M = Cr, is seen to be isolobal to a C_{2v} ML_3 fragment, M = $[Pt]^{2+}$, and a C_{2v} ML_4 fragment, M = Fe, is seen to be isolobal to a C_{2v} ML_2 fragment, M = Pt. However, the low-lying metal p orbital can become involved,

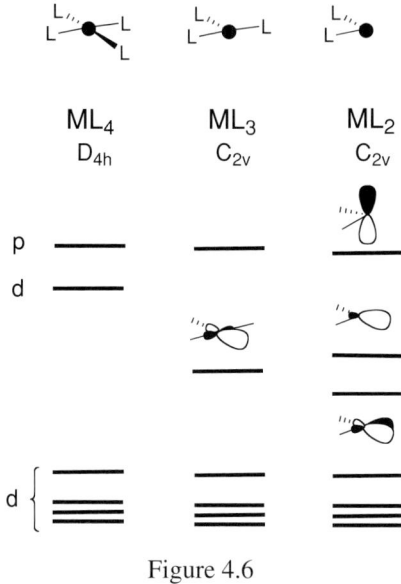

Figure 4.6

and Mingos has discussed situations in which this orbital plays an important role. For a more detailed analysis and discussion of ML$_3$ and ML$_2$ fragments including orbital symmetries and correlation diagrams, see Albright, Burdett and Whangbo.

Exercise 4.3. Compare the coordination of C$_2$H$_4$ to Fe(CO)$_4$ vs. Ni(PR$_3$)$_2$ using a fragment analysis.

Answer. As shown in Figure 4.2, the frontier orbitals of a C$_{2v}$ Fe(CO)$_4$ fragment are a$_1$ and b$_2$ symmetry functions (two electrons) suitable for interacting with the π and π* MOs of ethylene (two electrons). The former constitutes the primary ligand donor–metal acceptor σ interaction whereas the latter constitutes the secondary metal donor–ligand π acceptor interaction. In the d^{10} Ni(PR$_3$)$_2$ fragment (Figure 4.6) with the five lowest-lying orbitals filled, the two frontier orbitals with 2 electrons are analogous to those of the iron fragment. Hence, the Ni–ethylene interaction is qualitatively the same as the Fe–ethylene interaction.

An important ancillary ligand in organometallic chemistry is η^5-C$_5$R$_5$ (Cp, R = H, or Cp*, R = Me) which is usually considered to occupy three metal coordination positions on a metal center, i.e., a "tridentate" ligand. We can generate fragments containing CpML$_n$ fragments beginning with the 18-electron pseudo-octahedral CpML$_3$ complex with a group-7 metal center (Figure 4.7). Note that the "t$_{2g}$" set only contains six of the seven metal valence electrons as one must be used in binding the ancillary Cp ligand, i.e., a neutral Cp ligand brings only five electrons to the molecular dance rather than the six electrons of the three monodentate

4.1 Isolobal fragments

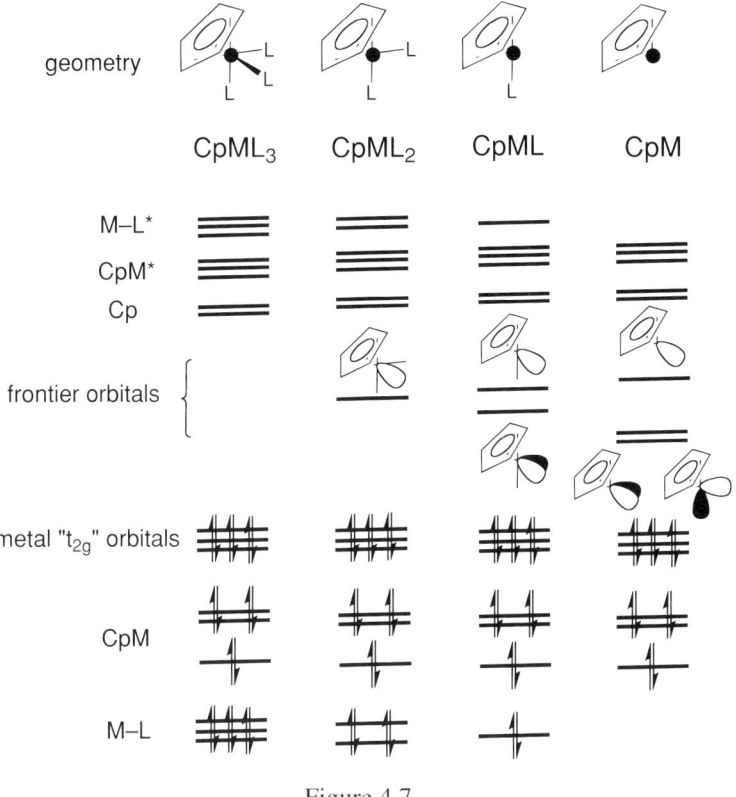

Figure 4.7

two-electron ligands of a M(CO)$_3$ fragment. If one considers the cyclopentadienyl ligand as a monoanion, [Cp]$^-$, it is a six-electron ligand; however, the metal oxidation state must be adjusted accordingly (a monocationic group-7 metal center is d^6). Take your choice! In either case, the diagram in Figure 4.7 is analogous to that in Figure 4.2.

Exercise 4.4. Suggest a CpM fragment isolobal with a BH fragment. Systematically replace one and two BH fragments in B$_5$H$_9$ with this metal fragment and enumerate all isomers.

Answer. A BH fragment possesses three orbitals (two of π and one of σ symmetry) containing two electrons. Hence, a CpM fragment with M = a group-9 metal is the appropriate choice (Figure 4.7). This leads to metal derivatives of B$_5$H$_9$ of the type (CpM)$_n$(BH)$_{n-5}$(H)$_4$ for M = Co, Rh or Ir. The square pyramidal *nido*-cluster possesses two types of vertices, one apical and four basal. Hence, for $n = 1$, two isomers are possible whereas for $n = 2$, three isomers are possible. The structures

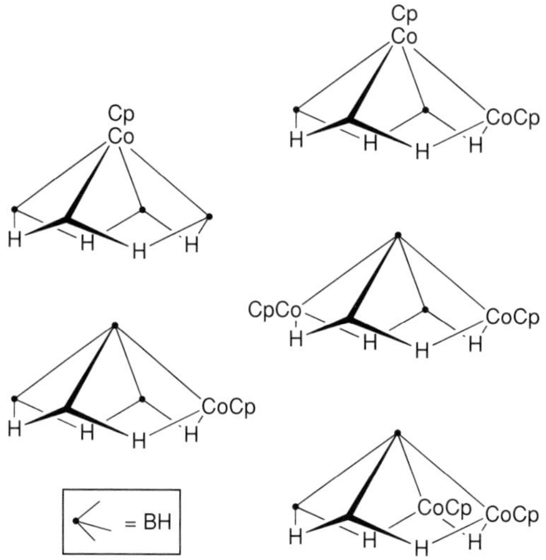

Exercise 4.4

are shown above and examples of all are known with selected group-9 metals – an impressive example of the usefulness of the isolobal approach.

4.2 Metal variation with fixed ancillary-ligand set

We saw in Chapter 3 that the structures of small main-group clusters were reproduced in transition-metal clusters of the same nuclearity. This provided our first look at metal-cluster bonding. However, we also saw that the more flexible bonding system of a transition metal led to greater complexity. There was correspondingly greater difficulty in evaluating a given cluster composition in terms of geometric and electronic structure. In terms of fragment analyses, a transition-metal fragment can adjust its frontier orbitals. This flexibility in bonding will be carried over to the mixed compounds. The new feature – not found in metal clusters alone – is that the preferred main-group tendencies will compete with the preferred transition-metal preferences. To explore this idea in the context of isolobal fragments we need to investigate how flexible electronic behavior of transition-metal fragments is expressed in the language of isolobal fragments.

All of the compounds discussed to this point in this chapter are accommodated by this assumption that the three filled "t_{2g}" orbitals do not constitute part of the frontier-orbital set. However, there are well-known organometallic examples where these metal d functions do participate in bonding. Hoffmann labels this behavior "into the t_{2g} set." Let's take a look at some dinuclear metal complexes that provide geometric evidence of the partial utilization of the "t_{2g}" set in bonding.

4.2 Metal variation with fixed ancillary-ligand set

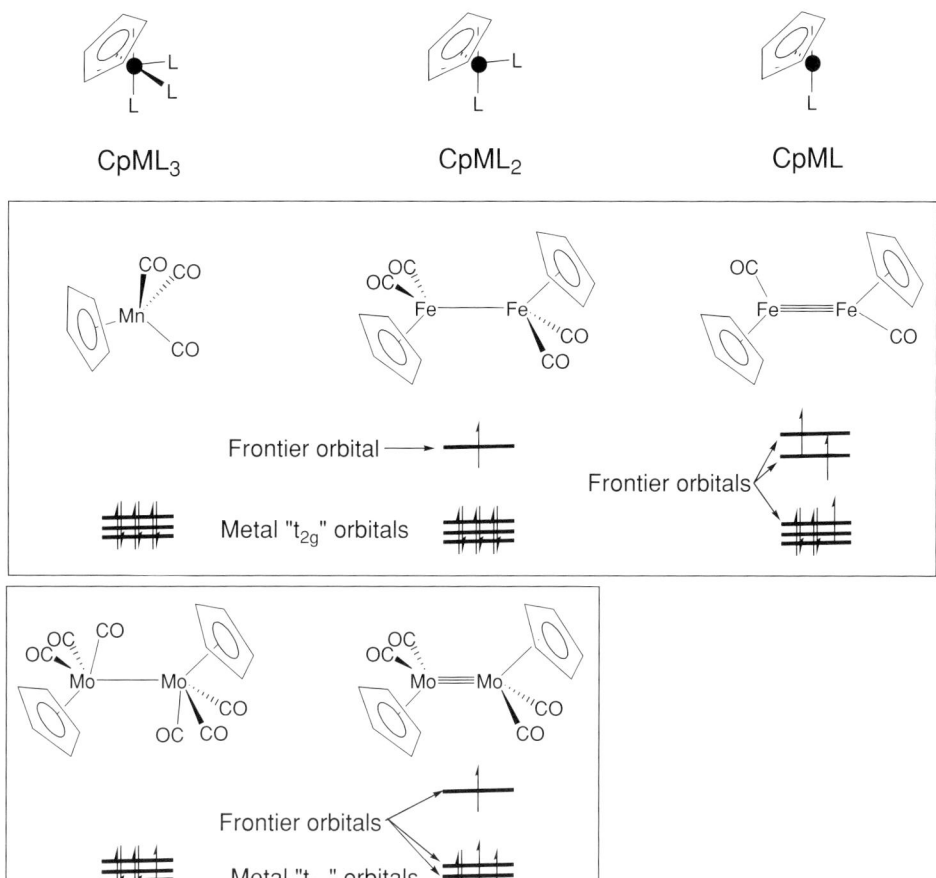

Figure 4.8

Consistent with our analyses above, CpMn(CO)$_3$ is a known 18-electron compound. The 17-electron CpMo(CO)$_3$ fragment is found as a dimer with an M–M single bond (Figure 4.8). In order to be isolobal with CH$_3$, the latter must use one "t$_{2g}$" orbital plus one electron in bonding. Seventeen-electron CpFe(CO)$_2$, with an "unused," filled "t$_{2g}$" set, is also isolobal with CH$_3$! Further, 15-electron CpMo(CO)$_2$ behaves as if it were isolobal with CH as it is found as a dimer with a short Mo–Mo bond attributed to a Mo≡Mo triple bond. Hence, two of the "t$_{2g}$" orbitals and three electrons must be used in the bonding. The Fe analog, 15-electron CpFe(CO) also forms a multiply bonded dimer which is isolated at low temperatures by photolysis of [CpFe(CO)$_2$]$_2$ in an inert matrix. It now must utilize one orbital from the "t$_{2g}$" set to be viewed as isolobal with CH.

In contrast to an E–H fragment, where variation in the main-group element E simply changes the frontier orbital populations, changes in metal identity for a L$_n$M fragment can generate changes in both the frontier-orbital set and population. We must be prepared to deal with this behavior in main-group–transition-metal clusters in spite of the fact that unambiguous geometric evidence for fragment electronic structure is rarely available.

Exercise 4.5. Count the skeletal electron pairs (or cluster valence electrons) in the organometallic complex shown below at the left. Is there a relationship to cyclobutadiene iron tricarbonyl shown next to it?

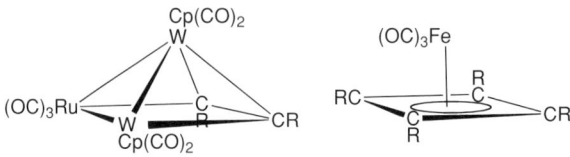

Exercise 4.5

Answer. A sep count approach is natural for an isolobal analysis. Taken as three orbital fragments, the CR fragments count as three each and the $Ru(CO)_3$ fragment most likely counts as two. In view of the discussion above, the $CpW(CO)_2$ fragment can be counted as a three-orbital–three-electron fragment. The total is $(6 + 2 + 6)/2 = 14/2 = 7$ sep. The structure should be based on an octahedron with one unoccupied vertex exactly as found. Viewed as a cluster analogous to B_5H_9, other isomers are possible depending on how the three different fragments are distributed over the framework. If the $CpW(CO)_2$ fragments are converted into their isolobal CR equivalents and the $Ru(CO)_3$ fragment is moved to the apical position by exchange with one $CpW(CO)_2$ fragment then one sees a relationship to $(CO)_3Fe(\eta^4\text{-}C_4H_4)$.

4.3 Metal–ligand complex vs. heteronuclear cluster

The ability to deconstruct a complex molecule into a set of fragments with compatible bonding properties along with an understanding of isolobal behavior places us in a position to explore main-group–transition-metal clusters in some detail. As described in the introduction to the chapter, the main-group entity can be viewed either as a complex ligand or a part of a cluster network in which both metal and main-group atoms participate. For example, isoelectronic $CpCoC_4H_4$ and $CpCoB_4H_8$ can be treated as mononuclear metal complexes or monometallic analogs of B_5H_9 (Figure 4.9) in order to rationalize structure by counting electrons. In terms of electronic structure, the metallaborane is better treated as a heteroatom cluster as the electronegativities of the B and Fe centers are of comparable magnitude. One practical consequence of this is that 2-$CpCoB_4H_8$ and 2,4-$\{Cp^*Co\}_2B_3H_7$ are known (Exercise 4.4) whereas the C analogs are not. Of

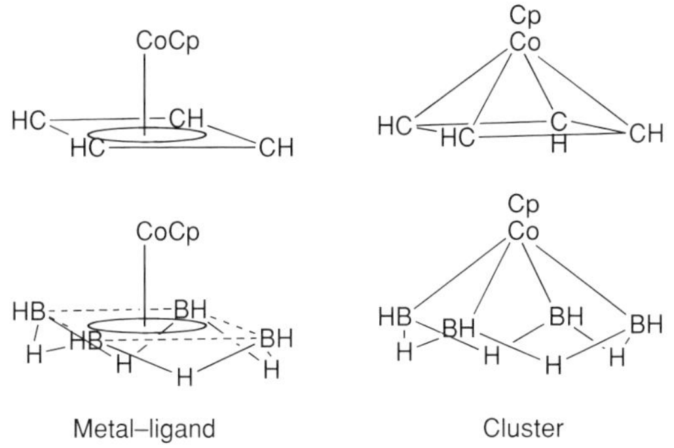

4.3 Metal–ligand complex vs. heteronuclear cluster   151

Figure 4.9

course, it is dangerous to base conclusions on negative evidence but approximate calculations suggest that hypothetical 2-CpCoC$_4$H$_4$ has little or no barrier to rearrangement to 1-CpCoC$_4$H$_8$.

The metal–ligand perspective is easy to appreciate when there is a single transition-metal center. How about multimetal clusters? In Chapter 3 we mentioned that a metal cluster such as Ru$_5$(CO)$_{15}$(PPh) (Figure 3.5) can be equally well considered a homonuclear *nido*-pentametallic cluster containing a μ_4-PPh ligand or a heteronuclear *closo*-hexanuclear cluster with five metal fragments and one main-group fragment. Which is preferable? It depends on the situation. In the example under consideration, the five metal atoms establish the identity of the compound as a metal cluster and for the purposes of correlating compound stoichiometry with electron count, neither method has an advantage. Certainly as the electronegativity of the main-group fragment becomes larger, its treatment as a bridging or capping ligand becomes more realistic than the cluster view.

Let's compare the two views with the real example in Figure 4.10. Reaction of CpRh(CO)$_2$ with alkyne yields a trinuclear complex with the alkyne coordinated parallel to a Rh–Rh bond (48 cve if the alkyne is a four-electron ligand). Shown immediately beneath it is a representation of the compound as a *nido*-cluster with sep = 7 and a square-pyramidal geometry analogous to that of, e.g., 2,3-C$_2$B$_3$H$_7$. Heating causes CO-ligand loss and formation of a trinuclear complex with μ_3-CPh ligands capping opposite sides (48 cve with CR a three-electron ligand). Apparently a C≡C triple bond has been cleaved. Again, a cluster representation is possible. This is a *closo*-2,3,4-Rh$_3$C$_2$ cluster with sep = 6 and a trigonal-bipyramidal geometry analogous to that of, e.g., 1,5-C$_2$B$_3$H$_5$. In dicarbon carboranes (Chapter 2), cluster isomers with non-adjacent carbon fragments are more stable than those with adjacent fragments, i.e., stability is driven by charge distribution. As 1,2-C$_2$B$_3$H$_5$ rearranges to 1,5-C$_2$B$_3$H$_5$ cleavage of the C–C interaction is a consequence of cluster electronic structure, i.e., the C–C bond cleavage is no longer spectacular.

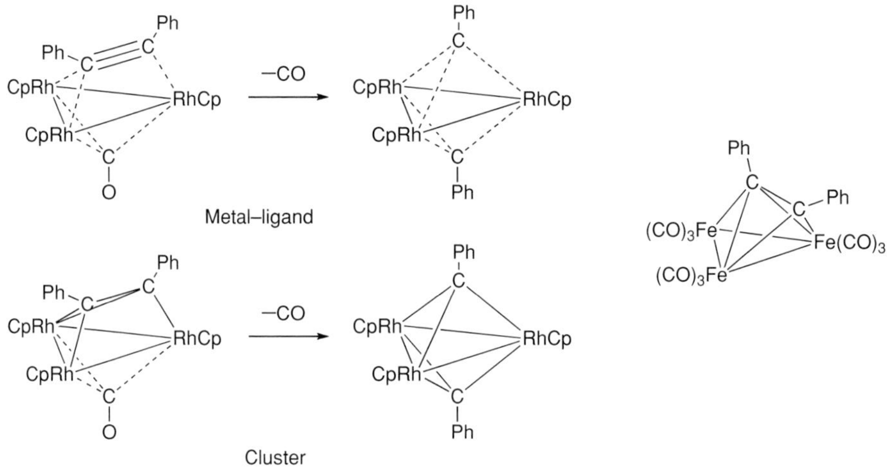

Figure 4.10

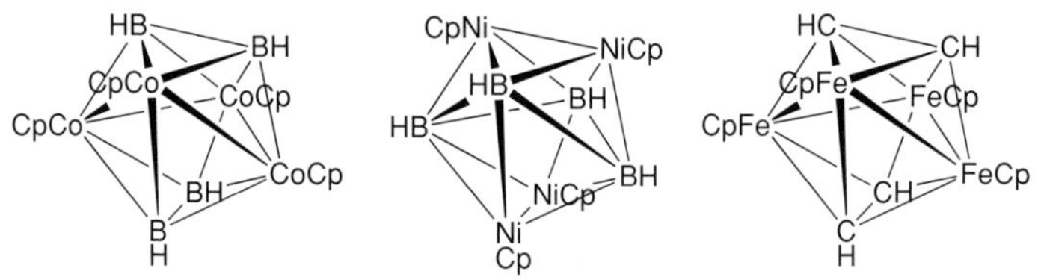

Figure 4.11

A similar conclusion is possible for the metal analogs. Note that metal analogs of 1,2-$C_2B_3H_5$ are known, e.g., $Fe_3(CO)_9(PhCCPh)$ (far right Figure 4.10).

When the number of metal and main-group fragments is nearly equal, the consequences of competition between preferred modes of cluster behavior should be most readily observed. Consider the $(CpM)_4E_4H_4$ clusters, E = B, M = Co, Ni, of Grimes and a C analog, E = C, M = Fe, of Okazaki. As shown in Figure 4.11 all have a dodecahedral cluster structure with an expected sep = 9. However, the cobaltaborane and ferracarbyne have eight sep and the nickelaborane has ten sep! These clusters are members of the cubane class of clusters, eight-atom M_4E_4 clusters or M_4 tetrahedra with E face-capping ligands, which exhibit large variations in electron count. They will be treated in detail in Chapter 5.

4.4 p-Block–d-block metal complexes

The isolobal idea has had considerable impact on organometallic chemistry. Closely related metal compounds containing p-block element ligands other than C also

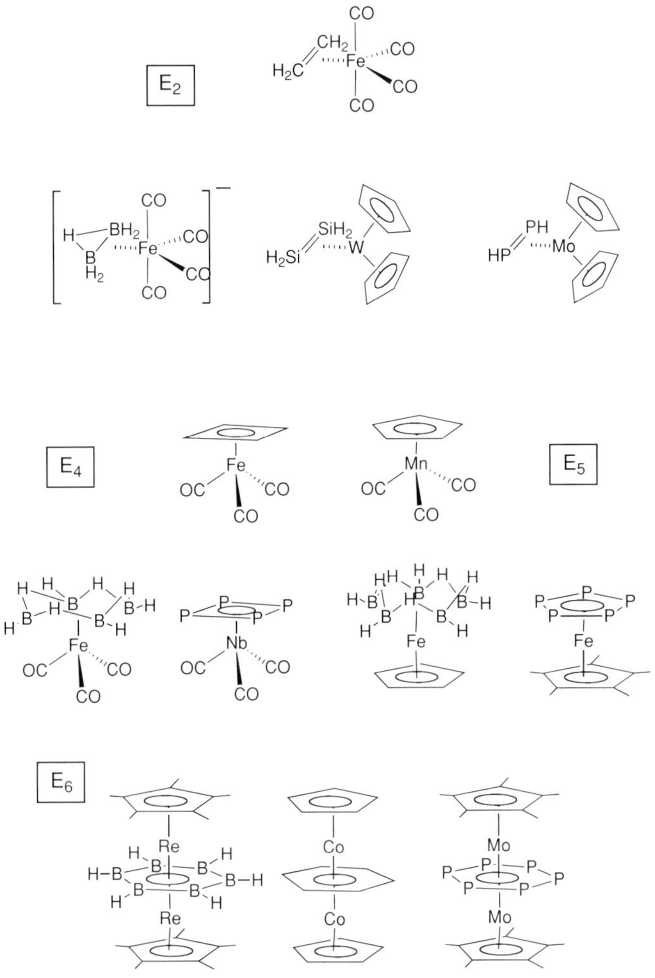

4.4 p-Block–d-block metal complexes

Figure 4.12

benefit from the isolobal concept. In Figure 4.12, representative metal complexes of C_n and E_n ligands, $n = 2$–6, are compared. For $n = 2$–5 the compounds are isolectronic; however, for $n = 6$ ("triple decker" complexes) they are not. This comparison has been highlighted by the term "inorganometallic" chemistry and a more comprehensive treatment will be found in a monograph of the same title. The focus is the coordination of the main-group entity to the metal center even though the main-group species may not have independent existence. In the case of the π complex of ethylene, the free ligand is available but others, e.g., cyclobutadiene, are not. Other p-block analogs of these ligands, if known, have more fleeting lifetimes. In some cases, bulky ligands permit a structural type to be characterized as a free entity, e.g., disilene.

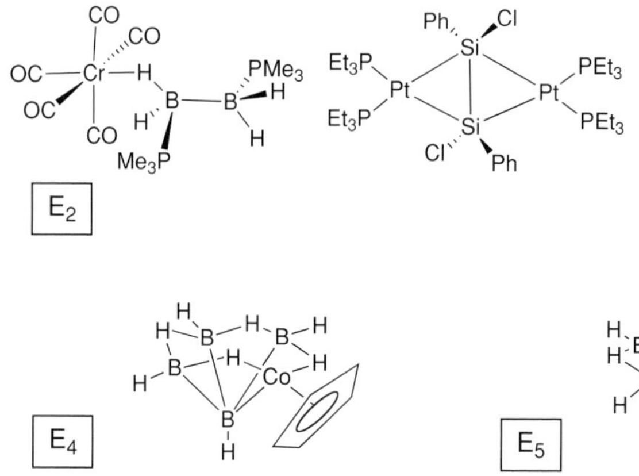

Figure 4.13

Like the borane analog of coordinated cyclobutadiene (Figure 4.9), inorganometallic analogs are found in isomeric forms not known for the organometallic analogs. A few are shown in Figure 4.13. An E_2 ligand has additional sites of Lewis basicity, i.e., lone pairs or hydridic H atoms, leading to different coordination sites. As shown, the cluster model provides a convenient explanation for the alternative forms of complexes containing four- and five-membered rings. Metal–ligand and cluster views can be understood as limiting models much like the two limiting models for olefin coordination, i.e., the Dewar model vs. the metallacyclopropane model. Each real compound lies somewhere in the continuum of possibilities between the two limits.

Group-13 and -14 analogs of the two C_1 complexes shown at the left side of Figure 4.14 are compared to the right. The differences between these p-block analogs and the organometallic compounds are clear. The binding of the $[BH_4]^-$ ion to a metal center is associated with one to three B–H–M bridges, and the example shown exhibits two. $[BH_4]^-$ is isoelectronic to CH_4, and $[(CO)_4Mo(BH_4)]^-$ may be considered a putative methane complex. The silane complex $Cp^*(CO)_2CrHSiR_3$ is another analog of a methane complex but now the bridging Si–H–Cr hydrogen is sufficiently unusual to be called "agostic" in the manner of C–H–M bridges.

The carbene complex contains a M=C double bond. A similar interaction in the group-13 analog must arise from interaction of an empty π-symmetry orbital on the main-group atom with a filled metal orbital of matching symmetry. Three possible scenarios are represented by the examples shown. With chelating oxygen ligands on the B atom, the orientation of the borane ligand permits significant interaction with a filled π-symmetry orbital on the Fe center. Hence, it is represented with a Fe=B double bond. In contrast, the same bonding network with phenyl groups on B exhibits a solid-state structure with the plane of the BPh_2 nearly perpendicular to

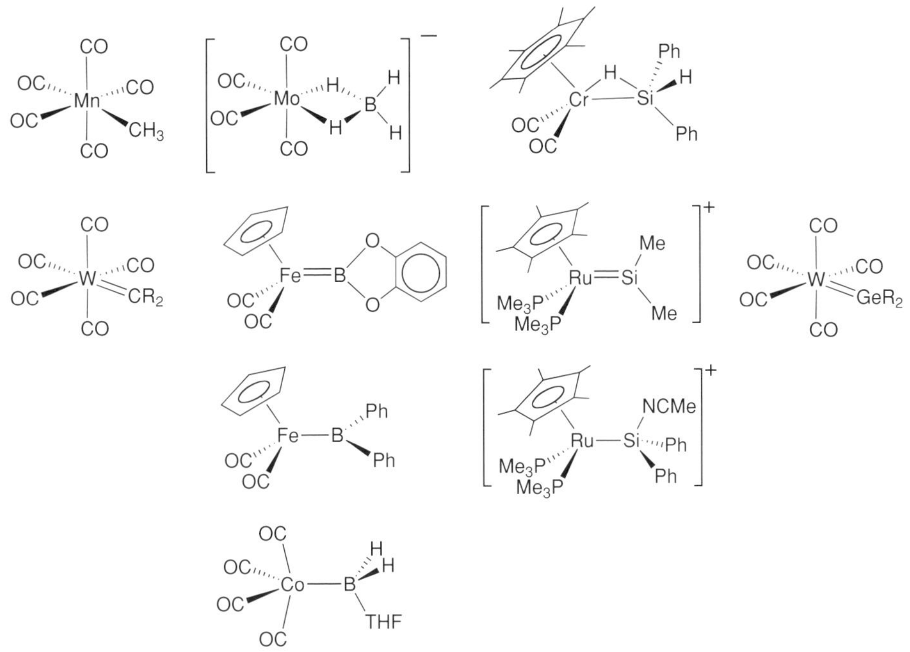

Figure 4.14

the plane of symmetry of the CpFe(CO)$_2$ fragment. Hence, no double bond with the principal π-symmetry orbital is possible and a weak π Fe→B interaction is implied. Finally, in the presence of Lewis bases, coordination to the empty valence orbital on the B atom precludes interaction with the metal center. Related observations are reported for Si, an element with an electronegativity similar to that of B. Thus, compounds with M=Si and M=Ge bonds have been characterized, but they were preceded in time by compounds containing a Lewis base coordinated to the Si center. Clearly this acid–base interaction is competitive with M=Si double-bond formation. Our major point is that element variation within a cluster framework should have equally large effects on cluster properties.

4.5 Carborane analogs of cyclopentadienyl–metal complexes

In Chapter 2, Section 2.6.1 we mentioned Hawthorne's analogy between the frontier orbitals of the [η^5-C$_5$H$_5$]$^-$ ([Cp]$^-$) ligand and those of the *nido*-[C$_2$B$_9$H$_{11}$]$^{2-}$ dianion (Figure 4.15). The former is a very common ancillary ligand in organometallic chemistry and the latter has developed into a potent ligand for metal coordination. In parallel chemistry, principally by Grimes, *nido*-carboranes such as C$_2$B$_4$H$_8$ were deprotonated to form *nido*-[C$_2$B$_4$H$_6$]$^{2-}$ ligands also presenting a five-atom–six-electron face to a metal center (Figure 4.15). These pentagonal-pyramidal clusters

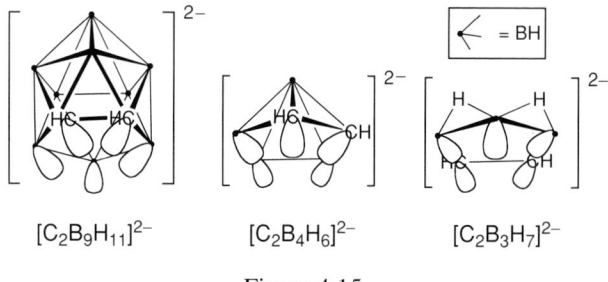

Figure 4.15

can be "decapitated" to form $[C_2B_3H_7]^{2-}$ thereby forming an isoelectronic analog of the $[Cp]^-$ ligand. Other planar five-membered rings, with or without heteroatoms, have been used with effect to mimic the $[Cp]^-$ ligand. For example, $[P_5]^-$ (Figure 4.12) as well as heteroatomic rings, e.g., $[SB_2C_2H_4]^-$ and mono- and tricarbon carboranes have been fruitfully employed.

This extensive adjunct to organometallic chemistry will be briefly illustrated to show that the metal–ligand description is particularly appropriate. That is, although the compounds can certainly be considered as clusters, important aspects of their behavior are readily accommodated in the context of coordination chemistry. The discussion will be limited to $[C_2B_4H_6]^{2-}$ and $[C_2B_9H_{11}]^{2-}$ analogs of the $[Cp]^-$ ligand.

4.5.1 Complexes of $[C_2B_9H_{11}]^{2-}$

A couple of compounds are illustrated in Figure 4.16, and a recent compilation may be found in a review by Jemmis. First, the numerous *commo*-$[(C_2B_9H_{11})_2M]^n$ complexes are analogs of metallocenes on the one hand and vertex-fused clusters on the other. Although we only discuss transition-metal complexes here, many examples containing main-group elements have been characterized as well, e.g., $[(C_2B_9H_{11})_2Si]$ with the same geometry as the Co complex in Figure 4.16. Similar main-group metallocene complexes are known, e.g., $(\eta^5\text{-}C_5Me_5)_2Si$ with a sandwich structure. Like the metallocenes, odd-electron species are often accessible for the same metal/ligand set, e.g., $[(C_2B_9H_{11})_2Fe]^{2-}$ and $[(C_2B_9H_{11})_2Fe]^-$ are analogs of ferrocene and ferrocenium. There is a difference, however, as now the carborane Fe^{III} complex is the more stable form.

Many complexes have the symmetrical structure illustrated for $[(C_2B_9H_{11})_2Co]^-$ in Figure 4.16; however, "slipped" structures are known as well for electron-rich metals. The Cp ligand can adopt η^1, one-electron and η^3, three-electron binding modes as well as the more common η^5, five-electron mode. So too, the carborane analog can adjust to the ligand electronic requirements of the metal. Although

4.5 Carborane analogs of cp–metal complexes

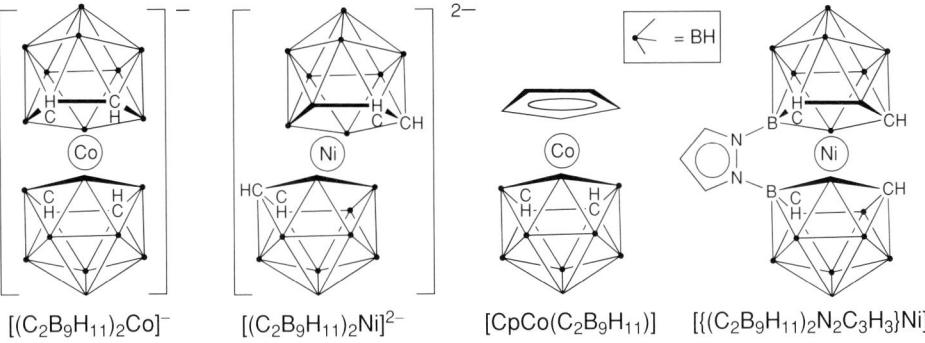

Figure 4.16

$[(C_2B_9H_{11})_2Ni]$ and $[(C_2B_9H_{11})_2Ni]^-$ exhibit symmetrical structures (the first with cisoid pairs of C atoms, the second with transoid C), $[(C_2B_9H_{11})_2Ni]^{2-}$ is found to adopt a slipped structure (shorter M–B and longer Ni–C distances). This first pair of compounds has been suggested by Hawthorne as the heart of a molecular-sized stepping motor driven by electron transfer.

Like the carboranes themselves, cluster rearrangements take place with adjacent C atoms moving to non-adjacent vertices on the icosahedral skeleton. Again there is a difference in that these rearrangements occur under much milder conditions than those for the metal-free $C_2B_{10}H_{12}$ clusters. As with the metal-free boranes, replacement of external cluster one-electron H ligands with two-electron Lewis bases changes the charge on the cluster. Hence, a neutral analog of $[(C_2B_9H_{11})_2Fe]^{2-}$ is $[(NEt_3C_2B_9H_{11})_2Fe]$ in which a triethylamine ligand has replaced one B terminal H on each of the two carborane cages.

Exercise 4.6. Test the compounds $[(C_2B_9H_{11})_2Co]^-$ and $[(C_2B_9H_{11})_2Ni]^{2-}$ for electronic saturation both as metallocenes and as vertex-fused clusters.

Answer. The $[C_2B_9H_{11}]^{2-}$ ligand is a six-electron ligand (four if considered neutral) and Co^{III} has six electrons (Co^{-I} has ten) giving 18 electrons for the metal center. Applying the Jemmis *mno* rule, we have $m = 2$ clusters, $n = 23$ vertices, and $o = 1$ shared single-vertex atom = 26 electron pairs required. The four CH groups contribute six pairs, the 18 BH groups 18 pairs, the Co $1\frac{1}{2}$ pairs (the 6 "t_{2g}" electrons are considered non-bonding) and the negative charge $\frac{1}{2}$ pair to give a total of 26. In the same manner, the Ni compound yields $6 + 6 + 8 = 20$ for the Ni center. The calculated *mno* is the same but there are now $6 + 18 + 2 + 1 = 27$ pairs available. Ligand "slippage" relieves the electronic situation from both metal–ligand and cluster perspectives.

Mixed-ligand systems such as CpCo($C_2B_9H_{11}$), shown in Figure 4.16, and CpCo($Et_2C_2B_4H_4$), shown in Figure 4.17, are also metallocene analogs. So-called "*ansa*" metallocenes have their analog in the "Venus fly-trap" of Hawthorne. As shown in Figure 4.16, joining two *nido*-carboranes together at the open faces generates a chelating ligand system that effectively acts as a hexadentate ligand. The higher negative charge on the carborane ligands relative to a cyclopentadienyl ligand plus the chelate effect makes this ligand a highly effective metal sequestering agent.

4.5.2 Multidecker complexes

Removal of the apical BH fragment from CpCo($Et_2C_2B_4H_4$) opens access to a large body of chemistry in which the borane Cp analog serves as the internal "bread slices" of multidecker sandwich complexes such as the two metal, tripledecker and three metal, tetradecker complexes shown in Figure 4.17. Continued use of this strategy leads to penta- and hexadecker complexes. Like their metallocene analogs multidecker complexes permit the incorporation of a variety of metals and metal oxidation states and the cluster count for these compounds is just one possibility out of many. For example, Siebert has described examples containing 30–33 valence electrons, i.e., CpM($C_3B_2H_5$)M'Cp; M M' = FeCo, CoCo, CoNi, and NiNi. Note that this valence count is a metal–ligand count, e.g., for CpCo($C_3B_2H_5$)CoCp: 2 Cp + 2 Co + central ring = $2 \times 5 + 2 \times 9 + 3 = 31$ ve. The compounds with FeCo, CoCo$^+$ and NiNi$^-$ are diamagnetic; those with FeCo$^+$, CoCo, NiCo$^+$ and NiNi are paramagnetic with one unpaired electron; and those with NiCo and NiNi are paramagnetic with two unpaired electrons.

A good example of the large variation in electron counts possible for the same qualitative shape is provided by tripledecker complexes, three of which are shown at the bottom of Figure 4.12. The valence electron counts for the Re, Co and Mo complexes are 24, 34 and 28, respectively. How does this compare with a cluster count? An example of a tripledecker complex following the counting rule is 30-ve (η^6–1,3,5-Me$_3$C$_6$H$_3$)Cr(1,3,5-Me$_3$C$_6$H$_3$)Cr(η^6–1,3,5-Me$_3$C$_6$H$_3$). It has a planar mesitylene ring sandwiched between two (η^6–1,3,5-Me$_3$C$_6$H$_3$)Cr fragments. Each of the latter is equivalent to (CO)$_3$Cr, a zero-electron–three-orbital fragment whereas each of the six R–C fragments is a three-electron–three-orbital fragment. Thus, it possesses nine sep which is the requirement for an eight-vertex *closo*-cluster. Note that the eight-vertex hexagonal bipyramid is not the more spherical eight-vertex dodecahedron of borane chemistry. The deviation of tripledeckers from a "normal" cluster count is not a problem if one chooses to view them as metal–ligand complexes. In fact, Jemmis showed how metal and central ring composition changes affect the positioning of the HOMO–LUMO gap and, hence, the valence electron count.

4.5 Carborane analogs of cp–metal complexes

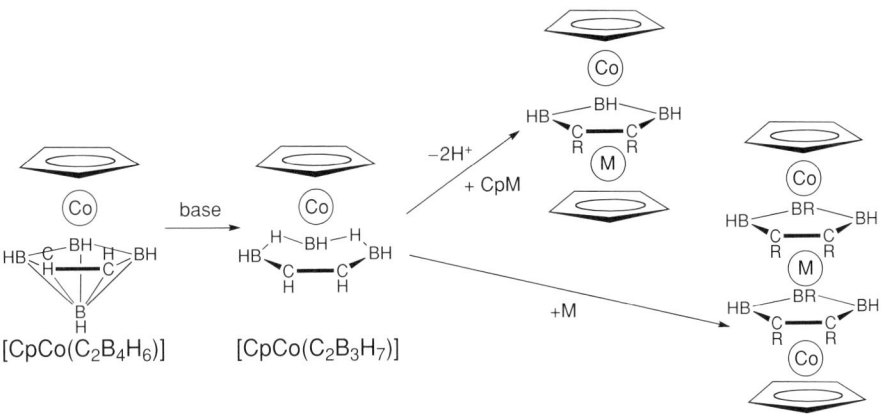

Figure 4.17

Exercise 4.7. How many valence electrons does [Cp*$_2$Co$_2$(Et$_2$C$_2$B$_3$H$_3$)] (structure in Figure 4.17) possess? Can one reasonably view this compound as an example of a *closo*-M$_2$E$_5$ pentagonal-bipyramidal cluster?

Answer. The valence electron count is: 2 Cp* + 2 Co + Et$_2$C$_2$B$_3$H$_3$ = 10 + 18 + 2 = 30. The sep count is: 2 Cp*Co + 3 BH + 2 CR = (4 + 6 + 6)/2 = 8 which is appropriate for a seven-vertex *closo*-deltahedron. Alternatively, the cve count is: 2 Cp* + 2 Co + 3 B + 2 C + 3 H + 2 R = 10 + 18 + 9 + 8 + 3 + 2 = 50 which is appropriate for a M$_2$E$_5$ pentagonal bipyramid (an all-metal cluster would have 14n + 2 = 100 yielding 50 for a M$_2$E$_5$ cluster).

4.5.3 Cluster aspects

These carborane ligands also demonstrate their cluster personalities in a number of ways. Just as the [C$_2$B$_9$H$_{11}$]$^{2-}$ ligand was made by base removal of a BH fragment from 1,2-C$_2$B$_{10}$H$_{12}$, so too the coordinated [C$_2$B$_9$H$_{11}$]$^{2-}$ is subject to degradation under basic conditions, e.g., 1-(CpCo)-2,3-C$_2$B$_9$H$_{11}$ on treatment with base followed by oxidation yields 11-vertex, 12-sep *closo*-1-(CpCo)-2,3-C$_2$B$_8$H$_{10}$ (Figure 4.18). Clearly, the view of 1-(CpCo)-2,3-C$_2$B$_9$H$_{11}$ as a cluster more easily accommodates this type of reactivity.

Hawthorne has described a rhodium carborane system in which the metal is capable of reversibly moving from an open face of the carborane to its external hydride surface where it "walks" around (Figure 4.18, bottom). *Exo-nido* species such as this one are potentially stable geometries, e.g., the Mn compound shown in Figure 4.18, upper right, is possible because B–H terminal H have Lewis-base character (recall (CO)$_5$Cr{η^1-B$_2$H$_4$(PMe$_3$)$_2$} shown in Figure 4.13).

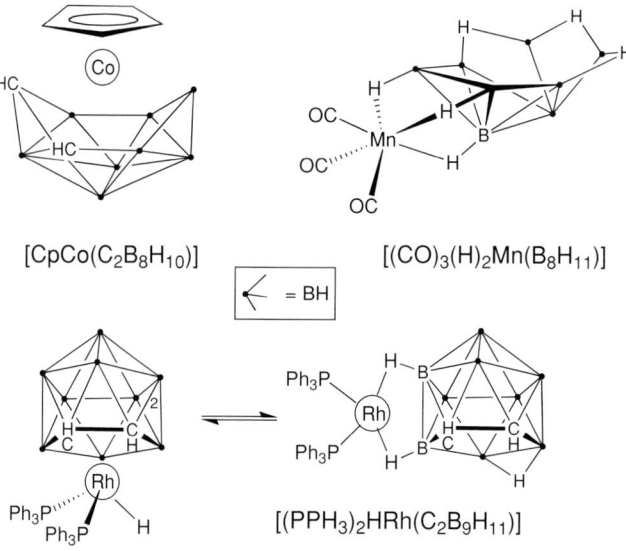

Figure 4.18

The coordinating ability of the B–H bond is also expressed in the chemistry of carborane complexes of the earlier metals. Stone has provided examples of the complex ways in which the carborane ligand can become involved in chemistry of metal carbyne complexes. As shown in Figure 4.19, the generation of M–H–B interactions serves to stabilize the Ru center. As with any such three-center–two-electron bridging interaction, the bridging H atom has protonic character and can be removed by a base to generate a M–B bond. These kinds of carborane cluster ligands can lead to beautifully complex species like the tungsten diiron complex shown at the bottom of Figure 4.19.

Exercise 4.8. Consider the WFe_2 complex at the bottom of Figure 4.19 as a trinuclear metal "cluster" and show that the number of cluster valence electrons = 48.

Answer. It is perhaps easier to let all the ligands remain neutral rather than trying to calculate metal oxidation states: $8\,CO + MeC + C_2Me_2B_9H_9 + 2\,B\text{–}Fe$ bonds $+\,W + 2\,Fe\,+\,$ charge $= 16 + 3 + 4 + 2 + 6 + 16 + 1 = 48$.

In this chapter we have strayed from the avowed focus on clusters. But it was judged worthwhile to emphasize the strong connections that exist with mononuclear organometallic chemistry via metallocene analogs. The treatment has been brief but also serves as a pedagogically useful exercise of the isolobal principle. With this preparation, we are now ready to tackle mixed main-group–transition-metal systems in a systematic fashion in the next chapter. Fragment analysis and the

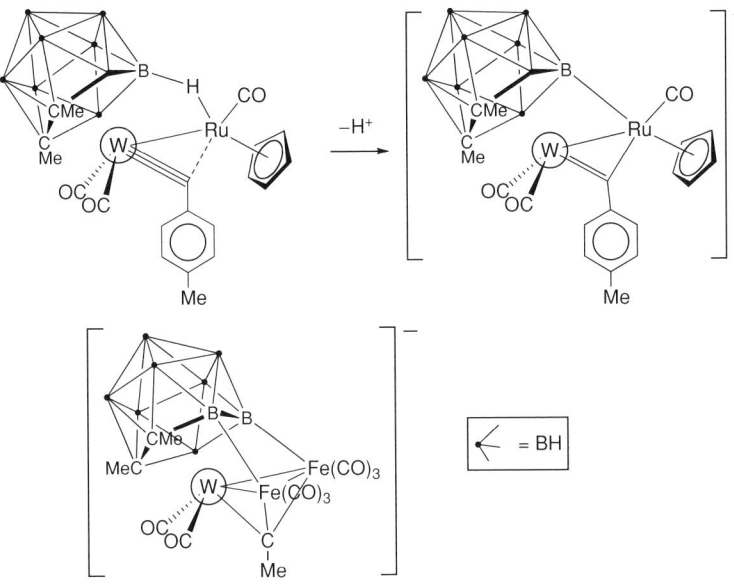

Figure 4.19

isolobal principle can be used to sort and organize the known compounds as well as search for geometric and electronic properties that reflect evidence of competition between the metal and main-group fragments.

Problems

1. How many different neutral square-pyramidal cluster compounds can you "synthesize" on paper using the isolobal principle, the cluster electron-counting rules, and SiH and CpRh fragments only.
2. The Fe(CO)$_4$ fragment can add PMe$_3$ or two H atoms to give (CO)$_4$FePMe$_3$ or the dihydride (CO)$_4$FeH$_2$, respectively. However, although BH$_3$ adds PMe$_3$ to give H$_3$BPMe$_3$, addition of two H atoms to give BH$_5$ does not yield a stable stoichiometry. Use the isolobal analogy to develop an explanation.
3. Using a fragment analysis and the isolobal analogy describe the bonding in the following molecules. As part of your discussion generate a main-group molecule isolobal to the inorganometallic species.

Problem 4.3

162 *Isolobal relationships between fragments*

4. Analyze the bonding of the following two molecules. On the basis of your analysis, account for the fact that the compound containing a single Cr atom absorbs at $27\,930\,\text{cm}^{-1}$ whereas the one with two Cr atoms absorbs at $16\,070\,\text{cm}^{-1}$.

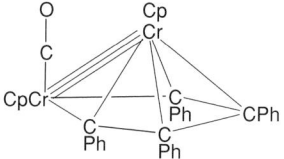

Problem 4.4

5. The organometallic compound shown below is represented with a Cr≡Cr triple bond in the literature. Describe it as a cluster compound. The measured Cr–Cr distance is 2.34 Å consistent with a triple bond. Comment on factors that favor localized Cr–Cr multiple bonding vs. delocalized cluster bonding as a mechanism for accommodating the electronic unsaturation.

Problem 4.5

6. The metal–ligand complex, $(CO)_4Fe-B(\eta^5-C_5Me_5)$ is reported as a "terminal borylene" metal complex, i.e., it possesses a B–R fragment coordinated to a single metal center. Formulate this compound as a metal cluster.

7. Consider the two WRu compounds at the top of Figure 4.19 as dinuclear metal complexes and show that they are electronically saturated 34-cve complexes. Now count the electrons at each metal center. Does this justify Stone's representation of the bonding in the WRuC three-membered ring?

8. Stone considers the compound shown below to be an electronically unsaturated 32-electron dinuclear metal complex. Justify this conclusion. The compound adds PMe_3 to yield the product shown. Is it now saturated?

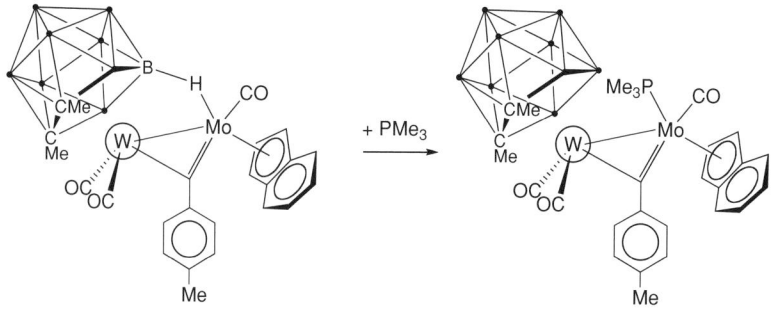

Problem 4.8

9. Consider the WFe$_2$ compound at the bottom of Figure 4.19 as a fused cluster as shown below. Does it obey the electron-counting rules? Hint: first calculate the count for an all-metal (or main-group) fused cluster system and then subtract ten electrons for each metal replaced by a main-group atom (or add ten for each main-group atom replaced by a metal).

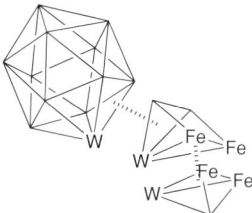

Problem 4.9

10. The complex [Pb$_5${Mo(CO)$_3$}$_2$]$^{4-}$ has been recently reported (Yong *et al.*, 2005) and exhibits the structure shown below. Does the compound obey the electron-counting rules for clusters? If not, where might one seek an explanation of its behavior? Consider explicitly tripledecker complexes as well as complications caused by separating the external cluster lone pairs from cluster bonding pairs (see Chapter 2, Problem 12). For your information, the measured distance between the Mo atoms is 3.216 Å, whereas twice the covalent radius of Mo is 2.90 Å, the Mo–Mo distance in [Mo$_2$(CO)$_{10}$]$^{2-}$ is 3.123 Å and the Mo–Mo distance in the 27-valence electron tripledecker complex CpMo(η^5-As$_5$)MoCp is 2.764 Å.

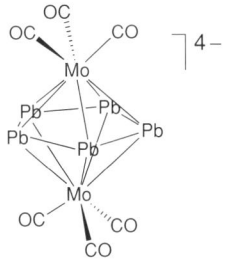

Problem 4.10

Additional reading

Section 4.1

Elian, M., Chen, M. M. L., Mingos, D. M. P. and Hoffmann, R. (1970). *Inorg. Chem.*, **15**, 1148.
Hoffmann, R. (1982). *Angew. Chem. Int. Ed. Engl.* **21**, 711.

Section 4.1.3

Seyferth, D. (1976). *Adv. Organomet. Chem.*, **14**, 97.
Nicholas, K. M. (1987). *Acc. Chem. Res.*, **20**, 207.

Section 4.1.4

Albright, T. A., Burdett, J. K. and Whangbo, H.-H. (1985). *Orbital Interactions in Chemistry*. New York: Wiley.
Evans, D. G. and Mingos, D. M. P. (1982). *J. Organomet. Chem.*, **240**, 321.
Mingos, D. M. P. and Wales, D. J. (1990). *Introduction to Cluster Chemistry*. New York: Prentice Hall.

Section 4.2

Hoffmann, R. (1981). *Science*, **211**, 995.

Section 4.3

Bowser, J. R. and Grimes, R. N. (1978). *J. Am. Chem. Soc.*, **100**, 4623.
Okazaki, M., Ohtani, T., Takano, M. and Ogino, H. (2002). *Inorg. Chem.*, **41**, 6726.

Section 4.4

Fehlner, T. P. (Ed.) (1992). *Inorganometallic Chemistry*. New York: Plenum.
Aldridge, S. and Coombs, D. L. (2003). *Coord. Chem. Rev.*, **248**, 535.

Section 4.5

Grimes, R. N. (2004). *J. Chem. Ed.*, **81**, 657.
Hawthorne, M. F. (1972). *Pure Appl. Chem.*, **29**, 547.
Hawthorne, M. F. (1975). *J. Organomet. Chem.*, **100**, 97.
Jemmis, E. D., Balakrishnarajan, M. M. and Pancharatna, P. D. (2001). *J. Am. Chem. Soc.*, **123**, 4313.

Section 4.5.2

Edwin, J., Bochmann, M., Böhm, M. C., Brennan, D. E., Geiger, W. E., Krüger, C., Pebler, J., Pritzkow, H., Siebert, W., Swiridoff, W., Wadepohl, H., Weiss, J. and Zenneck, U. (1983). *J. Am. Chem. Soc.*, **105**, 2582.
Jemmis, E. D. and Reddy, A. C. (1988). *Organometallics*, **7**, 1561.

Section 4.5.3

Stone, F. G. A. (1990). *Adv. Organomet. Chem*, **31**, 53.

5
Main-group–transition-metal clusters

A large number of mixed main-group–transition-metal clusters can be understood with isolobal ideas and the electron-counting rules. However, there is a growing number that cannot. These clusters are more closely related to metal clusters than main-group clusters. In this chapter we begin with a survey of the "rule-abiding" mixed-element clusters with emphasis on variety rather than comprehensiveness. Why a cursory survey? Because with a solid background of main-group and transition-metal cluster behavior under our belts, the isolobal analogy permits a ready understanding of mixed systems that follow the rules. Understanding permits prediction of possible stoichiometry and structure thereby generating goals for future synthesis.

It is the second compound type that constitutes an interesting challenge to our views of cluster electronic structure. With this compound type we encounter a structural response arising from main-group and transition-metal atom competition within the context of a molecular cluster. This competition generates both cluster shapes invariant to change in electron count as well as new cluster-structure types associated with unusual electron counts. Both pose a problem of interpretation. But the problem is a worthwhile one as structures that deviate from the electron-counting rules contain information on the electronic factors that underlie cluster chemistry in general. "Failure" of the rules actually constitutes a gateway leading to compounds with hybrid properties not accessible with either pure main-group or transition-metal clusters. Useful chemistry is all about properties and the more ways we develop to vary and control properties the better off we are.

5.1 Isolobal analogs of p-block and d-block clusters

The types of mixed p-block–d-block clusters that follow the isolobal principle and cluster electron-counting rules can be organized and compared in several different ways. Variables are cluster size and shape and main-group and transition-element

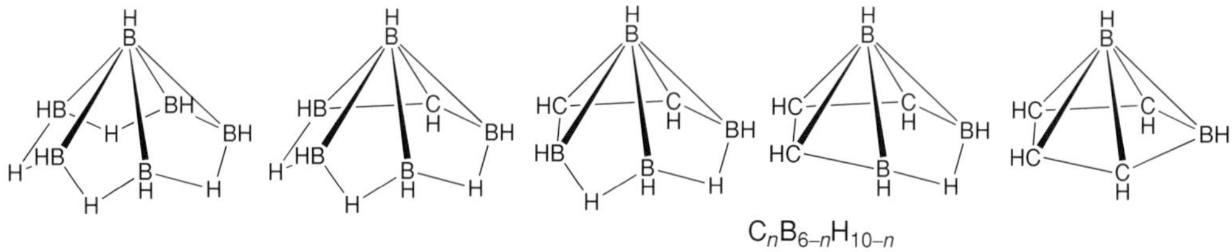

$C_nB_{6-n}H_{10-n}$

Figure 5.1

ratios and identities. We have selected systems that illustrate each of these variables thereby revealing relationships that would be hidden in the absence of the cluster connection. Metallaboranes predominate in this discussion simply because they constitute the most abundant class of four-connect and higher clusters. It is the intrinsic electronic structure of the B atom (electronegativity and less than half-filled valence shell) that fosters cluster generation and it continues to do so when combined with transition metals. However, examples containing a selection of the other p-block main-group elements will also be given at the end of this section. As might be expected, many of the latter examples are three-connect clusters.

5.1.1 A carborane model

We know from the work of the cluster pioneer, Williams, that carboranes follow the borane paradigm. For example, the series of five carboranes $C_nB_{6-n}(H_{10-n})$, $n = 0$–4, all exhibit *nido*-pentagonal-pyramidal cluster shapes in accord with a cluster electron count of eight sep. The difference between members of the series is the number of B–H–B bridges (Figure 5.1). We expect this conformity as the C and B atoms have the same number and type of valence orbitals. Do transition-metal fragments, with a larger valence set, follow the electron-counting rules as closely?

5.1.2 Closo-*clusters*

With group-8/9 metal fragments and acceptor ancillary ligands the answer is a cautious yes. In Figure 5.2 a series of clusters exhibiting a three-connect tetrahedral shape (sep = 6) is shown. For this series, the all-B member, B_4H_8, is a transient species only observed by mass spectrometry, but the metal analog, $H_4Ru_4(CO)_{12}$, is a known compound. A complete series with the same metal fragment is not known. Although Mn vs. Fe only changes the number of skeletal H, we have seen in the homonuclear cluster systems that the presence or absence of bridging H can be a significant perturbation on cluster structure for metal systems.

For the four-connect, seven-sep, *closo*-octahedral series shown in Figure 5.3 examples of the B, $[B_6H_6]^{2-}$, and metal, $H_2Ru_6(CO)_{18}$, clusters are known

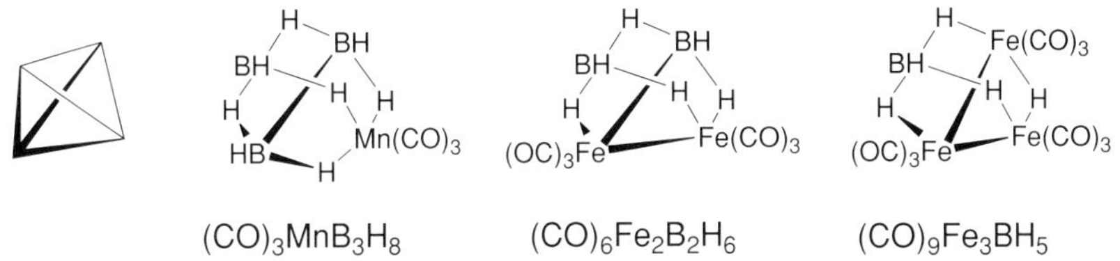

5.1 Isobal analogs of p-block and d-block clusters

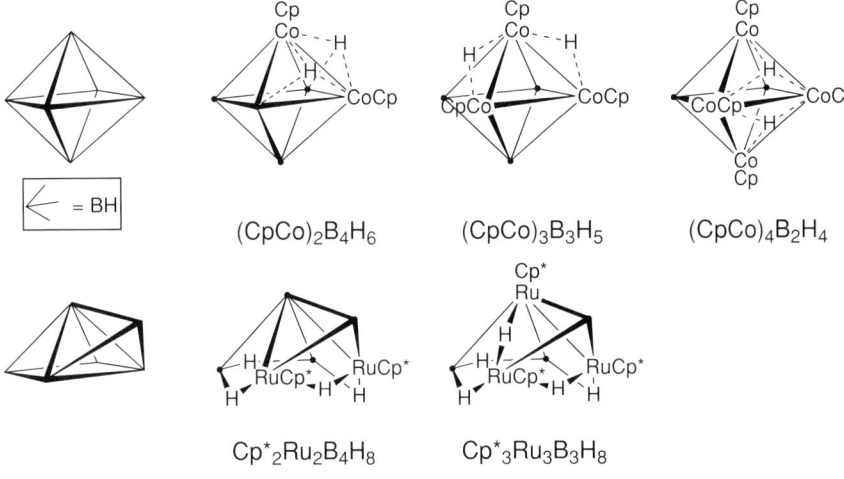

Figure 5.2

Figure 5.3

(Chapters 2 and 3). Pleasingly, examples with M/B > 1, < 1 and = 1 are seen to be direct structural analogs of each other as well as the homonuclear parents. If one moves from Co to Ru, however, a hint that the more versatile bonding properties of a metal fragment vs. a main-group fragment appears. For example, the sep = 7 ruthenaboranes shown at the bottom exhibit capped square-pyramidal rather than *closo*-octahedral shapes presumably due to the necessity of accommodating additional bridging H to meet the required cluster electron count with a group-8 Cp*M fragment. Note that the capping moiety is a B–H fragment.

One should not draw structural/stability conclusions from the absence of MB_5 and M_5B clusters in the first series. Quite possibly it is simply a question of finding the appropriate synthesis. On the other hand, there is reason to believe that there are problems with both structures. The two bridging H atoms of the known compounds are associated with M–M edges, and the MB_5 cluster with a single metal atom possesses no M–M edge for protonation. An anionic cluster would not have this problem. At the other compositional extreme, the M_5B system may not follow the main-group cluster scenario because a more stable alternative is accessible. For

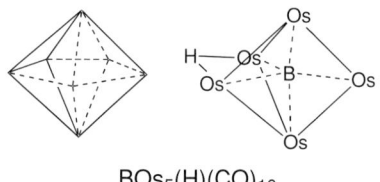

BOs₅(H)(CO)₁₆

Figure 5.4

example, the cluster Os$_5$(CO)$_{16}$BH has been reported by Shore and, in the absence of structural information, might be viewed as a member of the *closo*-series in Figure 5.3, i.e., five Os(CO)$_3$ fragments (analogous to CpCo fragments), one B–H and one extra CO equivalent to two bridging H atoms gives sep = 7. Contrariwise, its structure (Figure 5.4) shows an interstitial B and a separate metal framework H atom. Hence, rather than a two-electron B–H fragment, we have a three-electron B and a one-electron H. This gives a total of four electrons and a cluster count of sep = 8. As now there are only five cluster atoms, an *arachno*-geometry is required and observed. In fact, the cluster structure is analogous to that of COs$_5$(CO)$_{16}$, a compound with an interstitial C atom generated by the addition of CO to an Os analog of CFe$_5$(CO)$_{15}$ (Figure 3.7). This shape can be generated by removing two non-adjacent equatorial vertices from a pentagonal bipyramid as shown in Figure 5.4. Note that CRu$_5$H$_2$(CO)$_{15}$ has the same shape as COs$_5$(CO)$_{16}$ so the substitution of CO for 2H is not the cause of the non-octahedral shape. The observed cluster structure for BOs$_5$H(CO)$_{16}$ may well be Nature's response to the size mismatch between metal and main-group fragments – a response permitted by the more flexible bonding available to the metal centers.

5.1.3 Nido-*clusters*

What is the situation for more open clusters? One example was encountered in Exercise 4.4 and we now look at more metal derivatives of this seven-sep *nido*-square-pyramidal cluster (Figure 5.5). The end members of this series are B$_5$H$_9$ and H$_4$Fe$_5$(CO)$_{15}$. The latter is not known but Fe$_5$C(CO)$_{15}$ (Figure 3.7) with an interstitial C atom constitutes a satisfactory analog. For one- and two-metal fragments compounds representing all positional isomers are known. In addition, the dirhodapentaboranes shown exhibit evidence of balanced main-group–metal competition in that the 1,2-isomer exists in solution as an equilibrium mixture of two tautomeric forms. One is the direct analog of the borane parent, B$_5$H$_9$, whereas the other, with a Rh–H–Rh bridge, is not. Both 1,2-isomers rearrange to the 2,3-isomer which also exhibits a Rh–H–Rh bridge, but one which is now on the open square face of the cluster.

5.1 Isolobal analogs of p-block and d-block clusters

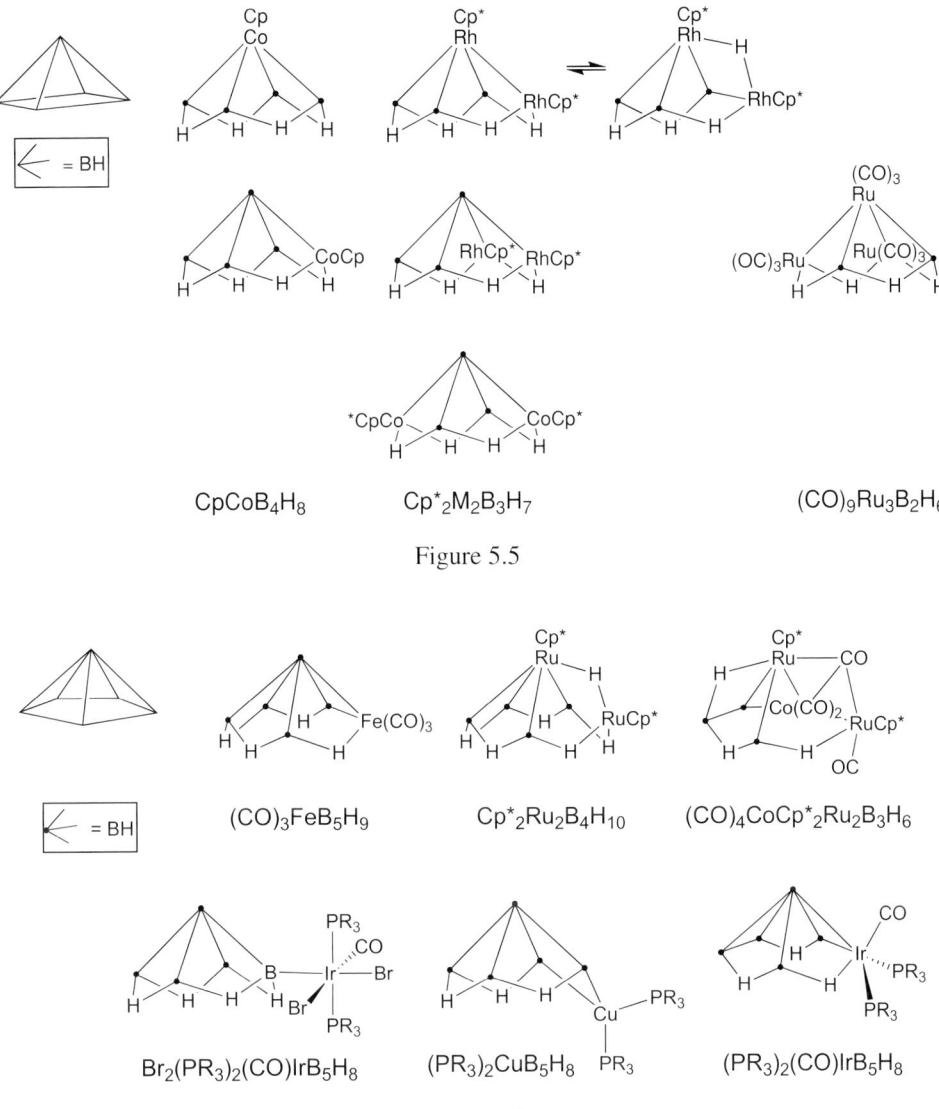

Figure 5.5

Figure 5.6

An eight-sep *nido*-series with six cluster atoms is shown in Figure 5.6. Although no metal-rich examples are known, the series supports the conclusions above. The all-B end member, B_6H_{10}, is known; however, the all metal structural analog is not. As pointed out in Chapter 3, open metal clusters analogous to the larger open boranes are not known.

This six-atom system provides an opportunity to understand how the greater bonding flexibility of the metal fragment can complicate the interpretation of a given molecular formula. The three MB_5 compounds at the bottom of Figure 5.6

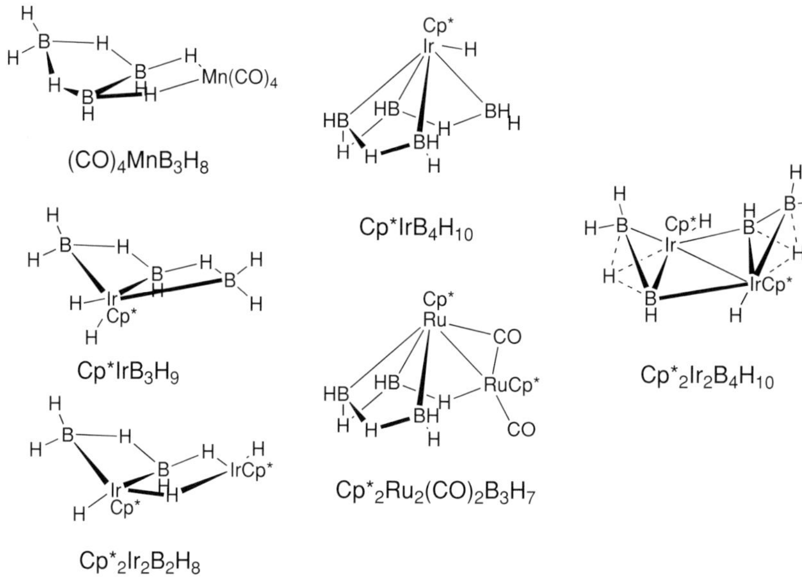

[Structures shown in Figure 5.7: (CO)₄MnB₃H₈, Cp*IrB₄H₁₀, Cp*IrB₃H₉, Cp*Ru₂(CO)₂B₃H₇ (labeled Cp*₂Ru₂(CO)₂B₃H₇), Cp*₂Ir₂B₄H₁₀, Cp*₂Ir₂B₂H₈]

Figure 5.7

illustrate three different bonding modes. At the far left we have a borane cluster with an external cluster metal substituent. In the middle is a borane with a metal fragment in a bridging position effectively replacing a bridging H atom. Finally, at the right is a true mixed cluster – a *nido*-metallahexaborane with the metal fragment incorporated into the cluster bonding network. In the first, the borane is a σ ligand of the metal or the metal fragment is a substituent of the borane cluster. In the second, the borane is a η^2-ligand to the metal or the metal fragment is a μ-bridging substituent to the borane cluster. Contrast these two with the third compound in which both metal and borane participate in a contiguous network.

5.1.4 Arachno-*clusters*

Arachno-metallaboranes are not as numerous but five related, mono- and dinuclear clusters are shown in Figure 5.7. The mononuclear open clusters are related to organometallic complexes. Thus, Cp*IrB₄H₁₀ is an analog of a metal butadiene complex and Cp*Ir(H)₂B₃H₇ is an analog of an allyl complex. Viewed as clusters they are analogs of B₅H₁₁ and B₄H₁₀, respectively. The *arachno*-Ir compounds do not lose H as readily as the Co analogs, e.g., eight-sep, *arachno*-1-Cp*CoB₄H₁₀ rapidly converts into seven-sep, *nido*-Cp*CoB₄H₈ whereas *arachno*-1-Cp*IrB₄H₁₀, shown in Figure 5.7, does not.

The dimetal species in Figure 5.7 are best compared with homonuclear borane clusters: eight-sep B₅H₁₁ for Cp*₂Ru₂(CO)₂B₃H₇ and nine-sep B₆H₁₂

5.1 Isolobal analogs of p-block and d-block clusters 171

for Cp*$_2$Ir$_2$H$_2$B$_4$H$_8$. Again the qualitative structural similarities are striking. *arachno*-Cp*$_2$Ir$_2$H$_2$B$_4$H$_8$, which is formed by the addition of borane to *arachno*-Cp*$_2$Ir$_2$B$_2$H$_8$, loses H to give *nido*-Cp*$_2$Ir$_2$B$_4$H$_8$ (Figure 5.8). The greater tendency for the iridaboranes to retain H carries over to the dimetal clusters. For example, Cp*$_2$Ir$_2$B$_2$H$_8$ is known but not Cp*$_2$Ir$_2$B$_2$H$_6$, whereas Cp*$_2$M$_2$B$_2$H$_6$, M = Co, Rh, are known but not Cp*$_2$M$_2$B$_2$H$_8$. Even though the electron-counting rules allow the ready interchange of equivalent metals, the structures of the clusters do reflect differences.

Likewise there are series in which the metal framework is constant and the main-group partner varies. For example, an all metal analog of B$_4$H$_{10}$ is known in the form of an isoelectronic set of clusters, HFe$_4$(CO)$_{12}$EH$_n$, E = B, C, N for n = 2, 1, 0, respectively, shown for E = B, n = 2 in Exercise 5.1 below. The formula, HFe$_4$(CO)$_{12}$BH$_2$, might suggest that this ferraborane is a member of a *closo*-(BH)$_n${Fe(CO)$_3$}$_{5-n}$H$_2$ series; however, the B atom is better viewed as interstitial and the BH fragment contributes four electrons to the cluster bonding of a four-, rather than five-atom cluster.

Exercise 5.1. Shown below are two organometallic clusters containing four metal atoms. One is Fe$_4$C(CO)$_{13}$ and the other [Fe$_4$(CO)$_{12}$CC(O)OMe]$^-$ made by the addition of [OCH$_3$]$^-$ to the former. Note the bridging CO ligand presumably migrates to the bare C atom before methoxide addition. Count the electrons of each in terms of five-atom M$_4$E *closo*-clusters. Then show that Fe$_4$(CO)$_{13}$C can also be considered as a four-atom *arachno*-cluster with an interstitial C atom. The experimentally determined dihedral angles of the metal "butterflies" in Fe$_4$(CO)$_{13}$C and [Fe$_4$(CO)$_{12}$CC(O)OMe]$^-$ are 101° and 130° and the Fe–C–Fe angles are 175° and 148°. Which is the more reasonable view of Fe$_4$(CO)$_{13}$C – an M$_4$ *arachno*-cluster or an M$_4$E *closo*-cluster? A closely related metallaborane cluster is also shown. The analogous "butterfly" angle in HFe$_4$(CO)$_{12}$BH$_2$ is 114° and the Fe–B–Fe angle is 162°. What can you conclude about the metallaborane?

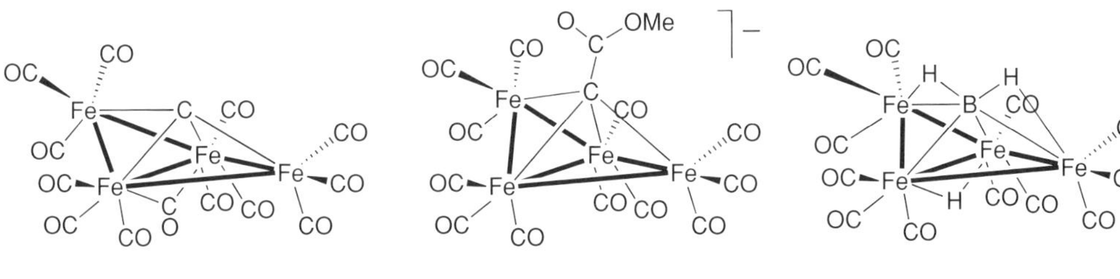

CFe$_4$(CO)$_{13}$ [Fe$_4$(CO)$_{12}$CC(O)OMe]$^-$ HFe$_4$(CO)$_{12}$BH$_2$

Exercise 5.1

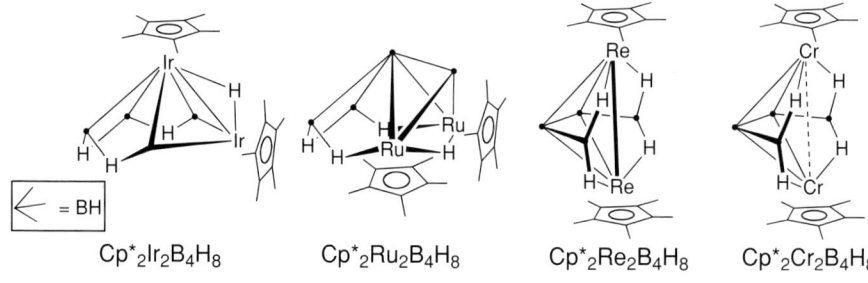

Figure 5.8

Answer. In an ideal four-atom *arachno*-"butterfly" cluster the dihedral angle is 109° whereas in an ideal five-atom *closo*-cluster the dihedral angle is 140°; a difference of 31°. The two organometallic clusters exhibit a difference in the "butterfly" dihedral angle of 29° consistent with considering the carbide cluster as a four-atom *arachno*-cluster and the methoxide adduct as a five-atom *closo*-cluster. The geometric parameters for the metallaborane are in between and insufficient in themselves to define the proper cluster description. However, considering that the difference between *arachno*-HFe$_4$(CO)$_{12}$CH and HFe$_4$(CO)$_{12}$BH$_2$ is one E–H–Fe bridging H, it is reasonable to describe the ferraborane as an *arachno*-four-atom metal cluster. Further evidence is the fact that addition of two Rh carbonyl fragments leads to 1,2-*closo*-[Rh$_2$Fe$_4$(CO)$_{16}$B]$^-$ that rearranges to the more stable 1,6-isomer. The latter is unambiguously characterized as an octahedral metal cluster with an interstitial B atom.

5.1.5 Metal variation

A particularly revealing series of compounds is Cp*$_2$M$_2$B$_4$H$_8$, M = Ir, Ru, Re, Cr (Figure 5.8). We have already seen in earlier chapters that the number of bridging H atoms as well as the nature of the metal and B ancillary ligands can result in a non-trivial perturbation of geometric structure. In this series, however, only the identity of the metal is varied; hence, variations in geometry and properties can be directly attributed to the participation of the metal. The first two compounds follow the rules. Cp*$_2$Ir$_2$B$_4$H$_8$ is a sep = 8 (cve = 48) six-atom *nido*-cluster with a geometry in accord with the electron-counting rules. The Ru compound is a capped *nido*-cluster with sep = 7 (cve = 46) which is permitted for metal but not pure borane clusters. In both cases we conclude that at the level of geometry, the metal t$_{2g}$ levels behave as if they are filled and cluster non-bonding.

The 44-cve Cp*$_2$Re$_2$B$_4$H$_8$ structure exhibits yet a different geometry from those of the Ir and Ru compounds. If one does not draw a Re–Re bond, the structure would be *nido* and an isomer of nine-sep Cp*$_2$Ir$_2$B$_4$H$_8$. For metallaboranes this

5.1 Isolobal analogs of p-block and d-block clusters

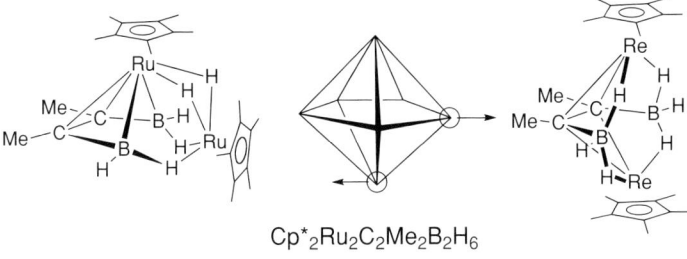

Cp*$_2$Ru$_2$C$_2$Me$_2$B$_2$H$_6$

Figure 5.9

type of isomerism is possible. A known example is shown in Figure 5.9 where two clusters of composition Cp*$_2$Ru$_2$B$_2$H$_6$C$_2$Me$_2$ (eight sep, 48 cve) are found to exhibit the two different *nido*-geometries that can be generated by removing one vertex from a pentagonal bipyramid. One mimics the structure of Cp*$_2$Ir$_2$B$_4$H$_8$ and the other the shape of Cp*$_2$Re$_2$B$_4$H$_8$. However, there is no M–M bond in the ruthencarborane to the right in Figure 5.9. Is there one in Cp*$_2$Re$_2$B$_4$H$_8$? The Re–Re distance is consistent with a Re–Re bond and on that basis rules out a *nido*-structure. With a Re–Re bond the cluster can be considered a bicapped tetrahedron with a B–H fragment capping each of the two Re$_2$B faces of the primary Re$_2$B$_2$ tetrahedron shown in bold in Figure 5.8. The required sep is six but note that it is the borane fragment that provides all six sep suggesting that the metal fragments contribute three orbitals each but no electrons. This is consistent with a filled, cluster non-bonding Re t$_{2g}$ levels, i.e., a simple isolobal relationship still holds.

A six-sep bicapped tetrahedron is the most condensed geometry possible for a six-atom cluster. But Nature doesn't recognize this logical limit and provides us with 42-cve Cp*$_2$Cr$_2$B$_4$H$_8$. We have the structure, now we must interpret it. Before doing so, let's consider the possibilities. Suppose we continue to keep three filled, non-bonding t$_{2g}$ levels on the two metal atoms. Each Cp*Cr fragment then is a −1 electron donor to cluster bonding! One way to generate an extra pair of electrons is to use one two-electron B–H fragment as a three-electron interstitial B atom and a one-electron metal hydride. A quick look at the structure shows that Nature does not adopt this option. Suppose we drop the condition of three filled, non-bonding t$_{2g}$ levels. In that case we might suggest a bicapped tetrahedron with a Cr=Cr double bond. There is a precedent. Forty-four-cve *nido*- Cp*$_2$Cr$_2$(CO)C$_4$Ph$_4$ (Chapter 4, Problem 5) exhibits a square-pyramidal structure with a Cr≡Cr triple bond. But the Cr–Cr distance in Cp*$_2$Cr$_2$B$_4$H$_8$ is longer than the Re–Re distance in Cp*$_2$Re$_2$B$_4$H$_8$! Clearly there is no localized multiple bond between the metal atoms.

Let's look at what does change in going from the Re to the Cr compound. True, the cluster structure is qualitatively like that of the Re compound, but is elongated

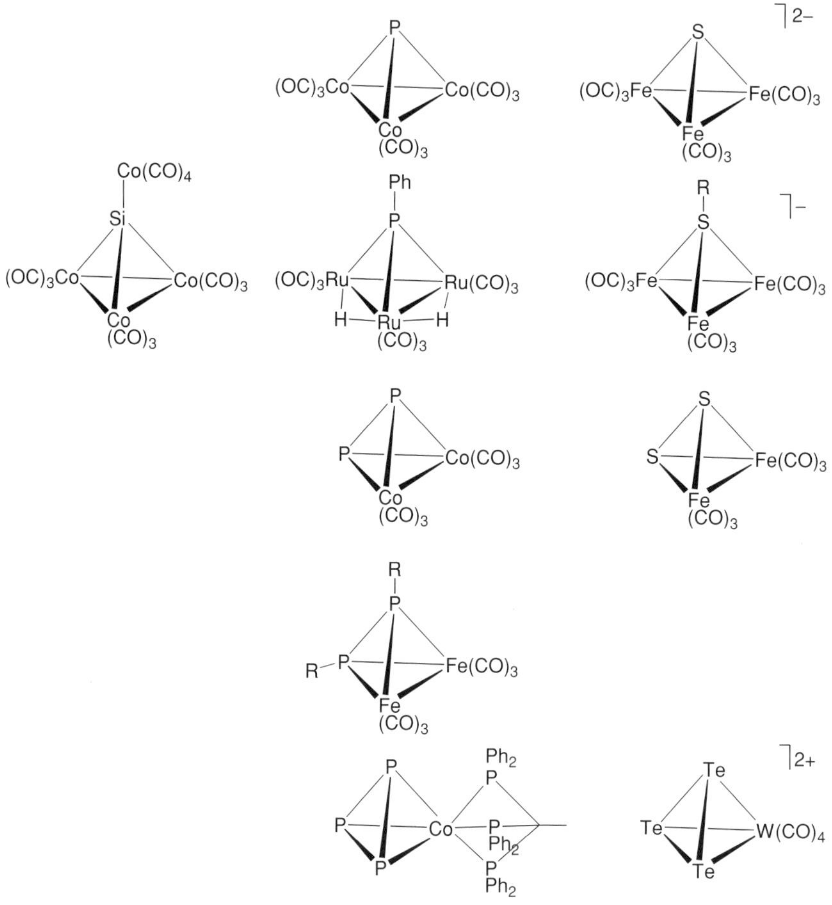

Figure 5.10

in the axial direction relative to the more compact Cp*$_2$Re$_2$B$_4$H$_8$. There is another important difference as well. The chemical shifts and ^{11}B–^{1}H coupling constants suggest large Re–H character for the "bridging H" in the case of Cp*$_2$Re$_2$B$_4$H$_8$, and large B–H character for the "bridging H" in the case of Cp*$_2$Cr$_2$B$_4$H$_8$. Definitely there are subtle electronic effects that are not explained by simply counting electrons. Approximate calculations suggest that stretching the cluster along the M–M axis and moving the bridging H from M to B generates a larger HOMO–LUMO gap for 42 rather than 44 cve. Reduced to the simplest terms, one of the largely metal-based orbitals, which is filled in the Re compound, is empty in the Cr compound. Whether one wishes to consider this another case of "into the t$_{2g}$ set" as discussed in Chapter 5 or not, it is clear that this compound is unsaturated in the same sense as Cp*$_2$Cr$_2$(CO)C$_4$Ph$_4$. The difference is that rather than exhibiting a localized M–M multiple bond, the unsaturation is expressed in distortions of the

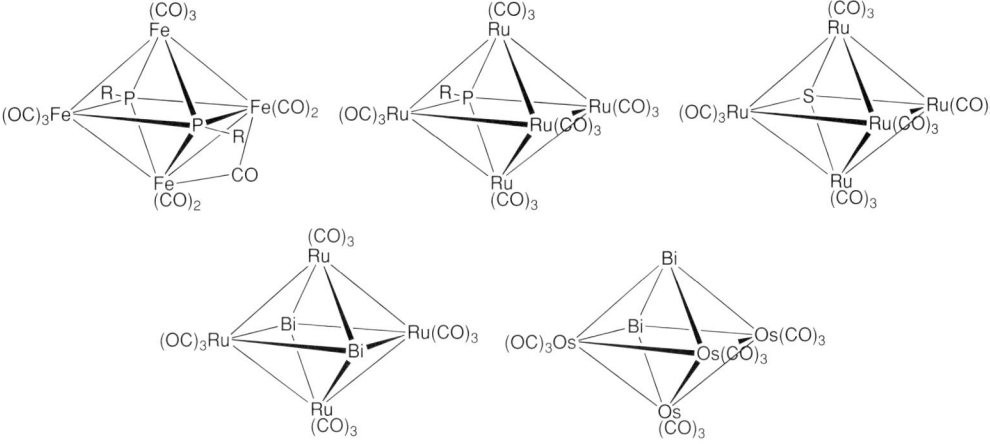

Figure 5.11

cluster structure itself. With this compound, then, we enter the realm of the rule-breakers – main-group–transition-metal clusters that exhibit geometric structures that differ from our expectations based on the cluster electron-counting rules.

5.1.6 Other p-block–d-block clusters

Before examining the rule-breakers, we need to take a more extensive look at p-block elements other than B – how do they conform to isolobal ideas in a cluster context? For the four-atom tetrahedral clusters there is a large selection of known compounds to peruse. In Figure 5.10 examples of E_nM_{4-n}, $n = 1$–3, containing representative elements from groups 14–16 are shown. Application of the isolobal principle allows us to consider $Co_3(CO)_9SiCo(CO)_4$ analogous to $Co_3(CO)_9SiR$. The $Co(CO)_4$ fragment is isolobal to a one-electron–one-orbital substituent and bound to the Si atom via a Si–Co two-center–two-electron bond. These electron-precise compounds lend themselves to a metal–ligand view as well, e.g., the 50-cve $[SFe_3(CO)_9]^{2-}$ M_3E tetrahedral cluster can be viewed as a 48-cve trimetal complex with a four-electron μ_3-S ligand.

Although the number of compounds is smaller, there is still a good selection of mixed main-group–transition-metal clusters with five and six vertices. In Figure 5.11, the sets chosen show that an octahedral, four-connect cluster geometry is supported by more elements than just B. In contrast to the situation with B, octahedral clusters containing more than three p-block elements other than B are not yet known. The 76-cve M_5E and 1,6-isomers of the 66-cve M_4E_2 octahedral clusters can also be viewed as 74-cve square-pyramidal and 62-cve square metal clusters with μ_4-bridging ligands. The two Bi derivatives provide an interesting example of isomeric forms reminiscent of 1,2-$C_2B_4H_6$ and 1,6-$C_2B_4H_6$ with which they are isolobal. The E_n, $n = 4, 5$, ring–metal derivatives, e.g., Figure 4.12, can be viewed as p-block element-rich *nido*-ME_n clusters. A more extensive compilation of this class of compound has been made by Housecroft where examples containing the heavier main-group elements as well as a variety of other cluster shapes including metal clusters with interstitial main-group atoms are described.

Exercise 5.2. Consider $Fe_3(CO)_9N_2R_2$, $Cp_2Co_2B_8H_{12}$, $Cp_4Co_4B_2H_2PPh$, $Co_3(CO)_9GeFe(CO)_2Cp$, $Cp_3Co_3(CO)B_3H_3$ and $Fe_2(CO)_6S_2R_2$ with structures shown below. Compute the sep or cve counts (choose the one that suits your convenience) and justify observed cluster shapes. Note: for the sep count, first divide the cluster into a set of metal and main-group fragments. How many of these compounds can be reasonably considered metal–ligand complexes?

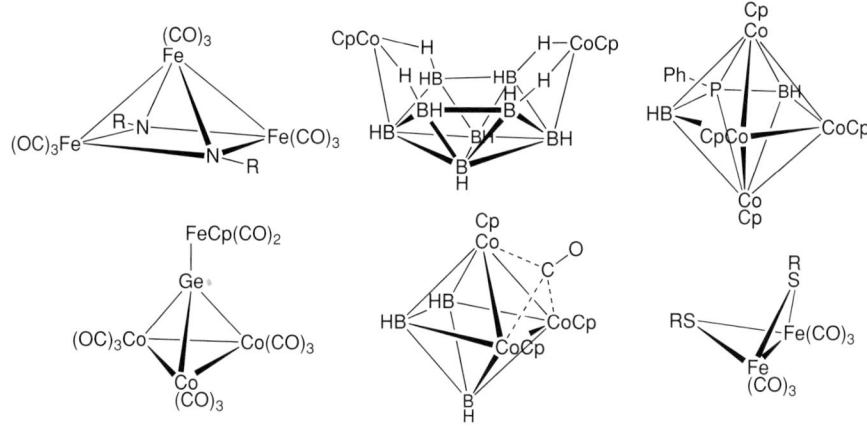

Exercise 5.2

Answer. $Fe_3(CO)_9N_2R_2$; sep: $3 Fe(CO)_3 + 2 NR = (3 \times 2 + 2 \times 4)/2 = 7$, $n = 6$, *nido*-octahedron with one vertex unoccupied; cve: $3 Fe + 9 CO + 2 N + 2 R = 24 + 18 + 10 + 2 = 54$ ($14n + 2 - 12 = 74$ cve for square pyramidal M_5 cluster – 20 for two M to E conversion = 54). This cluster may also be considered as a three-metal complex with triply bridging four-electron NR ligands ($24 + 18 + 8 = 50$) obeying the 18-electron rule ($2 \times 17 + 16 = 50$). $Cp_2Co_2B_8H_{12}$; two CpCo plus eight BH two-electron fragments and 4 bridging H, sep = 12, *nido* based on 11-vertex deltahedron with the six-connect vertex unoccupied. $Cp_4Co_4B_2H_2PPh$; four CpCo and two BH two-electron fragments, one PPh four-electron fragment, sep = 8, *closo* based on pentagonal bipyramid; $Co_3(CO)_9GeFe(CO)_2Cp$; three $Co(CO)_3$ three-electron fragments, one GeR three-electron fragment ($CpFe(CO)_2$ is a one-electron fragment like an R group), sep = 6, tetrahedral with six two-center edge bonds;

5.1 Isolobal analogs of p-block and d-block clusters

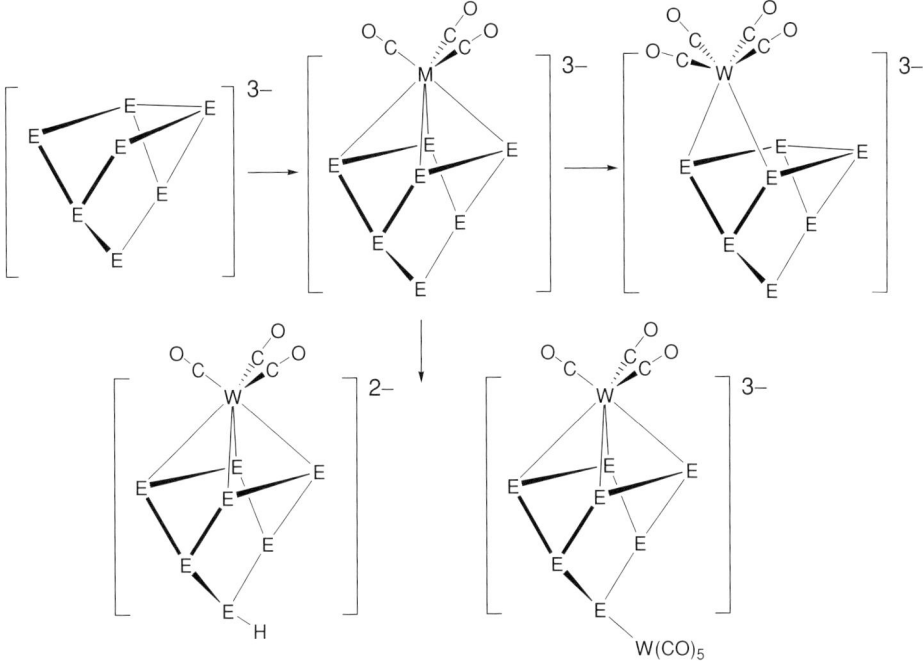

Figure 5.12

Cp$_3$Co$_3$(CO)B$_3$H$_3$; three CpCo and three BH two-electron fragments + a two-electron CO ligand, sep = 7, *closo*-octahedral; Fe$_2$(CO)$_6$S$_2$R$_2$; two Fe(CO)$_3$ two-electron and two SR five-electron fragments, sep = 7, *arachno* based on octahedron with two adjacent vertices unoccupied, a metal dimer with two bridging SR ligands (16 + 12 + 6 obeying the 18-electron rule (2 × 17 = 34).

5.1.7 Bare p-block–d-block clusters

Mixed cluster systems in which the main-group atoms lack external ligands are developing rapidly. Our first example from Eichhorn is a simple one and typical of a route that provides an alternative to the reaction of, e.g., P$_4$, with sources of transition-metal fragments. The Zintl ion used, [E$_7$]$^{3-}$, E = P, As, Sb, is a two- and three-connect cluster. A solution of [E$_7$]$^{3-}$ in an amine is used to displace a labile ligand from a transition-metal complex containing both labile and non-labile ligands. A set of clusters with bare group-15 networks incorporating a M(CO)$_3$, M = Cr, Mo, W, fragment is generated (Figure 5.12). These are most profitably viewed as metal complexes of bare group-15 ligands. Thus, the η^4-metal–ligand interaction becomes analogous to that of a rectangular [B$_4$H$_4$]$^{6-}$ interacting with [BH]$^{2+}$ (Figure 2.14). Formally, the ligand is a six-electron donor to

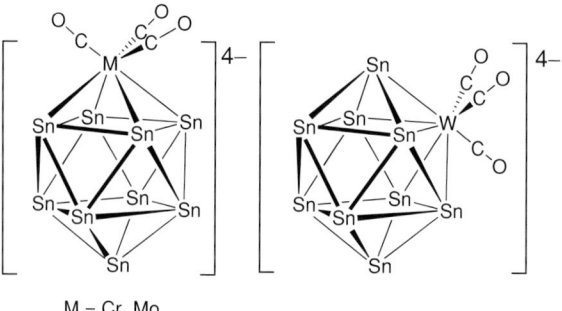

Figure 5.13

the 12-electron metal fragments. Increasing the electron count by a pair of electrons by addition of a CO ligand to the metal center causes two M–E bonds to rupture and one E–E bond to form. This generates the original structure of the $[E_7]^{3-}$ ion now acting as a simple bidentate ligand to the metal center. A high-lying lone-pair orbital associated with the two-connect P atom at the base of the $[(CO)_3WP_7]^{3-}$ complex is the site of electrophilic attack leading to P–H, P–R and P–M(CO)$_5$ derivatives.

Naked group-14 atoms are isolobal to BH fragments; hence, they possess a tendency to form cluster shapes with four-connect and higher vertices. The following example of the approach to metal complexes via Zintl ions illustrates the possibilities. The $[(CO)_3MSn_9]^{4-}$ clusters, M = Cr, Mo, W, are synthesized from $[Sn_9]^{4-}$ and exhibit bicapped square-antiprismatic shapes (Figure 5.13). In the solid state the Cr and Mo derivatives have the metal capping the square antiprism whereas for W the metal occupies a position in the square antiprism. Hence, these are 11-sep, 52-cve *closo*-cluster isomers. An advantage of Sn as a cluster atom is that ^{119}Sn NMR permits examination of the structural behavior in solution. It is found that all three clusters exhibit effective C_{4v} symmetry. The NMR experiments show that the frameworks are dynamic in solution and that there is an equilibrium mixture of the two different isomers in solution. This implies there is little energy difference between metal placement in the 4- and 5-connect vertices. The NMR observation of fast intra-molecular exchange of the metal position between η^4- and η^5-positions via diamond–square–diamond rearrangements also means that the two isomer structures are separated by a small energy barrier. Thus the potential energy surface describing this cluster bonding network must be a fairly flat one.

5.2 Rule-breakers

In Chapters 2 and 3 we examined clusters that violated the electron-counting rules for a variety of reasons: breakdown of the separation between external and internal

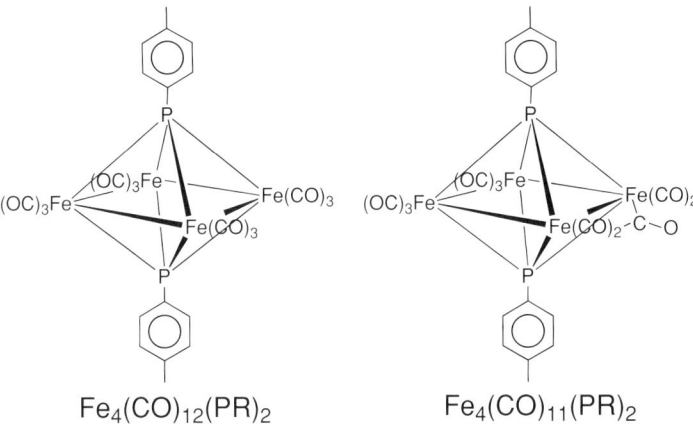

Figure 5.14

cluster bonding, e.g., strong mixing of external and internal cluster orbitals in bare clusters; idiosyncrasies of specific cluster geometries, e.g., clusters with threefold symmetry axes; intrinsically small HOMO–LUMO gaps due to weak skeletal bonding in metal clusters; and underutilization of tangential (surface) bonding orbitals. These factors apply here, and there is no need to repeat the discussion. Rather we focus on unexpected structures and compositions for deltahedral clusters caused by: the electronic flexibility of transition-metal fragments; fundamental changes in the MO ordering caused by ancillary-ligand properties; and mismatched electronic properties of main-group and transition-metal cluster fragments. Taken together, the examples that follow both review and exercise your understanding of the differences between main-group and transition-metal fragments in a cluster environment – differences implicit in the material of Chapters 2 and 3.

5.2.1 Variable electron count with constant shape

We begin with a straightforward example of rule-breaking that is representative of the problem. The two clusters shown in Figure 5.14 illustrate two octahedral *trans*-M_4E_2 clusters differing only by the presence or absence of one CO ligand. Verify for yourself that the cve or sep counts of the six-atom clusters are 66 and 68 or seven and eight. There are dozens of clusters of this type known with one of these two electron counts and *trans*-M_4E_2 cluster geometry. How does the same geometry support two electron counts?

At one level the explanation is trivial. As pointed out in the discussion of the differences between three-connect and four-connect clusters, it is possible to generate the same geometry with differing electron counts. Thus, if we consider the clusters in Figure 5.14 as M_4 metal clusters then one can generate a square shape

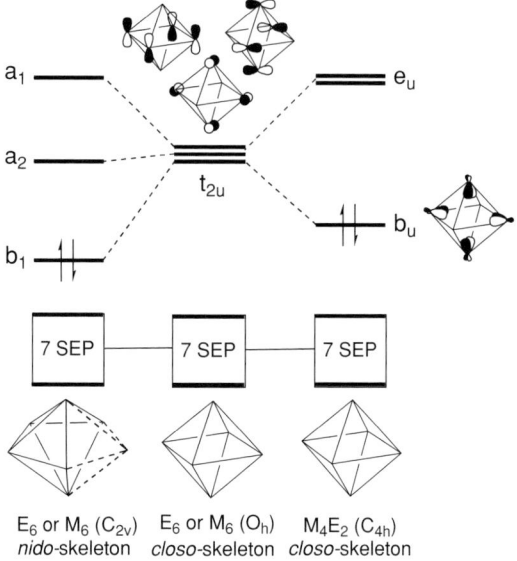

Figure 5.15

by considering a square M_4 ring that obeys the 18-electron rule ($16 \times 4 = 64$ cve) or one that is generated by removing two *trans*-vertices from an octahedral cluster ($86 - 24 = 62$ cve). If one considers the triply bridging PR groups as four-electron ligands, then the cve counts observed for four-atom clusters $Fe_4(CO)_{12}(PR)_2$ and $Fe_4(CO)_{11}(PR)_2$ are $4 \times 8 + 12 \times 2 + 2 \times 3 = 64$ and 62, respectively. But let's look at the problem more carefully from the perspective of an octahedral M_4E_2 cluster to look for a more satisfying reason.

First, go back to the MO diagram of a regular homonuclear E_6 or M_6 octahedral cluster which obeys the electron-counting rules. Take $[B_6H_6]^{2-}$, for example. The seven sep lie in bonding orbitals which are separated by a large energy gap from the antibonding t_{2u} LUMO, as shown in Figure 5.15. Each of the three components of this degenerate LUMO can be described as a π antibonding (π^*) orbital localized on one of the three squares of which the octahedron can be constructed. Have a look at the orbitals in Figure 5.15 where they are drawn for the case of E_6 clusters. Addition of two electrons to this seven-sep cluster induces a Jahn–Teller distortion. One octahedral edge opens to generate a *nido*-bipentagonal pyramid which satisfies the electron-counting rules (left side of Figure 5.15).

Why doesn't the same situation occur for the eight-sep $Fe_4(CO)_{12}(PR)_2$ cluster? It has the additional sep but there is no edge opening of its octahedral framework. The answer is found in the fact that the degeneracy of the LUMO is already removed in this species by the presence of two atom types in the cluster (see right-hand side of Figure 5.15). No additional distortion is needed. Well, you'll say that this is

5.2 Rule-breakers

obvious because a *trans*-M_4E_2 octahedron has less symmetry than an E_6 or an M_6 cluster. Indeed, orbital degeneracy splitting due to the deviation from the high ideal polyhedral symmetry is present in almost all the mixed main-group-metal clusters. The electron-counting rules work for most heteroatomic clusters and this splitting can be discarded since it does not change the cve count. OK, we agree, but what is important to notice here is that the splitting of the degeneracy is large. So, why is it so large in these *trans*-M_4E_2 octahedra? Have a look again at the drawing of the t_{2u} orbitals of the E_6 cluster in Figure 5.15 and then at the M_4E_2 cage. There are two identical M_2E_2 squares and one M_4 square which circle the octahedron. So, the degeneracy splitting will give two degenerate $\pi^*(M_2E_2)$ and one $\pi^*(M_4)$ orbitals. Due to the nature of the metal π AOs, the overlap in the latter orbital cannot be pure π^* on the M_4 square, but is intermediate between π^* and δ^*. Thus this MO is rendered weakly antibonding in contrast to the strongly antibonding $\pi^*(M_2E_2)$ MOs. The result is a single $\pi^*(M_4)$ lying in the middle of a large energy gap. We have a situation, encouted before, where two electron counts are possible. If $\pi^*(M_4)$ is occupied we have eight sep, 68 cve and if it is empty seven sep, 66 cve. In both cases, the HOMO–LUMO gap is large enough to provide stability to the octahedral structure. The energy of the $\pi^*(M_4)$ MO depends largely on the effective nuclear charge (electronegativity) of M. Without exception, the known examples with electronegative metals such as Co have the $\pi^*(M_4)$ at lower energy and an eight-sep count. The examples with electropositive metals such as Ru exhibit the seven-sep count. In the case of intermediate Fe, both electron counts are observed.

Thus, these compounds illustrate the limitations of the isolobal analogy which the electron-counting rules are based on. The take-home lesson is that the isolobal analogy, which focuses on the bonding similarities between fragments of different nature, sometimes fails because the differences override the similarities. This is what happens, for instance, with the PR and $Fe(CO)_3$ fragments in $Fe_4(CO)_{12}(PR)_2$. When mixing isolobal main-group and metal fragments one must be aware of the fact that the energy (and therefore size) of the orbitals on the different fragments may not match the same way as in the main-group model thereby leading to unexpected electron counts/cluster shapes.

But now you are scratching your head and wondering about M_6 octahedral clusters which, for group-8/9 metals and acceptor ligands, always have seven sep not eight. Why don't the three π^*/δ^* t_{2u} LUMOs lie at sufficiently low energy to be occupied? The answer lies in the fact that each of the ($\pi^*/\delta^* t_{2u}$) components is destabilized by the proper combination of the filled "t_{2g}" set associated with the metal atoms which cap the particular M_4 square considered. As a net result, these orbitals rise in energy and remain empty in the octahedral M_6 cluster. Thus, the "t_{2g}" sets participate to some extent in the cluster bonding, despite the fact that the isolobal analogy requires these orbitals to be considered non-bonding. So, we

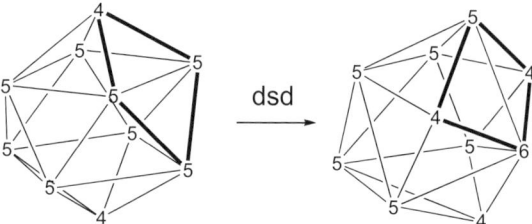

Figure 5.16

have the amusing situation (its better to smile than cry) that even though the isolobal $[B_6H_6]^{2-}/[Fe_6(CO)_{18}]^{2-}$ cluster analogy ostensibly relies on the frontier "t_{2g}" separation, it is the participation of these orbitals that makes it work! Perhaps Woolley's view (Section 3.1.3) is correct. The more complex examples which follow also emphasize the importance of all the metal valence orbitals.

5.2.2 Unusual electron count combined with unusual shape

Rather early in the history of metallaborane and metallacarborane chemistry, compounds with low formal electron counts and anomalous structures arose and provoked animated discussions. The Kennedy school identified *isocloso-*, *isonido-* and *isoarachno-*clusters. As illustrated in Figure 5.16 a ten-atom *isocloso-*shape is related to *closo* by a diamond–square–diamond, dsd, rearrangement which leaves the total connectivity of the deltahedron unchanged but changes the distribution of cluster vertex connectivities. Note that the greater the vertex inhomogeneity the greater the deviation of cluster structure from an ideal spherical shape. As a rule, in an *isocloso-*shape the metal atom occupies a site of higher connectivity than present in the equivalent borane deltahedron (Figure 2.13). The structural observations are fact. As always, the question is their implications for electronic structure in terms of metal participation in cluster bonding.

An important observation of Spencer shows that *isocloso-* and *closo-*geometries inter-convert on the addition and subtraction of a pair of electrons. As shown in Figure 5.17 this empirical inter-conversion expresses the deltahedral rearrangement illustrated in Figure 5.16. But this doesn't answer the question of whether the metal simply compensates for the loss of the electron pair by contributing another orbital containing two electrons ("into the t_{2g} set") or whether the geometric change raises the energy of a cluster MO such that it is emptied. The first explanation, equivalent to localization of the missing electrons at the metal center, is favored by Kennedy and described as an increase in metal oxidation state. The second, equivalent to delocalization of the missing electrons over the cluster framework (denoted by the term *hypercloso*), is favored by Baker and discussed in terms similar to the capping

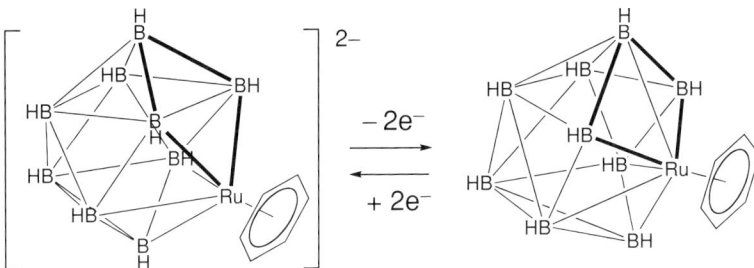

Figure 5.17

mechanism, i.e., a seven-atom capped octahedron possesses seven sep. Subsequent bonding analyses of Johnson and Mingos suggested that both mechanisms can be operative. A geometric adjustment is available to metal-containing systems which is not available to pure main-group clusters. The manner in which the metal matches the main-group fragment requires more than simplistic electron counting.

A related example, Cp*$_2$Cr$_2$B$_4$H$_8$, was mentioned in Section 5.1.5. As a member of the series Cp*$_2$M$_2$B$_4$H$_8$, M = Ir, Ru, Re, Cr (Figure 5.8), the geometric information is unambiguous as other factors (metal ancillary ligand, number of bridging H atoms) are invariant. The geometric change which accomplishes the electronic adjustment to reduced cve count is much more subtle than a dsd rearrangement, i.e., qualitatively the Re and Cr cluster shapes are the same but the quantitative differences tell the story. Shape is a complex reporter of the role of a metal in cluster electronic structure.

Even when cluster shape change is large, the electronic interpretation is a complex one. For example, consider the homologous series of rhenaboranes, Cp*$_2$Re$_2$B$_m$H$_m$, $m = 6$–10 (for $m = 6$, two adjacent BH fragments are replaced by BCl fragments in the structurally characterized derivative) in Figure 5.18. All of these clusters contain a Re–Re cross-cluster bond with total vertex connectivities equal to those of the corresponding borane deltahedron. Thus, the observed shapes are related to the borane shapes by dsd rearrangements. All the clusters are distinctly oblate and the borane fragments are ring-like. If the Cp*Re fragments have filled, cluster non-bonding "t_{2g}" sets, then each cluster is lacking three sep from the $n + 1$ required for a *closo*-cluster. Despite this difference, these clusters are thermally stable, insensitive to air and water, and exhibit no reversible redox waves by cyclic voltammetry. They have the electron count they need.

In the first member of the series, $n = 6$, the borane ring is hexagonal and planar allowing an alternate description of the compound as a 24-valence electron tripledecker complex of the type mentioned in Chapter 4 (Figures 4.12 and 4.17). The geometric/electronic mechanism for accommodating a varied electron count in a tripledecker complex can be applied to the $n = 7$–10 rhenaboranes with a

184   *Main-group–transition-metal clusters*

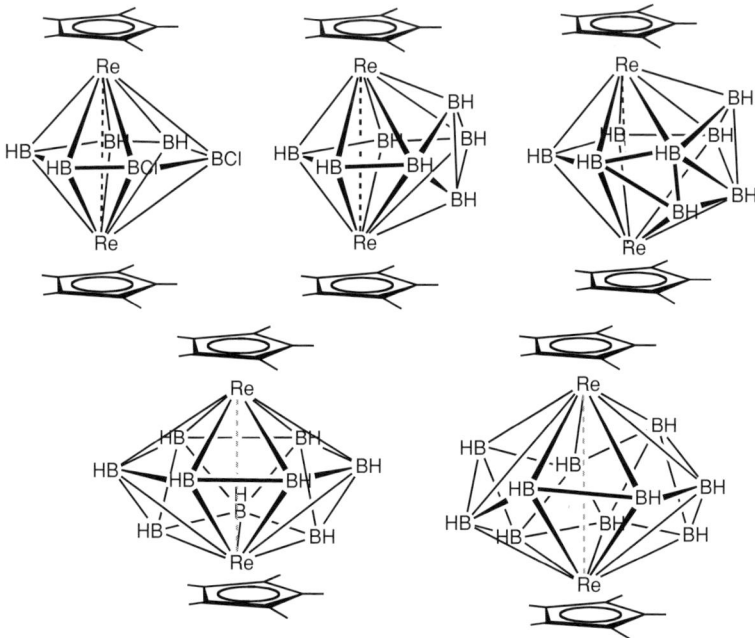

Figure 5.18

twist. Briefly, the strong bonding and antibonding interaction between the dimetal fragment and borane "ring" generates filled Cp*Re–ring bonding orbitals and three high-lying empty orbitals that in a late-metal analog would be filled (Figure 5.19). The "twist" is that the B_n fragment cannot have the same shape found in the equivalent borane. It must adopt a geometry that matches the frontier orbital set offered by the dirhenium fragment. In essence, the dirhenium fragment adopts a structure that generates the three pairs of electrons provided the borane fragment in turn adopts a shape to accommodate the metal-fragment orbitals. Clearly this electronic solution to the bonding problem involves extra metal orbitals plus cluster shape change!

If one retains formal sep count as an index, then the shape change on sep increase leading to *closo-*, *nido-* and *arachno-*clusters finds an inverse in metallaboranes. A series of known compounds containing ten occupied vertices that illustrates the new shapes available for mixed clusters is shown in Figure 5.20: $[(C_6H_6)RuB_9H_9]^{2-}$ (11 sep), $(C_6H_6)RuB_9H_9$ (ten sep), $(Cp*Ru)_2(C_6H_6)RuB_7H_7$ (nine sep) and $(Cp*Re)_2B_8H_8$ (eight sep). The electronic mechanisms are totally different. Instead of cluster opening generating low-lying orbitals to accommodate extra electrons, cluster flattening generates high-lying orbitals to accommodate the extra holes.

This can be carried further. As shown in Figure 5.21, a debor process, defined as a replacement of a BH vertex with a 2– charge, can be used to generate more

5.2 Rule-breakers

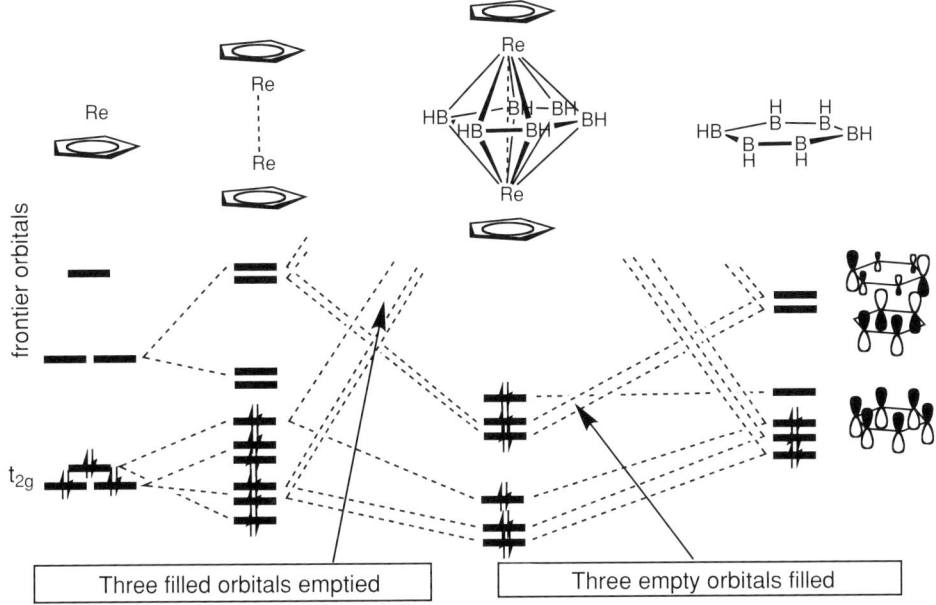

Figure 5.19

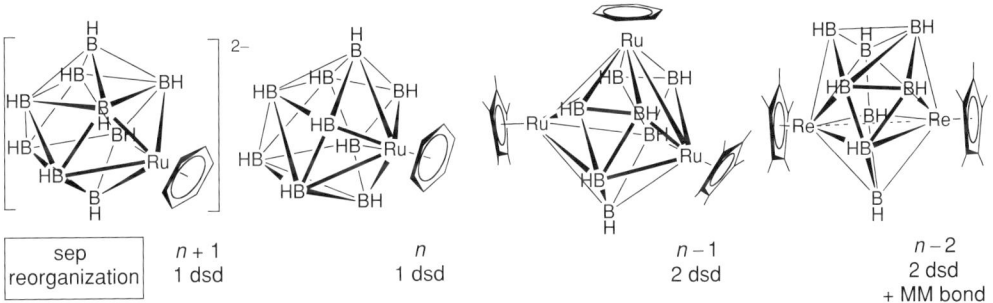

Figure 5.20

open structures all of which exhibit formal electron counts three sep lower than those prescribed by the electron-counting rules. The parent closed cluster in this series is the eight-vertex hexagonal bipyramid with a cross-cluster M–M bond. The series generated is represented by three structurally characterized compounds: $Cp^*_2Re_2B_6H_4Cl_2$, $Cp^*_2Re_2B_5H_2Cl_5$ and $Cp^*_2Re_2B_4H_8$ which exemplify the structure types and constitute six-sep clusters ($n - 3$ sep) with *closo-*, *nido-* and *arachno-*shapes. Presumably the same type of Re–B electronic interactions generate the three additional high-lying, empty orbitals required in each case. Alternative descriptions of the *nido-* and *arachno-*clusters as bicapped trigonal-bipyramidal and bicapped tetrahedral clusters match the formal sep counts; however this hides information.

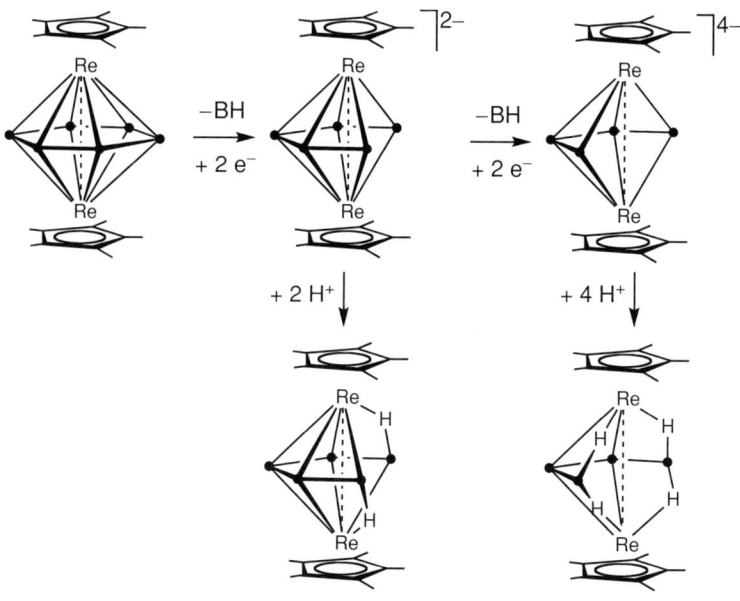

Figure 5.21

Exercise 5.3. Consider the observed geometric structure of $Cp^*_2Re_2B_7H_{11}$ with a cross-cluster Re–Re single bond and a formal sep count of nine. Derive its geometry from nine-sep $Cp^*_2Re_2B_9H_9$ by a debor process in order to see if it is an *arachno*-cluster and a member of the set of open clusters that can be derived from the homologous series shown in Figure 5.18.

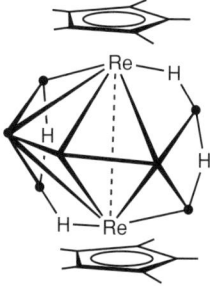

Exercise 5.3

Answer. The connectivity of $Cp^*_2Re_2B_7H_{11}$ cannot be generated from the observed structure of $Cp^*_2Re_2B_9H_9$ by removing two adjacent BH fragments (the two adjacent four-connect vertices cause the problem). However, if one first performs a dsd rearrangement as shown below, then the observed shape can be generated by two debor operations. Protonation on the open face produces the observed structure.

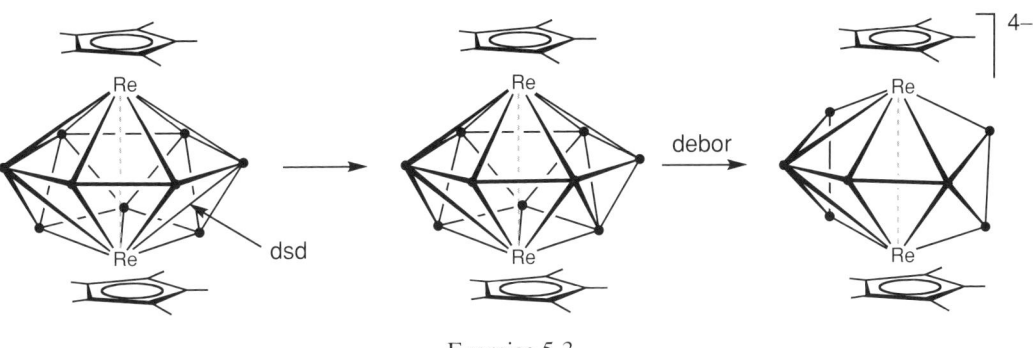

Exercise 5.3

In conclusion, transition-metal fragments do mimic isolobal main-group fragments in cluster chemistry but the transition-metal fragment can foster the existence of stable, non-canonical shapes as well. This is a satisfying conclusion simply because one of the important aspects of organometallic chemistry is the ability to modify the properties of bound C fragments by coordination to an appropriate transition-metal fragment. The fact that one can do likewise with other p-block elements suggests similar scope for new metal-modified main-group chemistry.

5.2.3 Clusters with internal atoms

Transition-metal clusters with single internal (interstitial) main-group atoms were discussed in Chapter 3. There we found that they could be dealt with at the level of electron counting by simply adding the valence electron count of the interstitial atom to the cluster electron count. The electronic problem created by sets of internal atoms, e.g., $[Al_{69}R_{18}]^{3-}$ (Chapter 2) or $[Pt_{38}(CO)_{44}H_m]^{2-}$ (Chapter 3), was found to be a much more difficult one to understand because of the fundamentally different ways in which inner and outer shells could interact. The two limiting cases were: radial bonding only between outer and inner cluster (electron count determined by the inner cluster count and the outer shell treated as external ligands); and radial plus tangential bonding for the outer shell (electron count determined by the outer cluster with the inner serving as "interstitial" atoms). Here, we illustrate this structural motif with two examples of mixed main-group–transition-metal clusters.

The first cluster follows the electron-counting rules for metal clusters containing an interstitial atom, but the replacement of a large metal atom with a small main-group atom has a dramatic geometric consequence. Two compounds (Figure 5.22) illustrate the point. For the structure at the left, to generate the observed electron count, take a B atom-centered trigonal-prismatic metal cluster (cve 90) and replace one metal fragment by an isolobal main-group fragment (cve 80). However, the observed structure has the cluster B–H capping a rectangular face of the trigonal

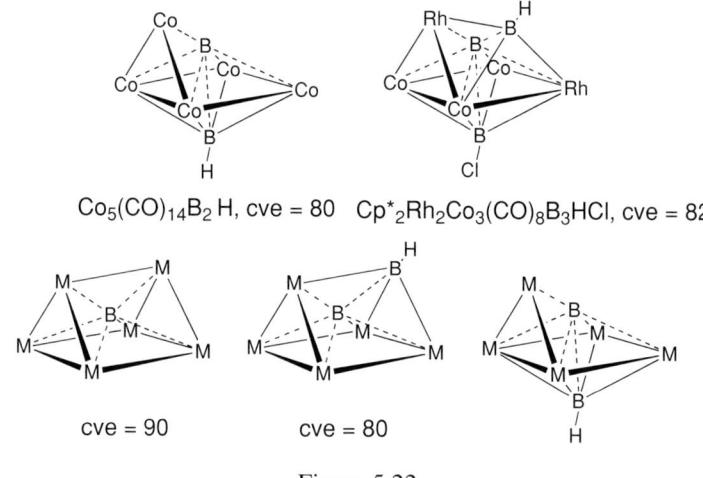

Figure 5.22

prism – perhaps you recognize this as a variant of the octahedron-capped square-pyramid relationship. The presumed driving force for the rearrangement is the presence of the interstitial B atom which would hold the corner B–H in the trigonal-prismatic geometry too far from the three nearest-neighbor metal atoms to form strong bonds. For the cluster at the right simply begin with a capped trigonal-prismatic metal cluster with 102 cve.

The second example by Eichhorn is described as "interpenetrating As_{20} fullerene and Ni_{12} icosahedra in the onion-skin $[As@Ni_{12}@As_{20}]^{3-}$ ion." The outer cluster layer is a three-connect cluster which is a pentagonal dodecahedron of I_h symmetry (Figure 5.23). The internal cluster is a centered metal icosahedron; hence, the outer and inner clusters are related in the same sense as a cube and an octahedron, i.e., switching faces and vertices generates the other shape. The As–As distances of the outer shell are bonding but are longer than typical As–As bonds. The Ni–Ni distances are similar to those found in centered icosahedral Ni clusters with carbonyl ligands (Chapter 3). The high symmetry of this cluster system makes it an outstanding example of the esthetic beauty possible in cluster chemistry.

The cluster $[As@Ni_{12}@As_{20}]^{3-}$ constitutes a problem of the type encountered for metal clusters except the outer cluster is now made up of main-group atoms only. The cve count is 268. If treated by the first limiting model (atoms of outer As_{20} shell act as ligands only) the count should be $170 + 40 = 210$ ($\Delta + 2n_s$) which is obviously not the case. The As_{20} shell is a three-connect cluster that meets the required cve count of 100 ($5n = 5 \times 20$). Hence, it needs no interstitial atoms and the second limiting model does not work either. However, the As_{20} pentagonal-dodecahedral cluster can act as a dodecadentate ligand to the internal cluster as it still has 20 lone pairs in radial functions. Hence, the predicted count is 170 (to

5.2 Rule-breakers

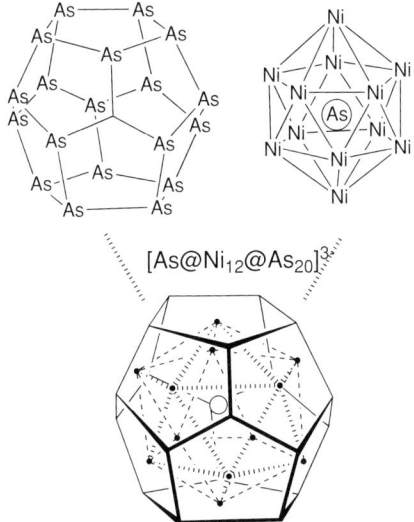

Figure 5.23

account for the internal icosahedral metal cluster bonding) + 40 (the radial outer to inner cluster bonding) + 60 (the 30 tangential bonds of the outer cluster) = 270. Based on this analysis the cluster is short of two electrons. Approximate MO calculations show the source of the problem is related to the high cluster symmetry. Although icosahedral Ni_{12} carbonyl clusters obey the cve = 170 count, the CO ligand envelope has symmetry lower than I_h and a proper match of ligand-donor to cluster-acceptor orbitals can be achieved. In the case of the As_{20} ligand a five-fold degenerate set of acceptor orbitals of the Ni_{12} cluster is excluded by symmetry but a four-fold degenerate set, allowed by symmetry, replaces it. Consequently only a total of 19 outer to inner donor–acceptor interactions are possible thereby reducing the count to 268, i.e., the count is $170 + 38 + 60 = 268$.

5.2.4 Cubane clusters

Behavior characteristic of transition metals but not p-block elements shows up in another guise in M_4E_4 cubane clusters. Two common types are illustrated in Figure 5.24. These clusters are characterized by versatile redox activity with little geometric change and a number of electron counts for the same cluster shape and connectivity. For example, cubane clusters with electron counts (treated as four-metal clusters) from 52 to 72 cve are known. As we have seen, this implies a small HOMO–LUMO gap for the structure type.

The classic work of Dahl provides the $[Cp_4Fe_4S_4]^{n+}$ clusters which have been structurally characterized for $n = 0$, 1 and 2. In addition, the structure of $Cp_4Co_4S_4$

190   *Main-group–transition-metal clusters*

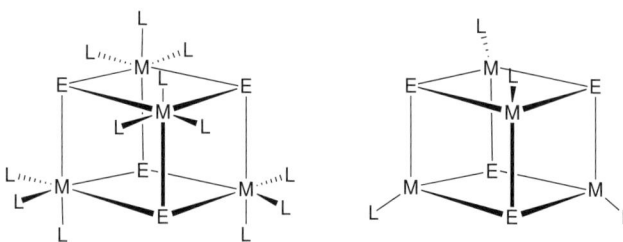

Figure 5.24

has been determined. These four compounds constitute a set of clusters with identical ancillary ligands and cubane cluster shapes but four different electron counts (66, 67, 68 and 72). Although the qualitative shape of the cubane cluster remains the same, the M–M distances reflect changes in cluster electronic structure accompanying the addition and loss of electrons. The purpose of this section is to examine the electronic origins of the multioxidation state/multielectron count behavior of these cubane clusters.

Given the importance of M–S cubane clusters as electron-transfer agents in living organisms, there is no lack of treatments of cubane clusters in the literature. We present only high points in the context of the cluster problem using cubanes with the metal centers in near octahedral coordination environments vs. metals in near tetrahedral environments (Figure 5.24). The following exercise gives a perspective of cubane clusters in terms of the electron-counting rules.

Exercise 5.4. Consider the two limiting cubane cluster geometries shown below. The first has a fully bonded metal tetrahedron with four triply bridging ligands whereas the latter has no M–M bonds and is a three-connect M_4E_4 cubic cluster. Calculate the expected cve and sep cluster counts of each and thereby illustrate the limits on electron count for this cluster type.

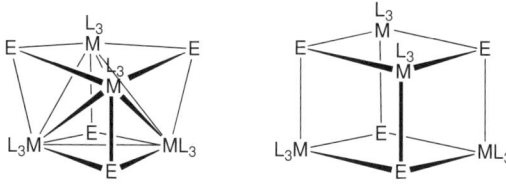

Exercise 5.4

Answer. A tetrahedral metal cluster is expected to exhibit a cve of 60 (sep of six); whereas if viewed as a tetracapped tetrahedron the sep remains six but the cve is 68 (tetra-M-capped M_4 tetrahedron is $60 + 4 \times 12 = 108$ from which we subtract 4×10 electrons in the change from four M to four E). A M_8 cube should have a

5.2 Rule-breakers

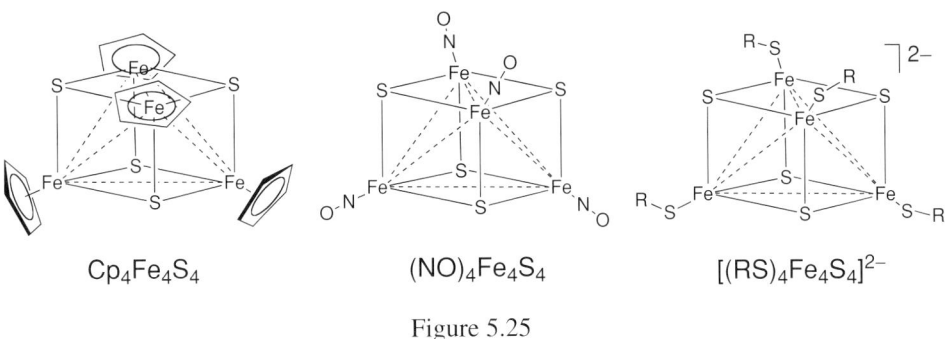

Figure 5.25

cve of $15 \times 8 = 120$: thus, M_4E_4 should have cve $= 120 - 40 = 80$ which is exactly $68 + 12$ (the number of electrons necessary to break all the two-center–two-electron M–M bonds). The sep count is 12.

Figure 5.24 shows the two structure types under consideration and Figure 5.25 shows three known compounds that will serve as our working examples. Two approaches have been used effectively: examination of the interaction of a M_4E_4 cluster core with an appropriate ligand (12 L or four L) set (Dahl) and examination of the interaction of four ML_xS_3, $X = 3$ and 1, fragments in the formation of a bridged tetramer (Harris). Of course, both generate the same MO diagram but the latter approach has two significant pedagogical advantages. Perturbation of the metal orbitals by the strong M–L interactions is treated first and perturbation by the weaker M–M interactions second. The first interaction is no different from the MO energy level diagrams for mononuclear metal complexes (octahedral and tetrahedral) well known to students.

Consider the formation of a cubane cluster by the sharing of each E ligand of an octahedral ML_3E_3 complex between three adjacent metal centers to form a three-connect cube. We begin with the splitting pattern of an octahedral complex and ignore the differences between the ligands L and E. This produces two sets of orbitals loosely defined as "e_g" (d_{z^2} and d_{xy} for the local coordinate system indicated in Figure 5.26) and "t_{2g}" (d_{xz}, d_{yz} and $d_{x^2-y^2}$). In addition there are six low-lying metal–ligand bonding orbitals. It is the "t_{2g}" set of each metal that is properly oriented for M–M bond formation. When interaction between the metal orbitals of four such units in a tetrahedral array is turned on, the "t_{2g}" set is split into bonding and antibonding blocks both of which lie below the "e_g" block. In addition to the metal orbitals shown, there are 24 low-lying occupied M–L(E) bonding orbitals and 16 high-lying M–L(E) antibonding orbitals (the eight "e_g" orbitals also have M–L(E) antibonding character). Hence, valence electrons in excess of 48 will occupy the metal orbitals shown in Figure 5.26 in the order M–M bonding, M–M antibonding and M–M non-bonding.

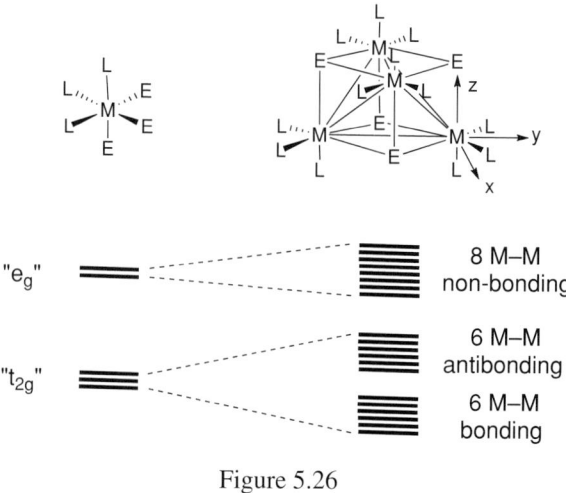

Figure 5.26

The $[Cp_4Fe_4S_4]^{n+}$, $n = 0$, 1 and 2, clusters with 20, 19 and 18 electrons in the metal MOs have fully occupied M–M bonding (12 electrons) and partially occupied M–M antibonding MOs (eight, seven and six electrons). Note that a Cp ligand is considered to occupy three metal coordination positions equivalent to three L ligands of an octahedral L_3MS_3 fragment. Consistent with this analysis the observed Fe–Fe distances are: $n = 0$, two bonding and four non-bonding; $n = 1$, two bonding, two non-bonding and two of intermediate length; $n = 2$, four bonding but long and two non-bonding. Symmetry prevents the geometric effects of removing electrons to be localized between pairs of metal centers; however, the overall increase in M–M bonding on reducing the valence electron count is clear. For $Cp_4Co_4S_4$ all four Co–Co distances are non-bonding consistent with 24 electrons in the metal orbitals and completely filled bonding and antibonding sets.

Is there a 60-cve four-metal cubane cluster with pseudo-octahedral metal centers and the six M–M bonding orbitals just filled? Of course! One example is $Cp_4Fe_4(CO)_4$ with four face-bridging CO ligands. Another is a metallaborane which connects these compounds to those discussed earlier in this chapter. This compound, $Cp^*_3Ru_3Co(CO)_2(BH)_3(CO)$ shown in Figure 5.27, possesses 60 cve and a cubane geometry with four M–M bonds based on observed M–M distances. Both the triply bridging BH and CO moieties are two-electron ligands. Viewing this metallaborane as a cubane rather than a tetracapped tetrahedron with six sep provides a pleasing solution to a long-standing anomaly in metallaborane chemistry – the existence of $Cp_4Co_4B_4H_4$ and $Cp_4Ni_4B_4H_4$ with eight and ten sep, respectively. The expected *closo*-cluster count for an eight-vertex cluster is nine sep. Both exhibit dodecahedral geometries (Figure 5.27) with Co occupying the four vertices of connectivity

5.2 Rule-breakers   193

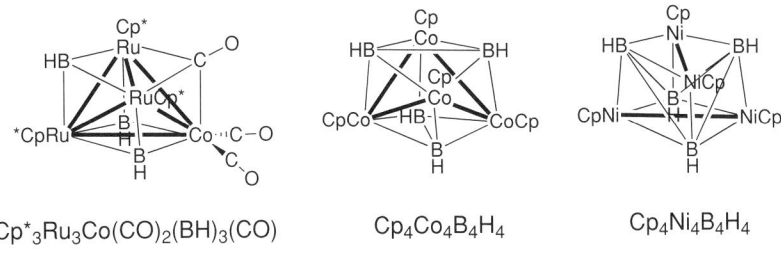

Figure 5.27

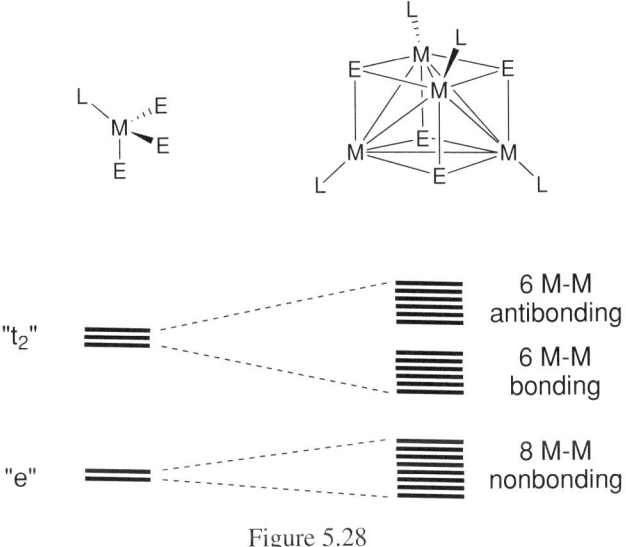

Figure 5.28

five and Ni the vertices of connectivity four. Note that counted as a four-metal cluster, the cobaltaborane has 64 cve and four Co–Co bonds (two less than a fully bonded tetrahedron) whereas the nickelaborane has 68 cve and two Ni–Ni bonds (four less than a fully bonded tetrahedron). Hence, these two compounds fit well the octahedral metal-fragment model of a cubane cluster. Other explanations have been published, but this one, due originally to Kennedy, has the merit of placing these compounds in the set of well-studied cubanes.

Move now to the cubanes with metals in a tetrahedral array of ligands. We begin with acceptor ligands. As illustrated in Figure 5.28, the splitting pattern is three over two which places the "t_2" set over the "e" set, i.e., the order of filling of metal-based orbitals is M–M non-bonding, bonding and antibonding. Again it is the "t_2" orbitals that generate the M–M interactions whereas the "e" orbitals are effectively M–M nonbonding. Keep in mind that this grouping is pretty crude as it ignores mixing possible in the lower symmetry of the real clusters. Now look at the chosen example, $(NO)_4Fe_4S_4$, with 60 cve. Of these, 32 occupy M–L bonding orbitals leaving 28 to

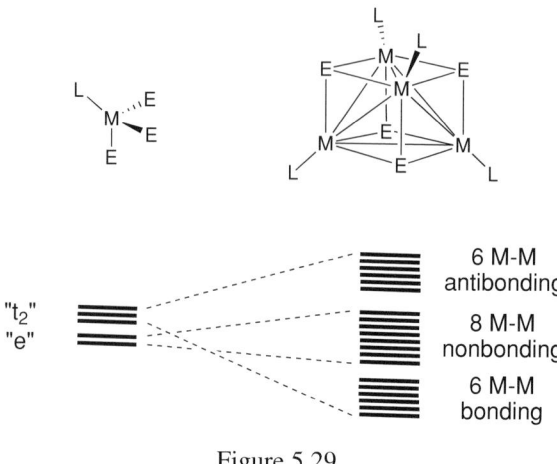

Figure 5.29

occupy the metal orbitals. Hence, the eight non-bonding orbitals are filled, as are the six bonding orbitals yielding a fully bonded Fe$_4$ tetrahedron consistent with the six observed Fe–Fe bonding distances.

The situation changes in going to $[(RS)_4Fe_4S_4]^{2-}$ with tetrahedral metal sites but π-donor ligands. As shown in Figure 5.29, the "e" non-bonding set is pushed to higher energy and falls in between the M–M bonding and M–M antibonding sets derived from the "t$_2$" sets. This 54-cve four-metal cluster again uses 32 electrons for M–L bonding leaving 22 for the metal orbitals. The six Fe–Fe bonding orbitals are filled, as are five of the eight Fe–Fe non-bonding orbitals. Six Fe–Fe bonding distances are observed. The fact that the highest occupied orbitals are M–M non-bonding has important consequences for reduction and oxidation. On addition or loss of electrons, the largest changes are in the M–L distances rather than the M–M distances (the orbitals derived from the "e" set have M–L antibonding character) consistent with the qualitative orbital diagram in Figure 5.29. Hence, if there is a situation where electron transfer without large structural change is required, this type of cubane cluster fits the bill. The model predicts that this cluster type can accommodate cve counts ranging from 44 to 60 while retaining a completely bonded metal tetrahedron. Quite a difference from expectations based solely on main-group cluster models but fully in accord with those developed in Chapter 3 where differences between clusters with acceptor vs. donor ancillary ligands can be large.

It should be clear that many other variations of cubane-cluster electronic structure are possible. Harris treats other metal-fragment geometries as well as mixed cubanes where the metal fragments have different ligand geometries. Like the cluster electron-counting rules, the beauty of this approach is that it not only allows one to make sense of existing compounds but it stimulates the imagination concerning

5.2 Rule-breakers

how one might design cubanes with desired electronic properties. Of course, the orbital diagrams are crude and as the cubanes become more complex one expects poorer agreement. In such cases one needs to go to real calculations for more precise information.

Exercise 5.5. In the cubane clusters, $Ti_4S_4Cp_4$ and $Cr_4S_4Cp_4$, the observed M–M distances are 2.93 (2) and 3.01 (4) vs. 2.83 (6) Å, respectively, where the number in parentheses is the number of M–M distances of that value. Provide a description of the electronic structure of these cubane clusters that rationalizes the different M–M distances.

Answer. Consider each cubane cluster as formed by the linking of four pseudo-octahedral $CpMS_3$ fragments via triply shared S atoms. Harris' Figure 5.26 applies. $Ti_4S_4Cp_4$ and $Cr_4S_4Cp_4$ have total electron counts of 52 and 60, respectively, of which 48 fill M–L bonding orbitals. This leaves four and 12 electrons, respectively, to populate the lowest lying set of six M–M bonding orbitals shown in Figure 5.26. Two and six M–M bonds result consistent with the observed patterns of M–M distances in the pair of compounds.

The cubane geometry is just one variant of this cluster type. The rich variety of cluster shapes possible is illustrated by Fe–S clusters, four examples of which are shown in Figure 5.30. Note that coordination geometry around the Fe centers in

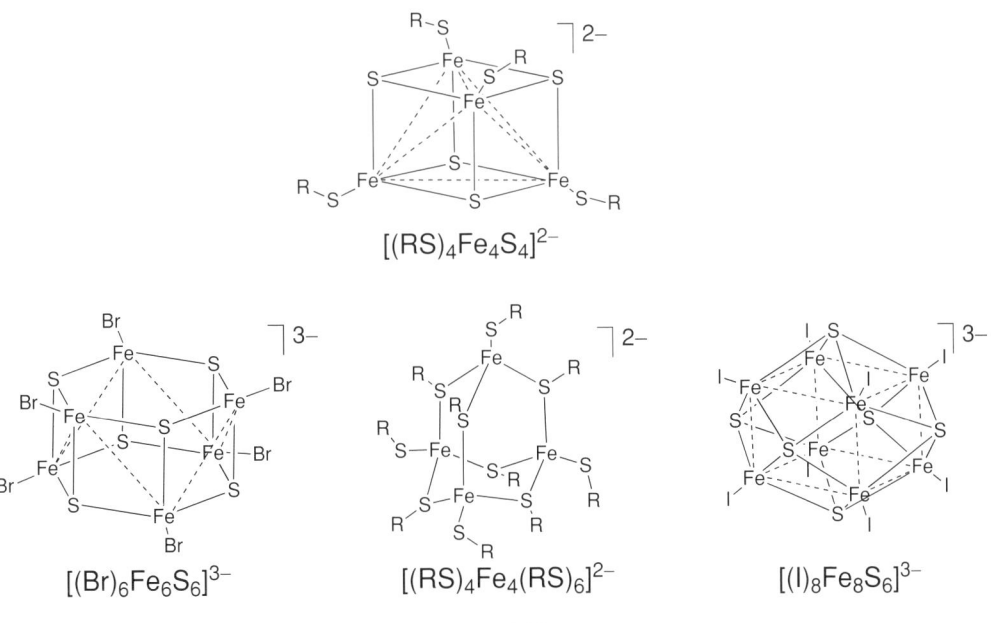

Figure 5.30

all of these compounds is tetrahedral and, hence, each can be thought of as being generated from LFeS$_3$ fragments.

Exercise 5.6. Consider the structure of [Br$_6$Fe$_6$S$_6$]$^{3-}$ as a hexagonal prism (Figure 5.30) and calculate the cve count required if it were a metal analog of the (RGe)$_{12}$ cluster which has the same shape (Chapter 2, Figure 2.3). Compare this number with that calculated for the observed composition of the Fe–S cluster. Comment on the possible source of the discrepancy. Try to develop an alternative description of the bonding if the Fe–S compound is considered as a six-metal complex with the S atoms triply bridging ligands. Note that the dotted Fe–Fe distances shown in the structure drawing are M–M bonding.

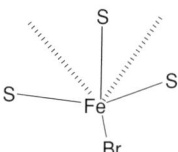

Exercise 5.6

Answer. The (RGe)$_{12}$ cluster has $5 \times 12 = 60$ cve so the M$_6$E$_6$ analog should have $60 + 6 \times 10 = 120$ cve. [Br$_6$Fe$_6$S$_6$]$^{3-}$ has $6 \times 1 + 6 \times 8 + 6 \times 6 + 3 = 93$. The comparison suggests the Fe–S cluster lacks valence electrons and possesses additional bonding not present in (RGe)$_{12}$. Note that the Fe atoms are arranged in a trigonally compressed octahedral array inter-penetrating a trigonally compressed S$_6$ octahedral array. This geometry provides six Fe–Fe bonding distances and six Fe–Fe non-bonding distances (Kanatzidis *et al.*, 1986). Let us consider the molecule as a hexamer of tetrahedral BrFeS$_3$ fragments with π-donor ligands. As shown in Figure 5.29 the five d functions are now close in energy. To form the Fe–Fe bonds across the diagonals of the rhoms on the circumference of the hexagonal antiprism (see drawing above where the dashes indicate the Fe–Fe bonding directions), we use two orbitals for each Fe center. This leaves three non-bonding orbitals and four low-lying M–L bonding orbitals. For the hexamer six M–M bonding, 18 M–M non-bonding and six M–M antibonding orbitals in order of increasing energy are generated. Below this will lie 24 M–L bonding orbitals containing 48 of the 81 valence electrons available (S = four-electron ligand). Of the remaining 33, 12 fill the 6 M–M bonding orbitals leaving 21 in the 18 non-bonding orbitals. As the highest occupied orbitals are non-bonding, redox activity is anticipated and observed (reversible redox between three oxidation levels).

A fragment approach to this cluster type is more than an academic exercise to make connections between known clusters. Recognition that the [CN]$^-$ ligand, which is

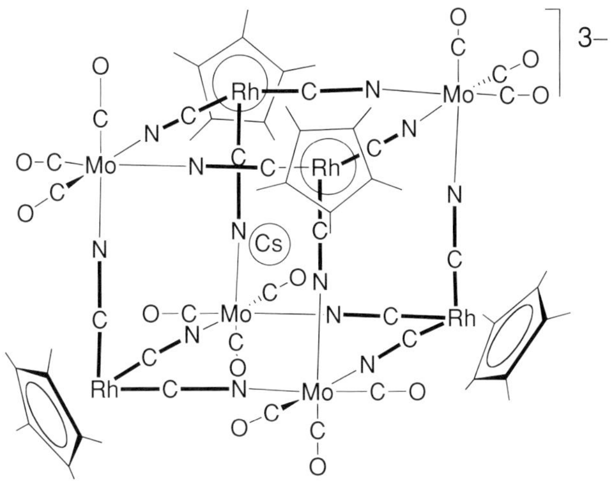

Figure 5.31

isoelectronic with CO, has significant metal-binding capabilities at both the C and N ends led to the design of isolatable building blocks for the construction of large cluster networks. It is this property of the [CN]⁻ ligand that generates extended metal–cyanide frameworks such as Prussian blue (Section 8.1.2). The extraordinary chemical and physical properties of this class of solid suggest that clusters based on the same motif will have equally interesting properties albeit modified ones. Of particular interest are the ion-exchange properties of the open networks and the magnetic behavior of high-nuclearity systems in which the metal centers have unpaired spin.

A synthetic approach utilized effectively by Rauchfuss is based on the utilization of two different, but complementary, types of cluster building blocks. Thus, to construct a cube, four inert [Cp*Rh(CN)$_3$]⁻ complexes are reacted with four labile (C$_6$H$_3$Me$_3$)Mo(CO)$_3$ complexes to yield a cube with four Rh and four Mo corners and cyanide-bridged edges (Figure 5.31). The Mo complex easily loses the arene π ligand to form an acceptor corner matched to the cyanide N atoms of the donor corner. Formed in the presence of [Cs]⁺, the cation is found in the center of the cubic cage and appears to provide a measure of stability. Alternatively, cationic complexes result if [Cp*M]²⁺ ions derived from Cp*MCl$_2$ are used for the acceptor corners rather than neutral (CO)$_3$Mo corners.

In a related approach by Zuo and Zhou the same type of donor corner, [TpFe(CN)$_3$]⁻, where Tp is a scorpionate ligand with a three-fold axis of symmetry (hydrotris(pyrazolyl)borate), is combined with a higher connectivity acceptor linker, [Cu]²⁺. As the Cu center adopts a square pyramidal (CN)$_4$Cu(H$_2$O) local coordination, the cluster structure observed is one with eight donor corners and

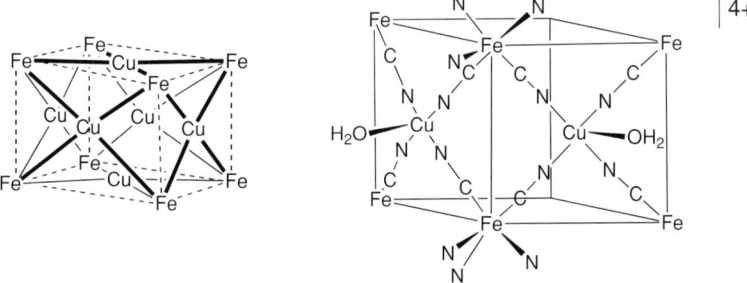

Figure 5.32

six face-capping Cu atoms (Figure 5.32, where the cubic core is shown on the left and the coordination environments of two Cu and Fe on the right). With $[Fe]^{3+}$ and $[Cu]^{2+}$ the cluster contains 14 paramagnetic metal centers and the magnetic behavior shows the presence of substantial magnetic anisotropy as well as single-molecule-magnet behavior.

5.2.5 Cubic clusters on the molecular-cluster–solid-state borderline

One of the distinct differences between clusters of metals and bulk metals is that the former have a significant HOMO–LUMO gap, whereas the latter have no band gap (Section 6.2.6). The former often give rise to defined electron counts for a given structure type leading to useful electron-counting rules. The latter permits different electron counts for the same structure or packing type. The cubane clusters with π-donor ligands and different electron counts for the same structure, discussed in the previous section, possess one rudimentary metallic property. In this section we explore this point further using larger metal clusters based on a cubic transition-metal core.

An example of the cluster type, $(CO)_8Ni_8(PPh)_6$ is shown in Figure 5.33 and you can show for yourself that with 120 cve it constitutes an example of an eight-metal cluster cube – a three-connect cluster system that can be adequately described with localized M–M bonds and the 18-electron rule. However, the attentive reader will have noted that an example of an Fe–S cluster of the same type, $[I_8Fe_8S_6]^{3-}$, with 99 cve has already been shown in Figure 5.30. In fact there are a dozen such clusters known with cve counts ranging between 99 and 120 possessing the same hexacapped-cubic metal-cluster structure. In addition, there are cubic metal clusters lacking a number of terminal ligands on the metal centers, e.g., $(PR_3)_4Ni_8Se_6$ with 112 cve illustrated in Figure 5.33.

There is general agreement on the MO structure of $(CO)_8Ni_8(PPh)_6$ shown as a block diagram in Figure 5.34. For contrast, we will use the Dahl fragment analysis

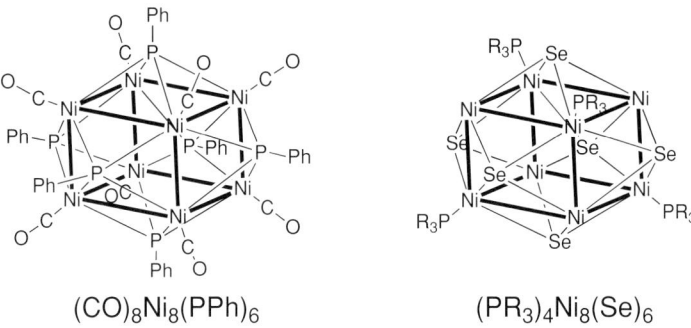

Figure 5.33

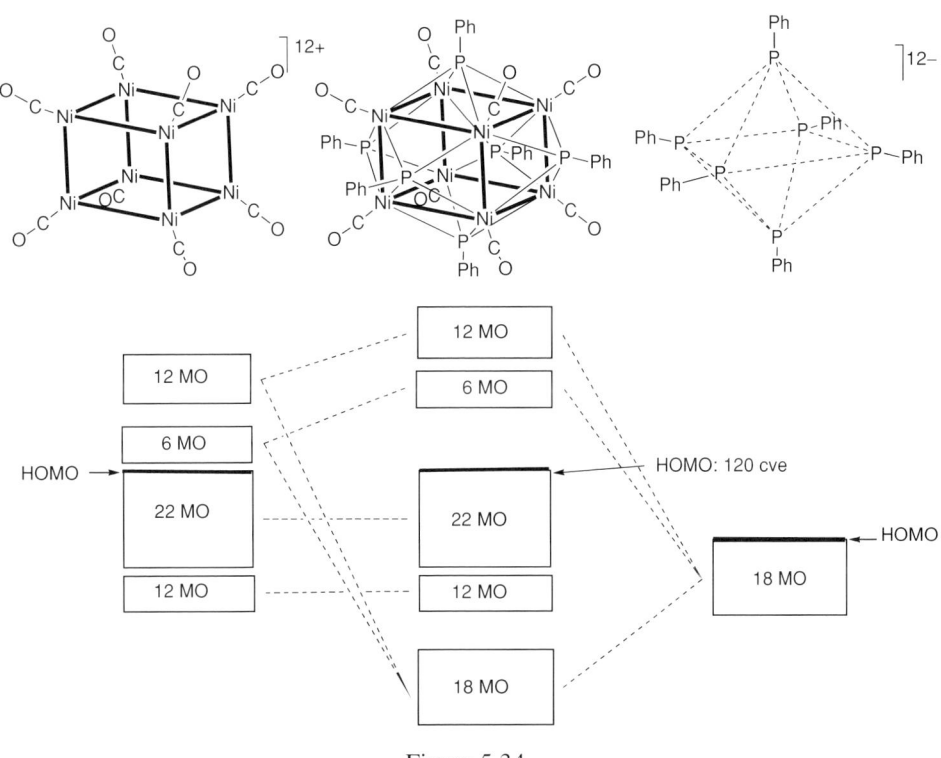

Figure 5.34

this time and consider a $[(CO)_8Ni_8]^{12+}$ cube interacting with an octahedral array of $[PPh]_6^{12-}$. A significant HOMO–LUMO gap for 120 cve is generated by the interaction of the eight face-capping PPh ligands with 12 high-lying empty metal sp acceptor orbitals plus six lower-lying empty metal d acceptor orbitals. It is the M–E interactions rather than the M–M interactions that generate the maximum cve count of 120. Larger electron counts would destabilize the cubic metal core by populating strongly antibonding orbitals. Experimentally, in terms of characterized compounds of this type, 120 cve is the upper limit and this count appears to be favored by metal ancillary ligands with π-acceptor character.

Of the two highest-lying blocks of filled MOs of mainly metal d character, the lower one with 12 MOs can be associated with the Ni–Ni bonding in the cube. The higher one consists of 22 MOs which are non-bonding or weakly antibonding. It is these orbitals that are emptied with only minor structural change for cve counts less than 120. A consequence of this model is the prediction that the lower limit of the cve count will be met when all 22 of these MOs are empty. This would correspond to a cve count of 76; however, to date the lowest observed is 99 for $[I_8Fe_8S_6]^{3-}$ which, you will note, has π-donor ligands.

There are many more aspects of the problem presented by cubic metal clusters of this type that we will not consider, e.g., larger clusters containing M_8 cubes, metal and main-group atom centered M_8 cubes, condensed cubic architectures leading all the way to cubic metal units in extended solid-state chemistry. The last serves as an appropriate point to end, as the next chapter begins our discussion of solid-state systems, and connections, such as this one, will be emphasized in Chapter 7. But before finishing, a comment on the origin of the behavior exhibited by the M_8 cubic clusters is in order. Most of the clusters with less than 120 cve exhibit open-shell electronic configurations and very similar geometries. Both properties are ultimately traceable to the high connectivity of the atoms of the cluster core which hinders structural distortion. High connectivity is a property associated with solid-state structures. So, in a real sense, these clusters do bridge to the solid state – they exhibit a fixed closed-shell electron count in the manner of "rule-abiding" metal clusters but also readily permit a range of lower electron counts with no significant gap between occupied and unoccupied orbitals in the manner of extended structures.

Exercise 5.7. Recall our serial discussions of large Al clusters in Chapters 2 and 3 and our attempts to match cve counts of $[Al_{77}R_{20}]^{2-}$ with large metal cluster models in Exercise 3.15. Do the observations on the cubane M_4 and cubic M_8 clusters discussed above provide yet another explanation of this failure and, perhaps, additional understanding? The structure is: $[Al@Al_{12}@Al_{44}(AlR)_{20}]^{2-}$ (Section 2.12.5).

Answer. In the manner of the cubic M_8 clusters, the external R ligands may impose an upper limit on cve count, but the highly inter-connected cluster core may also permit both lower counts and open-shell electronic configurations even without the five d functions of a transition metal. As shown in Exercise 3.15, if we invoke full radial and tangential bonding for the inner centered cluster but only radial bonding for the outer shell of $[Al_{77}R_{20}]^{2-}$, we obtain a cve count of 268 vs. an observed value of 253. If MOs associated with the "lone pairs" of the partially exposed second shell are lying near the HOMO–LUMO gap, some can be emptied. This appears to be the case for clusters like $[Ga_{19}R_6]^-$ discussed in Chapter 2. The resulting low electron count and the odd number of electrons would be analogous to those generated by the model obtained for $L_8M_8E_6$ cubic clusters, i.e., metal-like behavior. It also suggests any simple electron count for these giant clusters and related nanoparticles will only constitute a limit. Variable electron count must be expected for large clusters as a reflection of incipient metal properties.

Problems

1. Test the following clusters to see if each obeys the electron-counting rules. Use your choice of cve or sep count.

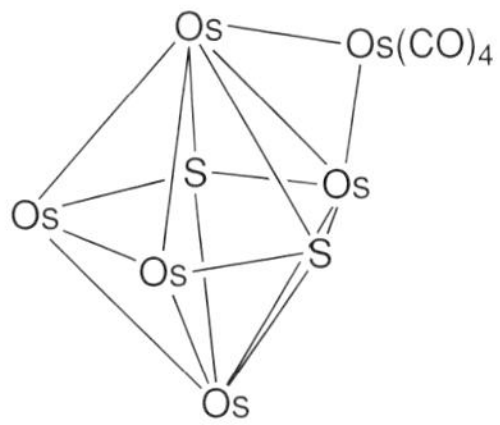

$Os_6(CO)_{17}S_2$

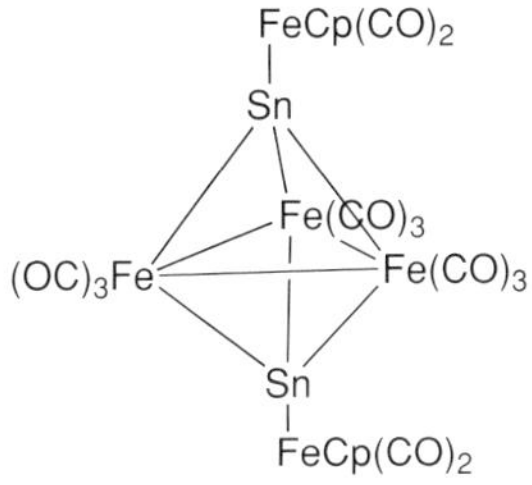

$Fe_3(CO)_9Sn_2\{CpFe(CO)_2\}_2$

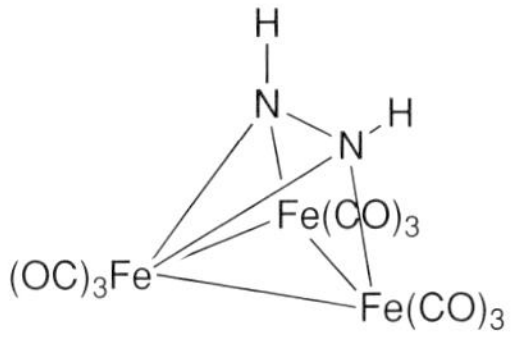

$Fe_3(CO)_9N_2H_2$

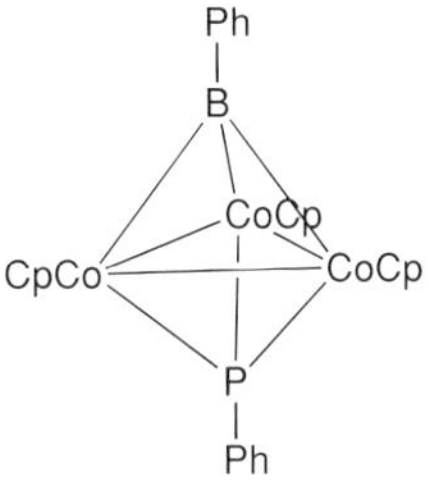

$(CpCo)_3(BPh)(PPh)$

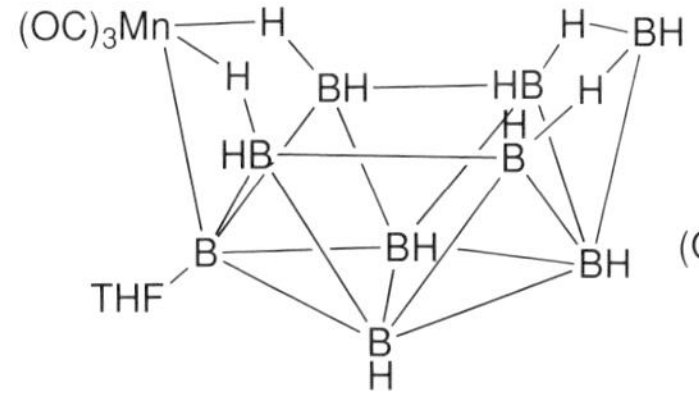

$Mn(CO)_3B_9H_{12}(OC_4H_8)$

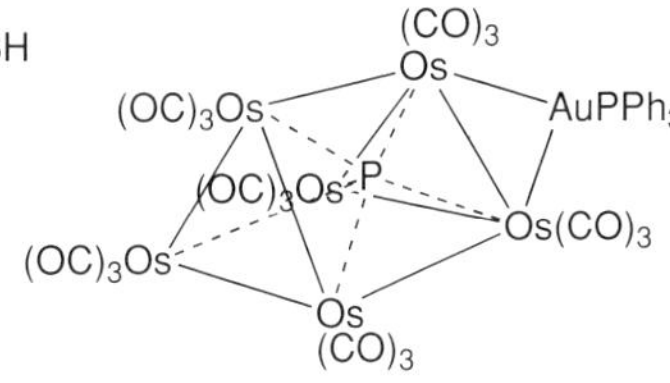

$Os_6(CO)_{18}P(AuPPh_3)$

Problem 5.1

2. Given the following molecular formulae, suggest a reasonable cluster structure for each: $[Rh_9(CO)_{21}P]^{2-}$, $Fe_3(CO)_9C_2BH_3$, $Co_3(CO)_9Bi$, $HRu_3Fe(CO)_{12}N$, $Fe_3(CO)_{12}(CH)As$.
3. The geometric structure of $Co_4(CO)_{11}Ge_2\{Co(CO)_4\}_2$ is shown below. Evaluate its shape using the electron-counting rules and discuss any discrepancies found.

Problem 5.3

4. Consider the cubane cluster $Cp_4Fe_4(CO)_4$ shown below. Count the electrons as a four-metal cluster (cve and sep) and then consider it as a cubane made up by the fusion of four pseudo-octahedral $CpFe(CO)_3$ fragments. (Okazaki et al., 1998) have shown that chemical reduction of this compound produces $Cp_4Fe_4(C_2H_2)_2$ (as they write it) possessing the dodecahedral structure shown below with the metal atoms in the five-connect vertices and the C atoms in the four-connect vertices. A dodecahedron may be considered as two inter-penetrating tetrahedra, one elongated (four-connect vertices of the dodecahedron) and one flattened (five-connect vertices). Now count the number of electrons (your choice) and the number of Fe–Fe and C–C bonds in the structure. Devise an explanation for the changes in structure (or number of Fe–Fe and C–C bonds) on reduction of the cubane cluster to the carbyne cluster.

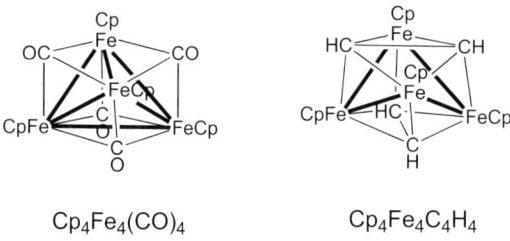

Problem 5.4

5. A cubane cluster containing a $Mo_2Ir_2S_4$ core (see structure below) has been reported (Masumori et al., 2000) to exhibit three M–M bonding distances (shown in bold lines) and three M–M nonbonding distances in the structure. Apply the Harris analysis and rationalize the number of M–M bonding interactions observed.

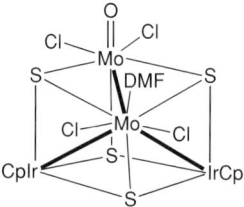

Problem 5.5

6. The structure of $[Sb_7Ni_3(CO)_3]^{3-}$ is shown below (Charles *et al.*, 1993). Discuss the composition and structure in light of metal and main-group cluster principles.

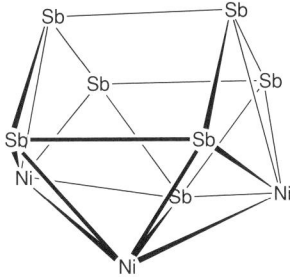

Problem 5.6

7. The cluster $Re_8In_4(CO)_{32}$ has the structure shown below (tetracapped tetrahedron). Justify its geometry based on the isolobal analogy and one form of the electron-counting rules.

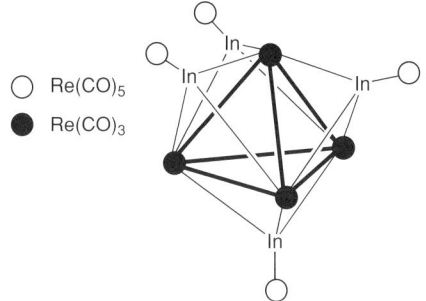

Problem 5.7

Additional reading

Section 5.1.1

Williams, R. E. (1976). *Adv. Inorg. Chem. and Radiochem.*, **18**, 67.

Section 5.1.2

Chung, J.-H., Knoeppel, D., McCarthy, D., Columbie, A. and Shore, S. G. (1993). *Inorg. Chem.*, **32**, 3391.

Section 5.1.6

Housecroft, C. E. (1992). In *Inorganometallic Chemistry*, Fehlner, T. P., (Ed.). New York: Plenum Press, p. 73.
Whitmire, K. (1988). *J. Coord. Chem.*, **17**, 95.

Section 5.1.7

Charles, S., Eichhorn, B. W., Rheingold, A. L. and Bott, S. G. (1994). *J. Am. Chem. Soc.*, **116**, 8077.

Kesanli, B., Fettinger, J. and Eichhorn, B. W. (2001). *Chem. Eur. J.*, **7**, 5277.

Section 5.2.1

Halet, J.-F., Hoffmann, R. and Saillard, J.-Y. (1985). *Inorg. Chem.*, **24**, 1695.

Section 5.2.2

Kennedy, J. D. (1998). In *The Borane, Carborane, Carbocation Continuum*, Casanova, J. (Ed.). New York: Wiley, p. 85.

Littger, R., English, U., Ruhlandt-Senge, K. and Spencer, J. T. (2000). *Angew. Chem. Int. Ed.*, **39**, 1472.

Baker, R. T. (1986). *Inorg. Chem.*, **25**, 109.

Johnston, R. L., Mingos, D. M. P. and Sherwood, P. (1991). *New J. Chem.*, **15**, 831.

Le Guennic, B., Jiao, H., Kahal, S., Saillard, J.-Y., Halet, J.-F., Ghosh, S., Shang, M., Beatty, A. M., Rheingold, A. L. and Fehlner, T. P. (2004). *J. Am. Chem. Soc.*, **126**, 3203.

Reddy, A. C., Jemmis, E. D., Scherer, O. J., Winter, R., Heckmann, G. and Wolmershäuser, G. (1992). *Organometallics*, **11**, 3894.

Callahan, K. P., Evans, W. J., Lo, F. Y., Strouse, C. E. and Hawthorne, M. F. (1975). *J. Am. Chem. Soc.*, **97**, 296.

Section 5.2.3

Moses, M. J., Fettinger, J. C. and Eichhorn, B. W. (2003). *Science*, **300**, 778.

Section 5.2.4

Trinh-Toan, Teo, B. K., Ferguson, J. A., Meyer, T. J. and Dahl, L. F. (1977). *J. Am. Chem. Soc.*, **99**, 408.

Harris, S. (1989). *Polyhedron*, **8**, 2843.

Klausmeyer, K. K., Wilson, S. R. and Rauchfuss, T. B. (1999). *J. Am. Chem. Soc.*, **121**, 2705.

Wang, S., Zuo, J.-L., Zhou, H.-C., Choi, H. J., Ke, Y., Long, J. R. and You, X.-Z. (2004). *Angew. Chem. Int. Ed.*, **43**, 5940.

Section 5.2.5

Burdett, J. K. and Miller, G. J. (1987). *J. Am. Chem. Soc.*, **109**, 4081.

Fenske, D., Ohmer, J., Hachgenei, J. and Merzweiler, K. (1988). *Angew. Chem. Int. Ed.*, **27**, 1277.

Halet, J.-F. and Saillard, J.-Y. (1999). *Metal Clusters in Chemistry*, Braunstein, P., Oro, L. and Raithby, P. R. (Eds.), Vol. 3. Weinheim: Wiley-VCH, p. 1643.

6

Transition to the solid state

The theme of this text, clusters as a bridge to solid-state chemistry, requires that we now consider the geometric and electronic aspects of substances that are solids. In doing so we will focus our attention initially on the nature of the atomic structures inside a bulk material; that is, we will completely ignore the surfaces. Towards the end of this chapter we will reincorporate surfaces into the problem and, in doing so, complete the bridge. The electronic-structure problem presented by periodic structures exhibiting extended bonding has been effectively dealt with in several earlier texts some of which are listed at the end of this chapter. These works go beyond what we need to establish our theme; however, the reader interested in more depth and breadth is referred to them.

6.1 Cluster molecules with extended bonding networks

As usual, let us begin with a discussion of geometric ideas relevant to a transition from molecular clusters to the solid state.

6.1.1 Surface vs. core atoms

In the structure of $[Al_{69}R_{18}]^{3-}$ (Figure 2.32) the number of nearest-neighbor Al atoms and bonding parameters changes in going from the outer shell made up of Al–R fragments deeper into the inner shells constructed from Al atoms alone. The internal cluster atoms display coordination numbers and inter-atomic distances more closely associated with bulk elemental Al than single-shell clusters. Is this reasonable? For the single-shell clusters discussed in preceding chapters the requirement for external ligands dominates the cluster stoichiometry/shape relationship. Except for the bare clusters, one nearest neighbor is a ligand. In a homonuclear cluster with more than one shell, those in the inner shells are surrounded by like

206   *Transition to the solid state*

atoms. Thus, if the atoms have, for example, more valence orbitals than electrons the situation is not directly moderated by external ligands. The atoms are left to deal with a local environment similar to that found in the bulk element so it is no surprise that they adopt a similar geometric/electronic strategy for dealing with it. Of course, the influence of the outer Al–R shell can extend more than a single shell deep into the cluster. Does it? The answer is not obvious. For the large Al clusters geometry suggests it does, i.e., the internal structure of $[Al_{77}R_{20}]^{3-}$ differs significantly from that of $[Al_{69}R_{18}]^{2-}$ albeit still displaying the same qualitative changes in going from outer to inner shells. But for the Au_{55} nanoclusters of Section 3.6 spectroscopic evidence suggests the first ligand-free shell has properties associated with the bulk. It is possible that the answer depends on metal type. Even so, as cluster nuclearity increases, a size will be reached for any metal where the electronic environment of the central atoms becomes effectively that of the bulk.

Flip the question of the size a cluster must be to exhibit bulk properties on its head and one realizes that clusters of lower nuclearity will express a gradation of the property as a function of size. The nature of the property in the bulk is one limit, whereas that in a single-shell cluster is another. Potential control of this size/property relationship justifies the excitement generated by the field of nanoparticles briefly touched in Chapter 3. Even if the nuclearity and geometry of a nanoparticle are known (a problem for nanochemistry), the connection between size and a given property is not a straightforward one. Think about it. For any given property the surface layers, first inner shell, second inner shell, etc., will make varying contributions. Even if we make the unjustified assumption that the contributions per atom in the inner layers are the same, the contribution from the surface layer will only be small for very large clusters. Plus, we can have different geometries at constant nuclearity – isomers. But even for large clusters or the bulk material, the difference between surface layers and internal layers produces a structural problem of considerable practical consequence, e.g., the so-called reconstruction of pure bulk Si surfaces. And for those concerned with reaction chemistry between solids and other phases, which must occur at the phase boundaries (chemistry in two dimensions), surface properties must be included. By their hybrid nature then, nanoparticles require a good understanding of both small cluster properties and those of large particles with extended bonding.

Exercise 6.1. A small cubic close-packed (the geometric structure is shown in the Appendix, Figure A1.12) crystallite of Al has a perfect tetrahedral shape and contains 120 atoms. (a) Calculate the number of surface atoms. (b) Estimate the fraction of surface atoms on a perfect cube of Al containing 6×10^{23} atoms.

Answer. (a) Starting from the top, we have 1 Al, 3 Al, 6 Al, 10 Al, 15 Al, etc. If we add them together then at the 8th layer with 36 Al a total of 120 atoms is

6.1 Cluster molecules with extended bonding networks

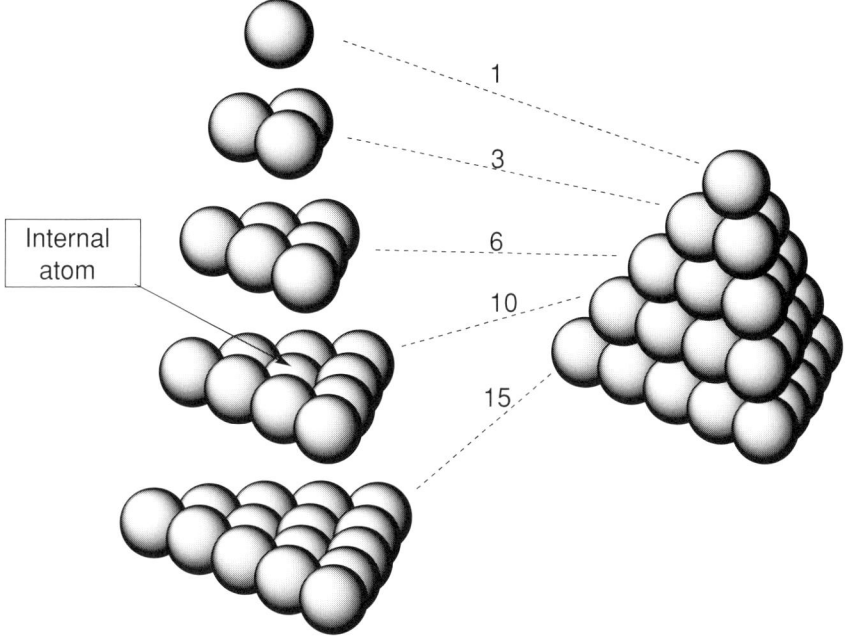

Exercise 6.1

reached. There are four corners, six edges and four faces. This gives four corner atoms, 6 × 6 edge atoms and 4 × 15 face atoms for a total of 100. Thus, there are only 20 internal atoms. Show for yourself that a tetrahedral cluster containing 35 atoms (five layers) is the smallest one with any internal atoms at all and then only one. Recall that the molecular cluster $[Os_{10}C(CO)_{24}]^{2-}$ (Figure 3.10) fits this motif and exhibits no internal metal atoms. The internal atoms start after the 4th layer and the internal pyramid builds up in the same progression as the larger external one. It is perhaps more understandable now that $[Al_{69}R_{18}]^{3-}$ with only 69 atoms plus some external ligands does not adopt a close-packed geometry but something intermediate. (b) The measured molar density permits the volume of an Al atom to be calculated (1.7×10^{-23} cm^3) as well as the volume of the sample. Assuming "cubic atoms" one can then calculate the number of atoms on a single face of the cube. It is a large number (about 10^{16} or about 10^{15} cm^{-2}); however, only about 10^{-7} of the total number of Al atoms.

6.1.2 The electronic-structure problem

Take a closer look at $[Al_{69}R_{18}]^{3-}$ to size up the problem. A couple of questions come to mind. How are the geometric and electronic structures of $[Al_{69}R_{18}]^{3-}$ connected? Are the polyhedral shapes of the various shells important or do other arrangements have similar energy? Is there a preferred valence electron count or, as we saw in Chapter 5 for the cubic clusters, is there a range of allowed values? Can we use an approach similar to those of the earlier chapters or do we need to seek a new approach? Let's see.

The core Al of $[Al_{69}R_{18}]^{3-}$ is bonded to 12 nearest neighbors so no simple localized bonding model is going to suffice to address these questions. How about the molecular-orbital model? The number of electrons and atomic orbitals required for the simplest MO treatment of $[Al_{69}(NH_2)_{18}]^{3-}$ (the ligands can be adequately mimicked by NH_2 to reduce the size of the problem) are 336 valence electrons and 384 functions yielding 384 MOs, 168 of which are filled. Doing the calculation is not the problem. In fact, quantum chemical investigations of $[Al_{69}R_{18}]^{3-}$ and related compounds have been published. The problem lies in the analysis of the 168 MOs in terms of useful concepts such as permitted stoichiometry or shape.

Clearly, these giant clusters are going to be a difficult problem if treated in the manner of small, molecular clusters. But if we can't handle the $[Al_{69}R_{18}]^{3-}$ cluster, you ask, what approach can we use to understand that of bulk Al? The molecular perspective we have taken thus far makes the problem look far worse. With 10^{23} Al atoms don't we need to consider 3×10^{23} valence electrons and 4×10^{23} atomic functions for the simplest MO treatment? Well, yes, but fortunately for substances in the form of single crystals translational symmetry provides a straightforward way around the problem. It is presented in the next section.

But first a geometric exercise is necessary to review the close-packed, hard-sphere model of metal structure. We will do it in a manner to reinforce the idea that there really are clusters in bulk Al! Bulk Al exhibits a ccp structure (ABCABC) with $d_{Al} = 2.86$ Å. Study the representation of the two close-packed layers shown in Figure 6.1. A triangular set of atoms is made transparent to show the lower layer. If you need greater clarity, make yourself three transparencies for an overhead projector with close-packed circles in three different colors and reproduce Figure 6.1. There is an octahedral hole, labeled "o", centered between the triangle of transparent atoms in the top layer and the triangles of three dark atoms immediately below. Moving outwards you should be able to find an octahedral array of tetrahedral holes, labeled "t", centered on the octahedral hole. At a larger radius a hexagon of octahedral holes is coplanar with the reference octahedral hole. Viewed from the side, the two layers of close-packed atoms contain a net of octahedral holes at $0.5\,d$, where d is the spacing between the two layers of close-packed atoms. Two nets of tetrahedral holes lie above and below at $0.225\,d$ and $0.775\,d$, respectively. There are twice as many tetrahedral holes as octahedral holes in this lattice.

The six atoms that define an octahedral hole, when excised from the lattice, constitute an octahedral cluster. Further, a hexa-capped octahedral cluster, generated by the fusion of six tetrahedral clusters, three up and three down, can also be cut

6.1 Cluster molecules with extended bonding networks

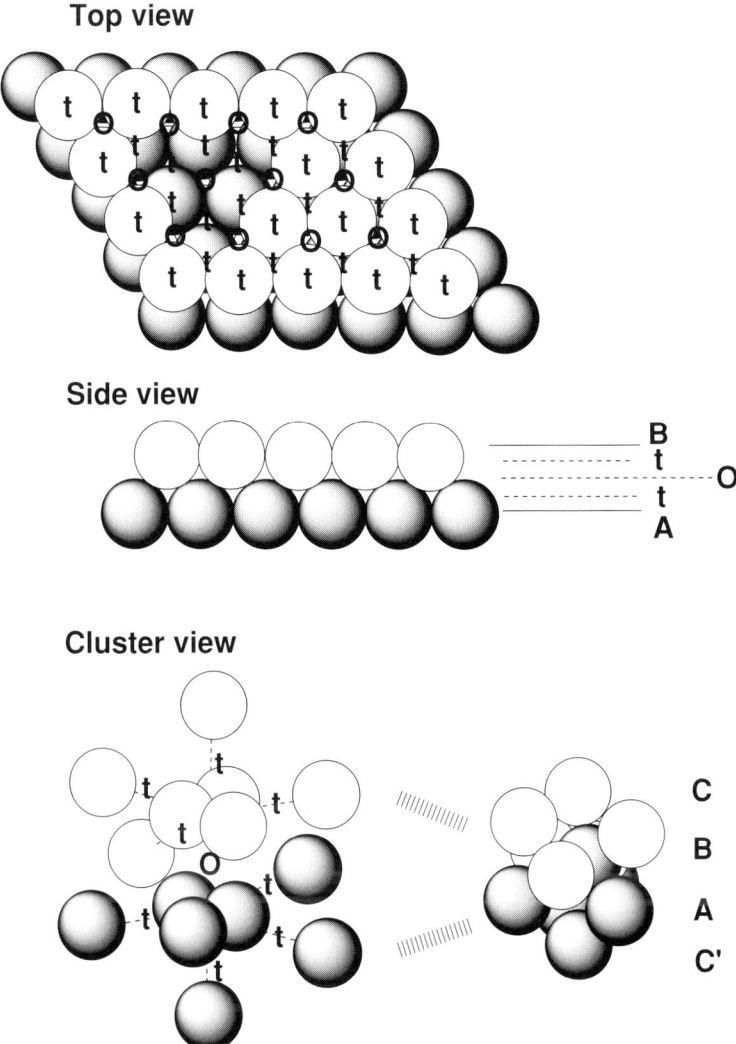

Figure 6.1

from these two layers. If these two layers are considered the A,B layers of a ccp lattice, inclusion of C layers generates a fully capped octahedron as shown in the "exploded" version in Figure 6.1. Here, then, is a geometric connection between single bare clusters and a close-packed solid. One implication of seeing clusters in a close-packed solid is the hint that between the limits of small bare clusters and a close-packed solid there exist a range of structures that can be formed by the fusion of small clusters. We know from the chemistry of molecular clusters that fusion is not restricted to faces. Vertex- and edge-fused clusters are well known. We will see a few examples of related solid-state compounds in Chapter 7. The most important

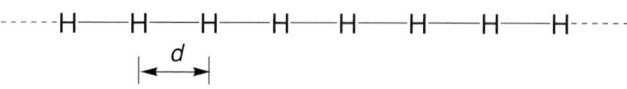

Figure 6.2

idea to take from this discussion of geometry is that of periodicity – these structures can be generated from a repeating template of atom, cluster or other unit.

In our discussion of main-group clusters, we emphasized the importance of developing a model of electronic structure capable of dealing with the major features of the problem at hand without being cluttered up with extraneous detail. We need such a model for crystalline materials and the majority of this chapter is devoted to illustrating and learning how to use an orbital-based approach in the manner of the molecular-orbital approach of Chapter 1. Connections to molecules and small clusters will be emphasized.

6.2 Outline of the electronic-structure solution in a one-dimensional world

The solution of the H atom problem of Chapter 1 provides us with the concept of atomic orbitals. Its extension provides a model for the electronic structure of the heavier atoms which can be developed into an MO model for molecules. Molecular orbitals formed from linear combinations of the same AOs provide a serviceable conceptual model for the electronic structures of molecules. The same AOs provide an approach to the electronic structure of extended systems with periodic geometric structures. It is useful because it avoids the nightmare of 10^{23} MOs dreamed above.

6.2.1 Crystal orbitals (COs) vs. molecular orbitals (MOs): the example of the H chain

In order to illustrate important aspects of the solutions in the least-confusing system, consider first a linear homonuclear chain of equispaced atoms with a single atomic function each. If it is a linear chain of H atoms as shown in Figure 6.2, then the atomic function is H 1s and there is one electron per atom for a neutral chain. This is a hypothetical species as under normal conditions of temperature and pressure this infinite H atom chain would revert to H_2 molecules. We will see why in Section 6.2.3. Purely hypothetical approaches, not necessarily limited to stable arrangements, give insights to electronic factors responsible for stability of a particular arrangement of nuclei and electrons.

As emphasized in Section 6.1, to model bulk properties a crystallite must be large enough so that bulk properties dominate over surface properties. So too the H-atom chain here must be long enough so the effects of the two end atoms are

6.2 The electronic-structure solution in a 1-D world

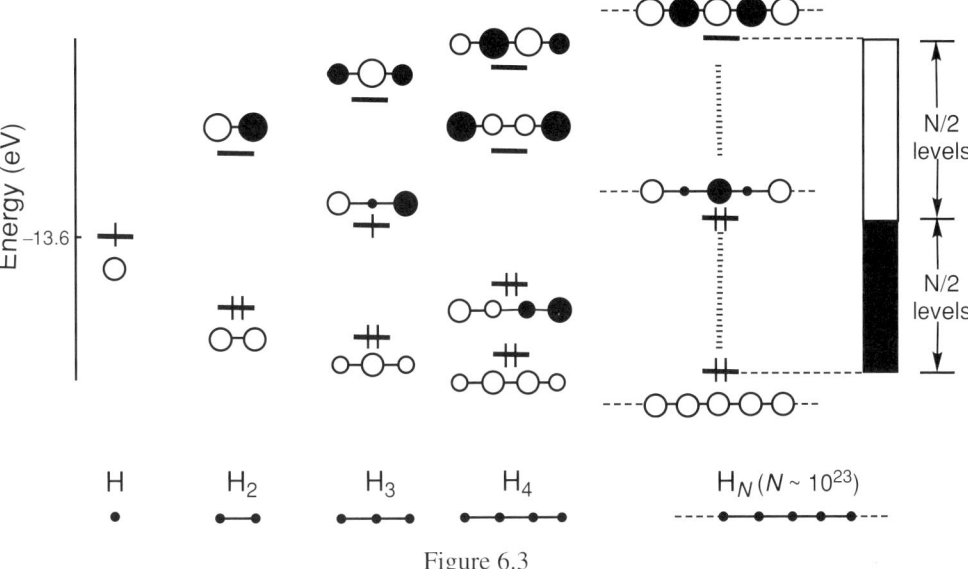

Figure 6.3

small. What we need are the MOs and relative energies. Let's bootstrap it. Consider the solutions to the MO problem for oligomers of H atoms as the chain length is progressively increased in size. The first few are shown schematically in Figure 6.3. Do you remember linear H_3 and butadiene from Chapter 1? In the limit as the number of H atoms approaches 10^{23} the spacing between the levels becomes very small and the solutions more complex. However the nature of the lowest- and highest-energy solutions should be clear by inspection, i.e., the lowest energy has no nodes and the highest has the maximum number of nodes. Note that even if the number of H atoms is infinite, the energies of the lowest and highest levels are finite, i.e., as the number of atoms increases in the oligomer, the maximum and minimum energies move asymptotically to finite values. We can use the number of nodes as an index, k, and order the functions in energy accordingly. Look at the bonding nature of the tetramer for example. For $k = 0$, there are three bonding interactions; for $k = 1$, two bonding and one antibonding interactions; for $k = 2$, one bonding and two antibonding interactions; and for $k = 3$, three antibonding interactions. For the extended chain, at the bottom (lowest energy) of the large set of levels, called a band, there will be a maximum number of bonding interactions. In the middle the number of bonding and antibonding interactions will be nearly the same and at the top there will be the maximum number of antibonding interactions.

Exercise 6.2. Instead of using linear chains use rings of H atoms and mimic the bootstrapping of Figure 6.3. There is no end effect in a ring. Did it disappear or has it been replaced by another effect? If another effect, how large a perturbation is it?

Answer. The solution of the H_3 triangle is given in Figure 1.13. The solutions for the larger rings mimic those of a planar $(CH)_5$ ring π system (Chapter 2, Problem 8). The main difference with the linear model is that the solutions for $k > 0$ are doubly degenerate but the variation in bonding/antibonding character from the bottom to the top of the band is reproduced. In a ring, the end effects of the linear chain are replaced by a bend in each three-atom segment. However, with only 100 atoms in the ring (two terminal atoms out of 100 for the linear chain) the H–H–H angle is 176.4°. As far as any set of three consecutive atoms is concerned, the ring for 10^{23} atoms is linear.

OK, fine you say. So far we are treating the extended chain like a giant molecule and we are still stuck with drawing in 10^{23} levels for a mole of H atoms. In the manner of Hoffmann, we would like to have a detailed and informative model that we can use in discussions of the electronic properties of solids without dealing explicitly with all these levels. In addition, we want to emphasize connections between clusters (molecules) and extended systems (solid state). So let's dig around in this band a little bit more and see if we can eliminate the necessity of talking about 10^{23} orbitals in order to discuss and rationalize properties of extended systems.

First ignore the ends of the chain (surfaces in the case of a three-dimensional solid) or adopt the strategy of Exercise 6.2. Recall that the point-group symmetry of a molecule can be used to obtain the symmetry-adapted linear combinations of, e.g., ligand orbitals, to generate MO diagrams of molecules. So too the translational symmetry of this H atom chain can be added to the solid-state problem to simplify the MO diagram. The one-dimensional translational unit here is an H atom – can't get any simpler than that. As was necessary for the regular H rings of Exercise 6.2, we have to solve the Schrödinger equation for a system in which all the atomic orbitals constituting the basis set are symmetry-equivalent. In the present case they are equivalent by translational symmetry. In this simple case the solutions of the Schrödinger equation (i.e., the coefficient of the MOs or COs) are fully determined by symmetry – no need to solve the Schrödinger equation! Simple symmetry tools generate an equation that expresses the translational symmetry-adapted linear combinations and the wave functions for H_N ($N \sim 10^{23}$) in Figure 6.3. It is called a Bloch function and is:

$$CO_k \, \alpha \sum_n \{\exp(iknd)\} \, [H1s(n)]$$

where the summation is over all the N atoms and CO = crystal orbital, n = the individual hydrogen atom label, d = H–H distance (unit cell parameter) and k = a translational symmetry index which can also be understood as a node counter. The coefficient of proportionality is the normalizing factor of CO_k which we need not

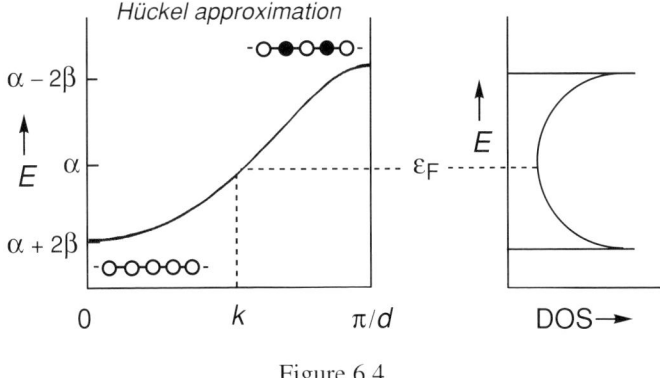

Figure 6.4

consider explicitly. The symmetry index k is just like the labels used in molecular problems, e.g., the label a_{1g}, denotes the totally symmetric ligand orbital linear combination for an octahedral coordination compound (Chapter 1). For $k = 0$, $CO_0 \propto \Sigma[H1s(n)]$ and for $k = \pi/d$, $CO_{\pi/d} \propto \Sigma(-1)^n[H1s(n)]$ thereby reproducing the lowest and highest energy COs in Figure 6.1.

So instead of having an index that runs from 1 to 10^{23} we have one with an absolute value that runs from 0 to π/d. In other words, the energy $E(k)$ associated with the CO_k orbital is a periodic function of k. This should not be a surprise given the periodic expression of CO_k by the Bloch function. A satisfying definition of the period range runs from $-\pi/d$ to $+\pi/d$ but $E(k) = E(-k)$ and each level is degenerate except for $k = 0$ (the ring model in Exercise 6.2 generates this representation, whereas the linear chain model does not). Outside of the range 0 to π/d the $E(k)$ vs. k equation repeats one of the functions already generated. In fact this range defines what is called the irreducible part of the first Brillouin zone.

If we now plot E vs. k for our H-atom chain we get a curve such as the one shown on the left side of Figure 6.4. There is a value of k for every translational unit in the crystal (i.e., as many values as H atoms) so the large number of points on the curve in Figure 6.4 makes it essentially a continuous function. Don't look, but you suddenly have been placed in k space – reciprocal space! Please note and remember that the sinusoidal type shape of the $E(k)$ vs. k curve originates from the fact that we are considering a very simple model in which all the atomic orbitals constituting the basis set are symmetry-equivalent by translation. Moreover, this particular $E(k)$ vs. k curve has been calculated within the Hückel approximation which neglects overlap between orbitals on different atoms. Within this approximation, the destabilization of the antibonding orbitals is exactly equivalent to the stabilization of their bonding counterparts. Although very approximate, this is the Hückel-type curve which is usually shown in introductory textbooks. However, in the following we will go beyond the Hückel approximation and consider overlap explicitly. When overlap is

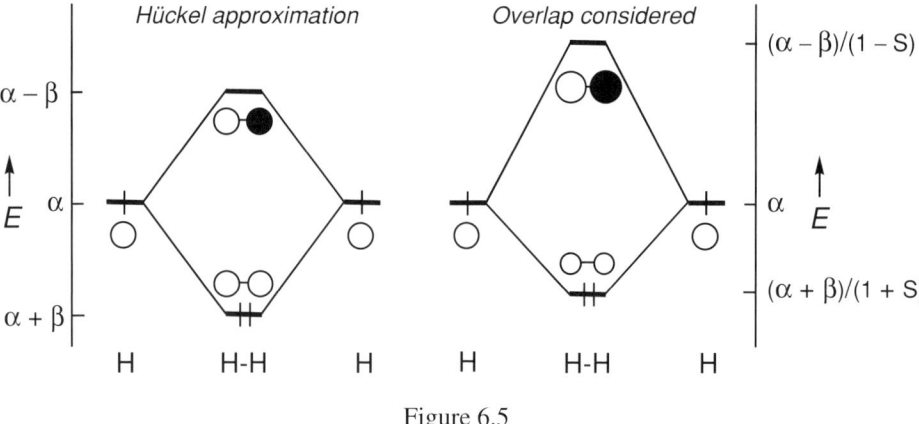

Figure 6.5

not neglected in the calculations, the antibonding effects are larger than the bonding ones, exactly as in the molecular H_2 example recalled in Figure 6.5, where σ^* is more destabilized than σ is stabilized if the overlap is considered. The result, in the case of the H-atom chain, is a stretching of the upper part of the $E(k)$ vs. k curve towards higher energies, giving rise to the distorted sinusoidal curve shown in the upper left of Figure 6.6. This, too, is ideal and, as we will see later, the band structure of a real solid shows fairly irregular $E(k)$ vs. k curves within the irreducible part of the first Brillouin zone.

Half of the levels lie in the lower branch of the curve and half in the upper. Note that in Figure 6.6, the lower half is concentrated on a smaller energy range than the upper half, in contrast to the Hückel band of Figure 6.4. Since in the H_N chain there are N electrons and N COs (or MOs), all the levels in the lower half are occupied and all in the upper branch are vacant. Thus, in this particular case, the HOMO energy corresponds to the inflexion point of the $E(k)$ vs. k curve. As the orbital corresponding to the inflexion point is non-bonding between nearest neighbors, it lies close to the energy of H 1s (i.e. -13.6 eV). In solid-state jargon, the HOMO energy is called the Fermi level and is labeled ε_F in Figures 6.4 and 6.6.

The plot of E vs. k in Figures 6.4 or 6.6 designates the band structure of linear H_N (now extrapolated to H_∞) and has characteristic properties. One is the band width or dispersion. This corresponds to the splitting between the bonding and antibonding MOs in, e.g., molecular H_2 (Figure 6.5). Good overlap between the 1s AOs in H_2 leads to a large σ^*/σ^* splitting (large $|\beta|$ value within the Hückel approximation). Similarly, good overlap between units in a chain leads to large band width. As with a molecule, good overlap depends in general on the value of the inter-nuclear distance, d, and AO type as well as symmetry. A very large value of d reduces the overlap between neighboring H 1s AOs to a negligible value and renders all the CO_k orbitals non-bonding and nearly degenerate. The resulting $E(k)$ vs. k

6.2 The electronic-structure solution in a 1-D world

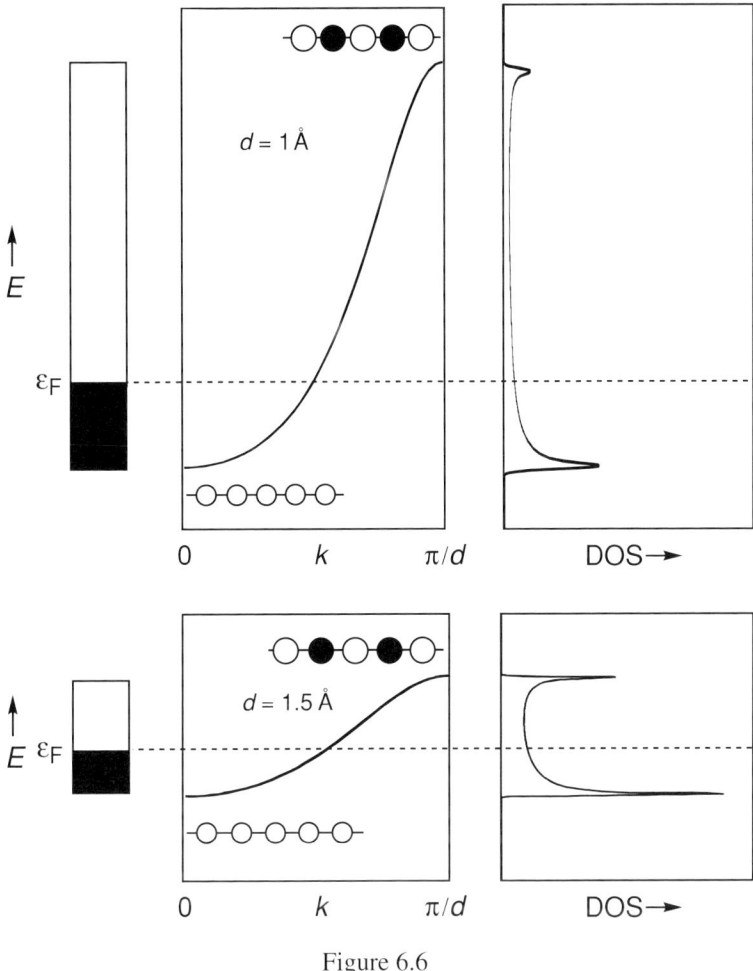

Figure 6.6

curve would be an almost horizontal line. The effect of two different inter-nuclear distances (top: short; bottom: long) on band width are illustrated in Figure 6.6.

Another feature of these bands is the density of states (DOS). In solid-state jargon, a state is an energy level $E(k)$. The density of states is nothing more than the concentration of levels as a function of energy and is shown at the right sides of Figures 6.4 (Hückel approximation) and 6.6 (overlap considered). Note that the density of CO energy levels as a function of energy is a strongly varying function that has its largest values (actually unbound for this case) at $k = 0$ and π/d. The value of DOS at any E is inversely proportional to the slope of the $E(k)$ vs. k curve so the smaller the band width the greater the density of states. This makes sense as the smaller the interaction (overlap) between H atoms, the smaller the energy over which the 10^{23} orbitals are distributed. Keep in mind that integral over the

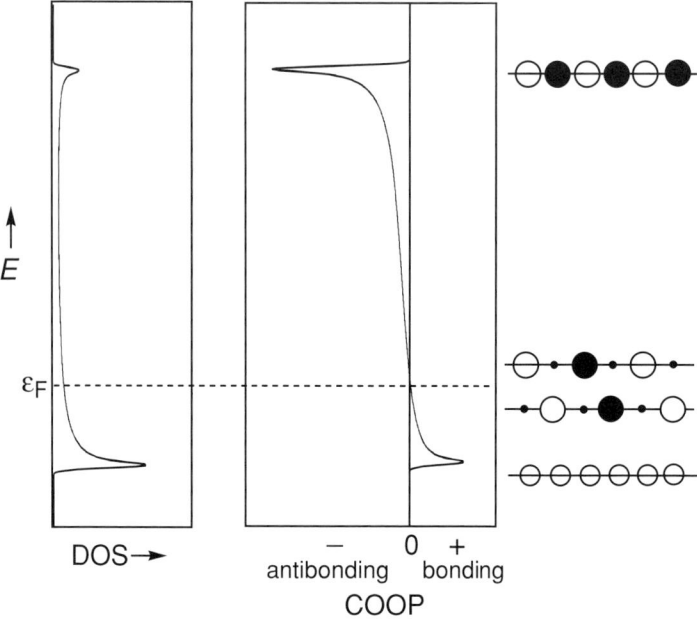

Figure 6.7

energy of the DOS curve *is* equal to the number of H 1s orbitals per repeat unit, i.e. one. As stated above, a change in the inter-nuclear distance d (thus varying the overlap magnitude) changes the band dispersion (width) and in turn changes the DOS curve. However, the value of the integrated DOS remains constant.

We have seen that the bonding/antibonding characters of the bands vary as a function of k and for the H_N (or H_∞) linear model the idea is straightforward. However, in more complex molecules we need a numerical parameter as a guide simply because a given MO could be bonding with respect to one pair of atoms and antibonding with respect to another. Likewise, we need a parameter for extended systems that mimics the overlap population used in the discussion of the bonding character of MOs. This parameter is the crystal orbital overlap populations (COOP). For H_N the orbital pictures can be "read" directly and, as shown on the right-hand side of Figure 6.7, the plot of COOP vs. E should be no surprise. The numerical value of COOP for any E depends on the magnitude of the overlap, the magnitude of the orbital coefficients and the value of the DOS at that E. Note that, as in the case of molecules, the antibonding character of the antibonding levels overrides the bonding character of the bonding levels. Within the Hückel approximation, both characters have exactly similar strength (see the molecular H_2 case in Figure 6.4). Just like Mulliken overlap populations for a molecule, which also depend on overlap and orbital coefficients, the COOP curves can be used to investigate the

6.2 The electronic-structure solution in a 1-D world

origins of bonding. The integral of the COOP curve up to the Fermi level gives the total overlap population which correlates with the strength of the bonding.

6.2.2 Consideration of a non-elementary repeat unit

Suppose now we take the same linear H_N (or H_∞) model in which all the H atoms are equidistant and consider it made up of H_2 repeating units rather than H atoms. You may think it stupid to disregard the existence of the elementary d translation and use $d' = 2d$. Yes, you are absolutely right but we have a pedagogical purpose for doing it. The process is similar to determining the MO diagram of the H_2O molecule within C_s symmetry rather than its actual C_{2v} symmetry. Even if ignored, the C_2 axis and one of the mirror planes of the H_2O molecule are still there and the final MOs and energy levels will be the same as those generated with full C_{2v} symmetry. By the same token, we expect the band structure and DOS of the linear H_∞ model to be equivalent whether we consider the repeat unit to be H, H_2, H_3 or H_n. Let's see how it works with a H_2 repeating unit.

First, the irreducible part of the Brillouin zone now varies from $k = 0$ to $k = \pi/d' = \pi/2d$. Indeed, doubling the parameter of the unit cell in real space halves the size of the Brillouin zone (or the reciprocal-space unit cell). Second, recall that orbital interactions are additive and that the final MO diagram (or band structure) is just the result of the sum of all the orbital interactions. Within each individual H_2 unit the interactions simply correspond to the bonding (σ) and antibonding (σ^*) MOs of each individual H_2 unit. There are three types of interactions involving the MOs of different H_2 units: interactions between all the σ orbitals interactions between all the σ^* orbitals and interactions between the σ and the σ^* orbitals. Since all the σ_n orbitals are equivalent by translational symmetry, their interaction is described by the Bloch function:

$$[CO^\sigma]_k \alpha \sum_n \{\exp(iknd)\} \sigma(n)$$

This equation is similar to the one we have seen when considering the H repeat unit, but this time it is expressed on the $\sigma(n)$ basis set, not on the [H1s(n)] one. Just by drawing the $[CO^\sigma]_k$ orbitals for $k = 0$ and $k = \pi/2d$, it is easy to see that the $E^\sigma(k)$ vs. k curve (lowest dashed band on the top part of Figure 6.8) has a pseudo-sinusoidal shape similar to that of Figure 6.6. Similarly, the interactions between the $\sigma^*(n)$ orbitals are described by the Bloch function:

$$[CO^{\sigma*}]_k \alpha \sum_n \{\exp(iknd)\} \sigma^*(n)$$

For $k = 0$, $[CO^{\sigma*}]_0 \alpha \Sigma \sigma^*(n)$ shows an antibonding relationship between the H_2 units. For $k = \pi/2d$, $[CO^{\sigma*}]_{\pi/2d} \alpha \Sigma (-1)^n \sigma^*(n)$ exhibits a bonding relationship.

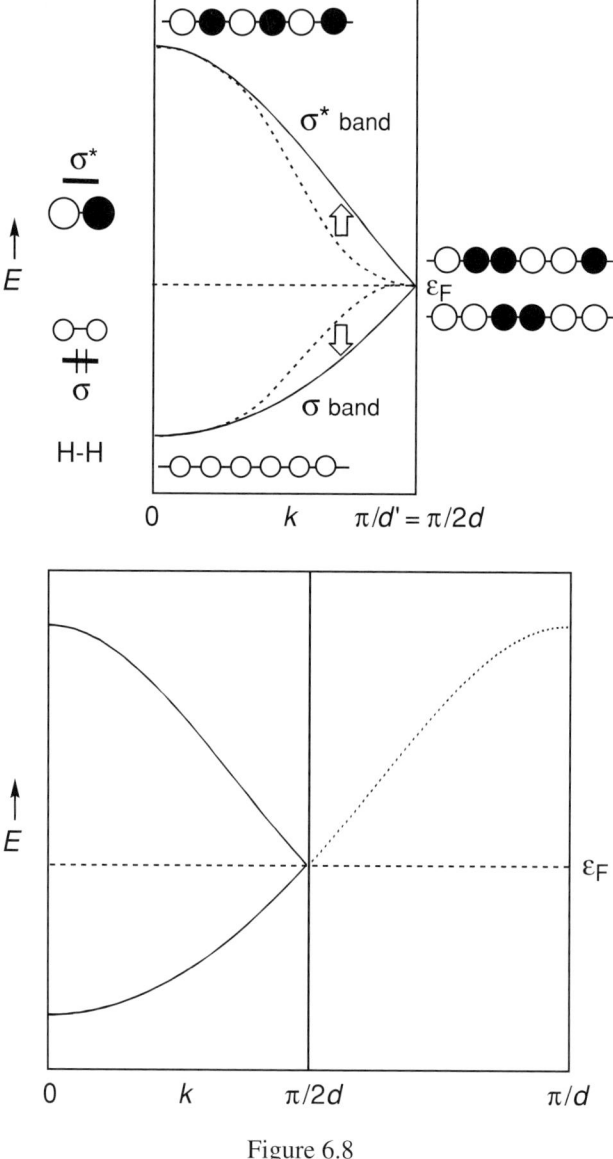

Figure 6.8

Unlike the $E^\sigma(k)$ vs. k curve, the $E^{\sigma*}(k)$ curve has a negative slope and a pseudo-cosinusoidal shape (e.g., highest dashed band on the top of Figure 6.8). An interesting feature of Figure 6.8 is that $[CO^\sigma]_{\pi/2d}$ and $[CO^{\sigma*}]_{\pi/2d}$ are degenerate. Indeed, the former is bonding inside the H_2 units and antibonding between these units, whereas the latter exhibits exactly opposite character. Since the H–H distances are the same inside and between the H_2 units, both COs have exactly the same non-bonding character. In a sense, considering this degeneracy means that we implicitly

6.2 The electronic-structure solution in a 1-D world

reintroduce the elementary translation d that was discarded in preferring H_2 repeating units rather than H atoms.

We have now to consider the σ/σ^* interactions, i.e., allow the $[CO^\sigma]_k$ and $[CO^{\sigma*}]_k$ combinations to interact. At first glance it looks extremely complicated since we have $\sim 10^{23}$ orbitals of each type. Symmetry saves us. Remember that k is a kind of symmetry index. This means that $[CO^\sigma]_k$ and $[CO^{\sigma*}]_{k'}$ can interact only if $k = k'$, otherwise they are orthogonal. Thus, for each k point we have two orbitals that interact. That is, at each individual value of k, $[CO^\sigma]_k$ and $[CO^{\sigma*}]_k$ mix and $[CO^{\sigma*}]_k$ becomes more antibonding whereas $[CO^\sigma]_k$ becomes more bonding. The net consequence is that the dotted curves repel each other. For a specific symmetry reason, there is no interaction (mixing) at the special k points corresponding to $k = 0$ and $k = \pi/2d$. Draw the corresponding COs in the same figure (it will help if you use two different colors) and you will see that they are orthogonal. The effect of the σ/σ^* interactions is sketched in the upper figure of Figure 6.8, which shows the final band structure (solid lines) of our linear H_∞ model considering H_2 repeat units. It is different from the one obtained for the same model, but considering H repeat units! Shouldn't they be the same? They are different because we have changed the reciprocal lattice vector but the DOS generated from this band structure (right-hand side of Figure 6.6) is the same. This is a relief as it is the DOS which is the observable. Thus, although the band diagrams are not the same they are equivalent. In fact, one can be generated from the other by folding the simple band in the lowest figure of Figure 6.8 at the point $k = \pi/2d$. To show this for yourself, take a piece of paper, draw the band shown in the top of Figure 6.6 and fold it at $k = \pi/2d$ to generate the bands defined by the solid lines in Figure 6.8 (top). The illustration at the bottom of Figure 6.8 shows the band structure before and after folding. This folding effect is an easy way to generalize a change in repeat unit size: $H_3, H_4, \ldots$, H_n repeat units lead to folding the band of Figure 6.4 into 3, 4, ..., n equal parts, respectively.

Exercise 6.3. Consider a linear H chain in which all the atoms are equidistant. Paint alternative atoms with different colors, e.g., red and yellow. The yellow H_n atoms correspond to n even and the red H_n atoms to n odd. We are left with two inter-penetrating subnets, of two different colors. Show that the interaction of these two subnets generates the band structure at the top of Figure 6.8. Hint: draw in color the band structure of each subnet on the same diagram and then color the COs at $k = 0$ and $k = \pi/2d$. Allow them to interact to generate the band structure at the top of Figure 6.8.

Answer. It is easy to see that each of the yellow and red subnets has an elementary translation of $2d$ which is the same as used in the double cell described just above. The band structure of the yellow chain is that of a regular chain, with one yellow

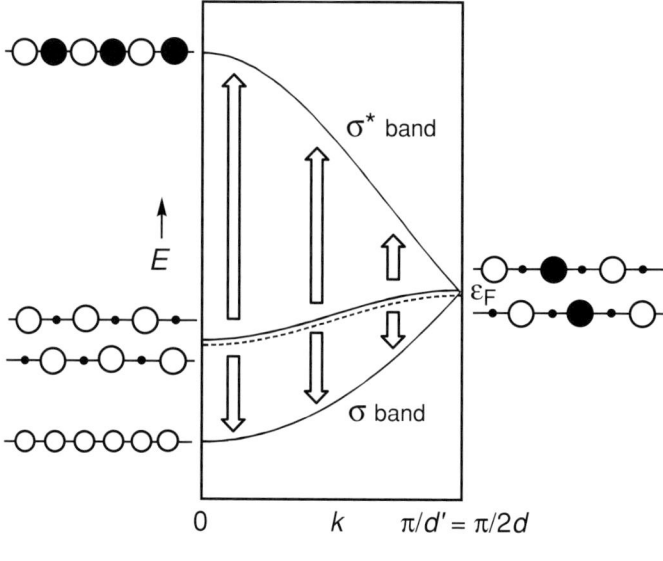

Exercise 6.3a

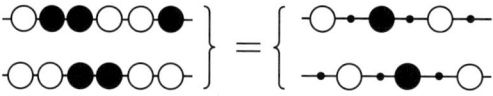

Exercise 6.3b

H per repeat unit, similar to that of Figure 6.4. However, it is important to keep in mind that the band width is very small since the $2d$ distance separating the yellow atoms is large (distance between second neighbors in the complete chain). Thus the band is almost horizontal, lying around $-13.6\,\mathrm{eV}$, the H 1s energy.

The red subnet is strictly equivalent to the yellow one and the red and yellow bands are degenerate. But thanks to the colors, you can easily distinguish them on your drawing even if it is less clear on our black and white diagram. At $k = 0$, the overlap between the yellow and red orbitals is very strong since all combinations have the same sign. As a result, the antibonding combination is strongly destabilized and the bonding combination is strongly stabilized. Once drawn, it is easy to see that these orbitals are the same as those shown for $k = 0$ at the top of Figure 6.8. At $k = \pi/2d$ the overlap between the yellow and red orbitals is zero since its contributions have alternative signs. By symmetry, there is no interaction at $k = \pi/2d$. Both orbitals are non-bonding in character, but you may be worried that they are different from those plotted at $k = \pi/2d$ in Figure 6.8. Again we have a situation where the two sets are equivalent. That is, any pair of degenerate orbitals can be replaced by a set

6.2 The electronic-structure solution in a 1-D world

of orthogonal, renormalized linear combinations. Taking the sum and the difference of two orbitals is a simple way to build orthogonal combinations. As shown above, the sum and difference of the orbitals at $k = \pi/2d$ are those of Figure 6.8. Finally, with the reasonable assumption that the strength of the yellow/red interaction varies monotonically between $k = 0$ (maximum) and $k = \pi/2d$ (zero), the two colored bands repel each other in such a way that we exactly get the band structure at the top of Figure 6.8.

In summary, we have developed three methods of generating a band structure: from atoms, molecular fragments, and nets. The first two should remind you of our approach to the MOs of molecules. The last is new but once the added complexity is reduced by the translational symmetry, the operations in developing the band structure are again similar to those for molecules. The choice of approach depends on the problem. Although the answer cannot depend on the approach chosen, the clarity and insight into a problem may well do so.

6.2.3 Connections between geometric and electronic structure: the Peierls instability and its relationship with Jahn–Teller instability

In molecular systems discussed in Chapter 1, connections between MO energies, electron count and geometric structure were described. For many systems, the properties of the HOMO serve as a useful guide to favored geometry. Recall the discussion of the simplification of Walsh (HOMO energy as a function of a pertinent geometric parameter) and Jahn–Teller effects. Are there analogous connections between geometric structure for extended chains and electronic structure as described by COs and DOS?

First, let's get a better feeling for molecules in solids before addressing the principal question of this section. In our H_∞ chain of Figure 6.3, shorten the distance between adjacent H–H pairs to generate alternating short, long, short, long, etc. spacing. The distortion of the regular chain that generates the alternating one is sketched in Figure 6.9. The formation of pairs (H_2 pairs in this case) is often called "dimerization" even though the "dimers" are not independent. Note that the translation d is lost and now the elementary repeat unit really is H_2 and the elementary translation is $d' = 2d$. Alternatively, the short and long distances can be represented by xd' and $(1 - x)d'$, where $0 < x < 1$.

This distortion will affect the bonding/antibonding strength of the COs shown at the top of Figure 6.8. Longer distances (between H_2 units) will stabilize COs having antibonding character and destabilize COs having bonding character. Shorter distances (inside H_2 units) will have the opposite effect. It is easy to see that the perturbation of the two COs at $k = 0$ is a result of destabilizing and stabilizing contributions that *virtually* cancel each other. Thus, the energies at $k = 0$ are not

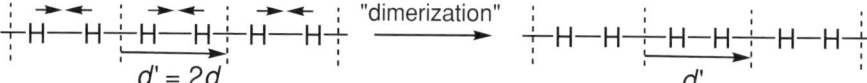

Figure 6.9

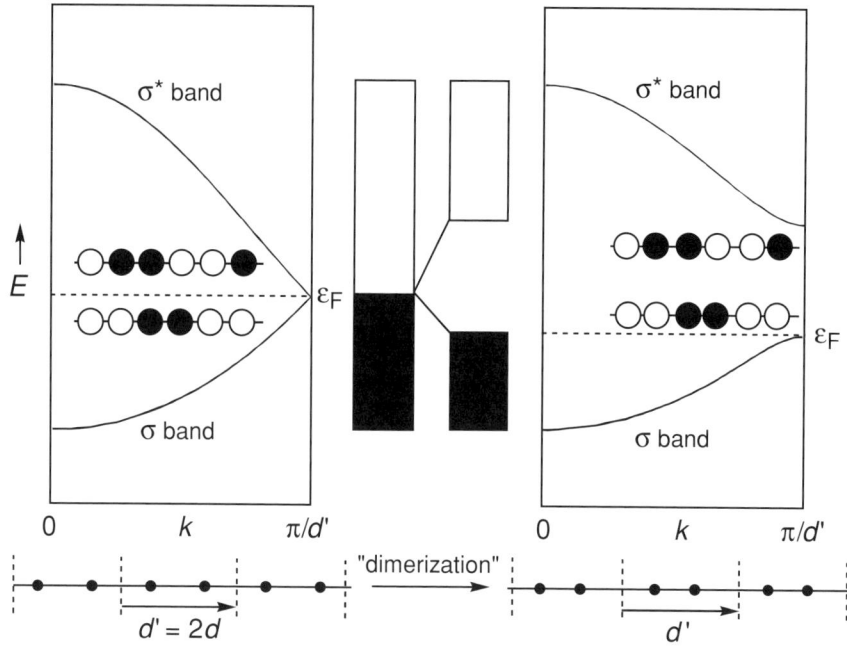

Figure 6.10

significantly perturbed by the "dimerization." A different result is found at $k = \pi/d'$ where both contributions add rather than cancel. $[CO^\sigma]_{\pi/d'}$ and $[CO^{\sigma*}]_{\pi/d'}$ undergo the sum of two stabilizations and two destabilizations, respectively, and no longer have equal energies. Hence, the degeneracy at $k = \pi/d'$ is broken. As in the case of molecules, a lowering of symmetry (in this case the loss of the translation d) can split orbital degeneracy. Between $k = 0$ and $k = \pi/d'$, the $E^\sigma(k)$ and $E^{\sigma*}(k)$ vs. k curves are distorted in a continuous way as illustrated in Figure 6.10. The band splits into a lower "σ" band, which is occupied, and a higher "$\sigma*$" band which is empty. The "dimerization" results in a net stabilization of the highest occupied levels (Figure 6.10) and the "dimerized" structure is expected to be more stable than the regular one. In other words, a half-occupied band (Figure 6.8) is expected to be unstable with respect to some kind of "dimerization" (Figure 6.10).

This effect is called Peierls instability and the astute reader will recognize that it is directly related to first-order Jahn–Teller instability in molecular chemistry. For

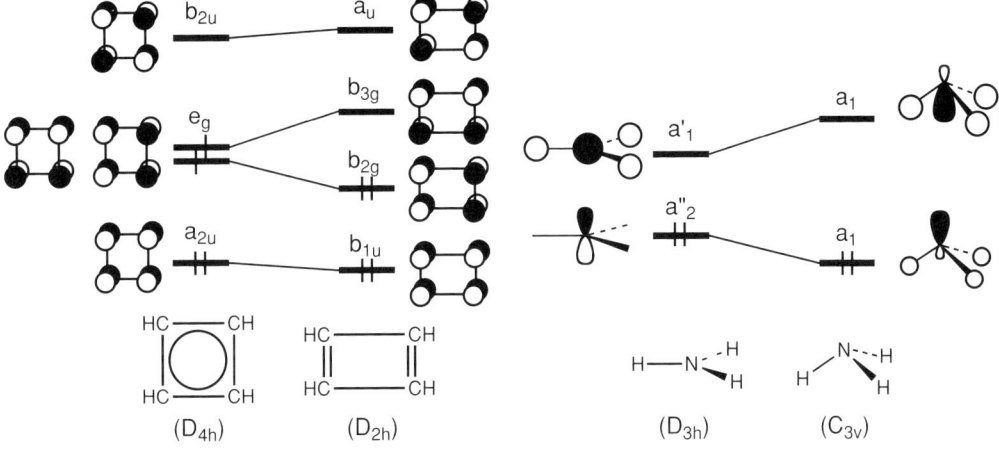

Figure 6.11

example, loss of symmetry accompanying the Jahn–Teller distortion of cyclobutadiene removes the HOMO/LUMO degeneracy (left side of Figure 6.11). As mentioned in Chapter 1, Jahn–Teller instability can also occur when a small but non-zero HOMO–LUMO gap occurs. The case of NH_3 is illustrated in the right side of Figure 6.11. Similarly, solid-state structures with very small band gaps above the Fermi level are subject to what is often called second-order Peierls instability.

In general, then, a band which is $1/3, 1/4 \ldots 1/p$ occupied is subject to Peierls instability with respect to a tri-, tetra-, p-merisation. It is important to remind you at this point that the possibility of a Jahn–Teller instability does not always mean the distortion occurs. The same applies for Peierls instability. There are several reasons. First of all, Peierls instability holds rigorously for ideal one-dimensional materials and reality is a more complicated three-dimensional space. Second, as in the case of molecules, when high-spin states are preferred, the distortion is not favored (see Problem 7 for an example). Third, even in the case of low-spin systems, the distortion may have additional consequences, e.g., weakening of spectator bonds. Coupled destabilizing effects can dampen or even prevent distortion. In the H_2 case described above, there is no effect opposing the distortion, and dimerization is complete, i.e., fully independent H_2 molecules result from our hypothetical chain.

Structural change resulting from a Peierls distortion can have dramatic effects on the physical properties. Look at the half-filled bands of Figure 6.6. There is no HOMO–LUMO gap. Such a situation depicts a metallic electrical conductor. On the other hand, after Peierls distortion there is a band gap at the Fermi level (Figure 6.10) and the material, depending on the width of the gap, is a semiconductor or an insulator. For a semiconductor, thermal excitation of electrons from the valence to conduction band leads to a conductivity that increases with increasing temperature, whereas for a metal conduction decreases with increasing temperature.

Before leaving this section, we need to tell you a very important point that beginners often forget. As in the case of Jahn–Teller instability, Peierls instability occurs for particular electron counts. For example, "dimerization" (a distortion leading to the doubling of the elementary unit cell parameter) is expected to occur only in the case of a half-filled band (or nearly half-filled if the material is not stoichiometric), i.e., Peierls instability depends on band population.

Exercise 6.4. Consider a crude, but useful, model of polyacetylene $(CH)_\infty$ consisting of the ideal zig-zag (all-*trans*) chain shown below in which all the C–C bond distances are equal. Draw its π-type band structure and show that it is Peierls unstable. Show that, unlike the H-chain case, the Peierls distortion does not correspond to a doubling of the unit cell.

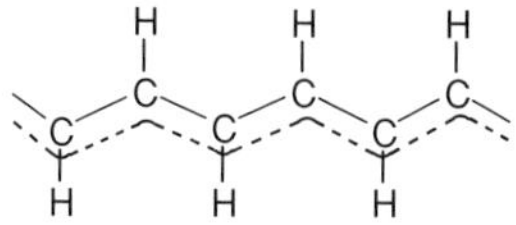

Exercise 6.4a

Answer. All-*trans* polyacetylene is an infinite planar ribbon. Although it has a certain finite width, it is a one-dimensional system since it extends infinitely (a very long distance in real life) in only one direction. First, set up the problem. What is the elementary repeat unit? This is the unit cell which is defined by the smallest translational vector. This vector is parallel to the horizontal axis in the drawing below, i.e., the direction of polymer growth. Translation vectors do not make zig-zags! The smallest translational vector is d and contains two CH units. The choice of the origin of the translation vector d is arbitrary and whether you choose a zig or a zag $(CH)_2$ repeat unit, the final result is the same. A simplification arises from the fact that the infinite ribbon is planar. Thus, as in conjugated planar molecules, the π and σ COs of polyacetylene do not interact by symmetry and may be considered independently. Just as in a molecule, the σ bonding levels will lie at low energy, the antibonding σ^* levels will lie at very high energy and the π levels will be situated in between irrespective of their bonding/antibonding character. Finally, there is one $2p_\pi$ orbital per C atom and, because C uses three electrons to form three σ bonds (two CC and one CH bond), the $2p_\pi$ AO contains one electron. Now we can develop the π-type band structure.

The π-type interactions within each individual $(CH)_2$ unit generate the bonding (π) and antibonding (π^*) MOs of each individual unit. The first one is occupied (two π electrons per unit cell), whereas the second is empty. Except for the fact that

6.2 The electronic-structure solution in a 1-D world

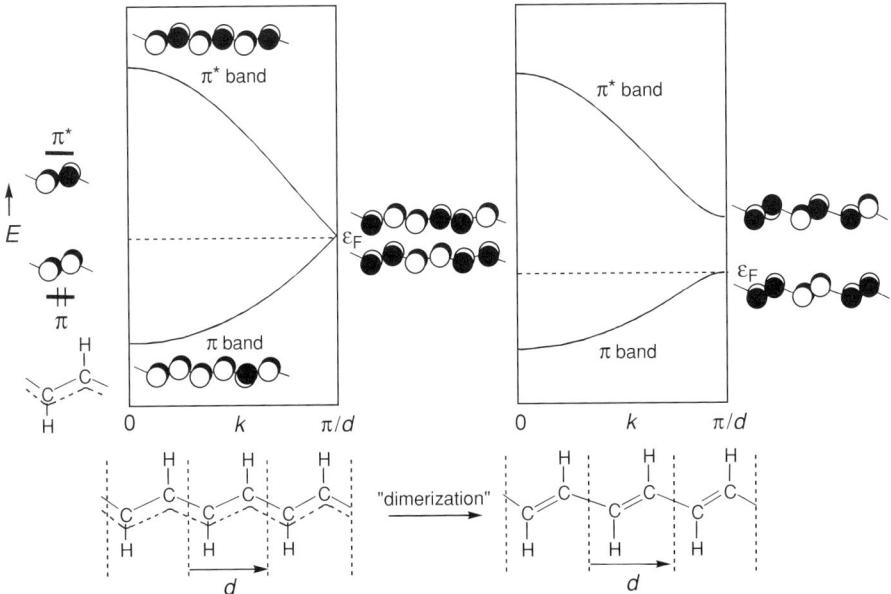

Exercise 6.4b

we are dealing with π and π^* COs in the place of σ and σ^* orbitals, the process of generating the band structure is the same as one we used for the H chain with H_2 units. The final band structure has a HOMO/LUMO degeneracy for $k = \pi/d$. Thus, as in the case of the H chain, Peierls instability occurs leading to a distorted structure of alternating single and double bonds. From this, one predicts a gap at the Fermi level which renders polyacetylene a semiconductor in agreement with experiment.

The unit cell parameter d is still the unit cell parameter of the distorted structure. No unit cell doubling occurs during the Peierls distortion! This time, the degeneracy splitting at $k = \pi/d$ is not due to the loss of an elementary translation but to the loss of a glide symmetry plane (combination of a mirror plane and a translation of $d/2$). Glide planes, as well as screw axes, have the same effect as multiplying the elementary translation vector by two, three, etc. They fold the bands at the edges of the irreducible part of the Brillouin zone and their loss leads to degeneracy splitting at these edges. Think about the kind of Peierls distortion a one-dimensional polymer having a 3_1 screw axis and a 1/3 filled band would be subject to.

6.2.4 Hypothetical one-dimensional homoatomic main-group element chain: the example of C

Consider now the problem of a regular chain of C atoms sketched in Figure 6.12. Unlike the H chain, such a one-dimensional structure is realistic. It is, after all, a

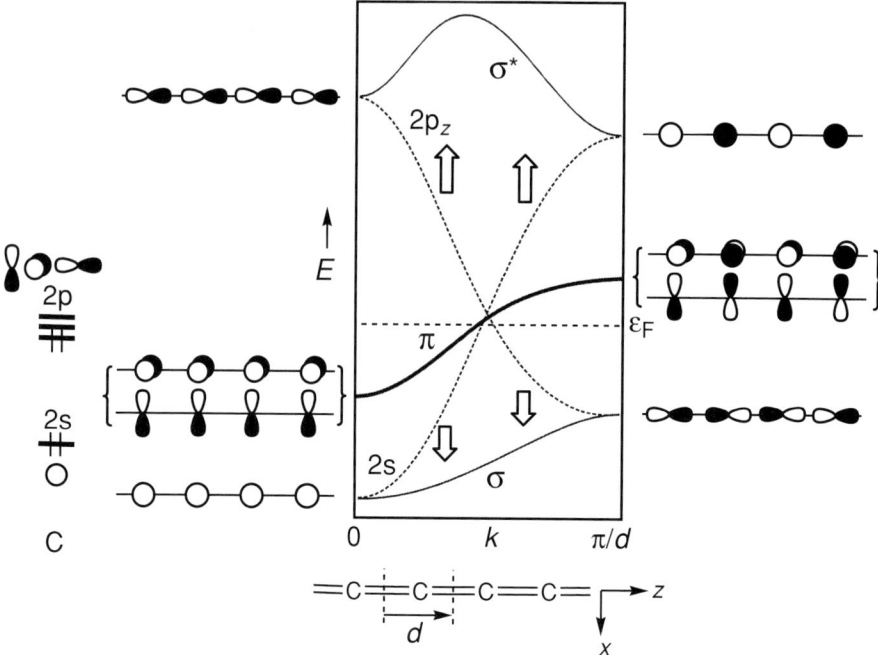

Figure 6.12

possible allotropic form of C and we will see in Chapter 7 that one-dimensional structures with sp-hybridized atoms satisfying the octet rule exist. Let's start with a regular chain in which all C–C bond distances are equal. These are, of course, double bonds. We now know how to tackle the band structure problem. First, identify the repeat unit (a single C atom) and the associated orbitals (2s, $2p_x$, $2p_y$ and $2p_z$). The next step is to build four Bloch functions associated with each of these AOs and then consider symmetry allowed interactions between COs of different Bloch functions. The four Bloch functions are easy to build, if we remember that: (i) for $k = 0$ the coefficients are all equal (i.e., with the same sign), whereas for $k = \pi/d$ their absolute values are equal but they exhibit sign alternation along the chain; (ii) they all have the same pseudo-sinusoidal or -cosinusoidal shape, depending if they are all bonding or all antibonding at $k = 0$ and vice versa at $k = \pi/d$; (iii) the width (or dispersion) of the band associated with a Bloch function depends on the overlap between the AOs (or MOs) on which it is expanded; (iv) their inflection point (i.e., $E(\pi/2d)$, roughly the middle of the band) is non-bonding between first neighbors and therefore should lie approximately at the energy of the AO (or MO) from which the Bloch function is generated.

It follows that the $2p_x$ and $2p_y$ bands are of π type, degenerate, and approximately centered on the energy of the C 2p shell (Figure 6.12). In similar fashion the bands associated with the 2s and $2p_z$ AOs can be generated (dotted curves in Figure 6.12).

6.2 The electronic-structure solution in a 1-D world

The band widths of the first two are less than the second two since π overlap is smaller than σ overlap. The 2s exhibits a positive slope and is roughly centered around the C 2s energy whereas the $2p_z$ exhibits a negative slope and is centered approximately on the C 2p energy. Once again, the negative slope of the $2p_z$ curve arises from the fact that the all-in-phase function at $k = 0$ is antibonding, whereas the all-out-of-phase combination at $k = \pi/d$ is bonding as shown in Figure 6.12. As a consequence, the two σ bands cross each other.

Although the π electronic structure of the C chain is well described by the π bands of Figure 6.12 (the π orbitals do not interact with the σ orbitals), this is not the case for the 2s and $2p_z$ Bloch functions. Both are of σ type and interactions between the corresponding two bands must be considered. As stated before, interaction between the $[CO^{2s}]_k$ and $[CO^{2p_z}]_{k'}$ functions is symmetry allowed only for $k = k'$. A series of two-orbital interactions remain, one for each value of k. Additional symmetry properties forbid interaction for $k = 0$ and $k = \pi/d$ (see Section 6.2.2 and Exercise 6.3). The consequence is that as k varies from 0 to π/d, the interactions go from zero to zero through a maximum roughly situated in the middle of the k range. Thus, in terms of the overlap criterion the two-orbital interactions should be stronger in the middle zone of the k range. However, an interaction between two orbitals depends also on their energy difference: the smaller the difference, the larger the interaction. As the energy difference is closest to zero near the point where the two bands cross, both the overlap and the energy difference parameters favor a strong interaction roughly in the middle zone of the k range. The resulting strong band repulsion is revealed in the solid line σ-type bands shown in Figure 6.12. Thus, the final band structure of the C chain is composed of the three solid line curves of Figure 6.12. Remember that the bold one (π-type) is doubly degenerate.

In contrast to the earlier examples, the σ-type bands have complex shapes since the various interactions at play are no longer simple. This complexity carries over to the DOS curve; hence DOS curves have shapes of little generality. In addition, although of σ-type, these bands have moderate dispersions as the repulsion in the middle of the k range flattens them. The lowest one is now bonding all along the k range with its character varying continuously from 100% 2s to 100% $2p_z$. Let's call it the σ band. The highest one is antibonding all along the k range with its character varying continuously from 100% $2p_z$ to 100% 2s. Let's call it the σ^* band. The σ band is occupied and the σ^* band is vacant (see why below). Clearly, these bands are associated with the σ component of the C=C double bonds of the chain.

There are four valence electrons on C which means that we have enough electrons to fill two of the four bands of Figure 6.12 (solid lines). The low-lying σ band will be fully occupied but it is not possible to fill just one of the two degenerate π bands. To do so would leave levels in the other π band at lower energy

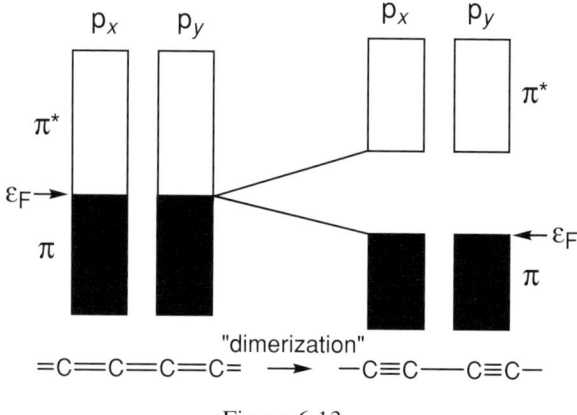

Figure 6.13

empty. Thus, the low-spin ground state corresponds to the filling of the two lowest halves of the degenerate $2p_x$ and $2p_y$ π bands as shown in the left hand side of Figure 6.13. All the bonding π states (or levels) are occupied, whereas all the antibonding π states are empty. Thus, we have two π-type bonding electrons per unit cell, in full agreement with a regular chain of C=C double bonds. A lot of work for such a simple result!

But don't get too happy and conclude that a regular C chain in which all the atoms are doubly bonded is a good candidate for a one-dimensional allotropic form of C. Such a structure is Peierls unstable since it has not one, but two, half-occupied π bands. A "dimerization" is expected, i.e. a distortion which would create alternating long and short bonds. By opening a gap at the Fermi level, such a distortion stabilizes the occupied π states (Figure 6.13). This distorted structure can be viewed as composed of alternating single and triple bonds which are conjugated. That is, the single and triple bonds are shorter and longer than standard ones, respectively. This one-dimensional C allotrope is named carbyne and it is presently unknown. It will be considered further in Chapter 7 where some of the problems in realizing this C form will be discussed.

If the process of generation of the band structure still seems strange go back and review Section 1.1.2 where the MOs of an E_2 diatomic molecule were developed. By consulting Figure 1.1 and Table 1.2 you will recall that the eight valence functions produce a set of 4 σ orbitals (2s, $2p_z$) and two pairs of orthogonal π orbitals ($2p_x$, $2p_y$). The complex σ manifold is a consequence of the fact that the 2s and $2p_z$ functions have the same symmetry and mix. As a result, the lowest energy orbital has dominant 2s character while the highest energy orbital has high $2p_z$ character. Parallels with the process used to develop the band structure for an infinite E_N chain should be clear. If you wish to further solidify the connections, generate a folded Figure 6.12 by using C_2 repeating units.

6.2.5 Hypothetical one-dimensional homoatomic transition-metal chain

The replacement of main-group atoms in clusters by transition-metal atoms generates a richer structural chemistry superimposed on the cluster basics illustrated by the p-block systems. A logical question arises here. What would a one-dimensional material containing a transition metal look like? Well, the d AOs will generate bands in a similar manner as the s and p orbitals. The major novelty will be the introduction of orbitals of δ symmetry. Let's look at a hypothetical chain composed of equidistant Ni atoms ($d = 2.5$ Å). The computed band structure, DOS and COOP are illustrated in Figure 6.14. The COs at $k = 0$ and π/d are drawn below. As a review of the previous section, we will reconstruct it starting from the Bloch functions associated with the nine Ni AOs.

Following convention, the z axis lies in the chain direction. Thus we have three AOs of σ type (4s, $4p_z$ and $3d_{z^2}$), two sets of degenerate AOs of π type ($4p_{x,y}$ and $3d_{xz,yz}$) and two degenerate δ-type AOs ($3d_{xy, x^2-y^2}$). Recall that orbitals of δ symmetry have two nodal planes containing the rotational axis, whereas the orbitals of π symmetry have only one nodal plane containing the rotational axis. Each of these AOs generates a Bloch function with corresponding $E(k)$ vs. k band widths that depend on the overlap between nearest neighbors. Overlap varies with distance, of course, but also it increases in the order $\delta < \pi < \sigma$. Thus, the two degenerate bands generated by the d_δ orbitals are the narrowest. For σ-type orbitals, the overlap also depends strongly on the quantum number n and follows the order $4p_\sigma \sim 4s > 3d_\sigma$. As a result the bands generated by the $4p_\sigma$ and 4s AOs are more dispersed than the one associated with the $3d_\sigma$ AOs. This is the simple part. We now must consider the s/p/d interactions. We saw with the C chain that the final σ-type band structure becomes complex due to an avoided s/p band crossing. The same occurs here as well as inter-mixing of the three metal AOs at any k point (except for the $4p_\sigma$ combinations which remain pure by symmetry at $k = 0$ and $k = \pi/d$). The net result is three σ bands of major 3d, 4s and 4p character for the lowest, intermediate and highest bands, respectively. The doubly degenerate bands generated by the $3p_\pi$ and $3d_\pi$ Bloch functions are weakly dispersed and well separated in energy. Note that the $3d_\pi$ band has a negative slope. Except at $k = 0$ and $k = \pi/d$, they repel on interaction and become flatter.

Compare the DOS in Figures 6.12 and 6.14. It is the larger number of orbitals in the repeat unit of the metallic chain that generates the additional complexity. You will recall that it was often useful to discuss a complex MO in terms of its AO components. Similarly, the DOS can be analyzed more deeply by decomposing it into its AO components, i.e., by looking at the contributions of the individual AOs to the total DOS. These contributions are called DOS projections and Figure 6.14 shows the 4s, 4p and 3d projections. For the sake of simplicity we have separately

230 Transition to the solid state

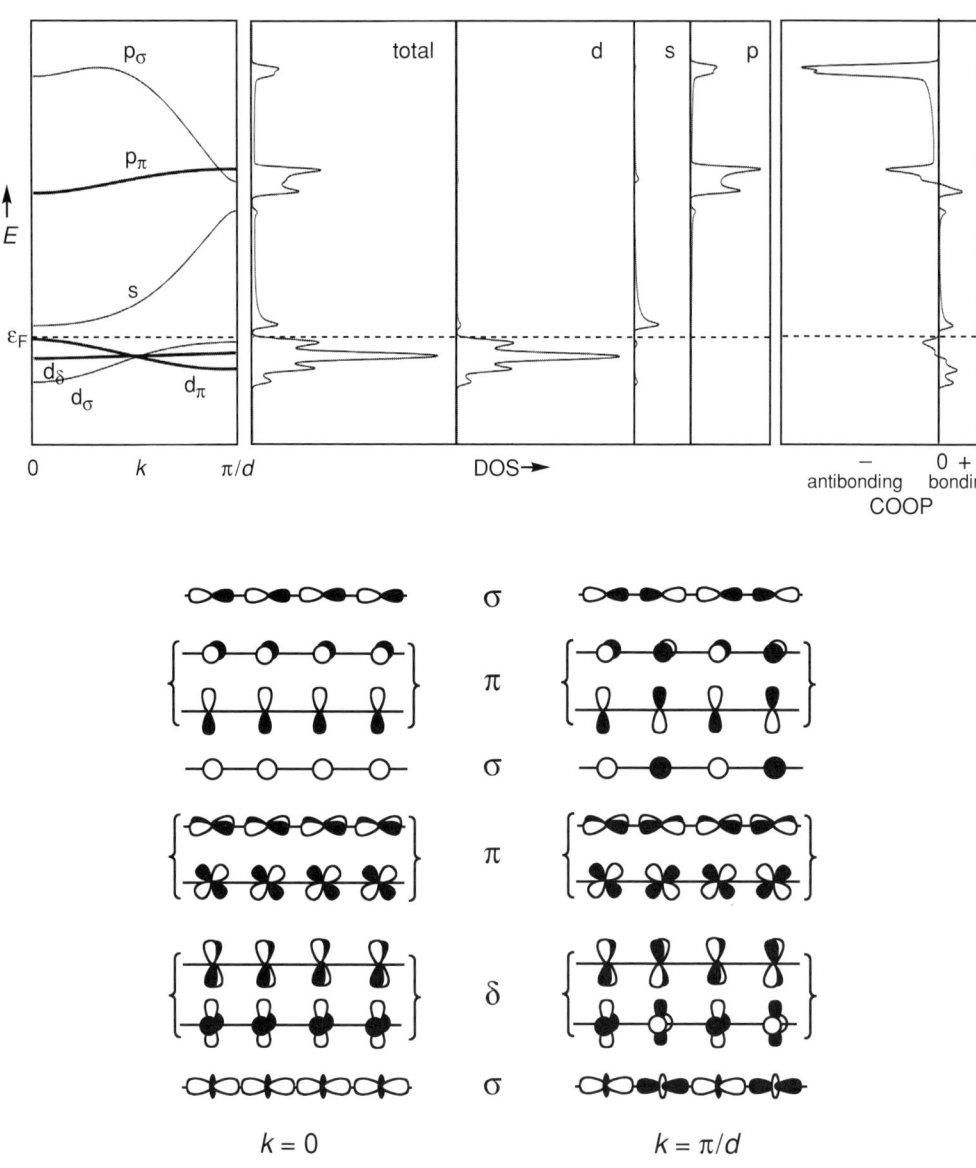

Figure 6.14

summed the three 4p AOs and the five 3d AOs. The 4s character is spread out over the whole DOS energy range so that its contribution is weak everywhere. A quite similar situation occurs for the 4p character in the upper part of the DOS. On the other hand, the projection on the more contracted 3d orbitals shows that their character is largely concentrated on the lowest part of the DOS, which is called the d-band. These results are sketched in a very crude way in Figure 6.15. We will take the same approach with the electronic structure of real three-dimensional metals.

6.2 The electronic-structure solution in a 1-D world

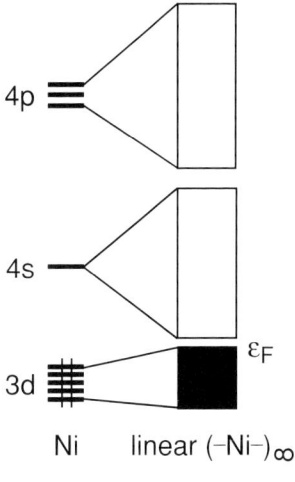

Figure 6.15

Note that this analysis is simplified. The 4p and 3d projections include orbitals of different symmetries whereas a separation between σ, π and δ is preferred for a careful analysis of the bonding in the chain.

The Ni–Ni COOP curve illustrated on the right side of Figure 6.14 reflects the band structure shown on the left side. Each individual band is bonding in its lower part and antibonding in its upper part. The bonding/antibonding character is more pronounced on the more dispersed bands. Flat bands are merely non-bonding at any k point. The COOP curve is the result of all these effects.

Finally, let's discuss the position of the Fermi level. The 10 valence electrons of Ni occupy the five lowest $E(k)$ vs. k bands all of which have dominant 3d character. There is a band gap at the Fermi level which separates the d band from the conduction band. Hence, this analysis suggests the hypothetical chain would be a semiconductor!

6.2.6 Hypothetical one-dimensional heteroatomic chains

In Chapter 1 we saw that in moving from homonuclear to heteronuclear diatomics a new factor enters – the atom characters are distributed differently over the filled and unfilled MOs. As only the filled orbitals contribute to the atomic charges, the Mulliken charge distribution reflects the polarity of the molecule. Similar information for the HOMO and LUMO permitted us to discuss properties such as Lewis acidity and basicity in terms of frontier-orbital characteristics. As we were able to unravel the DOS of the metal chain in terms of AO type, we can also interrogate the DOS of a heteroatomic system for information on the distribution of atomic character over the total DOS. That is, we can reveal the contributions or character of a chosen atom to the DOS. We can begin to appreciate the power of this tool by

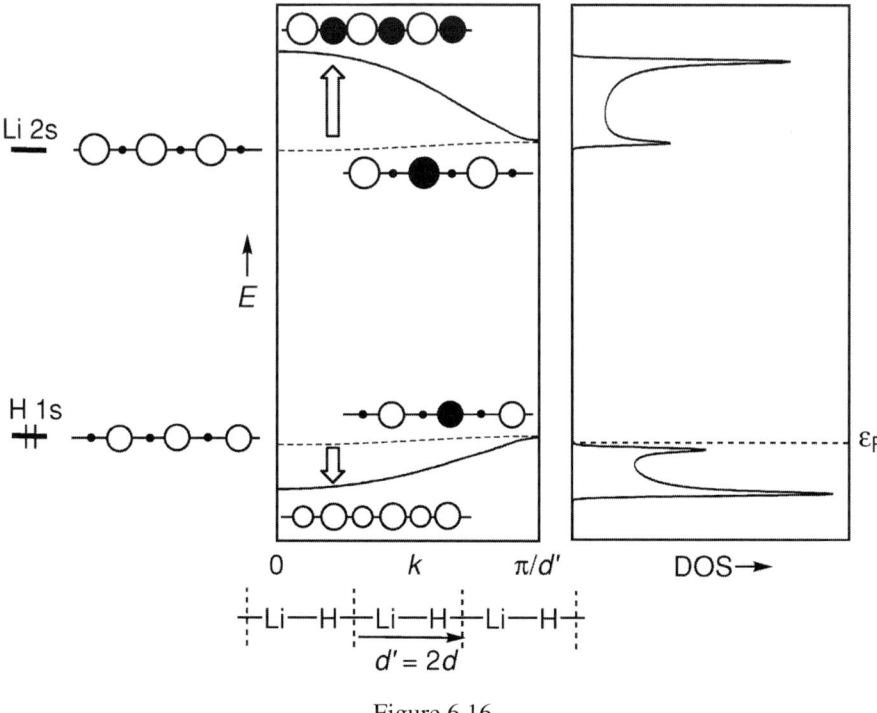

Figure 6.16

observing how an extended chain of alternating Li and H atoms behaves relative to the LiH diatomic molecule considered in Chapter 1. In this hypothetical model, we discard the Li 2p orbitals and consider only the Li 2s.

The first thing to be recognized is that a single atom can no longer serve as the repeating unit. This must be the LiH fragment even if all Li–H distances are equal. Hence, we have to consider the behavior of two orbitals under the translational symmetry of the chain with repeat distance $d' = 2d$, where d is the Li–H bond distance. The best way to solve the problem is to use the approach of Exercise 6.3 and consider two interpenetrating subnets of Li and H. As seen earlier for the H chain of Exercise 6.3, each subnet generates a flat (almost horizontal) band (dotted curve in Figure 6.16). The major difference is that now the two bands are not degenerate. The Li band is centered approximately at the energy of Li 2s which is significantly higher than the energy of H 1s, i.e., H is more electronegative (or less electropositive) than Li.

Now let the bands interact. The bands repel each other as they did in the H problem of Exercise 6.3. They mix in such a way that the lower band (valence band) will have dominant H character and the higher band (conduction band) dominant Li character. At $k = \pi/d'$, there is no interaction by symmetry. Unlike in the regular H problem these two orbitals are no longer degenerate and their linear combinations are not

6.2 The electronic-structure solution in a 1-D world

equivalent. Like the H chain, the lowest band is bonding overall and the highest is antibonding (solid line in Figure 6.16). The lower band is filled (two electrons per unit cell) but now we have a charge distribution of negative H and positive Li. There is a band gap at the Fermi level and the material will be a semiconductor or insulator depending on the size of the gap. So, unlike the isoelectronic regular H chain, the LiH chain is not subject to Peierls instability.

Exercise 6.5. Draw the band structure associated with the π orbitals of an infinite B=N chain which is isoelectronic to the regular C chain analyzed in Section 6.2.4.

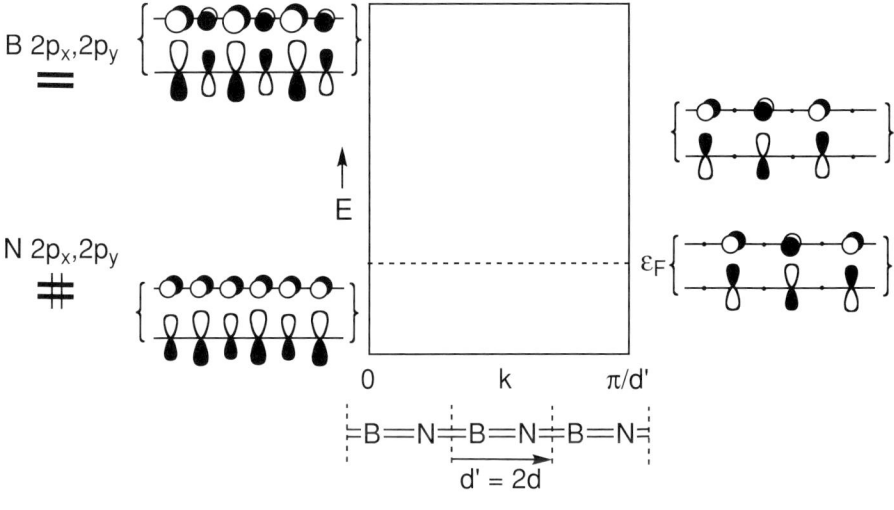

Exercise 6.5

Answer. The repeat unit in this system is B=N, i.e., the elementary translation is $d' = 2d$ (with $d =$ B–N). The π bands are generated by the $2p_x$ and $2p_y$ AOs of B and N. Again, the simplest way to solve the problem is to consider two interpenetrating networks of B and N subsystems. Each of these subnets generates a doubly degenerate π band, which is flat (no significant interaction) and approximately centered at the energy of B 2p and N 2p for the B and N bands, respectively. Both degenerate bands interact strongly at $k = 0$ and do not interact by symmetry at $k = \pi/2d$. The result is qualitatively similar to that of the LiH chain, except that the bands are doubly degenerate.

To summarize, we now have two ways by which the degeneracy at the Fermi level found for the C atom chain can be removed. Bond alternation along the chain is one way and substitution of the C atoms by two different alternating atoms, keeping the total electron count constant, is the other. Both require a doubling of the elementary translation (or unit cell).

Exercise 6.6. One answer for Problem 6 of Chapter 2 is *closo*-1-THF-2-PB$_5$H$_4$. It is a monomer from which one might construct a square joined by P–B donor–acceptor interactions. With *closo*-1-THF-6-PB$_5$H$_4$ an extended chain of clusters can be constructed. (a) Draw the chain, define the repeat unit and distance, and draw the pertinent orbitals of the repeat unit. (b) Develop qualitative band and DOS diagrams and predict whether the solid would have a band gap or not. (c) Partial reduction of the chain by adding 10 mole% Li is carried out. Assume the electron goes into a cluster-based orbital and the closed cluster structure is retained. Predict the change in the electronic properties.

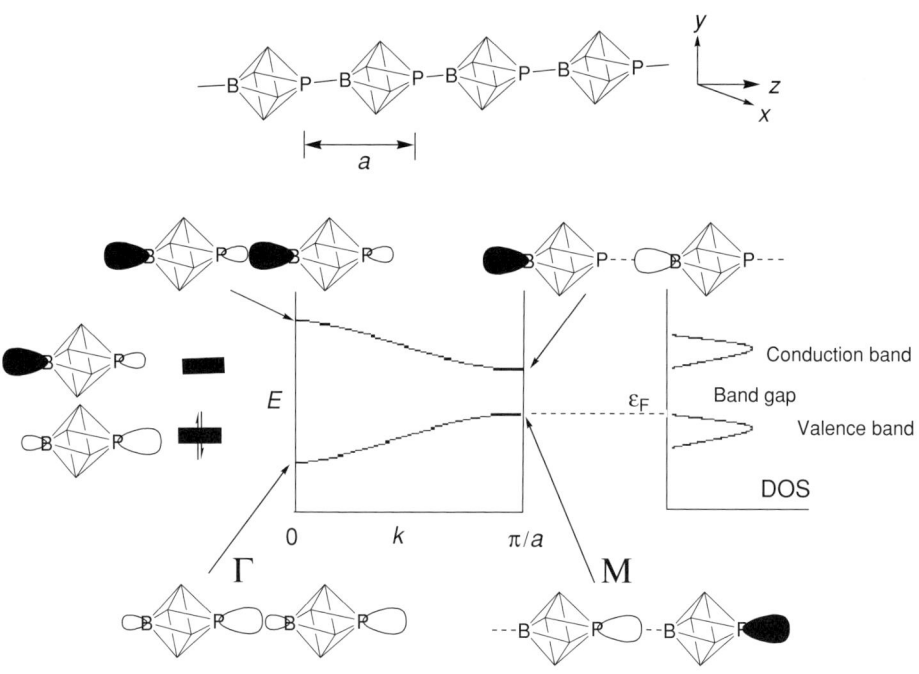

Exercise 6.6

Answer. (a) In the drawing above only the frontier donor and acceptor orbitals of the complex cluster repeat units are considered. The rest of the cluster orbitals are assumed to be only slightly perturbed by chain formation. (b) This problem is similar to that of the LiH chain, with a σ-type frontier orbital on each P and B atom in the 1,6-positions of the octahedral *closo*-cluster. These fragment orbitals are external cluster orbitals pointing in the direction of the next cluster. As shown below, this leads to two bands with the top of the lower one being largely of P character and the bottom of the upper one largely of B character. The lower part of the valence band should have high P character and P–B bonding character, whereas the top of the conduction band should have high B character and P–B antibonding character. The band gap should be significant. The lower band will be completely filled leading to semiconductor behavior. (c) Reducing 10 % of the repeating units will lead to population of the conduction band suggesting the doped chain should display metallic conductivity, i.e., behave as a "molecular wire."

6.2.7 What we have learned from one-dimensional models – a summary

Aren't you impressed by the elegant simplicity of this solution to our problem of 10^{23} orbitals? The molecular problem is of manageable size as the number of MOs depends on the number of basis functions used to describe the electronic structure of the atoms in the molecule. The periodic extended structure problem is of manageable size as we can partition the 10^{23} MOs formally required into a limited number of bands of COs using a periodic description of geometric structure. We now have all of the parts we need to mimic the parameters used to discuss molecules. They just have different names when applied to solid-state, extended systems. Here is a summary, our Rosetta stone, connecting small and very large systems:

molecular orbital (MO) ↔ crystal orbital (CO)
MO energy level distribution ↔ DOS
HOMO ↔ top of the valence band or Fermi level
LUMO ↔ bottom of the conduction band
HOMO–LUMO gap ↔ band gap at the Fermi level
Mulliken population analyses (charge and overlap) ↔ partitioning of DOS into atomic or fragment contributions and COOP.

Just as a MO diagram for a molecule is built, the band structure (the final COs on the diagram) of a one-dimensional compound is generated by summing all the orbital interactions. A systematic way to proceed is:

(1) Identify the elementary unit cell (the smallest repeat unit).
(2) Build the MO diagram associated with the contents of the unit cell (in the simplest case of an atom repeat unit these are its valence AOs).
(3) Build the $E(k)$ vs. k diagram of the Bloch functions associated with each of the unit cell MOs (or AOs). The shape of these bands is pseudo-sinusoidal/cosinusoidal.
(4) Turn on symmetry allowed interactions between the Bloch functions at each k point (repulsion and mixing between associated bands) to generate the final COs. In the general case, the shapes of the resulting bands will reflect characteristics of the specific compound.

Application of these simple tools to extended structures allows many to be partitioned into molecular fragments thereby drawing explicit connections between molecular and crystal electronic structures. We will use these concepts in the last part of this chapter to deal with the large cluster systems that lie somewhere in between small molecules and bulk materials.

6.3 Complex extended systems

The solid-state world we want to understand is not one-dimensional in general and the repeating unit is not as simple as an atom or diatomic molecule. To give you a sense of the problem, we will point out the directions in which these complicating factors take us and discuss, in qualitative terms, the solutions. Fortunately the concepts learned by treating a one-dimensional system provide the means to do so.

6.3.1 Two- and three-dimensional extended systems

What happens when we add dimensions? Well, as we found for clusters vs. small molecules, life becomes more complicated. Remember that k belongs to reciprocal space. In a one-dimensional space k is reduced to a scalar (see Section 6.3.2). In a three-dimensional space, k is a vector of components k_x, k_y and k_z and, of course, one now has to comprehend how the $E(k)$ energies depend on k_x, k_y and k_z. Even worse one has to convey the essential elements of band structure (E vs. k_x, k_y and k_z) in two-dimensional pictures. This is a common problem in science. If you think this problem is tough, consider a potential energy surface describing a reaction of polyatomic molecules where one needs a plot of E vs. $3N - 6$, where N is the number of atoms in the molecules! Hopefully, as in the one-dimensional case, the $E(k)$ functions are periodic. Unfortunately, the period (the first Brillouin zone) is no longer a segment in k space (or reciprocal space) but rather a volume centered at the origin of the k space and defined by a polyhedron whose shape and dimensions depend on the crystal system and cell parameters. The first Brillouin zone, which may have a complex shape, is nothing more than an equivalent representation of the reciprocal unit cell (reciprocal elementary repeat unit).

Simplification is necessary. We pick certain crucial points in k space, situated on the surface of the irreducible part of the first Brillouin zone, and see how the bands vary between the points. Recall we did a similar thing in the one-dimensional systems by focusing attention on the k points 0 and π/d. Remember at the $k = 0$ point a repeat unit function is taken $++++$ in the CO whereas at $k = \pi/d$ it is taken $+-+-+-$. Thus, for three dimensions we focus our attention on the points where k along x, y and z can have 0 and π/d. The four points that result, called symmetry points, are expressed in the form: *symbol* (k_x, k_y, k_z). They are: Γ (0, 0, 0), M (π/d, π/d, π/d), K (π/d, π/d, 0) and χ (0, 0, π/d). We will consider an extended cubic array of H atoms (the repeat unit is still an H atom), where d is the H–H distance. The points Γ (0, 0, 0), M (π/d, π/d, π/d) correspond to $k = 0$ and

6.3 Complex extended systems

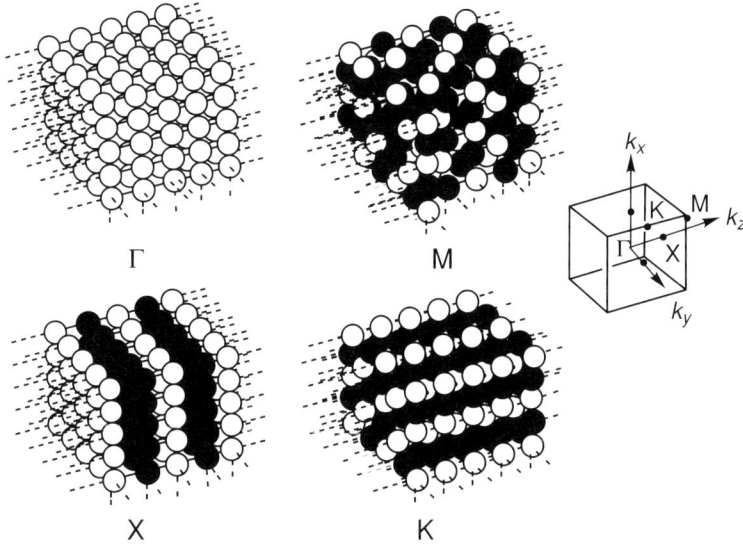

Figure 6.17

π/d in the one-dimensional case. The COs for a three-dimensional square lattice of H atoms at the four symmetry points are illustrated in Figure 6.17. Do you see the connection with the one-dimensional case?

Exercise 6.7. No doubt you see for K that the points at $(\pi/d, \pi/d, 0)$, $(\pi/d, 0, \pi/d)$, and $(0, \pi/d, \pi/d)$ have the same energy. The same degeneracy holds for X. Sketch them in the manner of Figure 6.17.

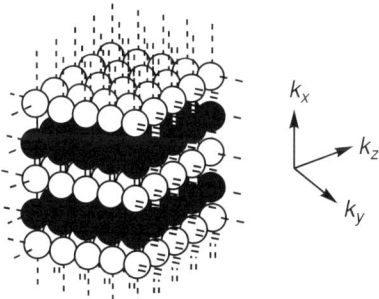

Exercise 6.7

Answer. For X $(0, 0, \pi/d)$ the sign of the H 1s contribution is unchanged in the k_x and k_y directions but alternates in the k_z direction (shown in Figure 6.17). For $(0, \pi/d, 0)$, the xz planes have the same sign and the sign alternates in the y direction. For $(\pi/d, 0, 0)$, the yz planes have the same sign and the sign alternates in the x direction as shown in the drawing above.

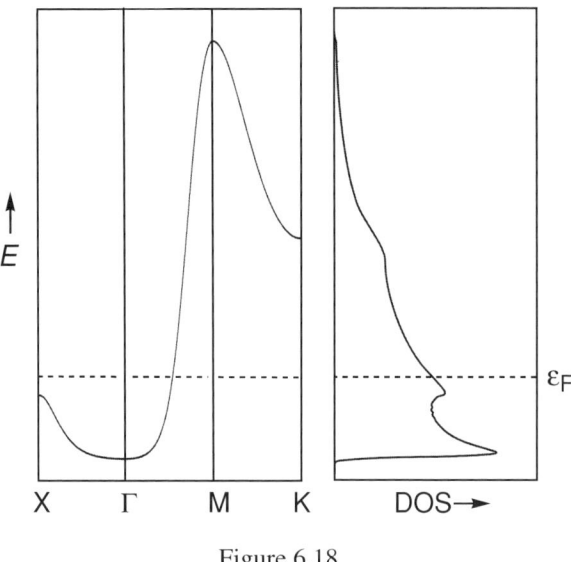

Figure 6.18

The band structure of cubic H_n is shown in Figure 6.18 in conventional form. Γ corresponds to the lowest energy as it is bonding along x, y and z (Figure 6.17), whereas M corresponds to the highest energy as it is antibonding along x y and z. The point K is antibonding along x and y, but bonding along z (each xy sheet is totally antibonding but the sheet–sheet interactions are bonding). Hence, at this symmetry point the CO is mostly antibonding. Point X is bonding along x and y, but antibonding along z (each xy sheet is totally bonding but the sheet–sheet interactions are antibonding). Hence, it corresponds to a mostly bonding CO. Both then will have energies between those of Γ and M. Finally, the four points are joined with a smooth curve, as shown on the left side of Figure 6.18. The corresponding DOS is shown on the right side. In a more complex structure, intuition alone is insufficient and calculations would be carried out at many points in k space and calculated energies plotted. A wiggly curve analogous to that in Figure 6.18 will result from each basis function yielding diagrams sometimes referred to as spaghetti diagrams because of the tangle of bands. Despite this, the resulting DOS and its interpretation employ the H-atom model as a guide in the same way as the H_2 model is used as a guide in molecular chemistry.

Exercise 6.8. Work out the band structure for a two-dimensional square array of H atoms in the xy plane separated by a distance d.

Answer. We need to find the COs for the points in k space (k_x, k_y); Γ $(0, 0)$, M $(\pi/d, \pi/d)$ and X $(0, \pi/d)$. You know that 0 means no change in sign of the function along the coordinate specified and π/d means an alternating sign. Hence,

6.3 Complex extended systems

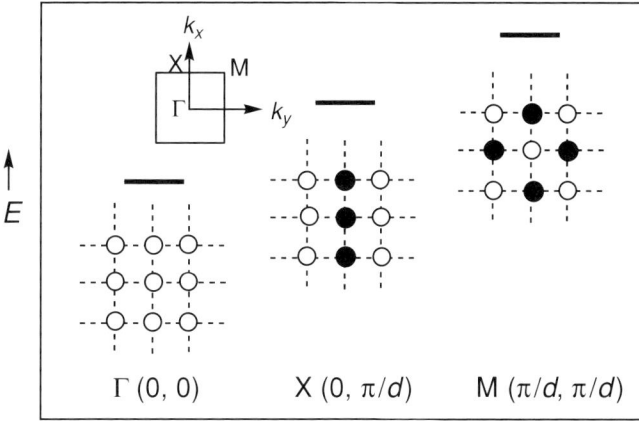

Exercise 6.8

Γ (0, 0) will have all H 1s functions in phase along x and y; M $(\pi/d, \pi/d)$ will have a change in sign for every lattice point along x and y; and X $(0, \pi/d)$ will have no change in sign along x and a change at every lattice point along y. These are shown in the drawing above. Note that Γ is all bonding, X is bonding within chains and antibonding between chains and M is completely antibonding between nearest neighbors. The ordering in energy shown should be obvious.

6.3.2 Complex periodic units

The H-atom chain is to solid-state structures as the diatomic molecule is to polyatomic molecules, e.g., clusters. The geometric-structure problems for H_2 and H_∞ are so simple that one can focus on the electronic structure problem exclusively. However, real solid-state structures, e.g., a solid with linked clusters or even bulk elemental Al, are not found in the form of a linear chain, square sheet or simple cubic structures so we need a way to treat solids with more complex structures, i.e., define a repeat unit that is more than a single atom.

What we need is a well-defined unit or building block that contains all the information necessary to completely describe the structure of the solid. This building block is the unit cell. With a unit cell we can generate any size crystal by translations of the cell, i.e., by stacking a sufficient number of unit cells together like the bricks in a building. The unit cell is defined by the pattern of atoms (or ions or molecules) in the material; however, the points that define the unit cell do not necessarily lie at atom centers. All that is necessary is that stacking the unit cell regenerates the crystal. Just as there is more than a single way of defining the repeat distance on wallpaper, so too there are several ways of drawing a unit cell in a three-dimensional crystal of a substance. Hence, conventions are required to avoid confusion in communication. We only need to know how to use defined

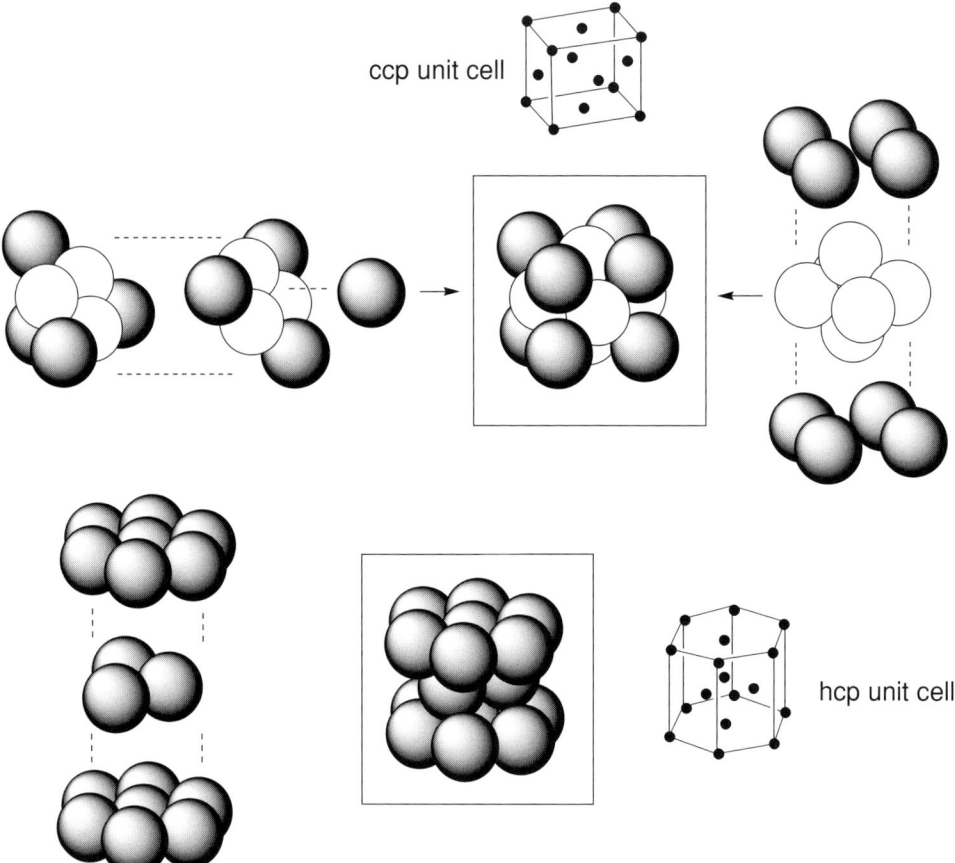

Figure 6.19

cells. Further discussion of existing conventions can be found in a standard text of crystallography.

Let us take a look at the unit cell descriptions of hcc and ccp (fcc) solids. In Figures 6.19 and 6.20 the unit cells are shown in "exploded" diagrams. The relationship between the hcc unit cell and the close-packed representation is easy to see. That for the ccp unit cell is a little more difficult as the planes of close-packed atoms now cut diagonally through the cubic cell. However, as can be seen, the unit cell can be constructed from fragments of close-packed planes as well as by face-capping (dark spheres) an octahedron (white spheres). Pay particular attention to the fact that the cell is defined by the lattice points (centers of the spheres) rather than the atomic surfaces so, e.g., in the ccp cell appropriate for Al only 1/8 of a corner atom resides in a single unit cell. Go back and take a look at Figure 6.1. The ccp cell is one of the capped clusters we excised from the bulk!

6.3 Complex extended systems

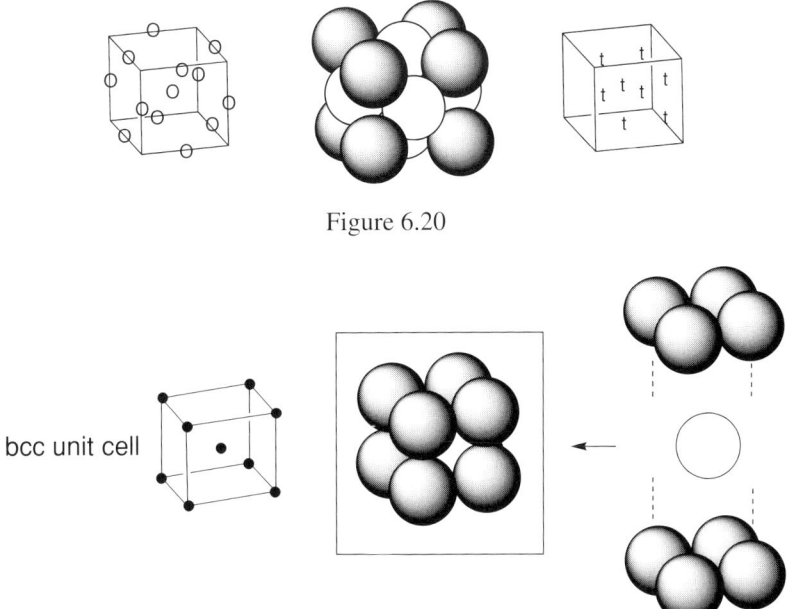

Figure 6.20

Figure 6.21

The beauty of the unit cell concept is that we now can describe a complex infinite array in terms of a small, finite entity with both characteristic geometry and, for compounds, composition. Thus, for example, the extended Al structure can be described by a ccp unit cell containing $1/2 \times 6 + 1/8 + 1/8 \times 8 = 4$ Al atoms (six face-sharing and eight corner-sharing atoms) per unit cell. The octahedral and tetrahedral holes can also be located in the ccp unit cell and are represented in Figure 6.20. Thus, one counts four octahedral holes ($1 + 1/4 \times 12 = 4$; one centered and 12 edge-sharing) and eight tetrahedral holes (eight within the cell). For more complex compounds the same type of exercise permits the composition of the solid to be obtained from its unit cell with ease: a not inconsiderable benefit.

Of course there are less efficient ways of packing hard spheres – the simple cubic lattice was dealt with already and group-1 s-block elements crystallize in a body-centered cubic (bcc) unit cell (Figure 6.21). In contrast to the ccp (fcc) and hcc structures above, each atom in the bcc structure only has eight nearest neighbors rather than 12. Because there are six next-nearest neighbors only 16 % further removed, space is still efficiently filled. The variations in structure are small and for metals with metal–metal bonding of low directional character, structural change with temperature (polymorphism) is common.

The differences between the structures of covalently bound elemental solids and that of Al, for example, are not as large as you might think. Thus, in Figure 6.22 the unit cell of the diamond structure (group 14) is compared with that of rhombohedral

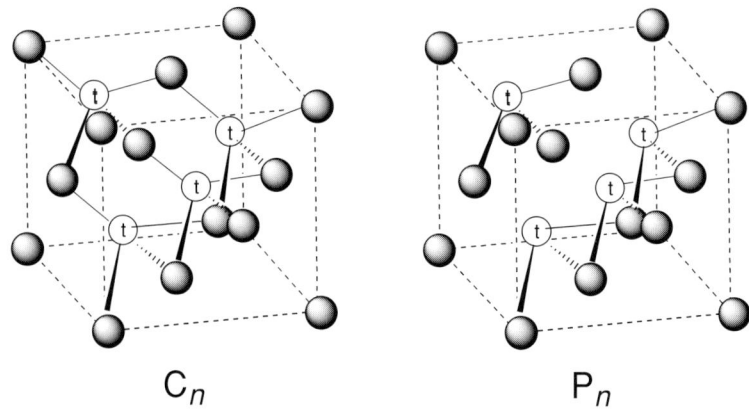

Figure 6.22

black phosphorus. Both may be described as ccp with half the tetrahedral sites (labeled with "t" in the drawing) filled with either C or P atoms. In an ideal close-packed fcc structure the tetrahedral sites can only accommodate a sphere 25 % the size of that situated on the lattice points. In the diamond structure all the atoms are of the same size and, consequently, the structure exhibits a considerably more open, less dense, lattice.

Exercise 6.9. A salt $[\text{cation}]_m[\text{anion}]_n$ possesses a structure with a ccp unit cell in which the anions occupy the lattice positions of the ccp cell and the cations occupy all of the octahedral holes. What are m and n?

Answer. A ccp unit cell contains $1/2 \times 6 + 1/8 \times 8 = 4$ atoms in the lattice positions and $1 + 1/4 \times 12 = 4$ octahedral holes. Hence, m and $n = 1$, i.e., the NaCl unit cell.

6.3.3 DOS for metals: examples of Al and Ni

To provide the other boundary point for the electronic structure of large clusters, e.g., our icon $[\text{Al}_{69}\text{R}_{18}]^{3-}$ and its Al_{77} partner, we need to generate the DOS for ccp Al metal. The repeating unit is the unit cell shown in Figure 6.19 and it must be repeated in three dimensions to generate the structure of bulk Al. We assume that the crystal size is large enough so that the exposed surfaces of the crystal constitute a small perturbation that can be ignored. But before dealing with the electronic structure, let's go a little bit deeper into the business of unit cells. You may be surprised now to learn that the ccp (fcc) unit cell is not the smallest repeat unit from which the whole crystal is generated by translation. In fact, the ccp (fcc) unit cell is the smallest unit containing all the symmetry properties of the crystal (it corresponds to what is named a Bravais lattice). The smallest unit cell

6.3 Complex extended systems

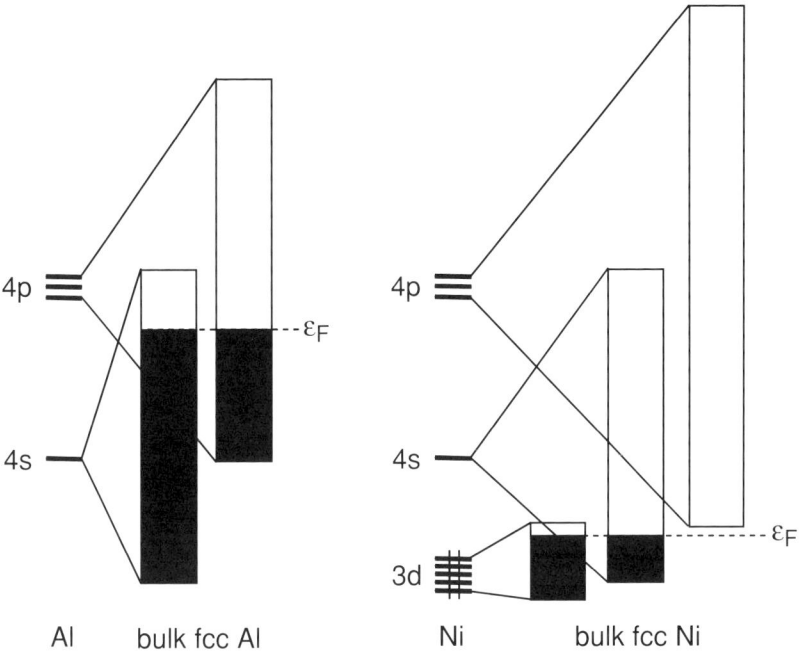

Figure 6.23

possible is rhombohedral and primitive, that is, it contains a single metal atom. For the definition of a Bravais lattice of a rhombohedral cell and for the connection between the fcc and rhombohedral cells in ccp crystals, the reader is again referred to a standard textbook of crystallography or solid-state physics. What is necessary to consider for the following is that it is a primitive lattice, i.e., it contains a single atom in contrast to the larger fcc cell which contains four atoms. With one atom per unit cell we are left with only four orbitals to deal with, i.e., Al 3s and 3p. For the following analysis, this is conceptually easier to handle qualitatively than a repeat unit made of an Al_4 cluster contained in the fcc unit cell.

Each of the four AOs of Al generates a Bloch function to which a $E(k)$ vs. k "band" is associated. Remember that k is a three-dimensional vector so that $E(k)$ vs. k defines a surface in the reciprocal space. Not simple to tackle qualitatively! Moreover, except for a few special k points situated on symmetry elements, all the Bloch functions interact at each k point and mix together. The best qualitative approach is to proceed as we did when building the crude diagram of the Ni chain sketched in Figure 6.15. Let's take into account the strong interaction between the 3s AOs on one side and that between the three 3p AOs taken together on the other side. This will generate two broad bands roughly centered at the energies of Al 3s and 3p, respectively (left side of Figure 6.23). A more precise description is given by the computed DOS and Al–Al COOP which are shown in Figure 6.24.

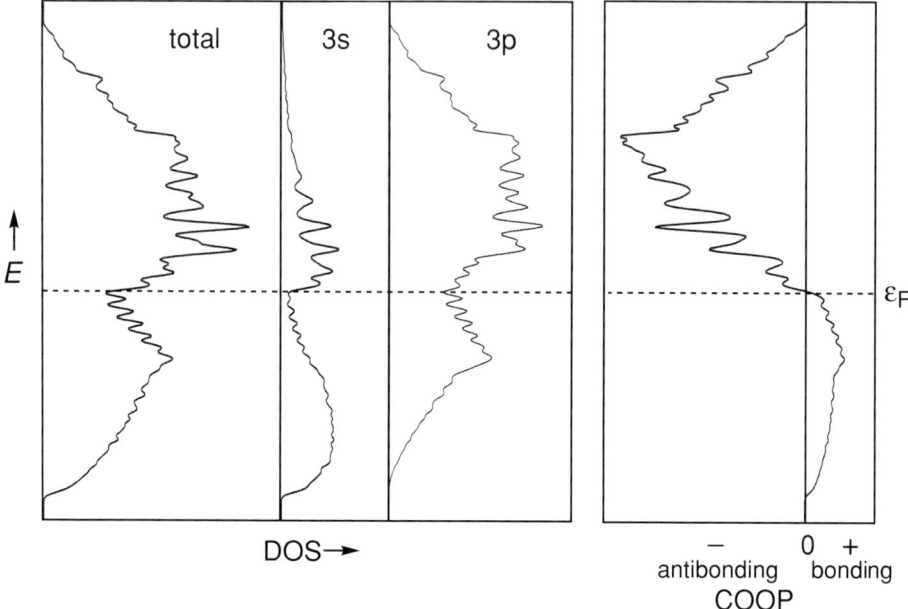

Figure 6.24

The COOP curve reflects the fact that the lowest parts of the 3s and 3p bands are bonding and their upper parts are antibonding. There are three valence electrons per Al atom so the band is 3/8 filled with the electrons in COs of Al–Al bonding character. Note that all the bonding levels are occupied whereas all the antibonding ones are empty. This is a mark of stability. There is no band gap at the Fermi level, consistent with the fact that bulk Al is metallic in character.

A similar approach is effective for ccp Ni. Considering a single Ni atom as the repeat unit, one is left with nine AOs which generate three bands, one for each valence shell, with multiplicities of five (3d), one (4s) and three (4p). From the calculated DOS and its projections (Figure 6.25), it is possible to draw the simplified Ni diagram of Figure 6.23 (right side). Note that because the 3d AOs are more contracted than 4s and 4p ones, the 3d band is much narrower. Compare this result with the mono-dimensional chain of Ni atoms discussed in Section 6.2.5. In the one-dimensional model, interaction (overlap) occurs only in one direction. In three-dimensional Ni, it occurs in the 12 directions of the 12 nearest neighbors of a given atom. The result is that the bands of three-dimensional Ni are much more dispersed than those of the one-dimensional model and they overlap. Thus, there is no band gap at the Fermi level. Unlike its one-dimensional model, three-dimensional nickel is metallic.

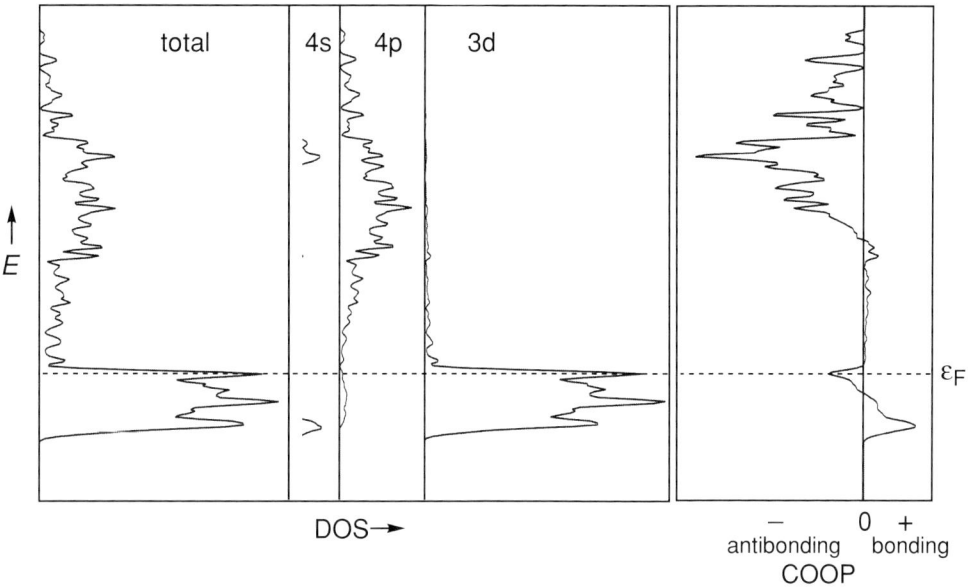

Figure 6.25

6.3.4 DOS for main-group elements: C graphite and diamond exemplars

A characteristic of the DOS of a metallic element is its large magnitude in the vicinity of the Fermi level. This feature, above and below the Fermi level, is associated with the highly delocalized character of the metallic bonding. However, most solid-state compounds are not metallic. Hence, we consider now the examples of graphite and diamond, two allotropic forms of elemental C. The chemically bonded network of the former is two-dimensional and that of the latter is three-dimensional. In Section 6.2.4 we presented the band structure of a hypothetical one-dimensional allotropic form of C. Zero-dimensional (molecular) forms do exist also; these are the fullerenes such as C_{60} which was mentioned in Chapter 2 and will again be discussed in Chapter 7. C nanotubes are intermediate between molecules and macroscopic solids and also will be considered further in Chapter 7.

In all these allotropic forms, the bonding is localized (two-center–two-electron bonds) and C satisfies the octet rule. In diamond, all the tetravalent sp^3 C atoms are equivalent (see Figure 6.22) and consequently the network is made of equivalent C–C single bonds. Reflecting this situation, the diamond DOS consists of a low-lying occupied σ band and a vacant high-lying σ^* band (Figure 6.26). That is, in the diamond structure the four C atomic orbitals mix together to lead to well-separated σ-bonding and σ^*-antibonding bands. This is analogous to the case of linear C where the C 2s and $2p_z$ AOs mix and repel each other to form the σ and σ^* bands

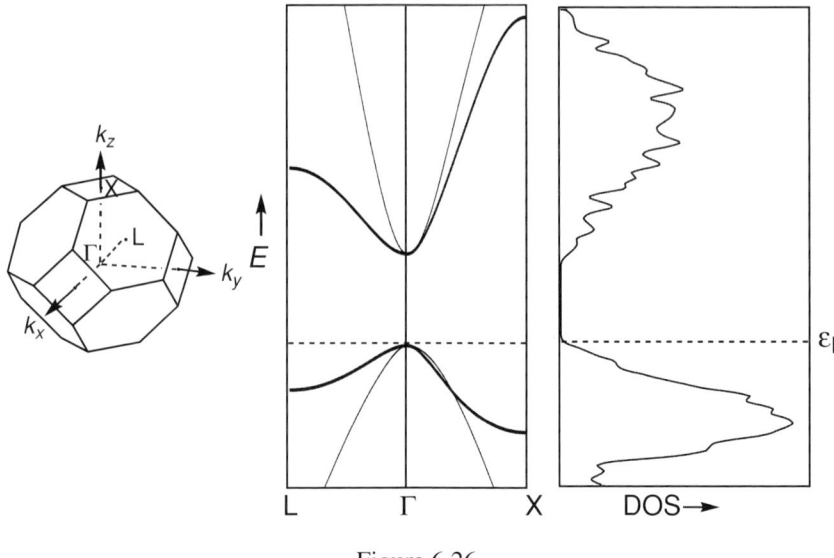

Figure 6.26

(see Section 6.2.3). The very large band gap at the Fermi level is indicative of the colorless and insulating properties of diamond.

The fusion of an infinite number of C_6 aromatic rings generates a sheet of graphite often called graphene. As with an aromatic molecule, the π-type and σ-type COs can be considered separately in this planar system and the π and π^* levels occupy energies intermediate between those of the occupied low-lying σ levels and vacant high-lying σ^* levels (Figure 6.27). Do you see the interesting difference? The occupied π and vacant π^* bands are not separated by an energy gap. They touch such that there is HOMO/LUMO degeneracy but the DOS is exactly zero at one unique point which is the Fermi level. This peculiar situation, imposed by symmetry, is intermediate between a semiconductor (non-zero band gap) and a conductor (no band gap and significant density of states at the Fermi level). Such systems are often called semi-metals. However, our model neglects the weak interactions between the planar sheets which slightly modify this ideal description.

6.4 From the bulk to surfaces and clusters

We finally arrive at a point where we can relate what we have learned to the principal theme of the text. Recall that small clusters essentially consist of surface whereas band theory applies to situations where the number of surface atoms is so small with respect to atoms in the bulk they can be ignored. In the Al_{77} cluster we have a shell-like structure of $Al_1@Al_{12}@Al_{44}@(AlR)_{20}$ or a ratio of $20/57 = 0.35$ surface to bulk. In addition, each of the 20 surface atoms is coordinated to

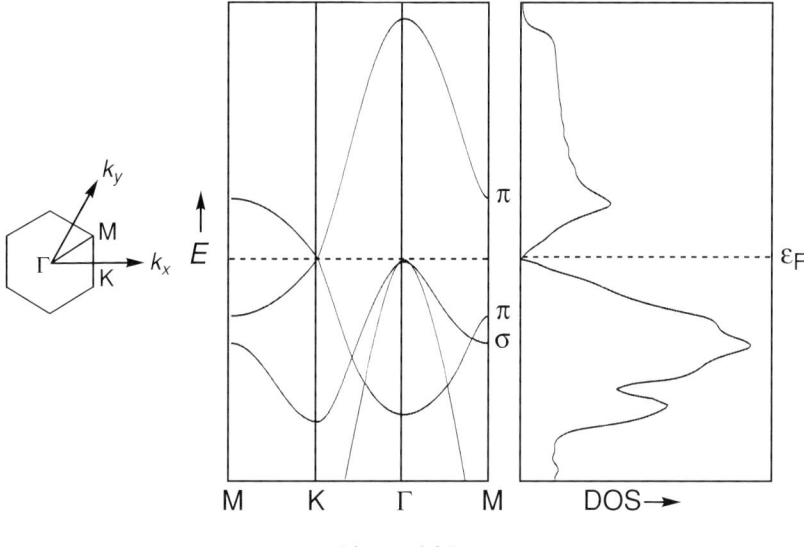

Figure 6.27

an external ligand. The Al_{69} cluster is similar. Now the bulk Al structure is ccp which can be considered to be built of fused Al_6 octahedra. So in this section let's compare the orbital structures of octahedral Al_6, related to $[(AlR)_6]^{2-}$, Al_{77}, related to $[Al_{77}R_{20}]^{2-}$ and ccp Al bulk metal. Note that although octahedral $[(AlR)_6]^{2-}$ is not known, icosahedral $[(AlR)_{12}]^{2-}$ is. Clearly a place to begin is to put the surface back into the bulk-metal problem because without it there is no connection.

6.4.1 Surface states

Surfaces can be conceptually generated by cleavage of a bulk material (which can be considered infinite) into two separated pieces, each of which exhibits a surface. The process is not so difficult to envision if you recall simple bond cleavage. Consider ethane H_3C-CH_3. In its MO diagram, the C–C bond is described by an occupied σ_{CC} bonding and a vacant σ^*_{CC} antibonding orbital (left side of Figure 6.28). Due to delocalization these MOs also have small H-atom character. The homolytic cleavage of the C–C bond leads to the formation of two non-bonding orbitals, one on each of the methyl radicals which are formed. Similarly, breaking the central bond of butane generates two non-bonding orbitals located mainly on the unsaturated C atoms (right side of Figure 6.28). In fact, whatever the length of a given linear alkane, when a C–C bond is broken, one always ends up with two non-bonding orbitals each mainly localized on *the* unsaturated atoms. Thus, breaking an "infinite" polyethylene chain in two pieces modifies its DOS as illustrated in Figure 6.29. The non-bonding level of one of the pieces, identified in our little exercise, is the "surface" state and the other levels are the "bulk" states. The "bulk"

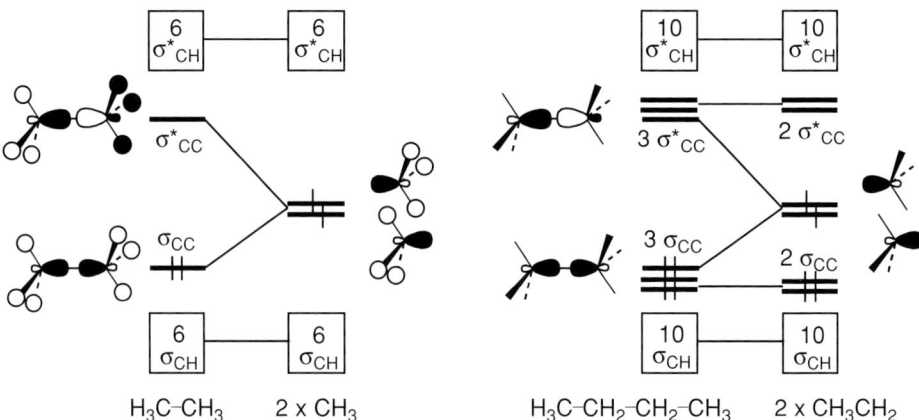

Figure 6.28

states are modestly perturbed by the bond cleavage but remain on various CC and CH bonds. There is one "surface" state for a very large number of "bulk" states.

The lesson taught by this exercise is that surface states are largely non-bonding in character and their number is extremely small as compared to bulk states for a solid sample of macroscopic size. The orbitals associated with the surface states are called "dangling bonds" in the jargon of solid-state scientists whether they are occupied or not. Since they are non-bonding this is an unfortunate choice of words. Despite their small number, these "dangling orbitals" are very important as they define the chemical properties of a surface. Depending on whether vacant, singly occupied or fully occupied, these orbitals induce electrophilic, radical or nucleophilic surface properties, respectively.

A "surface" made of a single atom is not very useful but a real two-dimensional surface can be generated by cleaving a three-dimensional extended structure into two pieces. Surface non-bonding orbitals will be generated on the atoms where the bonds have been broken. Since the surface is an extended two-dimensional array, the surface states will develop a band, i.e. the single surface orbital of Figure 6.29 is now a non-bonding band. Exercise 6.9 and Problem 1 are pertinent here. Although the non-bonding level of Figure 6.29 contains contributions from the "bulk" part of the molecule, delocalization is weak. The situation can be different in a real two-dimensional surface. In this case, the greater the magnitude of the overlap between nearest surface orbitals, the greater the dispersion of the surface bands and the more delocalized the surface states. The wider the surface bands the greater the mixing with the bulk states.

In fact, the surface of a transition metal is highly delocalized. Let's take a look at it. Focus on the d band where the Fermi level is situated regardless of the metal. For the sake of simplicity, we do not consider the overlap and mixing with the s

6.4 From the bulk to surfaces and clusters

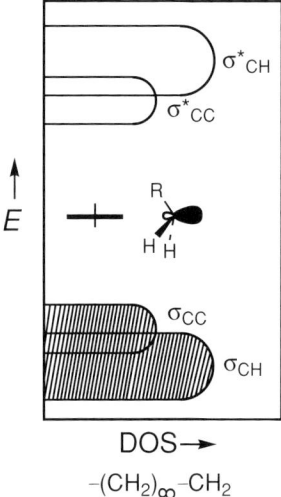

Figure 6.29

and p bands described in Section 6.2.5. But keep in mind that what we call the d band contains some supplementary levels provided by the overlapping s band (Figure 6.25). In a three-dimensional metal, the d AOs develop a band roughly centered at the energy of the metal d valence shell (non-bonding states). The lower part is bonding, the upper part is antibonding. So what happens to the surface states once the bulk has been cleaved in two half-infinite chunks? The atoms on the surface have lost several neighbor atoms. Delocalized bonds have been broken. Thus, all the orbitals with localization on the surface atoms have lost part of their bonding or antibonding character and their energy gets closer to the energy of the valence d shell. The associated density of states should be narrower. This is illustrated in Figure 6.30. For emphasis, the scale of the surface atoms projection is much larger than that of the total DOS. Remember that the surface states constitute only a tiny fraction of the total density of states in the case of a macroscopic solid. But because of the high delocalization of all the orbitals in the metal, there is no clear-cut separation of the surface and bulk states and a continuum is observed.

An interesting consequence of the narrowing of the surface DOS with respect to the bulk one is surface polarization. With late transition metals, the d band is more than half occupied, but even in the case of group-10 metals, this d band is not completely filled because of the supplementary levels provided by the overlapping bottom of the s band (see Figures 6.23 and 6.25). All the surface states are occupied, whereas some bulk states at the top of the d band are vacant. Consequently, the surface is polarized negatively. With early transition metals, there are fewer occupied surface states than bulk states. The surface is polarized positively. In both cases, the polarization of the bulk is diluted over so large a number of atoms that

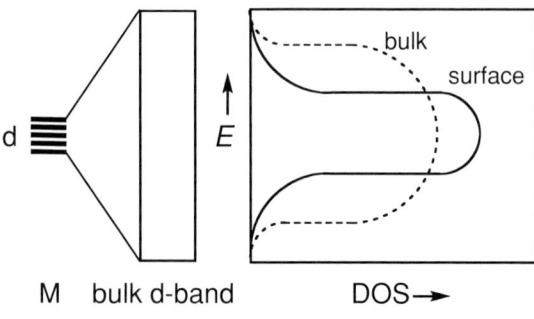

Figure 6.30

it is negligible. The bulk acts as an infinite electron reservoir/sink with respect to the surface and the level of the reservoir/sink (i.e. the Fermi level) is not affected by the existence of the surface.

Before returning to our correlation of clusters and bulk metals some caveats are in order. The important features of surface states have only been roughly sketched out here. Indeed, the detailed electronic structure of a surface depends on the details of its geometric structure which we have largely ignored. A surface can be irregular (with steps, kinks, islands ...) or regular. Regular surfaces of monocrystals are characterized by reticular (net-like) planes defined by their Miller indices (h, k, l). For the definition of such planes, please consult standard textbooks of crystallography. Surfaces with different Miller indices have different two-dimensional arrangements and therefore different electronic structures. The surface densities of states of Al (111) and Al (100) have different shapes and composition. For example, the former is more dispersed than the latter because the (111) surface is more compact than the (100) one. Besides that, one sometimes has to take into account rearrangements of the local two-dimensional structures of less compact surfaces to more compact ones, i.e., the surface reconstruction mentioned in Section 6.1.1. But these topics go beyond the scope of this text and the interested reader is referred to works on surface science.

6.4.2 Correlation between bulk-element COs and cluster MOs

Now that the qualitative aspects of surfaces are understood, we may use these ideas to address our problem of the connection between cluster MOs and bulk COs. Consider a crystalline film of a metal deposited on a planar support. Now the bulk, which constitutes the middle part of the film, can no longer be considered as infinite. Surface states will dominate the total density of states if the thickness of the film is not large. In fact, in a thin film consisting of a few layers of atoms the bulk no longer exists and the total DOS is composed of surface states. The DOS should be narrower than that of the bulk state and, if the surface is not dense, localized non-bonding orbitals will appear, generating "spikes" in the surface density of states.

Now, there is another way of conceptually generating a surface from the bulk material besides cleaving the infinite bulk to generate a pair of two-dimensional infinite surfaces. Consider extracting a piece of the solid from the bulk. We generate a particle of a certain size with a surface, which can be irregular or regular, and a bulk. The DOS of a metallic particle of a macroscopic size will look like that of Figure 6.30. Decreasing the size of the particle raises the surface/bulk atomic ratio. Down to a certain size, the DOS will look like that of a film. When the particle reaches the size of a cluster, discrete MO levels replace the (quasi) continuum of the DOS. Metallic nanoparticles lie at the boundary between macroscopic bulk particles (DOS) and molecular metal clusters (discrete MO levels).

We are now ready to consider a thought experiment in which a large crystal of fcc Al is subdivided again and again until one reaches the tiny Al_6 octahedral cluster (the isolation of a neutral Al_6 octahedron is an unlikely prospect; however, the beauty of MO models is that one is not restrained by reality). Each time a piece of the crystal is broken away, Al–Al bonding interactions are broken. Bonding and antibonding orbitals get more and more non-bonding and collapse to generate surface states closer to the center of the DOS. At the same time, more structure (spikes) will appear in the band as the particle becomes smaller and smaller. These spikes are on the way to becoming discrete energy levels. Ultimately, the MO levels of octahedral Al_6 emerge. This is illustrated in Figure 6.31 where calculated DOS of bulk Al and the MO levels of Al_6 are at the two sides of the diagram. The former was presented in Section 6.3.3 and the latter are easily derived from those of $[B_6H_6]^{2-}$ discussed in Chapter 2 (Figure 2.21). Ignore mixing and remove the six low-lying bonding B–H orbitals plus their six high-lying counterparts of $[B_6H_6]^{2-}$ (both sets of a_{1g}, t_{1u} and e_g symmetry), and replace them by six a_{1g}, t_{1u} and e_g combinations of non-bonding σ-type hybrids pointing outwards (dangling bonds) four of which are empty and lying in the HOMO/LUMO region (see also the comparison of $[B_6H_6]^{2-}$ and $[C_6]^{2-}$ in Figure 2.21). In the middle of Figure 6.31 we have inserted the DOS or MO levels (you choose) of an Al_{77} nanoparticle (the cluster on the cover). Well what do you think? The nanoparticle does have a large spike in the HOMO/LUMO or Fermi level region associated with the "surface" non-bonding orbitals. Some of these will become stabilized and bonding on addition of the 20 ligands to make up the experimentally isolated and characterized $[Al_{77}R_{20}]^{2-}$ nanoparticle cluster. Even so it is clear that the HOMO/LUMO gap will be small and that electronic-structure simplification to the level of a counting rule will not be valid (Section 1.3). Our attempts to apply various versions of cluster electron-counting rules failed for good reason.

In this chapter the electronic-structure model developed for extended main-group systems describes properties that do not depend on the surface states. It is simply

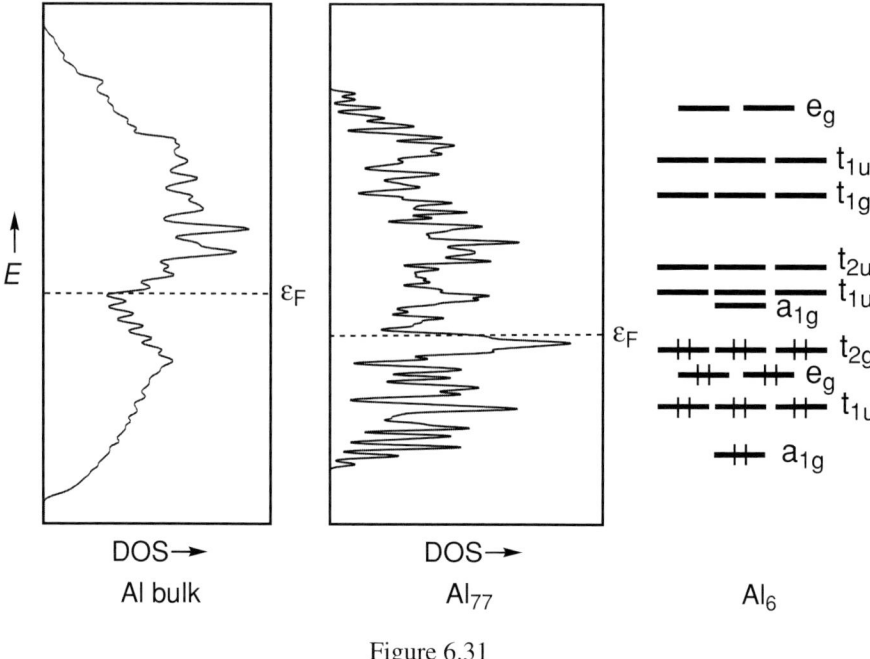

Figure 6.31

a case of the number of states associated with the bonding of the first couple of layers of atoms at the surface being a tiny fraction of the total DOS. On the other hand, the principal electronic-structure model developed in Chapter 2 for small clusters removes "surface states" from the problem by external-ligand interactions. The remaining cluster-surface layer of atoms is described by a model which can be simplified to the level of a useful electron-counting rule. It is not surprising then that when "surface states" are incompletely removed (Section 2.12.5) the paradigm must be modified. These, then, are the two ends of the bridge for main-group atoms. Based on earlier discussions, transition metals increase the complexity of the DOS/MO diagrams, but the general connection is the same. The analyses of Chapters 2, 3, 5, 6 and 7, which now follows, circumscribe a large uncharted area, exemplified by the Al_{77} cluster in which surface-like and bulk-like characteristics compete somewhat equally. These new hybrid structures are known to have unusual properties that deserve to be understood. Although we have not presented solutions to an active research problem, we hope that your understanding of both cluster and crystal (solid-state) electronic structure places you in a better position to address the problems presented by giant clusters and nanoparticles. In the same way that understanding the fundamentals of electronic structure fostered development of solid-state chemistry on the one hand and cluster chemistry on the other, so too the future development of the electronic structure of large clusters, such as $[Al_{77}R_{20}]^{2-}$, promises similar advances in nanoparticle chemistry.

Problems

1. Work out the band structure for the p_x, p_y and p_z functions of a two-dimensional square array of B atoms in the xy plane separated by a distance d. This might be a model for a material made up of weakly interacting sheets or of a very thin film or surface layer of a solid. Hint: carry out the operations used to generate the band structure in Exercise 6.9 with p functions rather than s functions.

2. Work out the band structure for the COs generated by the π MOs of N_2 for a two-dimensional square array with the molecules oriented vertically (along the z axis) relative to a flat surface in the xy plane to which they are axially coordinated. Assume all have the same surface–N bond distance and the net spacing is a uniform distance $a = 3$ Å. Plot the points Γ (0, 0), M (π/d, π/d) and X (0, π/d) and draw correlation lines between corresponding points. Assume the surface–molecule interaction does not involve any π-symmetry interactions. Do you expect the DOS to be broad or narrow?

3. Develop a crude model of the adsorption of an array of Lewis bases on a periodic Lewis acidic surface in the following way. First, work out the band structures for a square net of empty s orbitals and a square net of filled s orbitals having the same spacing parameter a. Assume a is larger than twice the van der Waals radius of each. Next consider a square array of donor–acceptor adducts formed by bringing the first two arrays together. Do so by separately working out the band structure for the donor–acceptor bonding MO and the donor–acceptor antibonding MO. Now make a correlation diagram in the manner of a MO diagram.

4. The sphalerite structure exhibits a unit cell with a ccp anion lattice with one type of tetrahedral hole occupied with cations. What is the compound stoichiometry?

5. H_3NBH_3 is isoelectronic with ethane, H_2NBH_2 is isoelectronic with ethylene, and HNBH is isoelectronic with acetylene. Derive the band structure and the DOS for planar poly-{–BHNH–} (isoelectronic to polyacetylene) with a single B–N distance and predict its conductivity and stability with respect to a Peierls distortion. Only consider the π electronic structure.

6. (a) Consider an infinite stack of benzene molecules, fully aligned with the planes of the rings parallel and separated by a distance d. Develop the one-dimensional diagram that arises from the highest-lying filled π MOs and lowest-lying unfilled π MOs of benzene. Will this stack exhibit metallic or non-metallic behavior along the stacking direction?
(b) Reduce 10 % of the benzene molecules to form the radical anions. Do you expect a change from metallic to non-metallic or vice versa? Can you now explain why the addition of K metal to graphite produces a material, C_8K, with a bronze appearance accompanied by an increase in conductivity of a factor of 30 and a conductivity that decreases with increasing temperature?

7. Tetracyanoethylene (TCNQ) shown below is a highly conjugated planar molecule. Its π-type LUMO (represented below) lies at a rather low energy so that the compound can be easily reduced by one electron, giving rise to the stable radical anion TCNQ$^-$.

 a. In several solid-state TCNQ$^-$ salts, the anions do not avoid each other as usually observed in ordinary salts (think of NaCl), but rather they stack as sketched below to make one-dimensional chains surrounded by cations. The reason for these rather weak anion–anion associations is π–π bonding between neighboring TCNQ⁻. The major structural features of these salts can be rationalized within a band model generated from the π-type SOMO (singly occupied HOMO) of TCNQ⁻. Two different phases of Rb(TCNQ), i.e., (Rb⁺, TCNQ⁻), have been characterized, one with regular spacing between the TCNQ⁻ anions (3.43 Å) and the other one with alternation of short (3.16 Å and 3.48 Å) and long TCNQ⁻...TCNQ⁻ spacing (i.e., "dimerization" along the stacks). Provide a rationalization for the existence of these two phases and predict some of their physical properties.

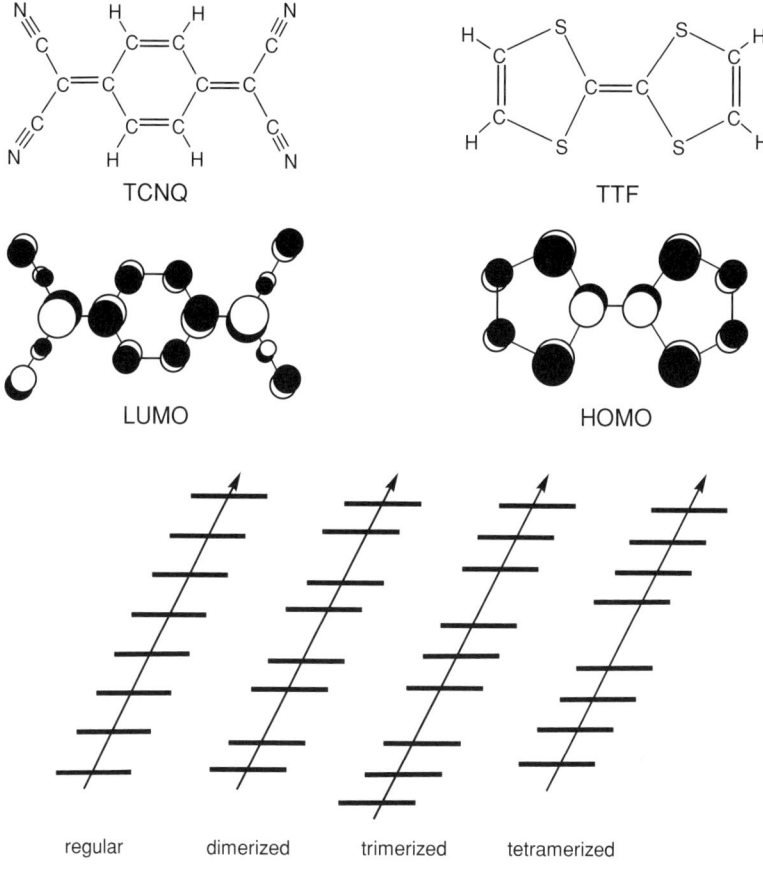

Problem 6.7

b. (NEt₄)(TCNQ)₂ and Rb₂(TCNQ)₃ also exhibit stacks of TCNQ units. In the former the TCNQ units associate into "tetramers," whereas in the latter they form "trimers." Explain.

c. Tetrathiofulvalene (TTF) shown above is also a highly conjugated planar molecule. Its π-type HOMO (represented above) lies at a rather high energy so that it is easily oxidized, giving rise to the stable radical cation TTF⁺. Solid-state TTF salts often give rise to stacks of TTF cations like the TCNQ salts. This is also the case in the (TTF)(TCNQ) salt which presents stacks of TTF as well as stacks of TCNQ moieties. It turns out that (TTF)(TCNQ) is a surprisingly good metallic conductor. Explain.

Additional reading

Section 6.1

Hoffmann, R. (1988). *Solids and Surfaces: A Chemist's View of Bonding in Extended Structures*. New York: VCH.
Albright, T. A., Burdett, J. K. and Whangbo, M.-H. (1985). *Orbital Interactions in Chemistry*. New York: Wiley.
Burdett, J. K. (1995). *Chemical Bonding in Solids*. New York: Oxford.
Dronskowski, R. (2005). *Computational Chemistry of Solid State Materials, A Guide for Materials Scientists, Chemists, Physicists and Others*. Weinheim: Wiley-VCH.

Section 6.3.2

Stout, G. H. and Jensen, L. H. (1968). *X-Ray Structure Determination*. New York: Macmillan.

Section 6.4.1

Muetterties, E. L., Rhodin, T. N., Band, E., Brucker, C. F. and Pretzer, W. R. (1979). *Chem. Rev.*, **79**, 91.
Somorjai, G. A. (1981). *Chemistry in Two Dimensions: Surfaces*. Ithaca, NY: Cornell University Press.

7

From molecules to extended solids

Of the millions of different chemical systems discovered since chemistry began, many are solids at room temperature. From the early days these solids have been classified in the four families, molecular, ionic, covalent and metallic solids, based on the nature of the forces which bind the atoms. Molecular solids are composed of groups of covalently bound atoms, i.e., molecules, held by weak charge-polarization (van der Waals) forces. In ionic solids, electrostatic attraction is the primary force binding cations and anions. Bonding in covalent solids is similar to that within molecules but extends over the whole crystallite. Metallic solids also exhibit extended bonding but, in addition, possess weakly bound, highly delocalized electrons easily moved by applied fields. Of course, this classification is somewhat artificial and many solids exhibit complex bonding in which more than one type of bonding is displayed. Molecular clusters in the solid state are naturally described nowadays with molecular-orbital models. Intermolecular interactions are weak. Although this is not true for solids with extended bonding networks, the solid-state machinery we developed in Chapter 6 shows that MO ideas smoothly transfer to crystalline solids. Hence, we have an analogous language for treating these more complex structures.

This is not a text of solid-state chemistry and the purpose of this chapter is to illustrate the use of the theoretical model of Chapter 6 with experimental examples. In doing so, we firmly establish the other foundation of our cluster bridge. There are many solid systems that could serve this purpose and the selection is somewhat personal – a collection of "portraits" that we find particularly educational. Most have connections to the clusters of the earlier chapters or simple coordination compounds. Enough review of pertinent molecules will be given to remind you of the molecular ideas. In doing so similarities between solid systems and molecules are highlighted as well as meaningful differences. Meaningful – are some not? Yes, as with bonding models some of the differences are more apparent than real, e.g., different ways of partitioning electrons before counting as well as the language of

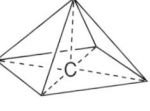

Fe₅C(CO)₁₅

Figure 7.1

discussion. These will be pointed out with pertinent comparisons. If successful, this approach will convince you that, despite some real differences, there are strong similarities between the bonding in molecules and that in extended solids. You can use many of the same tools, including relationships between the electron count and geometric structure, to effectively describe and understand the stoichiometry and bonding in solids.

Come on in! Take a look at the chemical portraits in the galleries that follow.

7.1 From a single atom to an infinite solid: the example of C

In Chapter 6 the "all-C" compounds of C, graphite and diamond, as well as new allotropic forms, such as C_{60} and C nanotubes were mentioned in the context of developing a bonding model for these structure types. The interesting way in which C_{60} accommodates the external cluster electrons in a delocalized system was also pointed out in Chapter 2 in the context of the "lone pair" problem of bare clusters. C_{60} and related species, some of which have been predicted theoretically but not yet observed experimentally, are thermodynamically less stable than graphite and diamond and provide a synthetic challenge. But small linear or cyclic C fragments can be prepared in molecular or infinite arrays in the solid state. Let's begin by looking at a few examples of fragments of hypothetical C allotropes.

7.1.1 Interstitial C atoms

The simplest C "allotrope" is a single atom. These are found in interstitial holes in metal lattices; hence, their name. They are also found in metal clusters as described in Chapter 3. The characterization of one of these can be seen with a historical perspective to be the first recognition of two ideas of great importance to this text. In 1962, Dahl and coworkers published the structure of the compound $Fe_5C(CO)_{15}$ (Figures 3.7 and 7.1) which contains a C atom embedded in a transition-metal carbonyl cluster. Think how unexpected a five-coordinate C atom was at that time when the icosahedral carborane clusters were only beginning to be structurally characterized! A qualitative MO description gave rise to the statement, "The representation of bonding for the apical iron [in $Fe_5C(CO)_{15}$] is formally analogous to

7.1 From a single atom to an infinite solid

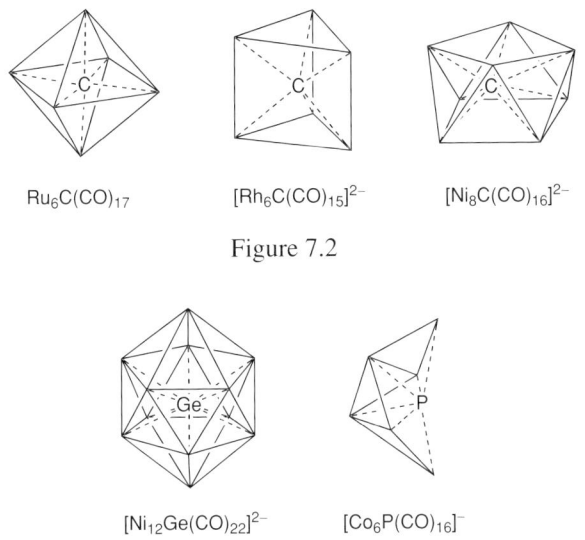

Figure 7.2

Figure 7.3

that for cyclobutadiene-iron tricarbonyl." Without mentioning isolobal – the word would not appear until 1976 – we see the beginnings of the isolobal analogy (Chapter 4). We read further, "The delocalized bonding of the carbide atom to the five irons in Fe$_5$C(CO)$_{15}$ is no doubt related to the Fe–C bonding in [cementite] Fe$_3$C." Dahl evokes a link between cluster and solid-state chemistry. More than 40 years later after this avant-garde work, these two concepts, albeit developed into robust tools over the years through the contributions of many others, facilitate our study.

For five years Fe$_5$C(CO)$_{15}$ was regarded as a curiosity until Ru$_6$C(CO)$_{17}$, and later others, such as [Rh$_6$C(CO)$_{15}$]$^{2-}$ and [Ni$_8$C(CO)$_{16}$]$^{2-}$, shown in Figure 7.2, were characterized. Respectively, these contain C atoms in octahedral, trigonal prismatic and square antiprismatic environments. The interest in these compounds is more in the possibility of stabilizing unusual metal geometries rather than the C atom itself. As shown in Chapter 3, the strong interaction between the metal atoms and the interstitial atom (Figure 3.8) increases the HOMO–LUMO gap and the stability of the metallic framework. A similar effect is observed in solid-state chemistry, e.g., upon adding C to Fe in the formation of steel. As good fit of atomic radius and cavity size is necessary, incorporation of second-row atoms requires larger clusters, e.g., the icosahedral [Ni$_{12}$Ge(CO)$_{22}$]$^{2-}$ cluster for Ge or more open structures, e.g., [Co$_6$P(CO)$_{16}$]$^-$ for P (Figure 7.3). Review the electronic structure and the cve/sep counts for these clusters in Section 3.3.1, if needed.

The variety of transition-metal carbonyl carbide clusters is mirrored in the large number of solid-state carbide compounds known. A difference is that the C atom in solid-state systems is nearly always six coordinate whereas the cluster systems can

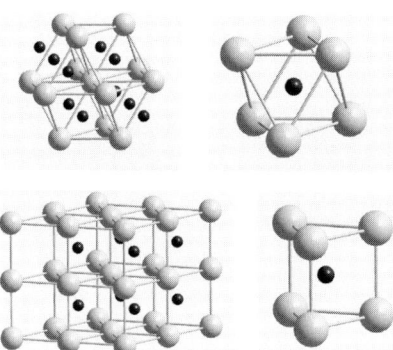

Figure 7.4

exhibit coordination numbers of four to eight. Let's compare a couple of cluster and solid-state systems. The structures of refractory carbides such as NbC and WC shown in figure 7.4 can be related to those of molecular clusters. On the left side of the Figure, the extended structures are portrayed (the small spheres represent the p-block elements and the large spheres the metal atoms) whereas on the right we have abstracted the immediate environment of a C atom. As in $Ru_6C(CO)_{17}$, the C atom in NbC shown at the top of Figure 7.4 is surrounded by six metal atoms in an octahedral array while in WC, shown at the bottom, the six W atoms form a trigonal prismatic array similar to that found for $[Rh_6C(CO)_{15}]^{2-}$. You may notice that in this chapter some structures are illustrated with both balls and sticks – the form favored by the solid-state chemist. Structure drawing, like other artificial representations of Nature, can generate strong emotional responses, e.g., the stick structures used in most earlier chapters were viewed with scorn by a solid-state chemist colleague. You have to get used to visualizing both!

The covalent bonding mode of the interstitial C in transition-metal carbide clusters is comparable to that observed in solid-state transition-metal binary compounds such as NbC or WC. With the language of bands and DOSs described in Chapter 6 we can make a side-by-side comparison of the bonding in the molecular cluster and the interstitial bulk carbide. The schematic diagram in Figure 7.5 juxtaposes the MO structure of the idealized cluster $[Ru_6C(CO)_{18}]^{2+}$ on the left (a mimic of $Ru_6C(CO)_{17}$) with the band structure of the solid metal carbide on the right. First, let us review the bonding in the octahedral cluster. The frontier orbitals of $[Ru_6(CO)_{18}]^{2-}$ are generated from six $Ru(CO)_3$ fragments and yield a "t_{2g}" set of 18 and a cluster bonding set of seven levels. The AOs of $[C]^{4+}$ interact primarily with four orbitals of the framework set but no new occupied orbitals are introduced by the interaction. In the solid carbide NbC, the metal AOs interact strongly with each other to give a relatively wide d band plus some framework orbitals, four of which point toward the cavity to be occupied by the C atom. As with the

7.1 From a single atom to an infinite solid

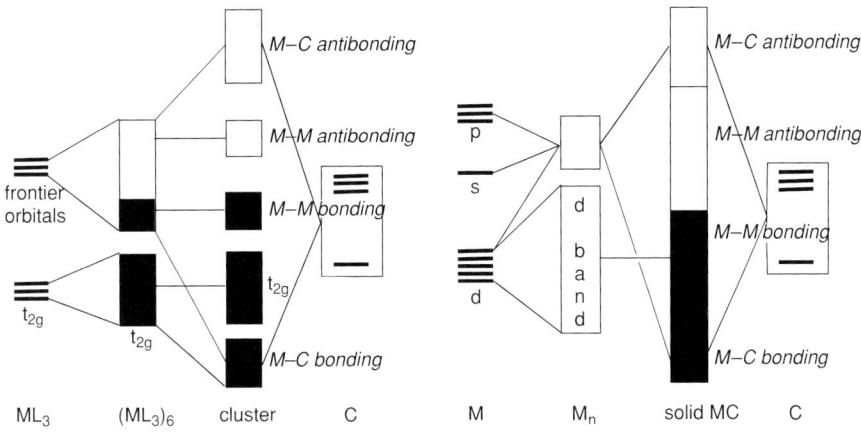

Figure 7.5

molecular cluster, the C AOs interact with these metallic orbitals to give bonding and antibonding combinations. So, in both cases M–C bonding is strong and the M–C bonding (either molecular or crystal) orbitals lie at low energies away from the HOMO (cluster) or the Fermi level (solid).

The similarity in the M–C bonding between the cluster and the extended solid is rather pleasing, isn't it? But maybe you have a niggling doubt – why didn't we compare $Ru_6C(CO)_{17}$ with an extended metal carbide of composition RuC. In fact monocarbides of the late transition metals are not known or found to be extremely unstable. They are too electron rich! Why? In the cluster carbide, for a d^8 metal all M–C and M–M bonding orbitals are filled and separated from the antibonding ones by a significant HOMO–LUMO gap. For the solid, the M–C and M–M bonding is maximized when only the bottom part of the metallic band is filled. This favors the lower electron counts produced by earlier metals such as d^4 (ZrC) or d^5 (NbC), as the higher counts of d^8 (RuC) would also populate M–M antibonding orbitals leading to weak bonding.

So, $Ru_6C(CO)_{17}$) is not a model for RuC but maybe it isn't clear yet why. We concentrated on the similarities of octahedral cavities but neglected the differences in the surrounding metal atoms. The molecular cluster is built of ML_3 fragments whereas if we were to derive a cluster to represent the environment around C in ZrC we would use six MC_5 fragments, i.e., ML_5 entities, around a central C atom. We saw in Chapter 4 that the frontier orbitals of ML_3 and ML_5 fragments differ considerably. Are there any molecular carbide clusters that can be derived from ML_5 fragments? Yes. The octahedral carbide cluster $[Zr_6C(\mu\text{-}Cl)_{12}Cl_6]^{4-}$ (Figure 7.6) with a cve of 74 rather than 86 as found for late metal-carbonyl clusters is one example. The reasons for this difference were discussed in Section 3.3.5. In accord with our theme, the relationship of the interstitial bulk carbide ZrC to $[Zr_6C(\mu\text{-}Cl)_{12}Cl_6]^{4-}$

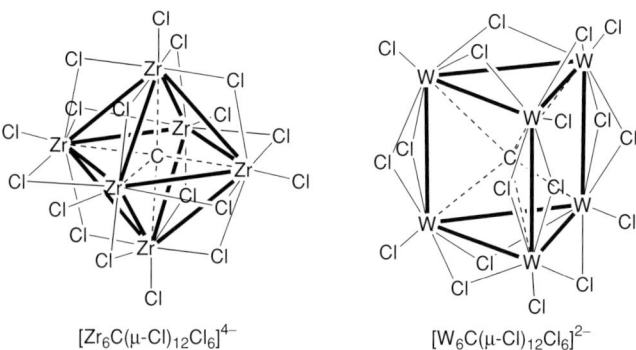

Figure 7.6

(Figure 7.6) is mirrored in the relationship of WC to the trigonal prismatic cluster $[W_6C(\mu\text{-Cl})_{12}Cl_6]^{2-}$.

Note that in both interstitial transition metal carbides and the molecular clusters with interstitial atoms, the octahedral metal arrangement changes to trigonal prismatic as the d-electron count increases. That is, octahedral carbides are found for d^4 or d^5 metals as exemplified by ZrC and NbC and trigonal prismatic carbides are found for d^6 metals as in WC.

Exercise 7.1. The solid carbide NbC has a melting point of 3900 °C whereas NbN has a melting point of 2600 °C. Develop an explanation using the band diagram in Figure 7.5.

Answer. The melting temperature of a crystal is one indicator of the quantity of energy required to destroy the regular atomic ordering, i.e., a measure of the energy necessary to break bonds sufficiently to form a liquid melt. If we assume this is related to the M–M bond strength as it appears to be for the metal elements (the maximum in melting point vs. metal is found for group 6), NbC is more strongly bonded than NbN. The total number of valence electrons in the unit cell will determine the position of the Fermi level. If we make another assumption that the Fermi level for NbC is close to the top of the M–M bonding band levels, then the additional electron per compositional unit goes into M–M antibonding levels. Hence, the overall bonding in solid NbN should be weaker. Now the caveats: you must continue to be aware that simple models that "work" hide many important complications. Melting temperature depends on the products of the melting process, i.e., are the products atoms (unlikely), clusters, etc. In the case of the nitride and carbide there is the possibility of M–E bond rupture in melting. Our answer assumes clusters in the melt retaining M–E bonds. In addition, we assume that the structures of the carbide and nitride are the same, whereas NbC is of the NaCl type and NbN adopts the NiAs type, but also is found with NaCl and WC types. Hence, although

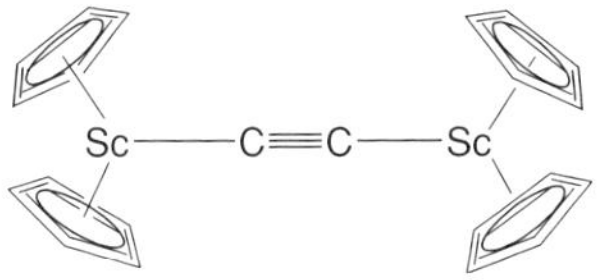

7.1 From a single atom to an infinite solid

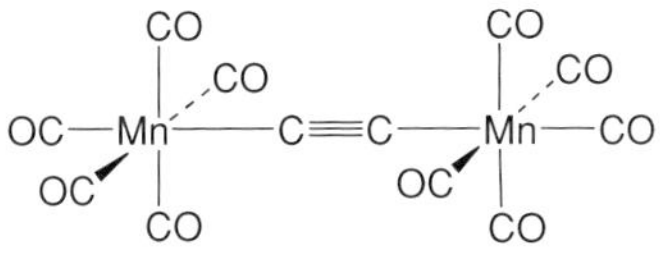

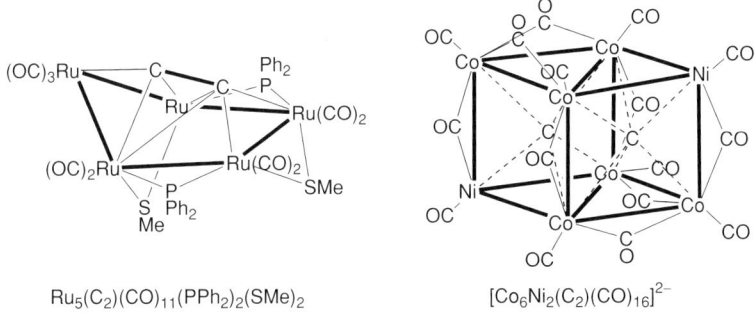

Figure 7.7

Figure 7.8

the simple argument given is satisfying, considerably more work would have to be done to see if it is a major factor causing the real 1300 °C difference.

7.1.2 C_2 "dumb-bells" in molecules and solids

The smallest molecular fragment of C, C_2, is a curiously interesting diatomic species. Although not found as a diatomic gas like the later first-row elements, it can be generated in a carbon arc, is found in comets and is responsible for the blue light we see in flames. It's a tiny dumb-bell in which the two nuclei are separated by 1.24 Å in the ground state. It can be stabilized by attachment to metal fragments such as in $[ScCp^*_2]_2C_2$ ($Cp^* = C_5Me_5$) or $[Mn(CO)_5]_2C_2$ (Figure 7.7) or by encapsulating it in a metal cluster. Five Ru atoms support an exposed C_2 fragment in the cluster $Ru_5(C_2)(CO)_{11}(PPh_2)_2(SMe)_2$ while C_2 is fully encapsulated in $[Co_6Ni_2(C_2)(CO)_{16}]^{2-}$ (Figure 7.8).

C_2 units are also found in solid-state compounds with C–C separations that depend on formal electron count. These are viewed as deprotonated ethyne, ethylene or ethane using a popular solid-state idea: the Zintl–Klemm concept. This concept is based on the simple idea that the metals transfer their valence electrons to the non-metal atoms thereby generating filled anion-centered bands at low energy, well separated from empty cation-based bands. Of course, this concept fails when the electronegativities of the metal and non-metal are not very different,

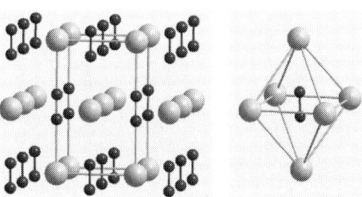

Figure 7.9

but it is tremendously useful in understanding a great variety of crystal structures. Think of $[C_2]^{2-}$ and N_2 for instance! These ten-electron species have qualitatively similar electronic structures. Thus, a triple bond is postulated for the C_2 moieties in CaC_2 (Figure 7.9) and the explosive powder Ag_2C_2 as the compounds are viewed as $Ca^{2+}[C_2]^{2-}$ and $(Ag^+)_2[C_2]^{2-}$. In Figure 7.9, the extended structure of CaC_2 is shown on the left and the octahedral cavity on the right. In the octahedral cavities of $Gd_2Cl_2C_2$ the C–C distance is 1.30 Å which corresponds to a double bond and the formulation $(Gd^{3+})_2(Cl^-)_2([C_2]^{4-})_2$. Finally, in the octahedral cavities of $Gd_{10}Cl_{18}C_4$ a C–C distance of 1.47 Å is consistent with a single bond and the formulation $(Gd^{3+})_{10}(Cl^-)_{18}([C_2]^{6-})_2$.

When a C_2 derivative includes a transition metal, the resulting covalent character requires a different approach. Of course, the C–C distance still serves as a useful guide to the electronic environment of the C_2 moiety but a number of interesting situations arise particularly with cluster and solid-state compounds. Let's take a look beginning with simple organometallic substituents on diatomic C_2. The bond representation of Figure 7.7 is substantiated by the C–C distances: $[ScCp^*_2]_2(C_2)$ (1.22 Å) or $[Mn(CO)_5]_2(C_2)$ (1.20 Å). A C–C length of 1.37 Å in $Ta[t-Bu_3SiO)_3]_2(C_2)$ is consistent with a C–C double bond. Intermediate C–C bond distances are common when the environment of the C_2 unit is the cavity of a cluster. For example it is 1.30 Å in $Ru_5(C_2)(CO)_{11}(PPh_2)_2(SMe)_2$ and 1.48 Å in $[Co_6Ni_2(C_2)(CO)_{16}]^{2-}$ (Figure 7.8). These are more difficult to interpret in the absence of quantum chemical calculations.

In the context of the problems to follow, it is useful to briefly review the major interactions derived from a fragment analysis of a simple compound like $[Mn(CO)_5]_2(C_2)$ in which the $[C_2]^{2-}$ unit spans two pseudo-octahedral ML_5 fragments (Figure 7.7). Although the metal fragments may be viewed as surrogates for, e.g., H atoms, in fact we already know that a metal fragment is much more versatile than a main-group one. Out-of-phase and in-phase combinations of the "t_{2g}" set and empty σ frontier orbitals of the two $[Mn(CO)_5]^+$ fragments give rise to a set of eight frontier orbitals, two of σ-type, four of π-type, and two of δ-type (Figure 7.10). These are available to interact with the σ_s and σ_p orbitals, and π and π^* (bonding and antibonding between C, respectively).

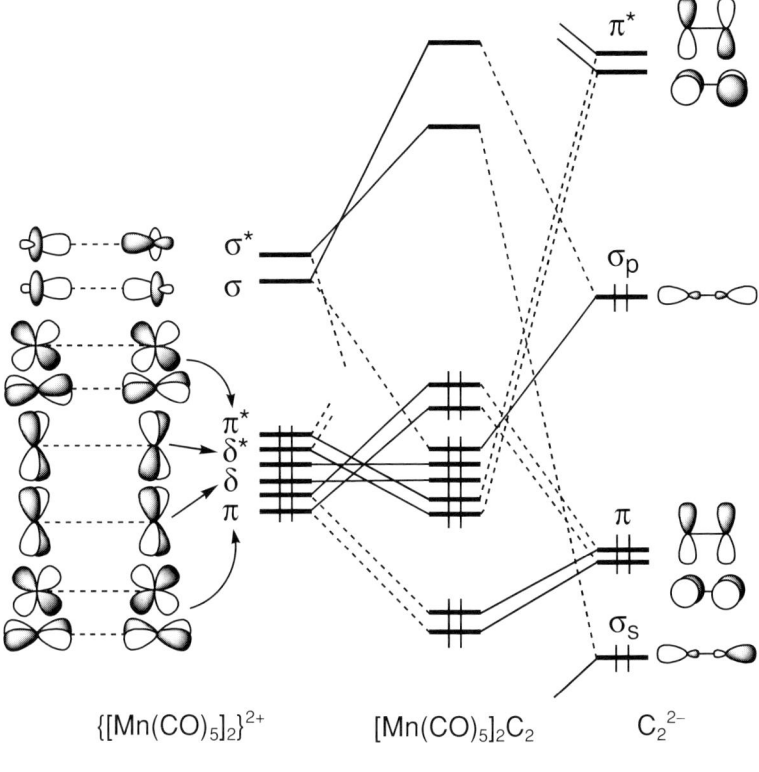

Figure 7.10

The principal bonding interaction occurs between metallic σ and σ^* frontier orbitals and the σ C_2 orbitals. You expect this because the $[Mn(CO)_5]^+$ fragment is isolobal with $[CH_3]^+$ and the product is isolobal with 2-butyne. However, the metal fragment provides the opportunity for weaker interactions between the π-symmetry orbitals. These are of two types. A backbonding interaction (occupied metallic π^* orbitals with unoccupied C_2 π^*) transfers charge from the metal to the C_2 unit, thereby reducing the C–C bond order. For this compound, the observed C–C distance is consistent with a bond order of three; hence, the π^*–π^* interaction must be small in this case. Another π-type interaction between filled orbitals generates the HOMO of the system. The HOMO is now delocalized over the Mn–C–C–Mn backbone rather than being largely of metal character. We conclude that $[Mn(CO)_5]_2(\mu\text{-}C_2)$ is correctly considered a metal-substituted acetylene. But you should be able to see that with an earlier metal or a different ligand set, the bond order of the C–C interaction in the $[C_2]^{2-}$ moiety can be reduced either by depopulating the C_2 π or populating the π^* orbitals. For example, $[W(t\text{-}BuO)_3]_2(C_2)$, which has an analogous structure, exhibits a C–C distance consistent with a single bond.

Exercise 7.2. The cluster $[Co_6Ni_2(C_2)(CO)_{16}]^{2-}$ consists of two trigonal prisms sharing a square face (Figure 7.8). The two C atoms, which lie at the center of the trigonal prisms, are separated by 1.48 Å, which corresponds to a single C–C bond. First, treat the compound as a fused cluster (Section 3.3.3) and compare total valence electrons with predicted cve. Second, consider the C–C distance and formulate the C_2 species as an anionic moiety as described above. Contrast the two views of electronic structure. What is the lesson learned?

Answer. Considering the C_2 unit as a fully encapsulated ligand, a count of 116 cve is obtained for $[Co_6Ni_2(C_2)(CO)_{16}]^{2-}$ (6×9 (Co) + 2×10 (Ni) + 8 (C_2) + 2×16 (CO) + 2 (anionic charge) = 116). This count corresponds to that expected if we consider that the cluster geometry can be generated from two trigonal prisms eliminating one square ($90 \times 2 - 64 = 116$). Treating the C_2 unit in this manner implies an interstitial $[C_2]^{8+}$ (Section 3.3.1). Alternatively, the C–C distance suggests a C–C single bond and a $[C_2]^{6-}$ interstitial anion! In terms of electron density, which is right? Neither, as both are conceptual models that partition the valence electrons to provide a useful tool for relating composition to structure. If one thinks about where the electrons actually are in, e.g., $Ru_6C(CO)_{17}$ (Figure 3.7), one should NOT imagine an empty octahedral cluster containing a $[C]^{4+}$ ion. In fact, there will be substantial electron density in the octahedral cavity. On the other hand, there will not be enough to make a $[C_2]^{6-}$ ion in the cluster considered. The interstitial C_2 fragment is particularly interesting as good theoretical analysis shows that not only is there some electronic charge transferred from bonding orbitals of the C_2 fragment to the metal, but also from occupied metal orbitals into antibonding orbitals of the C_2 fragment. This may remind you of the Dewar model in which the bond order of a coordinated olefin is reduced by ligand to metal and metal to ligand charge transfer processes (Section A1.4.7).

Another cluster type that contains C_2 fragments are the metallocarbohedrenes (met-cars) – cage-like clusters of transition metals decorated by C_2 dumb-bells, e.g., $Ti_8(C_2)_6$, which are found for Ti, V, Cr and other early transition metals. Formed in beams, they are species structurally characterized principally by theoretical studies. $Ti_8(C_2)_6$ was initially thought to have a T_h-symmetry pentagonal-dodecahedral structure (a Ti_8 cube hexacapped by C_2 units, Figure 7.11, left). Better calculations have shown that a tetracapped tetrahedron of metal atoms surrounded by the six C_2 units (T_d symmetry) is energetically preferred (Figure 7.11, right). A tetracapped tetrahedron requires $60 + 4(60 - 48) = 108$ cve whereas the total number of valence electrons in $Ti_8(C_2)_6$ is 80. This is far below the number required for a ligated cluster. In fact the calculations show a number of d orbitals are empty and the HOMO is triply degenerate, occupied with two electrons and with no HOMO–LUMO

7.1 From a single atom to an infinite solid 267

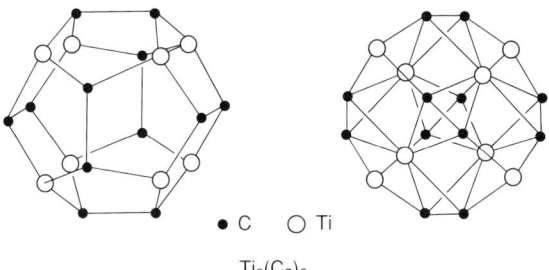

Figure 7.11

gap! The open-shell electron configuration rather than a closed-shell Jahn–Teller distorted configuration is attributed to the high connectivity of the atoms constituting the cluster and is reminiscent of that encountered for the cubic clusters treated in Section 5.2.5.

Now that we have glimpsed the variety of molecular-cluster environments in which we find the C_2 diatomic, let's move to the realm of solid-state chemistry. Solid-state carbides containing C_2 units are profuse in number. We have already mentioned an important compound in this family, calcium carbide, CaC_2. As shown in Figure 7.9 its structure consists of Ca and C_2 units arranged in a tetragonally distorted derivative of the NaCl structure with the C dumb-bells aligned along the c axis. It is often considered the archetype of ionic solids with C^{2-} anions encapsulated in octahedra composed of Ca^{2+} cations. Although a C–C distance of 1.19 Å is in agreement with a $[C_2]^{2-}$ unit, theoretical calculations suggest some covalency between the constituent elements.

Thus, to gain understanding of its electronic structure, use the Zintl–Klemm concept and fragment the compound into its ionic components Ca^{2+} and $[C_2]^{2-}$. The dicarbide sublattice can be constructed from an infinite number of non-interacting $[C_2]^{2-}$ "molecules." We can get some qualitative insight when $[C_2]^{2-}$ units are simply encapsulated in $(Ca^{2+})_6$ octahedra. Due to the large electronegativity difference between Ca and C, only weak interactions are observed, as illustrated on the left side of Figure 7.12. The C_2 orbitals are slightly stabilized by mixing with the Ca orbitals. Recall that if two levels differ greatly in energy, their covalent interaction is small even with favorable overlap. Hence, as shown on the right side of Figure 7.12, the bands associated with the sublattices only interact weakly. Hence, we find narrow, low-energy occupied bands centered at the discrete dicarbide energy levels and largely separated from high-lying empty Ca bands. Note that the relative area ratios of the generated bands are related to the number of orbitals per CaC_2 motif, i.e., 1 (C_2 σ_s) to 2 (C_2 π) to 1 (C_2 σ_p) to 2 (C_2 π^*) to 4 (Ca s and p).

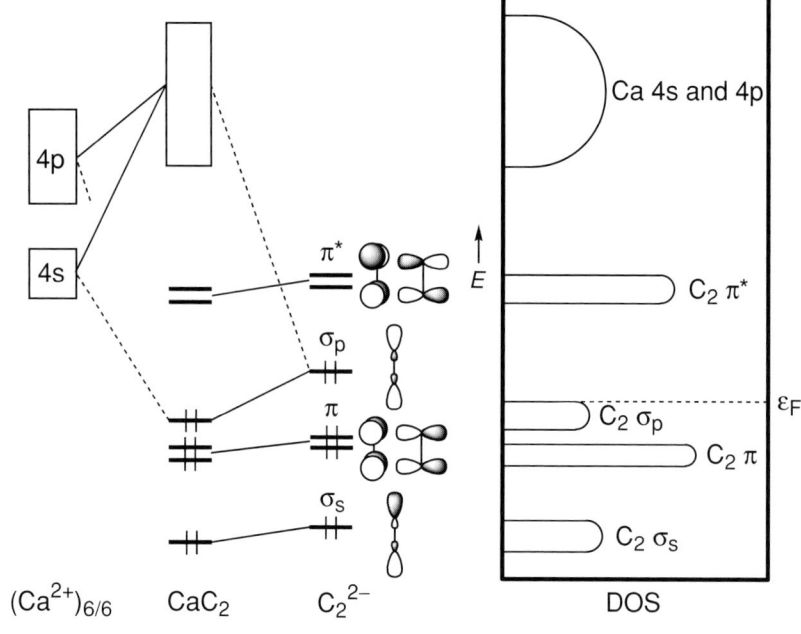

Figure 7.12

Exercise 7.3. The CaC$_2$ structure is adopted by most of the lanthanide and actinide analogs as well. However, the C–C distance measured in these compounds differs. It increases from 1.19 Å in CaC$_2$ to 1.28 Å in YC$_2$ to 1.30 Å in LaC$_2$. Explain.

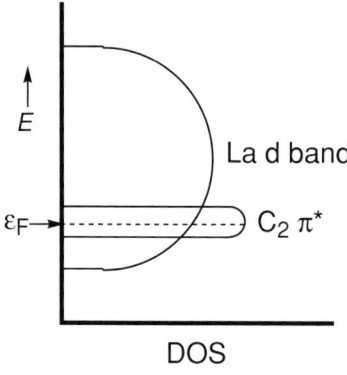

Exercise 7.3

Answer. Assuming an ionic Zintl–Klemm model and the common oxidation state of these metals, we have C$_2^{3-}$ ions locally in the octahedral cavities formed from the

7.1 From a single atom to an infinite solid

Y^{3+} or La^{3+} cations. Consequently, the additional electron populates an antibonding π_g^* orbital thereby reducing the bond order and increasing the C–C distance. Reality is a little more complicated since there are low-lying d orbitals on these metals. Consequently a metallic d band overlaps with the C_2 π_g^* band as shown below. For this reason, the electron is partly localized in the metal d band and partly localized in the C_2 band. Nevertheless the C–C distance in these materials is chiefly governed by the electron count of the metal.

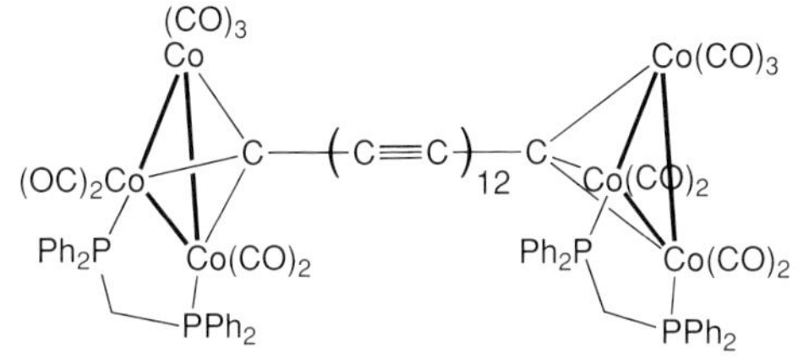

Figure 7.13
$[Co_3(\mu\text{-dppm})(CO)_7]_2(C_{26})$

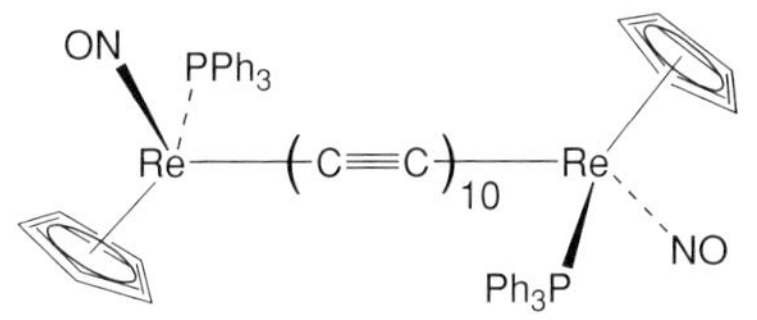

Figure 7.14
$[ReCp^*(PPh_3)(NO)]_2(C_{20})$

7.1.3 Finite C chains in molecules and solids

C chains with more than two C atoms can be isolated if stabilized with terminal metal complexes or clusters. The record number at the time of writing this book is 26 found in $[Co_3(\mu\text{-dppm})(CO)_7]_2(C_{26})$ and shown in Figure 7.13. The terminal C atoms cap metal triangles so alternatively one might view the stabilizing clusters as M_3C clusters with a chain of 24 C atoms. Another impressive example is $[ReCp^*(PPh_3)(NO)]_2(C_{20})$ containing 20 C with mononuclear terminal complexes (Figure 7.14).

One important characteristic that makes these compounds of interest is electronic configurations that permit "communication" between the metal termini. This is determined by the nature of the connections to the metal end groups. To illustrate, the bonding of $[RuCp(PPh_3)_2]_2(C_4)$ is analyzed. Its structure shows alternating

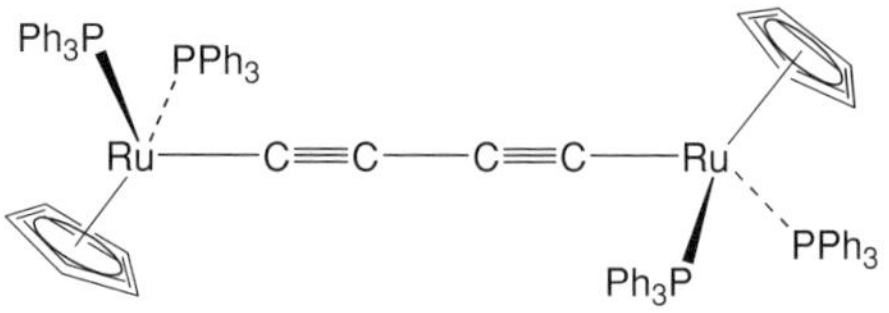

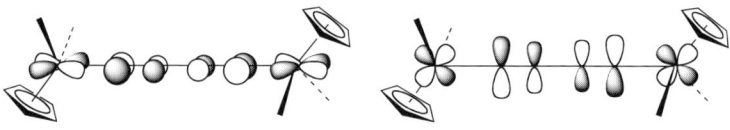

[RuCp(PPh₃)₂]₂(C₄)

Figure 7.15

HOMOs of [RuCp(PPh₃)]₂(C₄)

Figure 7.16

long and short C–C distances consistent with the representation in Figure 7.15. Consequently, the terminal complexes are saturated 18-electron species.

The M–C bonding in this species is analogous to that described above for [Mn(CO)$_5$]$_2$(C$_2$) in Figure 7.10. Particularly important is the π interaction that generates a pair of HOMOs. These result from a M–C antibonding combination and are well separated from the LUMOs and other occupied MOs. As shown in Figure 7.16, they have similar nodal properties, are antibonding between the metal and its adjacent C atom, bonding between the first and second, and third and fourth C atoms, and antibonding between the second and third C atoms. Oxidation is expected to depopulate these orbitals and affect the C–C distances in accord with their bonding and antibonding characters. Four reversible oxidation processes are observed, something of a record for such a small molecule. In going from neutral to dicationic species, a shortening of the M–C and central C–C bonds and a lengthening of the outer C–C bonds is observed in accord with the nodal properties of the HOMOs. Formally oxidation in two-electron steps converts the neutral diyndiyl complex (Figure 7.17(a) where the bars on Ru represent non-bonding electron pairs) into the cumulenic dicationic complex (Figure 7.17(b)) into an acetylene-bridged dicarbyne complex (Figure 7.17(c)). This provides a particularly nice connection between structure and electron count. Considerable experimental evidence shows that the same model applies to the compounds with longer C chains as well.

Metal carbide solids with long chains are not well characterized. In the Zintl–Klemm paradigm, this can be attributed to the fact that for $n \geq 3$ all C_n chains will have a 4− charge. The longest C chain has $n = 3$ as found in Sc$_3$C$_4$. As you

7.1 From a single atom to an infinite solid

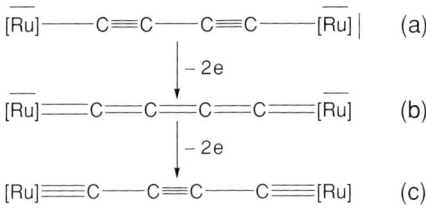

Figure 7.17

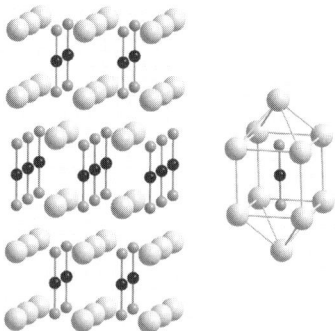

Figure 7.18

might expect from the formula, the structure is complicated. It has a large unit cell containing eight C_3 chains, two C_2 units and twelve isolated C atoms in holes created by the Sc sublattice and consistent with the formulation $Sc_{30}(C_3)_8(C_2)_2C_{12}$. The C–C distances are 1.34 Å in C_3 and 1.25 Å in C_2. These separations may be assigned to double and triple bonds, respectively, which leads to the charge assignments $[C_3]^{4-}$ and $[C_2]^{2-}$. You can draw the Lewis formula for such species can't you? Assuming C^{4-} for the isolated C atoms leaves a few electrons in the Sc d band to account for the metallic conductivity of the compound. In other words, the Sc atoms are formally in a slightly lower oxidation state than the usual 3+ found for rare-earth metal carbides. With 70 atoms and 430 valence orbitals per unit cell it is not easy to examine the bonding between the metals and the C groups in detail.

Consider instead the simpler solid-state arrangement in Sc_2BC_2 which contains triatomic BC_2 chains. We will see in a moment that these heteroatomic chains are isoelectronic to $[C_3]^{4-}$ so this compound allows us to explore the bonding interactions and compare it to what we find for the molecular compounds. The structure of Sc_2BC_2 is easy to understand if you recall that of CaC_2 (Figure 7.9). In the conventional tetragonal unit cell of CaC_2 the Ca atoms are at the corner and center positions whereas the C_2 dumb-bells are located in the middle of the edges and the middle of the lower and upper square faces of the Ca_6 octahedra. The structure of Sc_2BC_2 is shown in Figure 7.18 where the extended structure is shown

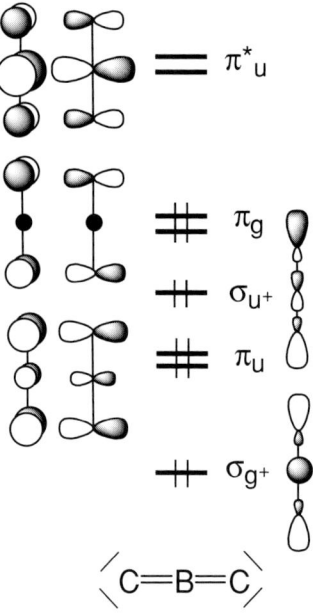

Figure 7.19

at the left and the environment of one BC_2 unit is shown on the right (Sc is light grey, C is medium grey and B is black). Thus, to obtain the structure of Sc_2BC_2 from that of CaC_2 we simply have to replace the Ca atoms by Sc_2 pairs and the C_2 dumb-bells by linear C–B–C units. Each BC_2 unit is encapsulated in a bicapped cube of Sc atoms. The B–C separation, 1.48 Å, corresponds to what is expected for a B–C double bond. But we need to look at the electronic structure more closely. You know what we are going to do. Yes, fragment the compound into metallic parts and main-group parts, examine each part separately and then recombine them to reveal their interactions.

Let us begin with the BC_2 units. Since they are isolated from each other, we can start by looking at the electronic properties of one BC_2 unit. Some of its molecular orbitals are shown in Figure 7.19. Above a set of σ-type MOs not shown, there is a set of π-bonding and σ and π non-bonding MOs well separated from the π^* molecular orbitals. Filling of the non-bonding and bonding MOs leads to the formal charge of 5– per BC_2 unit. Hey! $[BC_2]^{5-}$ is isoelectronic to $[C_3]^{4-}$, allene or CO_2 and consistent with the observed B–C distance. Indeed, except for differences associated with different atom electronegativities, their MO diagrams are pretty similar as well.

Where do the electrons for the 5– charge come from? The only possibility is from the Sc atoms. This leaves us with one valence electron per two Sc atoms per formula unit before the interaction between the $[BC_2]^{5-}$ and Sc sublattices. Partitioning of

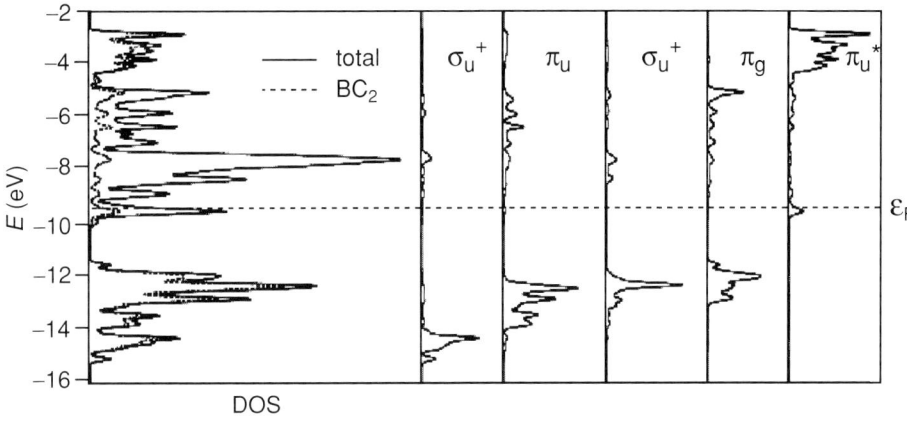

Figure 7.20

the electrons using the Zintl–Klemm paradigm is formal and ignores any covalent interactions between the chains and the metal atoms. With occupied orbitals at relatively high energy and vacant orbitals at relatively low energy, the $[BC_2]^{5-}$ chains are well prepared to act as both σ donors and π acceptors in the manner of a classical molecular organometallic complex. So let's explore the DOS of Sc_2BC_2 and see if we can find any evidence for such interactions.

The total DOS of Sc_2BC_2 is shown in Figure 7.20 with the solid line. The dashed line inside the total DOS is the % contribution of the BC_2 units to the total DOS. At the right are the projections of the frontier orbitals of the BC_2 sublattice after interaction with the metallic lattice. What do we see? First of all, the Fermi level cuts the bottom of a band mainly localized on the metal atoms. This is consistent with the compound's metallic conductivity. Secondly, there is a strong covalent interaction between the organic BC_2 units and the metals. That is, if the interaction were weak, the peaks corresponding to the frontier-orbital projection would be narrow in width. In fact, they are rather dispersed in energy. Some of the frontier orbitals of BC_2, σ_g^+, π_u, σ_u^+ and π_g, were occupied before interaction (Figure 7.19). The small peaks due to BC_2 in the metallic band result from the $[BC_2]^{5-}$ acting as a donor toward the metal atoms. On the other hand, the frontier orbital π_u^* was vacant before interaction with the metallic host. In the far right-hand plot, notice the peak below the Fermi level. This reflects $[BC_2]^{5-}$ acting as an acceptor. So, with evidence of electron donation from the $[BC_2]^{5-}$ chains to the metallic network supplemented by back-donation from the metallic lattice into the BC_2 units, we can view Sc_2BC_2 as a solid-state example of the molecular compounds containing atomic chains.

Although no chains longer than three C atoms have been stabilized in solid-state compounds, replacement of some C with B atoms permits chains with more than

274 *From molecules to extended solids*

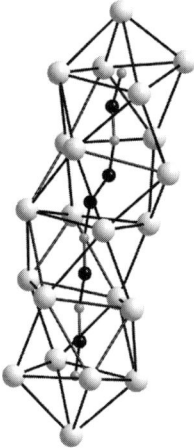

Figure 7.21

ten atoms to be observed. One example comes from the area of rare-earth B–C compounds known for their rich structural chemistry. In LaBC the La atoms form distorted square and puckered metallic layers. They stack in such a way that they form holes in which are encapsulated finite B_5C_5 chains in the sequence C–B–C–B–B–C–B–C–B–C as illustrated in Figure 7.21. In the scheme, La is light grey, C is medium grey and B is black. These worm-like entities buried in the solid are fairly linear with B–B, B–C and C–C separations corresponding approximately to coordinated double bonds.

Exercise 7.4. The solid-state compound, LaBC, contains C–B–C–B–B–C–B–C–B–C chains with B–C and B–B distances appropriate for double bonds. (a) Calculate an appropriate formal charge. (b) For this charge, predict the nature of the electrical conductivity of LaBC.

$$C=B=C=B=B=C=B=C=B=C$$

Exercise 7.4

Answer. (a) To obey the octet rule for each atom a charge of 9− is required; two lone pairs on the terminal C atoms and one negative charge for each B atom in the chain generate the Lewis formula shown above. This chain is isoelectronic to a hypothetical cumulenic $[C_{10}]^{4-}$ oligomer. (b) The charge is 9− on the chain in $(La)_5(B_5C_5)$ but La favors a +3 oxidation state so there are six electrons which remain in the metallic band. Hence, the compound should be an electrical conductor. As we have seen, there will also be some covalent interaction between the sublattices of B–C chains and the La atoms but the Zintl–Klemm idea plus electron counting provides a good estimate of electronic structure. There is a complication here that is incapable of being explained by a simple Lewis formula. Like O_2 or $[C_2]^{4-}$, $[B_5C_5]^{9-}$ with an odd number of B and C atoms would be a diradical if isolated.

At the time of writing, the record length in a solid-state compound for a linear chain containing C atoms is a B_5C_8 chain which also exhibits a cumulenic form. The infinite B–C rod with cumulenic double bonds shown in Figure 7.22 is a good model for the hypothetical metastable allotropic one-dimensional phase of C presented in Section 6.2.4. Called carbyne or chaoite, its preparation, although claimed by some, remains a challenge. As we saw in Chapter 6, such a chain with equispaced atoms would be subject to a Peierls distortion. Hence, it would be made up of alternating single and triple bonds (Figure 7.23) as we found with the organometallic analogs discussed in Section 7.1.3. We also learned in Chapter 6, that the necessity for a Peierls distortion in the infinite C chain is removed in an isoelectronic chain with alternating heteroatoms. As a $[BC]^-$ unit is isoelectronic with a (BN) unit, a linear –B–C–B–C– chain made of $[BC]^-$ units is equivalent to a linear –B–N–B–N– chain.

But there is an inorganic analog of infinite carbyne! Infinite B rods are found in LiB and, in the Zintl–Klemm paradigm, the compound is formulated (Li^+, B^-). B^-, of course, is isoelectronic with C. As shown in the two views in Figure 7.24, infinite B chains run in channels made by the Li atoms (large grey spheres). The view at the right is along the chains and that at the left shows a section of four chains. Given the discussion in Chapter 6 as well as above, you won't be surprised

$$=C=B=C=B=C=B=$$

Figure 7.22

$$\!=\!\!(C\!=\!\!C\!)_x\quad \text{vs.}\quad -\!(C\!\equiv\!C)_x\!-$$

Figure 7.23

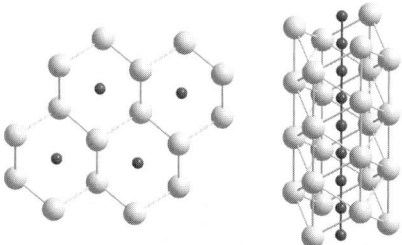

Figure 7.24

276   *From molecules to extended solids*

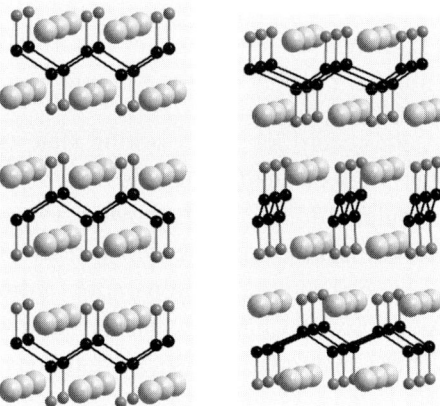

Figure 7.25

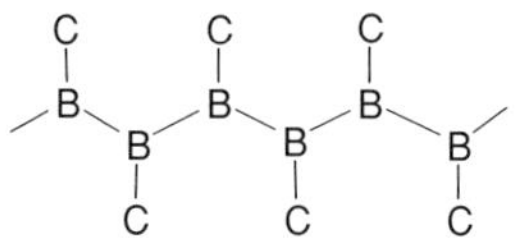

Figure 7.26

to learn that the B chains have alternating long (about 1.7 Å) and short (about 1.4 Å) B–B separations.

Let's test your understanding and look at a somewhat more challenging pair of systems in which Peierls distortion plays a role. YBC and ThBC are two more rare-earth boron carbides which have the same stoichiometry as LaBC but rather different structures. These are shown in Figure 7.25 with YBC on the left and ThBC on the right. The metals are represented by large grey spheres and the C and B by small grey and black spheres, respectively.

In both, the metal atoms form infinite channels containing zigzag B chains with C atom branches as represented schematically in Figure 7.26. The metal atoms and branched chains stack in one direction to give two-dimensional MBC slabs. In the structural pictures, the most obvious difference in the structures is the way these slabs stack in the third dimension. However, the crucial structural difference actually lies within the slabs themselves: the B–C zigzag chains differ. The B chain in YBC is regular but it is alternating in ThBC.

Why the difference? Apparently one is subject to a Peierls distortion but the other is not. If true, there must be a difference in the electronic structures. The first hint comes from applying the Zintl–Klemm model. With Y and Th, it is pretty clear, we have Y^{3+}, $[BC]^{3-}$ and Th^{4+}, $[BC]^{4-}$, respectively. That is, although the

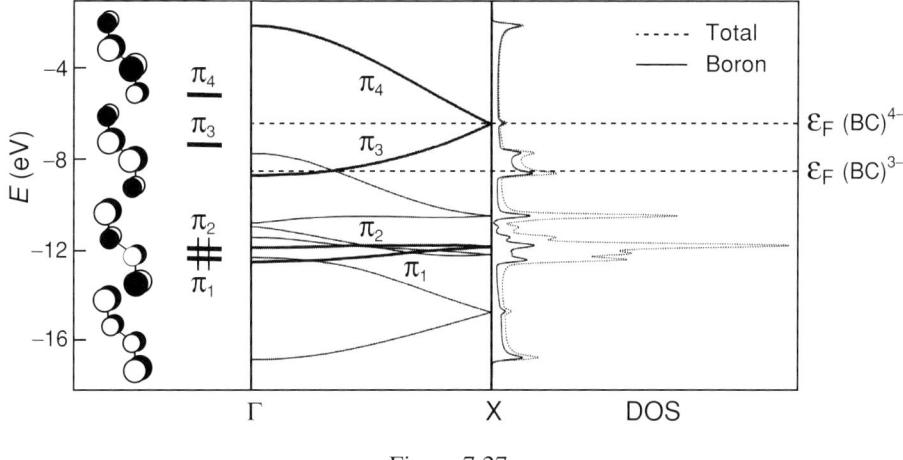

Figure 7.27

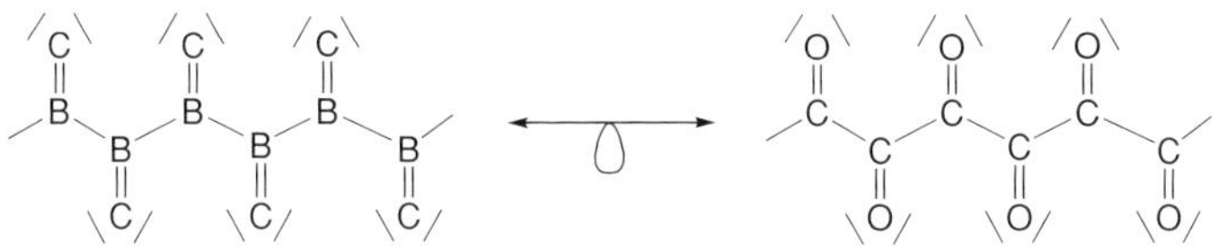

Figure 7.28

chain compositions are the same, the electron counts are different. Let's compare the electronic structure of an isolated regular C branched zigzag B chain with BC unit charges of 3− and 4−. The π levels of the −B(C)−B(C)− motif which constitutes the unit cell, the band structure and the DOS are shown in Figure 7.27. Note that the phases of the π levels of the −B(C)−B(C)− motif remind one of those of *trans*-butadiene.

For a charge of 3− the Fermi level (ε_F) cuts the bottom of a band which derives from the π_3 level associated mainly with the B chain. This corresponds to an important peak in the DOS. Thus, there is no reason for a distortion of the B chain and the full occupation of the two lowest π bands is consistent with the Lewis formula on the left side of Figure 7.28. To keep you in the mood of thinking isoelectronic, note that this polymer, which is embedded in an ionic matrix, is isoelectronic to a polyketone, $(CO)_\infty$ shown on the right of Figure 7.28. Calculations suggest that $(CO)_\infty$ is only slightly unstable with respect to carbon monoxide formation. In fact few examples of $(CO)_n$ oligomers exist.

With two more electrons per unit cell, i.e., $[B_2C_2]^{8-}$, the π_3 band becomes fully occupied. But the π_3 and π_4 bands are degenerate at the special point (zone edge) X leading to a first order Peierls instability. Distorting the branched B chain to generate alternating long and short separations leads to the band structure and

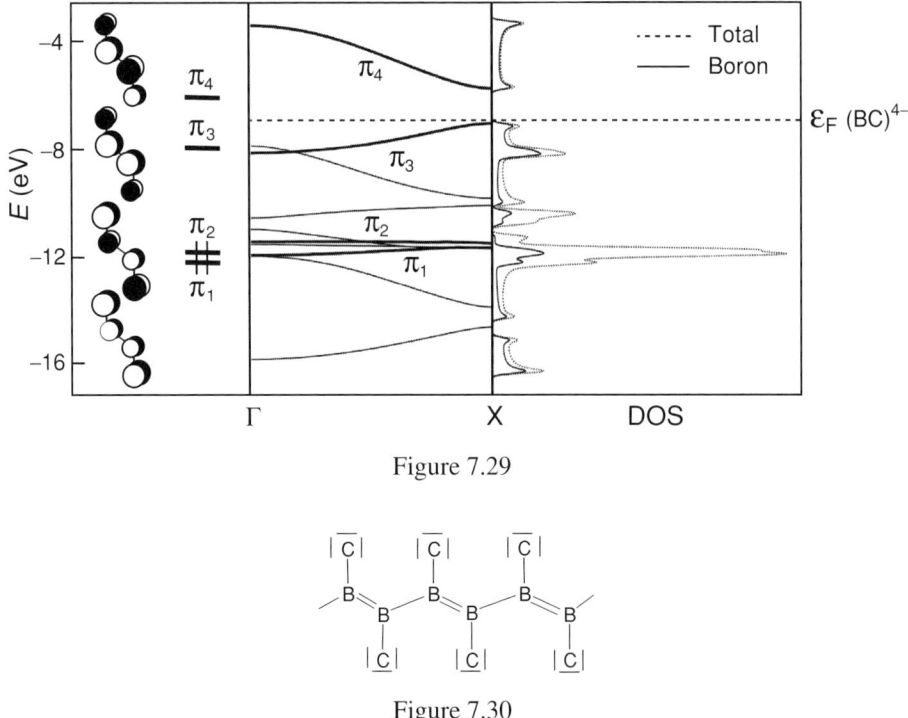

Figure 7.29

Figure 7.30

DOS diagrams in Figure 7.29. As expected, a gap opens between the π_3 and π_4 bands with concomitant lowering of the energy for this electronic configuration. As the π_3 band is filled and separated by an energy gap from the empty π_4 band, the Lewis formula in Figure 7.30 is now appropriate. It turns out that the B–B "single" bonds are very long in the "real" solid. The main reason for this property is that interactions with the surrounding metals were ignored. In spite of this, the qualitative observation, regular vs. alternating B–B distances, is rationalized by a simple electron-counting argument in the context of the DOS model of electronic structure.

7.1.4 More C than M: fullerenes and nanotubes

The naked C_{60} cluster (a fullerene) was briefly mentioned at the end of Chapter 2 as an example of a three-connect bare cluster with a delocalized system containing the external cluster electrons. Here we will discuss some properties of fullerene-derived solids (there is C_{60}, but also C_{70}, C_{84}, etc.). At room temperature, solid C_{60} adopts an fcc structure with weak van der Waals interactions between the C_{60} molecules. Look more closely at the structure of C_{60} itself. It has the highest symmetry possible for a molecule, I_h point group, and consists of a polyhedron with 20 hexagonal

7.1 From a single atom to an infinite solid

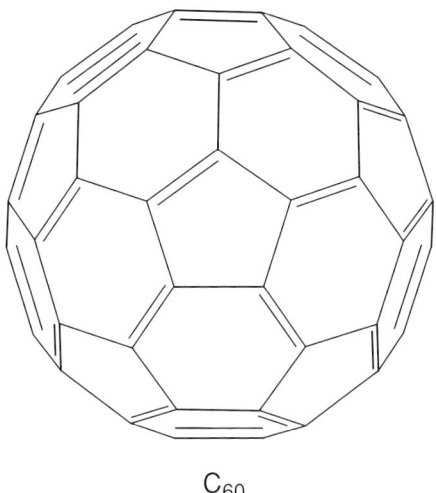

C$_{60}$

Figure 7.31

and 12 pentagonal faces (Figure 7.31). It is the presence of the pentagonal faces that requires the spherical shape as a 60 C-atom molecule made of solely regular hexagons would be planar similar to a small piece of graphite.

The spherical geometry and high symmetry of C$_{60}$ gives it a singular electronic structure. As the three-connect bonding in the surface of the sphere is analogous to the σ-bonding network of a planar aromatic molecule, the main interest lies in the curved π system, i.e., each of the external orbitals of this cluster contains a single electron rather than a lone pair. An approximate description of this electronic feature can be generated from the symmetry-adapted combinations of the radial p orbital of the 60 C atoms and the result is shown in Figure 7.32. A substantial HOMO–LUMO gap separates the 30 occupied orbitals from a group of 30 vacant orbitals. However, like aromatic hydrocarbons, C$_{60}$ easily accommodates extra electrons in the low-lying LUMO of t$_{1u}$ symmetry.

Now consider the solid. It is easy to understand why pure bulk C$_{60}$ is semiconducting. The clusters are about 3.1 Å apart and the interaction between clusters must be small. Therefore, the discrete levels in the HOMO–LUMO region of C$_{60}$ give rise to narrow or flat occupied bands separated from flat unoccupied bands as shown by the band structure diagram in Figure 7.33 where the first Brillouin zone is shown at the right. Solid C$_{60}$ is a molecular solid.

So why is solid C$_{60}$ of any more interest than the molecular cluster itself? Addition of alkali metals such as K, Cs or Rb generates salts of the composition $[M^+]_x[C_{60}]^{x-}$. These have metallic character as shown by the electrical conductivity which increases by many orders of magnitude. The K$_3$C$_{60}$ solid, for example, exhibits an fcc structure with K atoms in all the octahedral and tetrahedral holes,

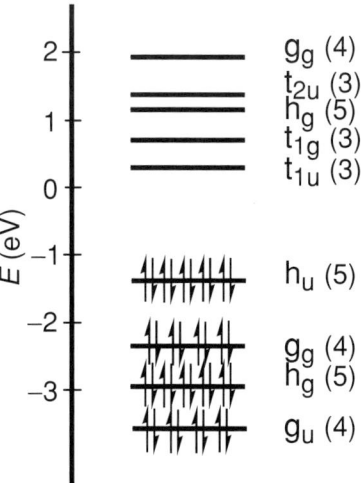

Figure 7.32

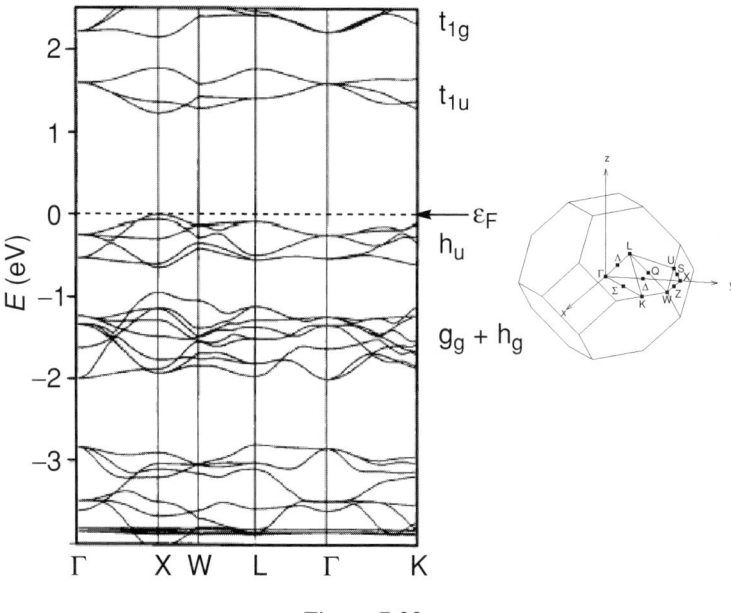

Figure 7.33

i.e., outside of the C_{60} clusters. As x increases, the conductivity increases, reaching a maximum for $x = 3$. This corresponds to a half-filled t_{1u} band in Figure 7.33. A further increase in the alkali-metal content causes the conductivity to decrease until for $x = 6$ the material becomes an insulator. Now the t_{1u} band is filled. Of great interest is the observation that for $x = 3$ these salts become superconducting at surprisingly high transition temperatures (T_C). Rb_2CsC_{60} has a T_C of 35 K whereas it is 12 K for K_3C_{60}. This difference is significant. We can use the DOS to explain this difference because physicists have shown that the T_C varies simply with the DOS at ε_F: the larger the DOS at ε_F, the higher the T_C. Now recall that band width depends in part on overlap which decreases with an increase in distance between the interacting centers. As the alkali metals lie outside the clusters, the C_{60} units are farther apart for Rb_2CsC_{60} than for K_3C_{60}. Hence, the t_{1u} band width is smaller for the former. Since the total integrated DOS is constant, the value of the DOS at the Fermi level must be larger for Rb_2CsC_{60} than for K_3C_{60} consistent with the relative magnitudes of the T_Cs.

With C_{60}, as well as the larger analogs, atoms can be introduced into the internal cavities to form main-group versions of transition-metal clusters containing interstitial atoms. Entities such as main-group atoms like N or a rare gas, molecules like H_2, rare-earth metals and others can be encapsulated. As with external metals, the maximum conductivity occurs for internal metals which are able to transfer three electrons to the radial t_{1u} band of solid C_{60}.

Recall from Chapter 6 that a single sheet of graphite is called a graphene. It is a two-dimensional net that can be rolled into cylinders but lacks the five-membered rings required by the spherical shape of C_{60}. This cylindrical form, called a nanotube, constitutes another allotrope of C. Two types are known: a single-walled nanotube, SWNT, is formed from a single graphene sheet whereas multiwalled nanotubes, MWNT, possess a number of concentric rolled graphenes. A close look at the geometry of these species reveals considerable geometric complexity which implies complex electronic structure. To describe the structures, the parameters, a_1 and a_2, given in Figure 7.34 and called the primitive vectors, are defined. The diameter of a SWNT varies from 1.2 to 2 nm and is given by the roll-up vector $C_h = na_1 + ma_2$ more simply designated by (n,m). By following the pattern across the diameter, three variants of SWNTs, which depend on the manner in which the graphitic sheet is rolled up, can be recognized. These are: achiral armchair (n,n), achiral zigzag (n,0), and chiral (n,m) ($n > m$ and $m \neq 0$). For (n,m) SWNT, a chiral angle, θ, is defined as the angle between the roll-up vector C_h and the (n,0) zigzag a_1 direction. In Figure 7.34 the example of the roll-up vector for the SWNT is (4,1). Below it is drawn a (5,5) armchair tube in its cyclindrical form.

Although these types of nanotubes differ only slightly in geometry, the differences have a significant effect on properties. As briefly described in Section 6.3.4, the simple geometric structure of graphite supports a complex electronic structure. It is called a zero-gap semiconductor (a semi-metal) because even though the π and π^* bands touch (no gap) the DOS at the Fermi energy is zero. Considering the close relationship with graphite, it's no surprise that the band structure of a SWNT is dominated by the dispersion of the π and π^* bands close to the Fermi level. The

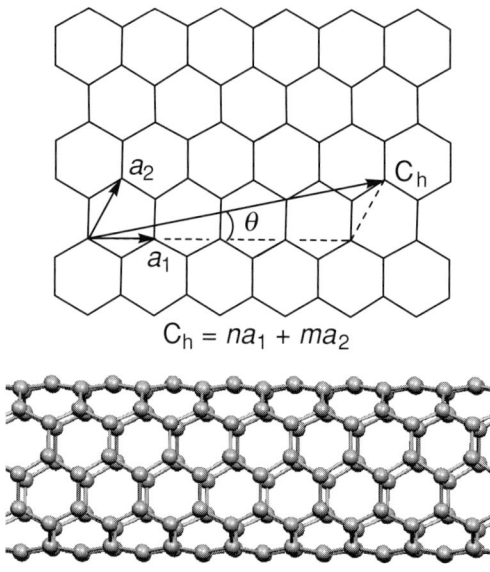

Figure 7.34

actual dispersion is a sensitive function of structure type (armchair, zigzag or chiral) and curvature as well as interactions with the internal tubes in the MWNTs. It follows that the physical properties will similarly differ. In fact, some nanotube types are conductors, whereas others are semiconductors with substantial band gaps.

As the tube ends may be open, the possibilities for encapsulating guests are large. For instance, they have been stuffed with water, main-group elements like O_2, and metals such as Cu. They are capable of storing up to 65% of their own weight in H_2 making them of interest for H_2 storage and, they easily encapsulate Li ions which are important charge carriers in a class of battery. Amusingly, the first molecule ever reported inside C nanotubes was C_{60} and the filled nanotubes were appropriately named "peapods." C_{60} acts as an electron-withdrawing dopant and perturbs the electronic properties of the nanotube that contains the clusters.

7.2 B clusters in solids: connections with molecular boranes

In Section 2.3.2 a little history of the prediction of $[B_{12}H_{12}]^{2-}$ as the composition and charge of an icosahedral borane molecule was related. The "dangling bonds" of a B_{12} icosahedral fragment "cut" from solid elemental B were terminated with H atoms and the charge required for stability derived from the assigned HOMO–LUMO gap of an MO calculation. The cluster electron-counting rules can be traced to this exercise. At this point in our studies we can look more closely at solids containing B polyhedra and seek more similarities and differences between molecular and solid compounds. Although B icosahedra are present, the geometric structure of

7.2 B clusters in solids 283

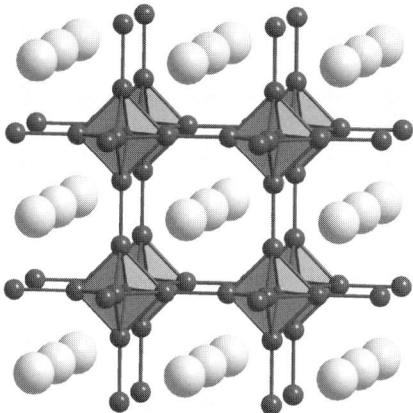

Figure 7.35

elemental B is complicated (the simpler form is shown in Figure A1.11). Even with today's powerful computers, no complete theoretical study of the electronic structure of elemental B has been done! Thus, we must consider a couple of simpler examples.

7.2.1 Octahedral B clusters in solids: analogy with $[B_6H_6]^{2-}$

Metal hexaborides (metal = alkali metal, alkaline earth, rare earth, actinide) such as CaB_6 are examples which can be related to octahedral borane clusters. These compounds exhibit a three-dimensional lattice of B_6 octahedral clusters linked to each other through inter-cluster B–B bonds in three directions. As shown in Figure 7.35, this forms a simple cubic arrangement in which the Ca atoms are embedded. In the figure, the metals are large grey spheres, the B atoms are small black spheres and the octahedra are shaded. Applying the Zintl–Klemm idea, we have $[Ca^{2+}][B_6]^{2-}$. The $[B_6]^{2-}$ moiety has 20 cve and six more come from the shared inter-cluster B–B bonds giving a total of 26, appropriate for a main-group cluster of order six. Alternatively, if you consider six of the 20 participating in external bonds, then the sep = 14/2 = 7 is also consistent with expectations. Thus, going from the $[B_6H_6]^{2-}$ molecule to the $[B_6]^{2-}$ network found in CaB_6 changes the localized B–H bonds into inter-cluster B–B bonds just like those found in *conjuncto*-$[B_5H_8]_2$ shown in Figure 2.18.

Electron counting hides interesting details of electronic structure so let's see how the band structure of CaB_6 compares to the MO structure of $[B_6H_6]^{2-}$. The MO diagram of $[B_6H_6]^{2-}$ is shown in Figure 7.36. To review, there is a set of six outwardly-directed MOs below seven cluster molecular orbitals separated by a large HOMO–LUMO gap from 11 antibonding MOs. Thus, we expect the $[B_6]^{2-}$ sublattice to exhibit a comparable pattern, i.e., a sequence of low-lying occupied

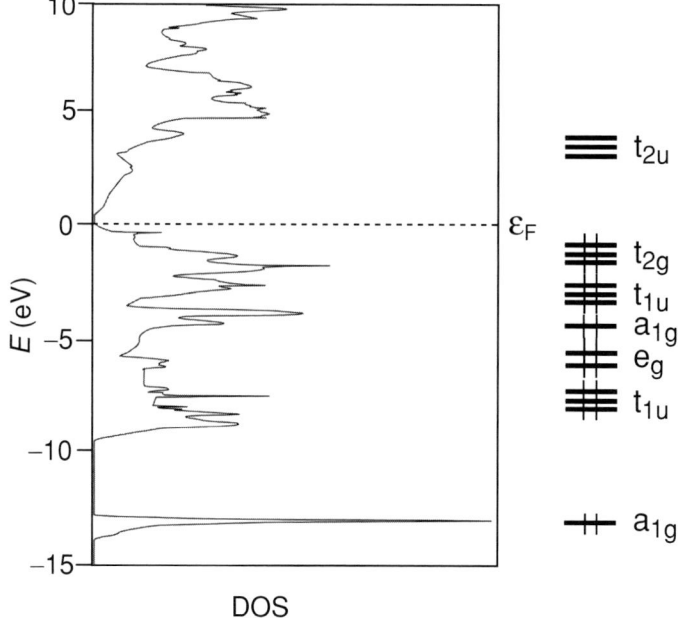

Figure 7.36

inter-cluster bonding bands, occupied intra-cluster bands and unoccupied antibonding bands. And the calculations on MB_6 (M = alkaline earth) demonstrate this. For example, the DOS for CaB_6, shown in Figure 7.36 at the left, can be compared with the MO diagram of $[B_6H_6]^{2-}$ at the right. By now you realize these solid-state compounds will not be purely ionic in character. Some weak covalent bonding occurs between B and M and the corresponding bands may pick up some metal character. But the qualitative properties are generated. We have a semiconducting solid with rather flat occupied bands separated from vacant bands.

You may be willing now to believe that CaB_6 is a close relative of $[B_6H_6]^{2-}$ in the sense of electronic structure. However, MB_6 is also found for M = lanthanides. LaB_6, for instance, is well characterized and also exhibits the structure of CaB_6. Given the preferred oxidation states of +3 rather than +2, this causes a problem. Applying the Zintl–Klemm approach, we would have $[B_6]^{3-}$; however, molecular $[B_6H_6]^{2-}$ has not been successfully reduced. But wait, we've encountered this situation earlier in this chapter. Even if the cation can provide an additional electron, the extra electron need not be transferred to the B network. It can occupy the bottom of a La-based band which will lie between the low-lying occupied B-based bands and the high-lying vacant B-based bands. This will impart metallic conductivity, and LaB_6 is known to be metallic. The important new idea is that the metal electrons can serve as a reservoir which can be fully or partially used to supply the electron requirements of the main-group partner.

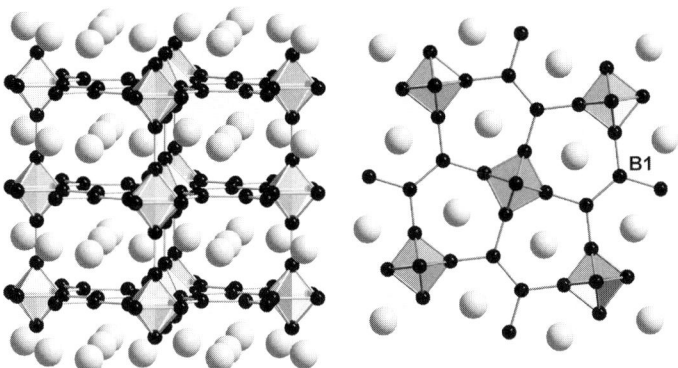

Figure 7.37

Many other linked polyhedra similar to molecular boranes are found for B-rich solid compounds of alkali and alkaline-earth metals. There are octahedra in Li_2B_6, octahedra and pentagonal bipyramids in Na_3B_{20}, dodecahedra and bicapped square antiprisms in Li_3B_{14} and icosahedra in Na_2B_{29}. The cluster connection is particularly clear for compounds of this type.

7.2.2 Localized and delocalized bonding in B solid-state chemistry: the example of GdB₄

Rare-earth tetraborides such as GdB_4 exhibit a more complicated architecture than metal hexaborides. Two representations of the structural arrangement of GdB_4 are shown in Figure 7.37. That at the left is a view perpendicular to the stacking axis while that at the right is down the same axis. Here Gd is represented by the large grey spheres and B by the small black spheres. Although GdB_4 contains B_6 octahedra (shaded in Figure 7.37) the octahedra are directly connected in one rather than three directions. In the two other directions, the link is made through B_2 "units," one B of which is denoted B1 in Figure 7.37. The connecting units are formed from sp^2-hybridized B atoms and each is bonded to two B_6 octahedra and a partner. The B–B distances are about 1.8 Å. Thus, GdB_4 is viewed as $Gd_2(B_6)(B_2)$.

The first question posed by this structure concerns the formal charge distribution between the two Gd, B_6 and B_2 units. If we naively apply the Zintl–Klemm concept we arrive at $[Gd^{3+}]_2[B_6, B_2]^{6-}$. The $[B_6]^{2-}$ cluster with six external bonds obeys the cluster electron-counting rules. Consequently, the B_2 fragment must have a charge of -4. This corresponds to saturated eight-electron B centers, and requires non-planar (tetrahedral) B centers. This does not agree with the observed planar B centers. But we know the metal does not need to be fully oxidized. Consider sp^2-hybridized B atoms which satisfy the octet rule. This would lead to $[Gd^{2+}]_2[B_6^{2-}][B_2^{2-}]$ and suggests a B=B double bond. OK, but planar B is

Figure 7.38

also found with only six electrons in, for example, planar BCl_3. A neutral B_2 unit with a single B–B bond would give us $[Gd^+]_2[B_6{}^{2-}][B_2]$. This leaves the metal in an unrealistically low oxidation state given its strong preference for a trivalent state. The Lewis structures for the three scenarios for the B_2 moiety are shown in Figure 7.38. So, which of these three scenarios is the best description? No doubt you would like an additional hint. But we don't have any! Only good calculations can help us. These have been done and indicate the bonding is best described by the $[Gd^{2+}]_2[B_6{}^{2-}][B_2{}^{2-}]$ charge distribution but with a weak coordinated B–B double bond.

Exercise 7.5. The structural arrangement of the ternary solid-state compound $Gd_5Si_2B_8$ is represented below. At the left it is shown perpendicular to the stacking axis and at the right down the stacking axis. The large grey spheres are Gd, the small black and dark grey spheres are Si and B, respectively, and the octahedra are shaded. You can see that the structure is related to that of GdB_4 in that layers of GdB_4 are intercalated by Gd_3Si_2 sheets. In the process one set of B–B bonds between B_6 octahedra are broken. A reasonable initial partitioning is $(Gd)_5(Si_2)(B_6)(B_2)$. Knowing that the Si atoms form singly bonded isolated pairs and the B–B distance suggests a double bond in the B_2 units linking the octahedra together, assign charges to the Gd atoms, Si_2 pairs, B_6 octahedra and B_2 units. Predict the electrical conductivity of this material.

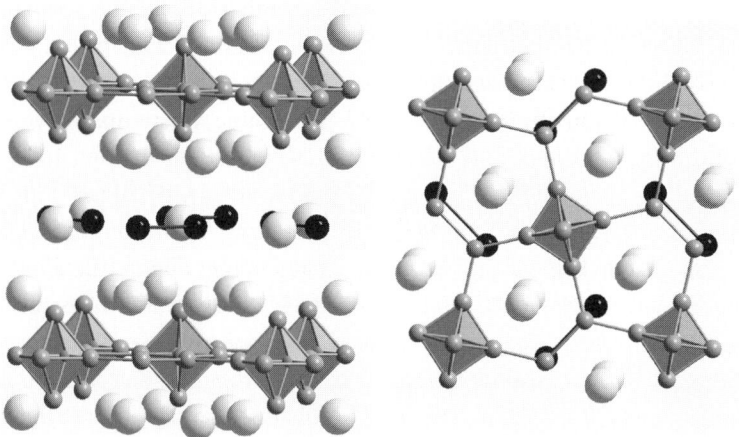

Exercise 7.5

Answer. Let the isolated Si_2 pairs satisfy the octet rule. This gives $[Si_2]^{6-}$ with a Si–Si single bond. The B_6 octahedra require 26 cve. As there are now only four external B–B bonds, this gives 22 for the neutral unit. Hence, the charge must be 4− leading to $[B_6^{4-}]$. With a B–B bond of bond order two, the B_2 units must bear a charge of 2−. Thus, we have $[Gd^{2.4+}]_5[Si_2^{6-}][B_6^{4-}][B_2^{2-}]$! Don't let the partial oxidation state of the metal bother you. It simply means that once the electronic requirements of the main-group moieties are satisfied, a total of three electrons per formula unit, i.e., per Gd_5, remain on the metal atoms. Since the metal is not fully oxidized, we expect the compound will be a metallic conductor. It is.

7.3 Molecular transition-metal complexes in solids

To give more breadth to our discussion of the connections between molecules and solids we now consider examples of systems in which one can "see" transition-metal complexes. They are of pedagogical interest because their simple electronic structures can be used to good effect in elementary treatments of the electronic structure of coordination compounds.

7.3.1 Ternary hydride solids

Ternary metal hydrides of formula A_2MH_6 (A = Mg, Ca, Sr, Eu and M = Fe, Ru and Os) are known. Full characterization requires knowledge of the distribution of the H atoms within the structure of the A_2M host. Using what we have learned, let's try to deduce their arrangement employing Mg_2FeH_6 as an example.

The metallic lattice is of antifluorite type. Recall that fluorite (CaF_2) can be described as an fcc lattice of Ca atoms with F atoms in half of the tetrahedral holes. Replacing Ca by Fe atoms and the F by Mg atoms leads to the structure of the Mg_2Fe host. In other words every Fe atom lies in a cube of Mg atoms. What about the H atoms? Perhaps the methods we have developed up to this point can be used to generate a reasonable answer. Let's see. The greater electronegativity of H and Fe relative to Mg invites application of the Zintl–Klemm principle. The $[FeH_6]^{4-}$ anions have an electron count at Fe of 18! Compare $[FeH_6]^{4-}$ to $Cr(CO)_6$ (Section 1.1.5) and you will find the solid-state hydride an analog of the molecular metal-carbonyl complex. Both have an octahedral arrangement of ligands around the metal (formally Cr^0 and Fe^{II}) with filled t_{2g} and empty e_g orbitals. But is this solution correct? Yes, neutron diffraction measurements confirm the octahedral structure shown in Figure 7.39. Although you will not find this example in textbooks, we cannot think of a simpler ML_6 octahedral complex with which to begin a discussion of the MO structure of an octahedral complex. The $[H]^-$ ligands have no π-donor or acceptor orbitals to complicate the analysis!

288   *From molecules to extended solids*

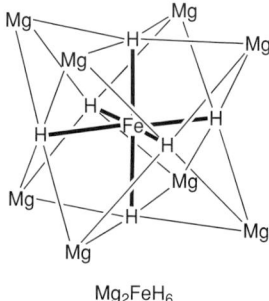

Figure 7.39

But with your experience of the earlier examples, you are probably wondering if the perturbations by the electropositive Mg^{2+} cations invalidate this simple model. Band-structure calculations indicate that, although the valence orbitals of Mg reinforce stability through covalent interactions with the Fe and the H atoms, the "molecular properties" essentially survive. That is, the occupied t_{2g} and vacant e_g levels of the $[FeH_6]^{4-}$ anions give rise to separated bands of low dispersion (flat) in the solid in agreement with the Zintl–Klemm analysis and the behavior of the material as an insulator.

Exercise 7.6. The solid-state compound Mg_2CoH_5 has been prepared and exhibits the same metal lattice as $MgFeH_6$. Using the Zintl–Klemm approach and your knowledge of molecular coordination chemistry, suggest a logical geometry for the H atoms around the Co center.

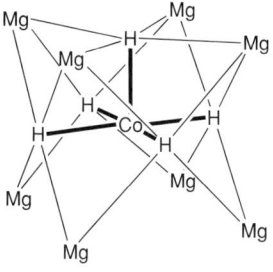

Exercise 7.6

Answer. As we deal again with Mg^{2+} ions, the anion is formulated as $[CoH_5]^{4-}$ with Co^I and five $[H]^-$ ligands. Thus, it is an 18-electron compound. The limiting geometries adopted by five-coordinate complexes are trigonal bipyramidal (D_{3h} symmetry and most common) for which $Fe(CO)_5$ provides an example and square pyramidal (C_{4v} symmetry) for which $[Ni(CN)_5]^{3-}$ provides an example, albeit slightly distorted. These, and distorted structures in between, have similar stabilities.

Interestingly, as shown in the drawing below, the square-pyramidal structure is favored in Mg_2CoH_5 and presumably is one imposed by the stabilizing Mg_8 cube which surrounds it.

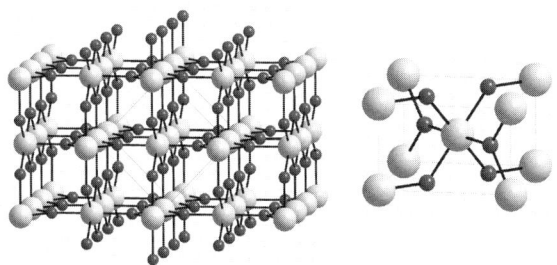

Figure 7.40

7.3.2 Shared ligands: the example of rutile TiO_2

Can all solids be fragmented into "transition-metal molecular ions" separated from each other by "spectator" counterions which satisfy charge balance and generate ionic interactions which stabilize the solid? If so, the problem is a simple one and we can move on. But it is not always possible to usefully identify "molecular pieces" in solids. Further, even when it is, complications arise that make understanding a little more challenging than with the hydrides above. Although bridging ligands are common in molecular chemistry, the process of fragmenting a solid-state structure with shared atoms is a good exercise. Let us look at a deceptively simple example, that of rutile TiO_2.

To describe the structural arrangement, the solid state chemist highlights the planar coordination of O by three metal atoms plus the infinite chains of edge-sharing TiO_6 octahedra which run in a single crystal direction. This is shown at the left side of Figure 7.40 where Ti is represented by the large grey spheres and O by the small black spheres. On the other hand, as shown on the right side of Figure 7.40, the molecular chemist will highlight the octahedral environment of the Ti^{IV} ions formed by six shared $[O]^{2-}$ ligands leading to the stoichiometry ($TiO_{6/3}$). Both viewpoints are right; they just differ in focus: the three-dimensional framework or the local environment of the atoms. Can the latter be helpful?

As usual, we begin with electron counting. For convenience consider the $TiO_{6/3}$ "complexes" to be constructed of d^0 $[Ti]^{4+}$ and $[O^{2-}]$ ions. Each O atom is in bonding contact with three Ti atoms; hence three of its four pairs form donor bonds with the three metal atoms leaving one lone pair. It follows that every TiO_6 octahedral "molecular complex" reaches the count of 12 electrons rather than 18! Although the 18-electron rule often fails if applied to early metal complexes, counting electrons in this case certainly doesn't provide understanding. Ti–O covalent character

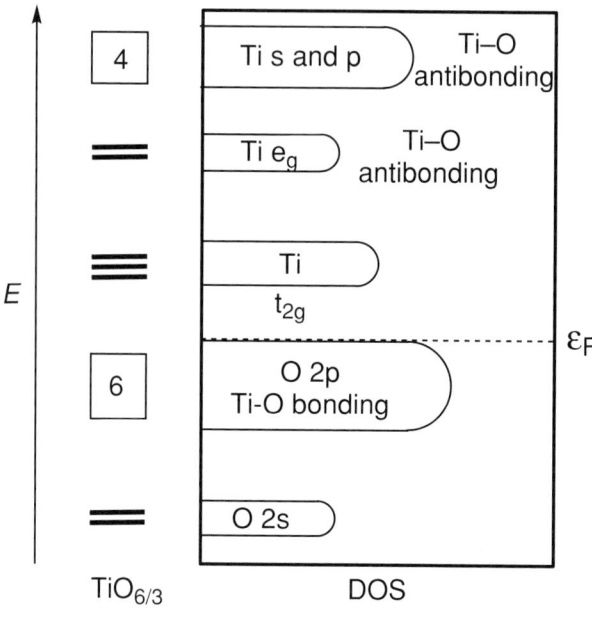

Figure 7.41

is important and any totally ionic model will fail. Examine the covalent character using a MO model.

First consider the orbital interactions in a single "isolated" TiO$_{6/3}$ octahedron. The 3d orbital set of each octahedral Ti atom splits into t_{2g} and e_g combinations as found for any octahedral molecular complex. For a count of 12 electrons both sets are empty. Below the t_{2g} orbitals lie the ligand O 2s and 2p occupied combinations associated with the Ti–O bonds and with the O lone pairs. Above the e_g orbitals are the empty Ti–O antibonding MOs with major Ti 4s and 4p character. This description is that of a typical octahedral transition-metal complex and describes the major interactions in rutile as covalent Ti–O interactions. Next we need to add to it the interactions between TiO$_{6/3}$ octahedra in the solid (interactions are additive). Although there is no effective Ti . . .Ti and O . . .O overlap in the solid, There is some through-bridge overlap between MOs of neighboring octahedra. These interactions are not strong so the bands generated by the localized MOs are only moderately dispersed. As a result, an approximate DOS for TiO$_2$ should look like that shown in Figure 7.41. The levels broaden into bands but the memory of the octahedral splitting remains and well-separated O 2s, O 2p, Ti t_{2g}, Ti e_g and Ti 4s and 4p bands are observed. Notice that these bands have the same relative area ratios as the number of orbitals per unit cell, i.e., 2 to 6 to 3 to 2 to 4. The presence of covalent bonding is reflected by Ti character in the oxygen bands and oxygen in the Ti bands consistent with the MOs of the TiO$_{6/3}$ "complex."

Exercise 7.7. At room temperature VO$_2$ exhibits the rutile structure of TiO$_2$ but at low temperature (< 340 K) it distorts so that the V atoms pair up along the edge-sharing chains. This leads to alternating short and long V–V distances. Explain this distortion.

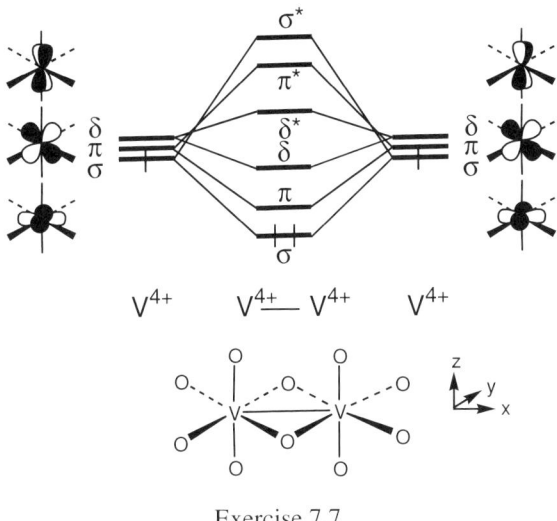

Exercise 7.7

Answer. VO$_2$ has one more electron than TiO$_2$, so the t$_{2g}$ band will contain one electron. Within the scheme given above, you can see that the t$_{2g}$ set of each metal contains one σ (d$_{x^2-y^2}$), one π (d$_{xz}$) and one δ (d$_{yz}$) type orbitals relative to the M–M axis. When two metals in the chains get close to each other their t$_{2g}$ orbitals interact. The orbital interactions decrease in the order σ > π > δ leading to the splitting pattern shown above. On moving the metal atoms closer, VO$_2$ with a d^1 V^{IV} configuration has the possibility of forming a V–V bond by filling the lower σ combination but not the σ* combination. This is not possible for TiO$_2$. The small energy separation between the filled σ and the empty π bands in the DOS makes the solid a semiconductor at low temperature. Note that for undistorted VO$_2$ the Fermi level cuts the V t$_{2g}$ band (see Figure 7.41) and that the observed behavior is typical of a Peierls distortion.

7.4 Molecular vs. solid-state condensed octahedral transition-metal chalcogenide clusters: rule-breakers again

In each of the molecular-cluster chapters we eventually came to the rule-breakers. We are at that point again. Our example deals with metal clusters of a familiar geometry that are found in the Chevrel phases. Young readers will not appreciate

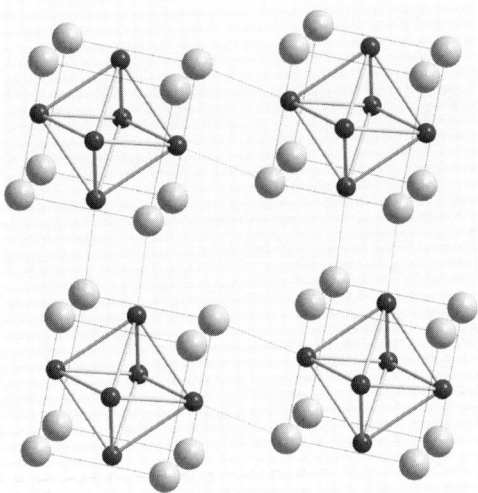

Figure 7.42

the impact the discovery of these ternary molybdenum chalcogenide materials of formula AMo_6X_8, where A is a monovalent or divalent metal and X usually a chalcogenide, had on solid-state chemists and physicists in the 1970s. The excitement arose from the fact that these phases are both high-temperature and very high-field superconductors. But high temperature then meant 15 K not 125 K as found later for some copper oxides! The fundamental structural unit for the prototypical example, $PbMo_6S_8$, is an octahedral cluster, Mo_6X_8, analogous to molecular $[Mo_6Cl_8L_6]^{4+}$ discussed in Section 3.3.5. In $PbMo_6S_8$ (Figure 7.42, where Mo is represented by the small black spheres and S by the large grey spheres) face-capped clusters are packed so that the face-capping chalcogens form terminal bonds to Mo atoms on an adjacent cluster. Within a cluster the Mo atoms have square-planar chalcogen coordination, but, on including coordination to neighboring clusters, have square-pyramidal coordination. The presence of the inter-cluster Mo–S bridge produces the three-dimensional net in the solid. In addition, and not shown, every cluster is encased in a cube formed by Pb atoms.

Formulation of this compound is not obvious; however, if we apply the Zintl–Klemm paradigm to Pb, then $[Pb]^{2+}[Mo_6S_8]^{2-}$ results. The three-dimensional extended network is made up of cluster units $[Mo_6S_8]^{2-}$ with 82 cve (six Mo, eight triply bridging S ligands and six terminal S ligands from adjacent clusters). However, solid-state chemists focus on the M_6 core electron count in $[Mo_6^{14+}(S^{2-})_8]^{2-}$ i.e., 22. Both ways were used in Section 3.3.5 and, if you check, you will find that both counts are two lower than found for the molecular cluster $[Mo_6Cl_8L_6]^{4+}$. As you have now sufficient information and experience to solve this conundrum, it

7.4 Molecular vs. solid-state

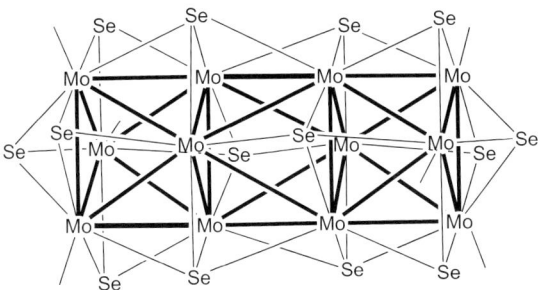

Figure 7.43

makes an excellent homework problem and you will find it as problems 2 and 3 at the end of this chapter.

This family of compounds includes more complex cluster units such as the $Mo_{12}Se_{14}$ entity present in $Cs_2Mo_{12}Se_{14}$ shown in Figure 7.43. These condensed clusters are linked together via the metal atoms in the outer layers much like for $PbMo_6S_8$ above. The finite units themselves are considered face-sharing octahedra surrounded by bridging chalcogenide ligands. In fact, this face-fused cluster unit is a member of the series $Mo_{3n}X_{3n+2}$, i.e., Mo_6S_8 is the smallest, that can be generated by successive insertion of Mo_3X_3 planar fragments. Oligomers with X = S, Se and Te, $n = 2$–8, 10 and 12, as well an infinite chain with stoichiometry $(Mo_3X_3)_\infty$ as found in $Tl_2Mo_6Se_8$, have been characterized.

Because the Mo atoms in the terminal Mo_3X_3 triangles are capped as well as linked to adjacent cluster units as found in $PbMo_6S_8$, the metal atoms are electronically similar. But note that the Mo atoms in the internal Mo_3X_3 triangles lie in a distorted C_{2v} square ligand environment. Consider their electron counts. As these clusters result from the fusion of octahedra through triangular faces, perhaps the treatment of Section 3.3.3 would be useful. Let's see what it looks like using the solid-state chemist's cluster count. Assuming the electron count of $[Mo_6Cl_8L_6]^{4+}$, i.e., 24, is an appropriate model, a fused cluster will have a count equal to 24 times the number of fused octahedra minus six times the number of shared triangular faces, six being the number of metal electrons per triangle. It means that we should get 42 ($24 \times 2 - 6$), 60 ($24 \times 3 - 6 \times 2$), 78 ($24 \times 4 - 6 \times 3$), etc. for Mo_9X_{11}, $Mo_{12}X_{14}$, $Mo_{15}X_{17}$, etc. oligomeric units. Electron counts of 34 for $[Mo_9Se_{11}]^{2-}$, 46 for $[Mo_{12}Se_{14}]^{2-}$ and 59 for $[Mo_{15}Se_{17}]^{3-}$, encountered in $K_2Mo_9S_{11}$, $Rb_2Mo_{12}Se_{14}$ and $Rb_3Mo_{15}Se_{17}$ solid phases, respectively, are much lower. Even if we use both the lower count of 22 observed for $[M_6S_8]^{2-}$ and note that when fusing late metal-carbonyl clusters, 50 rather than 48 cve are subtracted, the counts are too high. This is puzzling.

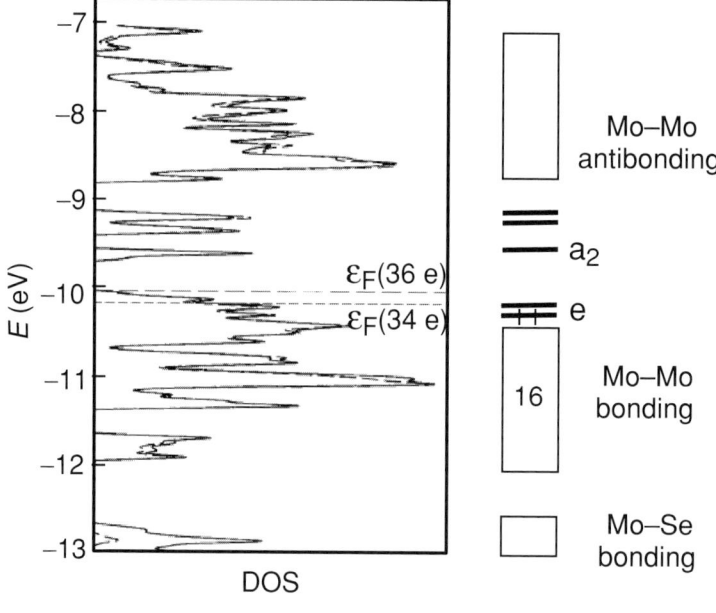

Figure 7.44

When counting electrons leads to nonsense, you know we need to move to a different paradigm or adopt another method. So let's analyze the bonding in $K_2Mo_9S_{11}$ to see if the information gleaned can be extrapolated to the other members of the series. A qualitative molecular orbital diagram of characterized $[Mo_9Se_{11}]^{2-}$ is shown in Figure 7.44 at the right. The overall distribution of energy levels shows a rough separation of the chalcogen p and Mo d orbitals with a region of predominant metal character in the HOMO/LUMO region. Further, there is a substantial energy gap separating the bonding and antibonding Mo d-based molecular orbitals. An open-shell electronic configuration is found for this oligomer. That is, for 34 electrons the degenerate e levels are partially occupied. Look at the calculated DOS at the left of Figure 7.44 and note the close correspondence with the MO diagram, even though the DOS includes inter-cluster interactions. The Fermi level crosses a partially filled narrow band which mainly comes from two e symmetry MO level sets. This means the solid compound should be a metallic conductor. It is a superconductor at very low temperature as well.

Look again at the DOS diagram. If two extra electrons could be added per Mo_9S_{11} unit, they would fill the lower e symmetry MO of the isolated cluster or fill the band at the Fermi level. The latter would render the compound semiconducting. A way to do this is to add a dopant. Experimentally this is done by insertion of Cu into $K_2Mo_9S_{11}$ by a redox reaction which leads to $Cu_2K_{1.8}Mo_9S_{11}$. This is successful as

7.4 Molecular vs. solid-state

crystalline $K_2Mo_9S_{11}$ has holes available around the clusters for the Cu atoms. Some K is lost upon reaction but the result is achieved. The compound is a semiconductor. Reduction without major structural change is a characteristic of high-connectivity clusters.

Exercise 7.8. Interestingly the condensation of octahedra via common faces is also exemplified in molecular chemistry. $Co_9Se_{11}(PPh_3)_6$ is an example which adopts the same structural arrangement as the $[Mo_9Se_{11}]^{2-}$ unit $K_2Mo_9S_{11}$ (see below). Count the cluster metallic electrons (using the "solid-state chemist's" way) and compare the value to the 34 electron count of $[Mo_9Se_{11}]^{2-}$.

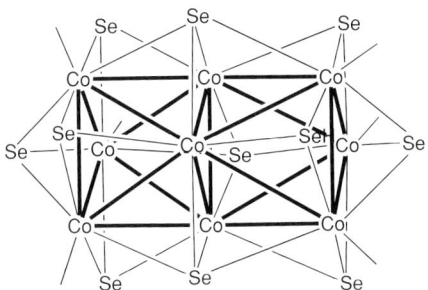

Exercise 7.8

Answer. The metal count is 59 (9 × 9 (Co) − 11 × 2 ($[Se]^{2-}$) = 59). This is 25 more than for $[Mo_9Se_{11}]^{2-}$! Indeed a large gamut of electron counts is possible for these condensed octahedral clusters, either molecular or part of extended solids. They are nice examples of compounds which can display variable electron count with constant shape. This situation is somewhat reminiscent of the transition-metal elements which exhibit a metallic d band which is gradually filled as we go from the left to the right of the periodic table without substantial change in their structure. Also review the discussion of cubic clusters in Section 5.2.5.

Exercise 7.9. Bicapped-octahedral M_6 clusters can be observed in solid-state chemistry as in $La_2Mo_{16}O_{28}$ and $Nd_2Mo_{16}O_{28}$ for instance. The former contains a well-ordered mixture of *cis*- and *trans*-bicapped "Mo_8O_{24}" clusters in equal proportion, whereas the latter shows only *cis*-bicapped "Mo_8O_{24}" clusters (see below). Count the cluster metallic electrons per bi-capped octahedral Mo_8 cluster. Try to explain why $La_2Mo_{16}O_{28}$ is semiconducting and $Nd_2Mo_{16}O_{28}$ metallic in character. Hint: in contrast to what we learned for molecular clusters, e.g., Exercise 3.6, permit the isolated *cis*- and *trans*-bicapped "Mo_8O_{24}" clusters to accommodate different closed-shell electron counts without large change in HOMO energies.

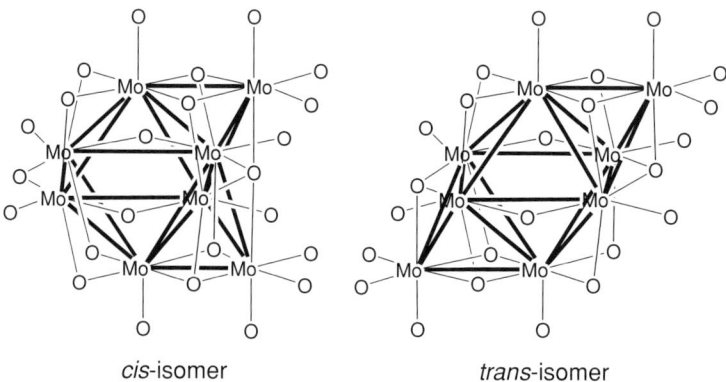

cis-isomer   trans-isomer

Exercise 7.9

Answer. La and Nd strongly prefer a +3 oxidation state leading to the formulation $[La(Nd)^{3+}]_2[Mo_{16}]^{50+}[O^{2-}]_{28}$. Hence, we have 46 metal electrons to distribute over two Mo_8 clusters. In $Nd_2Mo_{16}O_{28}$ there are two identical cis-bicapped isomers and hence each must have a count of 23 electrons. With an odd number of electrons, the HOMO of the cluster is singly occupied and will generate a half-filled band and metallic behavior is predicted for the solid. On the other hand, there are two possibilities for $La_2Mo_{16}O_{28}$ and we have different isomers. If both cis- and trans-bicapped Mo_8 isomers have the same electron count of 23, metallic conductivity in the solid is again predicted. This cannot be the case as $La_2Mo_{16}O_2$ is found to be a semiconductor. Filled bands at the Fermi level are thus required and therefore the cis- and trans-bicapped Mo_8 isomers must have even electron counts. Consequently the two isomers must have different electron counts. Indeed, calculations indicate that the cis- and trans-bicapped Mo_8 isomers satisfy closed shell requirements for 22 and 24 metallic electrons and exhibit HOMOs at roughly the same energy. Note that in this particular case, the capping principle described in Section 3.3.2 is not satisfied since two different bi-capped clusters display two different electron counts. This is due to some mismatch between the frontier orbitals of the octahedron and the frontier orbitals of the capping units. This mismatch differs for cis- and trans-capping.

7.5 Cubic clusters in solids

Molecules and solids based on cubic rather than octahedral motifs are less abundant in chemistry. The former were treated in Section 5.2.5 and now we consider a few solid-state analogs. Similar cubic architectures are found in $(Fe/Co/Ni)_9S_8$, or synthetic Co_9S_8 transition-metal sulfide minerals called pentlandites. In a binary

7.5 Cubic clusters in solids

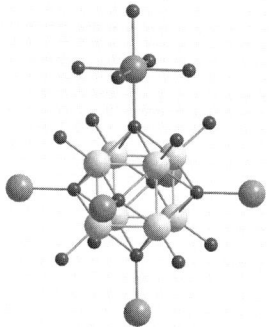

Figure 7.45

compound, e.g., Co$_9$S$_8$, hexacapped cubic Co$_8$S$_6$ clusters can be distinguished linked together by octahedral CoS$_6$ units as shown in Figure 7.45. In this scheme the Co$_8$ cube (cluster Co atoms are represented by large, light grey spheres, mononuclear Co atoms by large, dark grey spheres and S atoms by small dark spheres) is surrounded by an octahedron of mononuclear Co complexes where its S ligands also cap the square faces of the cube. For clarity, only one full ligand envelope of the mononuclear linkers is shown.

You know counting electrons is not going to be helpful here as the cubic cluster systems treated in Chapter 5 had counts ranging from 120 to 99 cve! Plus, we have a mononuclear complex to deal with in the same lattice and we can't even be sure it will be an 18-electron species. On the other hand, we can use an electron count to evaluate some boundary conditions for this system. With eight S^{2-} ligands we have (Co$_9$$^{16+}$), i.e, 65 metal electrons are left for the nine cobalt atoms. The solution to the assignment of these electrons to the two different metal types is: $8x + y = 65$, where x and y each represent a different d-electron count per metal. Still x and y cannot differ greatly since we deal with a single metal type, Co, albeit in two different environments. Consider the mononuclear center and the value of y. A d^6 electron count is expected for an octahedral Co center with π-acceptor ligands, i.e., a filled t$_{2g}$ set and empty e$_g$ set. However, we have π-donor ligands so the e$_g$ set may be partially filled (d^7 or d^8). For example, the 20-electron (Ni^{2+}, d^8) octahedral complex [Ni(H$_2$O)$_6$]$^{2+}$ models the latter. With $y = 6$, 7 or 8, we are left with $8x = 59$, 58 and 57. Magnetic susceptibility measurements would be helpful in going further but keep in mind that we have both the cluster and the complex to consider. In any case, none have been measured for this compound.

So we must proceed from the "molecular approach" to a solid-state approach. Look at the t$_{2g}$ and e$_g$ projections of the mononuclear cobalt in the computed DOS of the material. This is shown in Figure 7.46 as the hatched areas. As expected,

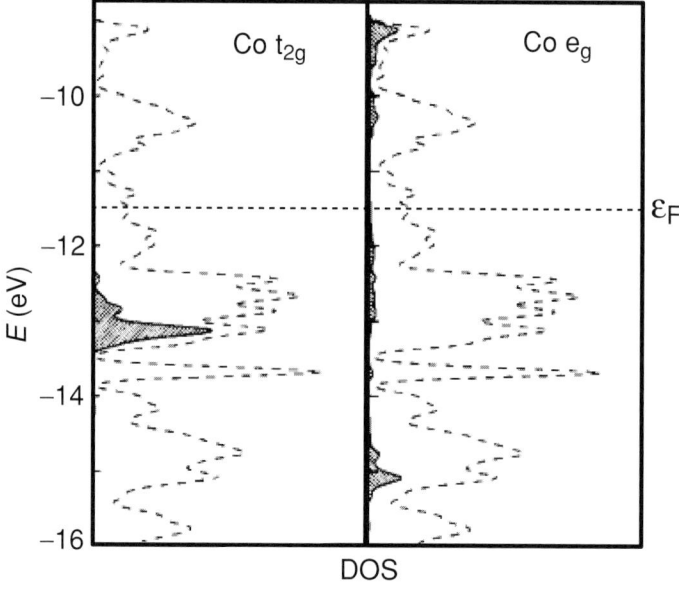

Figure 7.46

the t_{2g} band is rather narrow and completely occupied whereas the e_g band is spread out above and below the Fermi level due to mixing with S ligand orbitals. It means that the Co–S bonding and antibonding states are occupied and empty, respectively. This suggests a Co^{3+} d^6 octahedral cobalt with its t_{2g} set filled and its e_g set empty. As already said this gives 59 metal electrons for the Co cubes or, alternatively, 111 cve for the "isolated" cluster $[Co_8(\mu_4\text{-S})_6S_8]^{15-}$. To get the latter we must "cut" a molecular cubic cluster from the solid and saturate the S ligands surrounding the metal cube. The assignment of the oxidation state of the octahedral cobalt is not unambiguous. Nevertheless, it is clear the Co_8S_6 clusters in Co_9S_8 are electron-poor with respect to the optimal count of 120 cve discussed in Section 5.2.5. Compare it with cubic molecular clusters with π-donor ligands such as the 108 and 109 cve analogs $[Co_8(\mu_4\text{-S})_6(SPh)_8]^{4-/5-}$. It is worth reemphasizing a point made in Chapter 5. The cubic clusters present an open-shell electronic structure which might translate into a metallic character in solids such as Co_9S_8. Indeed, the DOS does show the Fermi level crossing partially filled bands (Figure 7.46). This solid is a metallic conductor. In the same way that molecular cluster cubes show a wide range of cve counts, so too, it is possible to play with the nature of the metal and synthesize pendlandite-type compounds with different electron counts.

We have come to the end of our gallery tour of solid-state systems with useful molecular connections. What we have tried to do in this chapter is to explore a link between molecules and extended solids by grafting simple band structure perspectives onto "popular" chemical thinking. Thus, we have interpreted these delocalized band structures from a very chemical point of view – via frontier-orbital considerations based on the same interaction diagrams used for small and large (clusters) molecules. We found the treatment of electronic structure in extended systems is no more (nor less) complicated than in discrete molecules. With the tools developed in Chapter 6, we have been able to build the electronic structure of complicated three-dimensional solids from their unit cell contents. Many similarities between molecules or clusters and extended structures have emerged, as well as some novel effects as the result of extensive delocalization in solids. By parting the curtains of the delocalized picture of Bloch functions, we see the essential chemical bonds that determine the structure of extended solids. Both molecules and extended solids deal with the fundamental questions: Where are the electrons? Where are the bonds? There is value in considering the solid as a molecule, a big one, yes, but just a molecule. To finish this chapter, now test your own skill in applying this approach with the problems below.

Problems

1. In Section 7.1.2 the solid-state compound $Gd_{10}Cl_{18}C_4$ was mentioned. Its structure, see below, consists of two Gd_6 octahedra sharing an edge with 18 bridging Cl ligands and C_2 dumb-bells in the octahedral holes. If the compound $Gd_{10}Cl_{18}B_4$ were synthesized, what B–B bond distance would be expected assuming it adopted the same structure as the known C derivative?

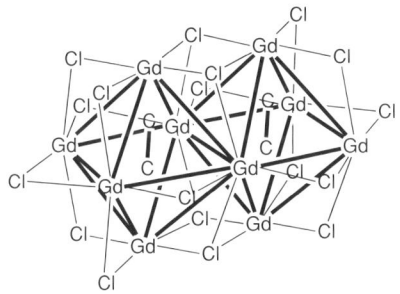

Problem 7.1

2. Propose geometries for the NiH_4 and PtH_4 moieties which are found in Mg_2NiH_4 and Na_2PtH_4, respectively. Construct corresponding qualitative MO diagrams.
3. The octahedral molecular cluster anion $[Mo_6(\mu_3\text{-}Cl)_8Cl_6]^{2-}$ with O_h symmetry shown below is an analog of the molecular cluster $[Mo_6Cl_8L_6]^{4+}$ described in Section 3.3.5.

300   *From molecules to extended solids*

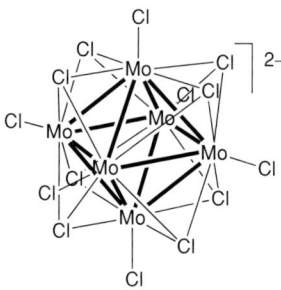

Problem 7.3

a. Calculate the oxidation state of the metal atoms and show that each Mo atom obeys the 18-electron rule. Calculate the number of metal electrons of the Mo_6 core. How many of these electrons are Mo–Mo bonding and Mo–Mo non-bonding?

b. The MOs of predominant metal character for $[Mo_6(\mu_3\text{-}Cl)_8Cl_6]^{2-}$ are shown below. Show that the diagram agrees with the cluster metal electron count.

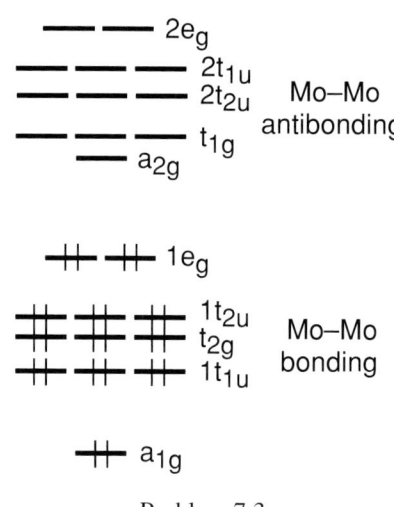

Problem 7.3

c. The cluster $Mo_6(\mu_3\text{-}S)_8(PEt_3)_6$ adopts the same structure. Count the number of metallic electrons for the Mo_6 core. Assuming that its MO diagram is similar to that of $[Mo_6(\mu_3\text{-}Cl)_8Cl_6]^{2-}$, rationalize the difference in their metal electron count. Do you think that other electron counts should be possible?

4. Do question 3 first.

a. The octahedral building unit Mo_6S_8 in the Chevrel phase $PbMo_6S_8$ described above in Section 7.4 possesses 22 metal electrons. Make a schematic drawing of the DOS for $PbMo_6S_8$ assuming substantial inter-cluster interactions. Indicate the Fermi level on the DOS diagram and predict the type of electrical conductivity that should be observed.

b. The solid-state compound $Mo_2Re_4S_8$ possesses a structure closely related to that of $PbMo_6S_8$. Comment on its electrical properties.

Additional reading

Cox, P. A. (1987). *The Electronic Structure and Chemistry of Solids*, Oxford: Oxford University Press.

Section 7.1

Braye, E. H., Dahl, L. F., Hubel, W. and Wampler, D. L. (1962). *J. Am. Chem. Soc.*, **84**, 4633.
Wijeyesekera, S. D. and Hoffmann, R. (1984). *Organometallics*, **3**, 949.
Halet, J.-F. in M. Gielen (Ed.) (1992). *Topics in Physical Organometallic Chemistry*, vol. 4. London: Freund Publishing House, p. 221.
Ouddaï, N., Costuas, K., Bencharif, M., Saillard, J.-F. and Halet, J.-F. (2005). *C. R. Chimie*, **8**, 1336.
Rohmer, M.-M., Bénard, M. and Poblet, J.-M. (2000). *Chem. Rev.*, **100**, 495.
Long, J. R., Hoffmann, R. and Meyer, H.-J. (1992). *Inorg. Chem.*, **31**, 1734.
Szafert, S. and Gladysz, J. A. (2003). *Chem. Rev.*, **103**, 4175.
Bauer, J., Halet, J.-F. and Saillard, J.-Y. (1998). *Coord. Chem. Rev.*, **178–179**, 723.
Haddon, R. C. (1992). *Acc. Chem. Res.*, **25**, 127.
P. Avouris (2002). *Acc. Chem. Res.*, **35**, 1026.
P. Lambin (2003). *C. R. Physique*, **4**, 1009.

Section 7.2

Albert, B. (2000). *Eur. J. Inorg. Chem.*, 1679.

Section 7.3

Miller, G. J., Deng, H. and Hoffmann, R. (1994). *Inorg. Chem.*, **33**, 1330.
Burdett, J. K. (1995). *Acta Crystallogr.*, **B51**, 547.

Section 7.4

Hughbanks, T. and Hoffmann, R. (1983). *J. Am. Chem. Soc.*, **105**, 1150.
Simon, A. (1994). *From Theory to Applications*, Schmid, G. (Ed.). Weinheim: Wiley-VCH, p. 373.
Svensson, G., Köhler, J. and Simon, A. (1999). *Metal Clusters in Chemistry*, Braunstein, P., Oro, L. and Raithby, P. R. (Eds.), vol. 3. Weinheim: Wiley-VCH, p. 1509.

Section 7.5

Burdett, J. K. and Miller, G. J. (1987). *J. Am. Chem. Soc.*, **109**, 4081.
Halet, J.-F. and Saillard, J.-Y. (1997). *Struct. Bond.*, **87**, 81.

8
Inter-conversion of clusters and solid-state materials

The conceptual connection between cluster and solid-state chemistries is the unifying theme of the first seven chapters. Complementary empirical connections between cluster and solid-state chemistries are emphasized in this final chapter. That is, the synthesis of solid-state materials from molecular precursors including clusters permits the strengths of molecular synthesis to be used in the development of new materials. On the other hand, the utilization of Zintl clusters as novel reagents in solution permits the advantages of thermodynamically driven solid-state synthesis to be transferred to the production of clusters in solution. Most of the examples discussed could have been included in earlier chapters, but are gathered here to serve as a review as well as a stimulus to creative thought for future research in cluster and materials chemistries.

8.1 Cluster precursors to new solid-state phases

In this section we give examples of molecular clusters used as precursors to new dense phases or to new porous networks.

8.1.1 III/VI Semiconductor synthesis

Traditional solid-state syntheses at high temperatures are guided by thermodynamics expressed in phase diagrams in distinct contrast to much of molecular chemistry that utilizes kinetics to guide synthesis. We viewed clusters as fragments of bulk solids stabilized by ligands; however, not all clusters can be viewed as building blocks of known bulk structures, e.g., icosahedral clusters. Hence, metastable phases not accessible by conventional solid-state synthesis might arise from cluster precursors. In other words, the structure of a cluster building block could determine the nature of the first-formed solid phase. To do so, a conversion technique that operates far from equilibrium conditions must be employed and the external

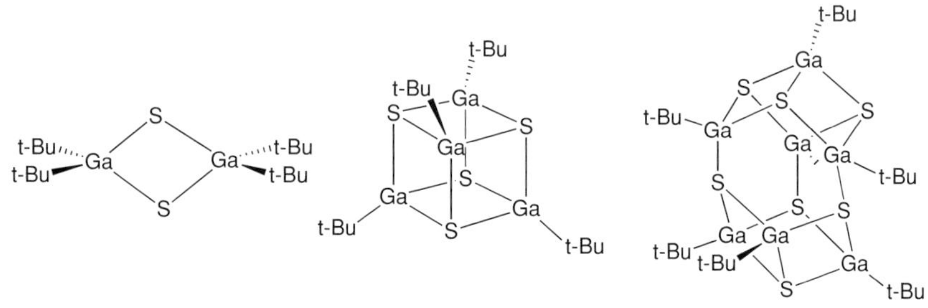

304   *Inter-conversion of clusters and solid-state materials*

Figure 8.1

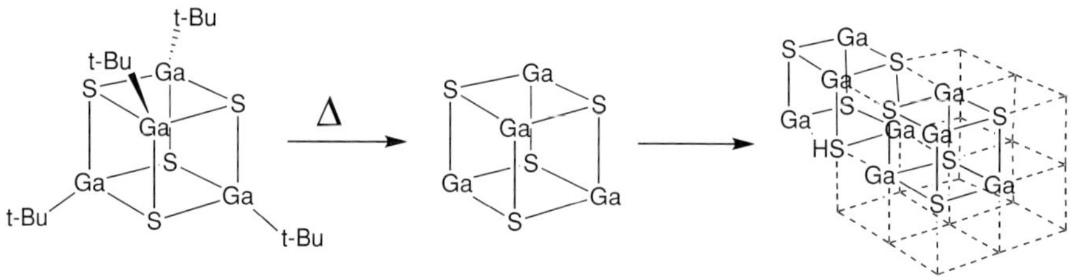

Figure 8.2

cluster ligands cannot be bound so strongly that disruption of the cluster bonding network is likely.

The thermodynamically stable phase of GaS exhibits a hexagonal (hcp) structure. Three volatile precursors for chemical vapor deposition were synthesized to generate metastable cubic GaS phases. (MacInnes *et al.*, 1993). The structures are shown in Figure 8.1. The first is a bridged dimer; the second, a cubane cluster and the third, a dimer of two partial cubane clusters. Upon removal of the external t-butyl ligands all have stoichiometry GaS but different core Ga–S connectivities, i.e., each Ga is connected to two, three and three S atoms, respectively. Contiguous, high-purity films of composition GaS, which are featureless by SEM, can be grown at about 400 °C on a variety of substrates from all the precursors. The dimer yields hexagonal GaS, the cubane yields a new cubic GaS phase, and the dimer of partial cubanes yields amorphous material. At lower temperatures, the cubane yields an amorphous film and at higher temperatures an amorphous S-rich film with crystalline needles – clear evidence of kinetic control in the formation of the new phase.

The new cubic phase can be generated as shown in Figure 8.2. As the external ligands on Ga are lost, the cubic building blocks generate a fcc NaCl lattice as verified by the diffraction properties observed. A crucial assumption in this interpretation is retention of the cluster core structure. We have emphasized earlier in the text that the energetics of external cluster–ligand bonding can match or even exceed that of the cluster core. Hence, it is difficult to argue a priori for cluster structure-retention based simply on a building-block correspondence between cluster core and solid-state structure.

Subsequent work by the same group addressed this point (Cleaver *et al.*, 1995). First, vaporization of the molecular solid containing the cubane cluster generated $[(^tBu)GaS]_4$ in the gas phase. Its structure determined by electron diffraction has parameters that are chemically equivalent to those of the solid-state structure. Hence, vaporization does not disrupt the cluster. As gas-phase thermal pyrolysis can be mimicked by infrared laser-powered photolysis, this technique was used to investigate whether the cubane cores could survive loss of the alkyl ligands on Ga. Evidence for the generation of $(t\text{-}Bu)_xGa_4S_4$, $x = 0-3$, suggests the GaS cubane core is sufficiently stable to account for the generation of the new cubic phase of GaS observed. The fact that the other precursors do not generate the new phase is significant. Presumably the Ga_7S_7 core of the partial cubane dimer also remains intact during CVD deposition. However, it has a C_3 axis and there is no possibility of ordered close packing as postulated for the cubane precursor. In similar manner, the results for the Ga_2S_2 dimer core supports this mechanism as, if the cubane core was degraded into, e.g., Ga_2S_2 dimers, it would lead to the known hexagonal phase. With appropriate properties, a molecular cluster controls the stoichiometry and structure of the solid-state product generated from it.

8.1.2 Cluster-expanded solids

Porous materials are technologically important in separation technology and as substrates for catalysts. One approach to the synthesis of porous solids is to expand a network solid by replacing one of its components with a larger one having the same functionality. Unfortunately, the larger pores can lead to the generation of two interpenetrating lattices, and when template molecules are used to avoid this problem, the resulting lattice may not survive the removal of the templates. Usually this strategy involves the increase of a single dimension of the extended network. Long proposed that both problems can be avoided if the component replaced increases the lattice dimensions isotropically (Shores *et al.*, 1999), e.g., replace an octahedral atomic connector by an octahedral metal cluster as illustrated next.

Prussian blue, $Fe_4[Fe(CN)_6]_3 \cdot xH_2O$, consists of a cubic lattice of alternating Fe^{2+} and Fe^{3+} ions connected with cyanide bridges (Figure 8.3). The $[Fe(CN)_6]^{4-}$ sites are 75% occupied and the exposed Fe^{3+} ions are coordinated with H_2O. Dehydration and hydration by moisture is reversible; hence, Prussian blue is a porous solid albeit one with very small pores. To increase pore size, the $[Fe(CN)_6]^{4-}$ complexes were replaced with octahedral $[Re_6Te_8(CN)_6]^{4-}$ clusters (Figure 8.4). This cluster displays an octahedral array of N donors on a sphere about 4 Å larger in diameter than the mononuclear complex. Reaction of Fe^{3+} with the cluster in

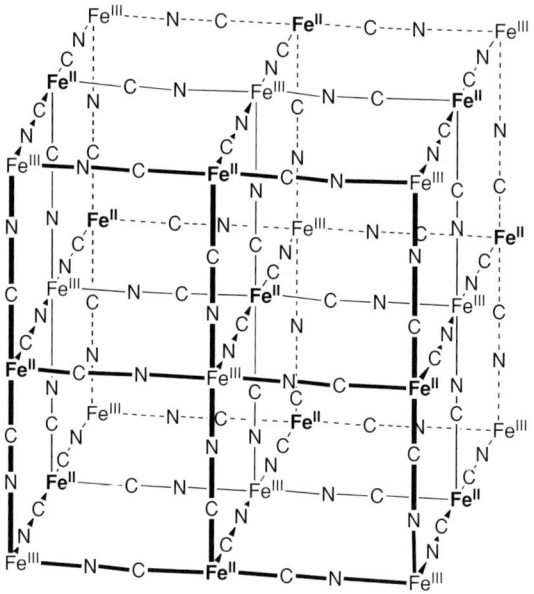

Figure 8.3

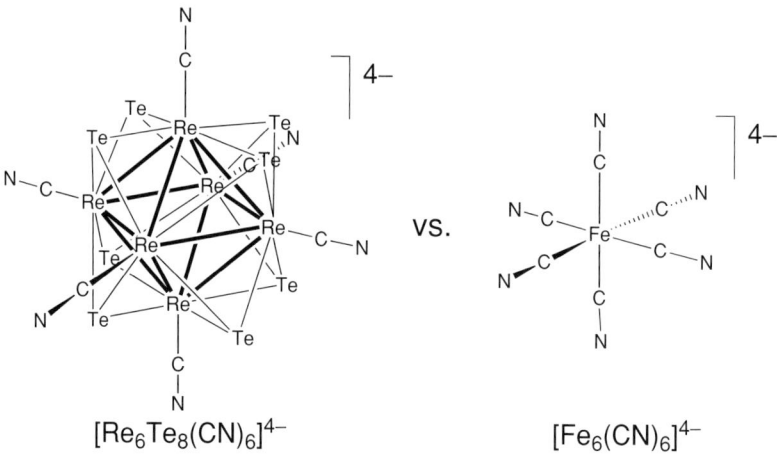

[Re$_6$Te$_8$(CN)$_6$]$^{4-}$ vs.   [Fe$_6$(CN)$_6$]$^{4-}$

Figure 8.4

a 4:3 ratio yields Fe$_4$[Re$_6$Te$_8$(CN)$_6$]$_3 \cdot x$H$_2$O which possesses a structure analogous to that of Prussian blue with 75 % occupancy of the [Re$_6$Te$_8$(CN)$_6$]$^{4-}$ sites. The environment of a single [Re$_6$Te$_8$(CN)$_6$]$^{4-}$ cluster is shown in Figure 8.5. Compare it to Figure 8.3. Simple calculations show that the pores in the cluster-based material are cubes of about 10 Å dimensions vs. about 6 Å in Prussian blue.

The cluster-expanded solids exhibit increased capacities for uptake of water, methanol and ethanol. Even better uptake results were obtained with cluster solids

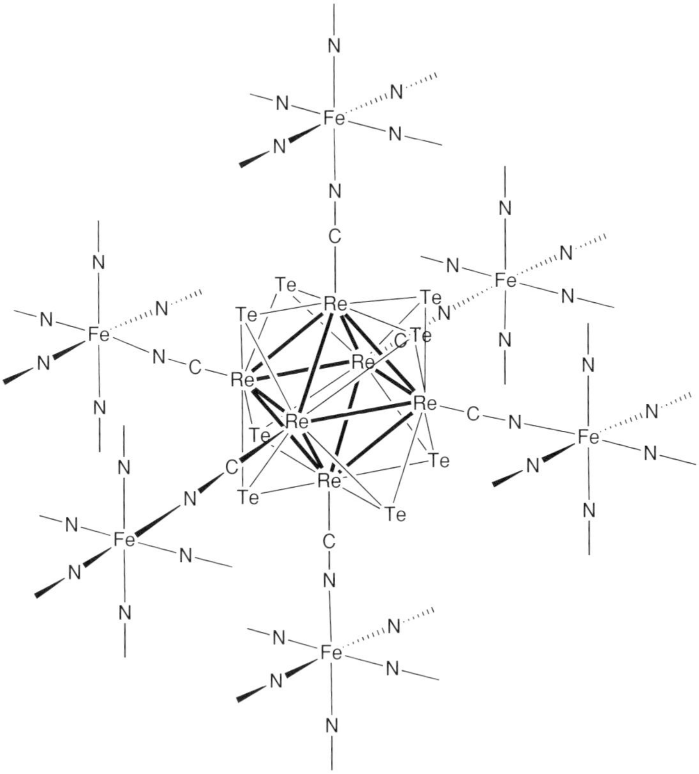

8.1 Cluster precursors to new solid-state phases

Figure 8.5

generated by replacing Fe^{3+} with Ga^{3+} (and Se with Te). By using Co^{2+} aquo ions instead of Fe, related cluster-expanded networks were obtained which could be dehydrated and which would take up small organic solvents in the pores. These compounds changed color on uptake of, e.g., diethyl ether, orange to blue, in a reversible manner. The color change is attributed to a change in coordination at the Co centers from octahedral to tetrahedral geometry thereby changing from a weakly absorbing coordination center to a more strongly absorbing one (recall that solutions of $[Co(H_2O)_6]^{2+}$ are light pink whereas those of $[CoCl_4]^{2-}$ are deep blue). With visible sensitivity to the presence or absence of guest molecules, this material is a solid chemical sensor (Beauvais et al., 2000).

8.1.3 Novel arrays from polyfunctionalized clusters

Not only can existing structural motifs be modified by cluster "substitution" but completely new materials can be synthesized. (Zhou et al., 2004) employed an early transition-metal octahedral cluster with terminal cyano ligands, $[Nb_6Cl_{12}(CN)_6]^{4-}$ (Figure 8.6, see also Section 3.3.5) in combination with a metal linker complex

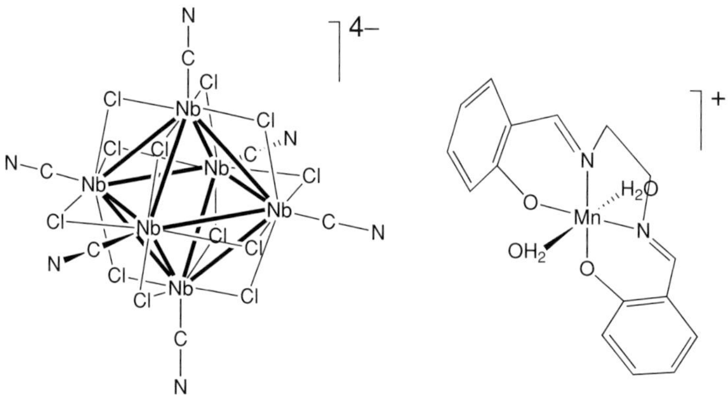

Figure 8.6

[Mn(*salen*)]$^+$. The *salen* ligand is a tetradentate ligand that occupies four equatorial sites of the six octahedral positions of the Mn ion leaving two axial sites available for coordination (Figure 8.6). The [Mn(*salen*)]$^+$ complex is the acceptor, hence, the axial ligands must be labile to permit facile assembly of the extended structure. In addition to isotropic expansion of a lattice, the use of cluster components permits, e.g., building-block charge to be varied without large structural change.

More than one framework can be generated from the two building blocks depending on conditions and other factors, e.g., in this case the size of the counterion is important and suggests a templating effect. Further, the dimensionality of the network generated need not utilize the full dimensionality of the cluster building block. In this work a dimer was observed (zero dimensions), a net (two dimensions) and a chain (one dimension). The net-like compound, $(Me_4N)_2[Mn(salen)]_2[Nb_6Cl_{12}(CN)_6]$, is shown in Figure 8.7.

8.1.4 Zintl clusters from molecular precursors

The synthesis of a Zintl compound from a molecular precursor makes a connection between molecular species and Zintl cluster compounds (Beswick *et al.*, 1998). $[Sb_7]^{3-}$ can be prepared by the addition of a sequestering agent to an intermetallic alloy of Sb. The high-temperature method and the cost of most sequestering agents is a limitation. A heterometallic alkali-metal/Sb(III) phosphinidene, $\{[Sb(PCy)_3]_2Li_6 \cdot 6HNMe_2\}$ generates $[Sb_7]^{3-}$ (Figure 8.8) at room temperature and is driven by formation of the coproduct $(PR)_4$. The molecular precursor constitutes an inexpensive route to gram-scale quantities of the $[Sb_7]^{3-}$ ion. One of the two forms isolated is $[Sb_7Li_3 \cdot HNMe_3]$ which contains a volatile ligand. Lusterous

8.2 Solid-state phases to molecular clusters

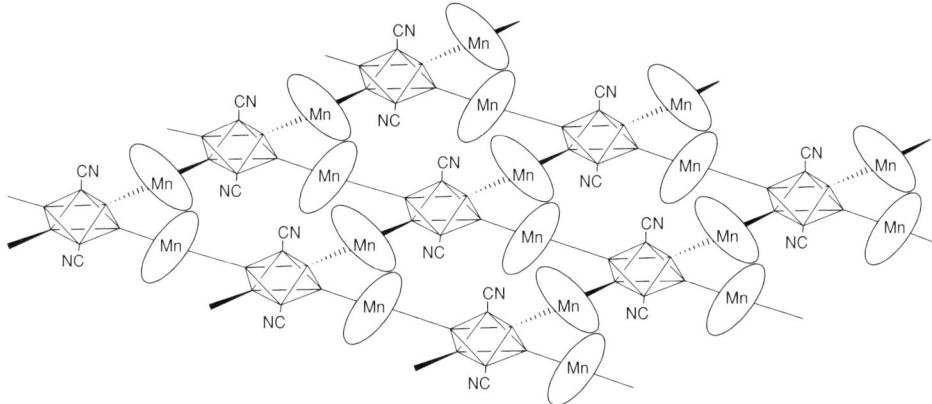

(Me$_4$N)$_2$[Mn(*salen*)]$_2$[Nb$_6$Cl$_{12}$(CN)$_6$]

Figure 8.7

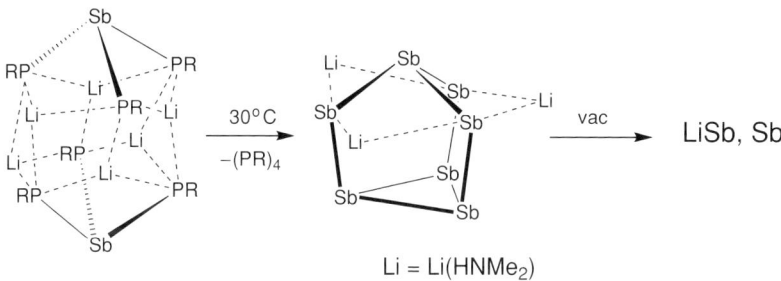

Li = Li(HNMe$_2$)

Figure 8.8

inter-metallic Sb/alkali-metal films are produced from this precursor under mild conditions. These films are potentially of value in the manufacture of photodiodes which requires the composition-controlled deposition of metallic Sb and alkali metals from the vapor.

8.2 Solid-state phases to molecular clusters

NMR studies of solutions of inter-metallic phases led to recognition of the potential use of polyatomic Zintl ions as reagents for the generation of new cluster compounds (Eichhorn *et al.*, 1988). Examples of transition-metal derivatives have been discussed in Chapter 5 and the structures of the ions themselves were used as examples of ligand-free main-group clusters in Chapter 2 (Corbett, 1985; Fassler, 2001). The following examples illustrate recent developments in the chemistry of these clusters derived from solid-state syntheses.

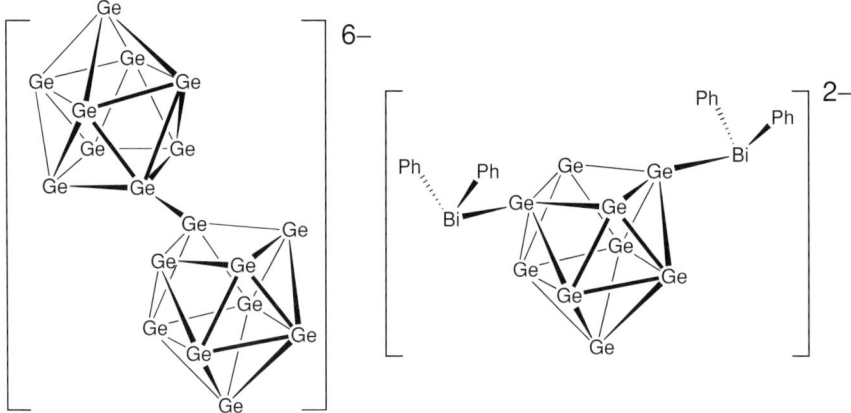

Figure 8.9

8.2.1 Cluster derivatives and oligomers

The bare $[Ge_9]^{4-}$ cluster, a monocapped square-antiprismatic cluster found both in a Zintl phase as well as a solution species, has been functionalized on the external cluster surface. Rather than capping a rectangular face with a transition metal (Chapter 5) external bonds to other clusters or substituents are formed. A dimer, $[Ge_9-Ge_9]^{6-}$ is shown in Figure 8.9 (Xu and Sevov, 1999). The two nido-Ge_9 clusters are joined by a single Ge–Ge exopolyhedral bond between atoms on the open faces. Consistent with its electron count (each $[Ge_9]^{3-}$ cluster has 11 sep, eight Ge atoms with lone pairs and one Ge atom with a single external orbital containing one electron which forms a Ge–Ge bond with the other $[Ge_9]^{3-}$ cluster) the dimer is analogous to $[B_5H_8]_2$ (Figure 2.18). Similarly, reaction of Ph_3Bi with an ethylenediamine solution of K_4Ge_9 (the major phase of the precursor contains both $[Ge_9]^{4-}$ and $[Ge_4]^{4-}$ clusters) leads to $[nido\text{-}6,8\text{-}(Ph_2Bi)_2\text{-}Ge_9]^{2-}$ also shown in Figure 8.9. The one-electron Ph_2Bi ligands replace one cluster negative charge each. The sep count of the cluster is unchanged.

With Ph_3Sb, not only is $[nido\text{-}6,8\text{-}(Ph_2Sb)_2\text{-}Ge_9]^{2-}$ formed, but $[nido\text{-}6\text{-}(Ph)\text{-}8\text{-}(Ph_2Sb)\text{-}Ge_9]^{2-}$ and $[6,6'\text{-}conjuncto\text{-}\{8,8'\text{-}(Ph_2Sb)_2\text{-}(nido\text{-}Ge_9)_2\}]^{4-}$ shown in Figure 8.10 are isolated (Ugrinov and Sevov, 2002). Recognition of both the nucleophilic displacement character of the reaction and the fact that $[Ge_9]^{4-}$, $[Ge_9]^{3-}$ and $[Ge_9]^{2-}$ are in equilibrium in solution permitted $[nido\text{-}6,8\text{-}(Ph_3E)_2\text{-}Ge_9]^{2-}$, E = Ge, Sn, $[nido\text{-}6\text{-}(R_3Sn)\text{-}Ge_9]^{3-}$, R = Me, Ph, and $[6,6'\text{-}conjuncto\text{-}\{8,8'\text{-}(Ph_3Sn)_2\text{-}(nido\text{-}Ge_9)_2\}]^{4-}$ to be prepared from R_3ECl, E = Ge, Sn, R = Me, Ph. Reaction with ^{t}BuCl leads to alkyl functionalization and the product $[6,6'\text{-}conjuncto\text{-}\{8,8'\text{-}(^tBu)_2\text{-}(nido\text{-}Ge_9)_2\}]^{4-}$. Finally, reaction with the appropriate sequestering agent leads to dimerization and the formation of a solid containing infinite chains of

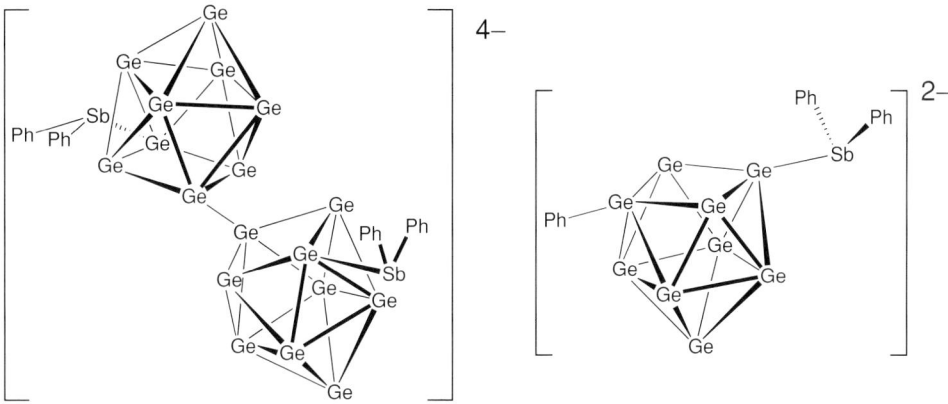

Figure 8.10

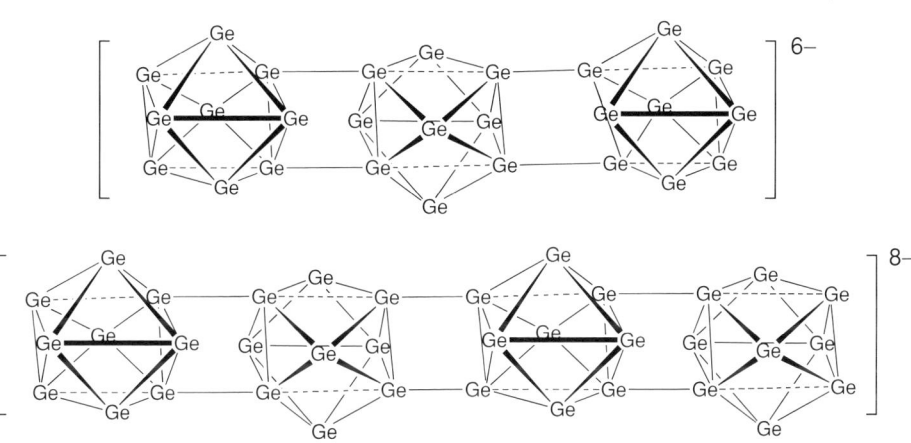

Figure 8.11

($-Ge_9{}^{2-}-$)$_\infty$ where each [Ge$_9$]$^{2-}$ unit is bonded in the 6,8-positions via Ge–Ge bonds to two similar clusters.

Not all derivatives isolated follow simple main-group cluster ideas. For example, oligomers of the Ge$_9$ cluster system, [Ge$_9$=Ge$_9$=Ge$_9$]$^{6-}$ and [Ge$_9$=Ge$_9$=Ge$_9$=Ge$_9$]$^{8-}$, are shown in Figure 8.11 (Ugrinov and Sevov, 2003). These clusters are connected by two inter-cluster Ge–Ge interactions and the Ge$_9$ cluster units exhibit tricapped trigonal-prismatic cluster shapes elongated along two prismatic edges (dotted lines in the drawing). Similar distortions have been discussed in Section 2.11.2 in connection with bare nine-atom clusters.

Each inter-cluster two-center–two-electron bond reduces the charge of the clusters joined by one unit each. Thus, if the clusters in Figure 8.11 are connected by two inter-cluster Ge–Ge bonds of the same type per pair, the charge of each end

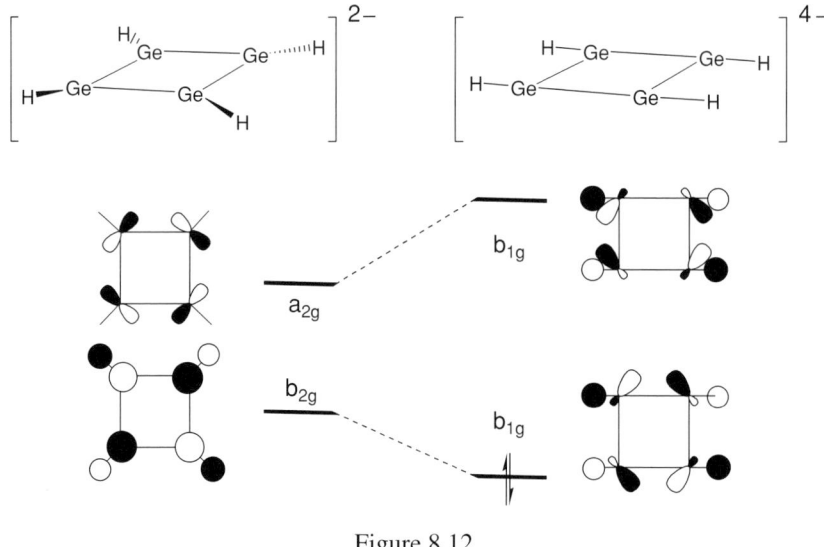

Figure 8.12

cluster would be 2− and that of each central cluster would be 0. However, the observed charges for the trimer and tetramer are 6− and 8− not 4−. Somehow the double-linked inner clusters accommodate an extra 2− charge each. Consistent with a different cluster–cluster connectivity, the inter-cluster Ge–Ge distances are significantly longer than those in $[Ge_9-Ge_9]^{6-}$. A key observation is that the four external cluster Ge–Ge bond vectors are not oriented in a radial direction but rather lie parallel to the elongated trigonal-prismatic edges of the Ge_9 clusters. A simple model shows how bending of the exocluster bonds away from the radial direction of the cluster lowers the energy of a lone-pair-type orbital so that it is filled.

The central cluster is modeled by square D_{4h} *arachno*-$[Ge_4H_4]^{2-}$ shown in Figure 8.12. The two pertinent empty orbitals are of b_{2g} symmetry (cluster and ligand antibonding) and a_{2g} symmetry (cluster antibonding) and cannot mix. Bending the Ge–H bonds to model the doubly connected Ge_9 cluster units in the dimer and trimer lowers the symmetry to D_{2h} and these two orbitals now are of b_{1g} symmetry and can mix. One is stabilized and the other destabilized as shown. The former is now available at low energy to accommodate the "extra" lone pair. An important consequence of this exocluster orbital mixing is that the oligomers are not viewed as delocalized clusters connected by localized bonds but single delocalized entities.

8.2.2 Extraction of Zintl clusters from transition-metal cluster solids

Solid-state centered-Zr cluster compounds with compositions $M^I_x[Zr_6(Z)Cl_{12}]Cl_n$ are known. For $n < 6$ the structures exhibit inter-cluster bridging by shared Cl atoms

8.2 Solid-state phases to molecular clusters

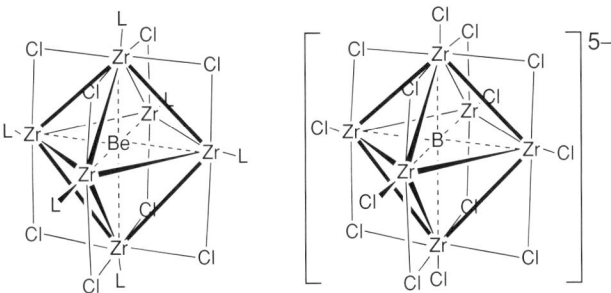

Figure 8.13

with the number of bridges increasing with decreasing n. For these compounds, excision of soluble cluster units results from the addition of Lewis bases which displace and open up the Zr–Cl–Zr bridges. Thus, addition of $[Cl]^-$, neutral amines or phosphines generates metal clusters in solution (Rogel and Corbett, 1990). For example, reaction of ethylamine with $K_3[Zr_6(Be)Cl_{12}]Cl_3$, $Na_4[Zr_6(Be)Cl_{12}]Cl_4$ and $Li_6[Zr_6(H)Cl_{12}]Cl_6$ leads to M^ICl and $Zr_6(Z)Cl_{12}(EtNH_2)_6$, $Z = Be$, H. The structure of the neutral Be-centered cluster is shown in Figure 8.13 at the left where it is seen that the six axial Zr positions through which inter-cluster bonding in the solid takes place are now occupied by the added base L.

An alternative solvent system permits access to a greater variety of clusters (Tian and Hughbanks, 1995). This redox-stable system employs a room-temperature molten salt formed from 1-ethyl-3-methylimidazolium chloride $(ImCl)/AlCl_3$ mixtures. The $Rb_5Zr_6Cl_{18}B$ cluster solid ($n = 6$ so there are no Zr–Cl–Zr bridges between clusters in the solid state), which yields a one-electron oxidation product when solubized in conventional solvent systems, generates $[Zr_6(B)Cl_{18}]^{5-}$ on treatment with the $ImCl/AlCl_3$ ionic liquid (Figure 8.13 on the right). The diamagnetic cluster in solution is now amenable to all the spectroscopic tools of molecular chemistry, e.g., the highly downfield shifted ^{11}B NMR resonance corresponds well with that of octahedral $[(CO)_{16}Rh_2Fe_4B]^-$ which also contains an interstitial B atom.

8.2.3 Cluster synthesis via ligand-arrested solid growth

The synthesis of large clusters such as $[Al_{69}R_{18}]^{3-}$ (Chapters 2 and 3) proceeds by Al atom cluster-core build up. Cluster-core growth is terminated at some point by external ligands. The method of Schnöckel is a variation of metal-atom vapor-deposition techniques and relies on: (a) the reversibility of the equilibrium between the liquid metal, e.g., Al, and gaseous metal halide, e.g., $AlCl_3$, with gaseous subhalide, e.g., AlCl; (b) the shift in equilibrium position with temperature and (c) competitive rates at similar temperatures of subhalide disproportionation to metal

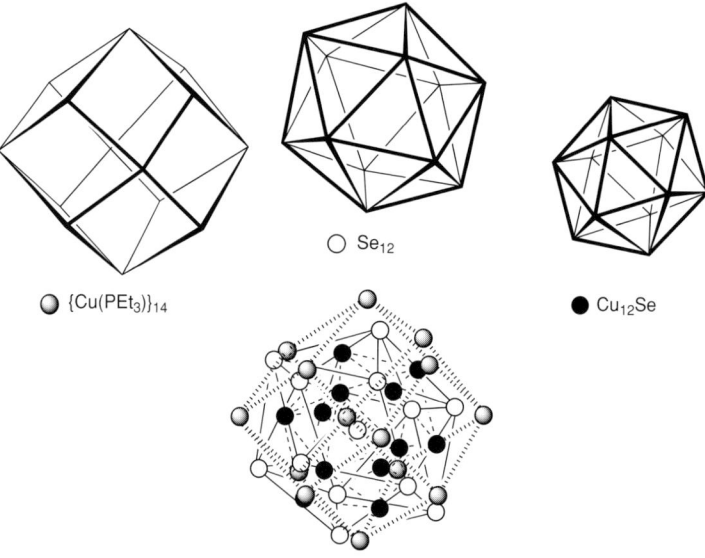

Figure 8.14

and halide (metal atom generation and core growth) and the metathesis of subhalide with a metalated ligand, e.g., LiN(SiMe$_3$)$_2$, to terminate core growth (Schnepf and Schnöckel, 2002). The reaction takes place in an appropriate solvent and reaction cluster size is controlled by reaction temperature, e.g., [Al$_7$R$_6$]$^-$ at $-7\,^\circ$C vs. [Al$_{69}$R$_{18}$]$^{3-}$ at 60 $^\circ$C. The barrier for disproportionation of AlCl is larger than that of metathesis so the core growth rate/ligand-trapping rate ratio increases with increasing temperature.

The same principles apply to the deposition of vapor of transition-metal chalcogenides in the presence of PR$_3$ ligands (Crawford et al., 2002). Depending on transition metal, small, e.g., M$_4$S$_4$(PR$_3$)$_4$, M = Fe, Ni; medium, e.g., M$_8$S$_8$(PR$_3$)$_6$, M = Cr, Co and large, e.g., Cu$_{26}$Se$_{13}$(PEt$_3$)$_{14}$ (Figure 8.14), clusters were isolated as major products. The large Cu$_{26}$ cluster can be described as an Se-centered Cu$_{12}$ icosahedron within a Se$_{12}$ icosahedron surrounded by a (CuPR$_3$)$_{14}$ rhombic dodecahedron with the PR$_3$ ligands in axial orientations. Just as the core structures of the large Al clusters differ from that of bulk Al metal, so too the core structure of Cu$_{26}$Se$_{13}$(PEt$_3$)$_{14}$ differs from that of the fcc lattice exhibited by Cu$_2$Se.

8.3 Clusters to materials

The emphasis of the first two sections was the generation of novel substances. Here we note that there are advantages to the utilization of cluster precursors for solid materials for which conventional syntheses are well established.

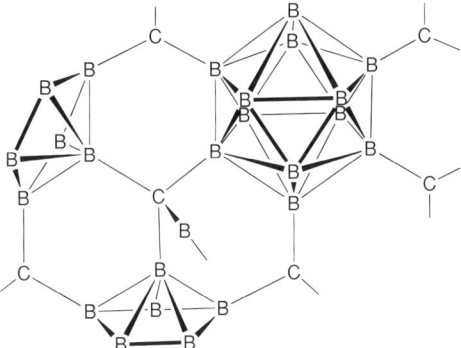

Figure 8.15

8.3.1 Borane clusters to boron carbides

Non-oxide ceramic materials, such as boron carbide, are technologically important. The stoichiometry of boron carbide is represented as B_4C, i.e., $B_{12}C_3$, with a structure based on $B_{11}C$ icosahedra and inter-icosahedral C–B–C chains as schematically illustrated by the structure of $B_{13}C_2$ in Figure 8.15. The structure is related to that of β-rhombohedral B and, like this allotrope of elemental B, the complexity of the structure is difficult to represent in two dimensions. It suffices for our purposes to point out some important differences. In contrast to elemental B, the links between deltahedral clusters are C–B–C units rather than larger B_{10}–B–B_{10} units. This results in closer packing of the icosahedra and the generation of C_2B_4 rings between icosahedra. These links utilize six B atoms in each icosahedron and the other six form direct B–B bonds to adjacent icosahedra. As limited substitution of B and C atoms within both cluster and chains is possible, the C content can range from 8.8 to 20 atom %. The limits correspond approximately to $B_{10.5}C$ to B_4C.

Borocarbide powders can be synthesized by the direct reaction of the elements at high temperatures. The properties that make boron carbide of interest also inhibit processing into useful forms other than powders. A way around this problem is to develop a polymer capable of being processed into shapes and converted into a ceramic while retaining shape. The requirements: a high-yield polymer precursor synthesis; a high-yield polymerization; a polymer with useable properties and conversion to a ceramic in high yield are demanding ones. Cluster precursors to solid-state materials syntheses will never be viable cost-wise for preparing bulk materials in quantity; however, for specialized applications requiring materials with designed properties, these tailored precursors offer considerable scope to the materials engineer. Hence, the illustration discussed is the generation of boron carbide nanostructures (Sneddon *et al.*, 2005).

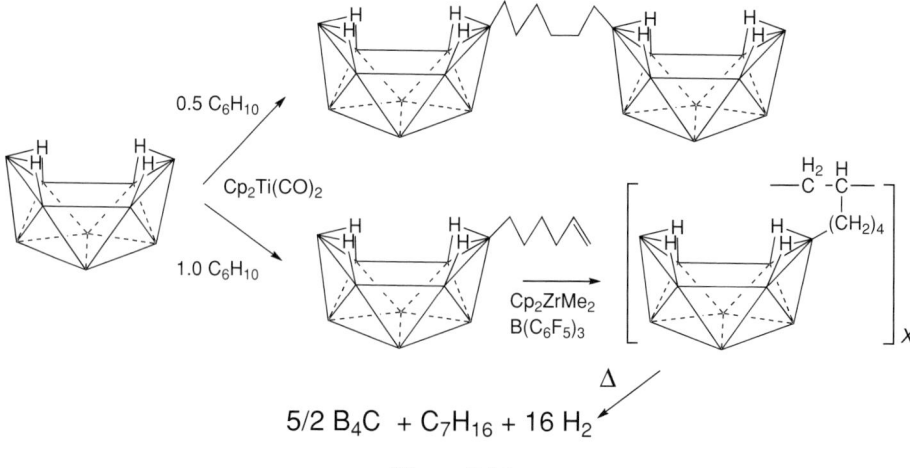

Figure 8.16

The attachment of *nido*-$B_{10}H_{14}$ clusters to a saturated hydrocarbon polymer produces a processable precursor for a boron carbide ceramic. Two steps are required: functionalizing the cluster and polymerizing the functional group to generate a linear polymer with pendant clusters. As shown in Figure 8.16, a $Cp_2Ti(CO)_2$ catalyst is used to generate linked cluster dimers or mono-cluster substituted alkenes using 1,5-hexadiene. 6-hexenyldecaborane is readily polymerized with another early transition-metal complex catalyst system to yield a cluster/organic hybrid polymer composed of a polyolefin backbone with dangling decaborane clusters. Poly(hexenyldecaborane) is soluble and can be converted into a bulk ceramic in 60 % yield (68 % theoretical).

Poly(hexenyldecaborane) can be spun into fibers and the green fibers can be heated to generate boron carbide fibers. Interest in smaller-scale fibers requires different techniques. In this case the 6–6'-$(CH_2)_6$-$(B_{10}H_{13})_2$ dimer, Figure 8.16, has found use in the synthesis of nanostructured ceramics. For example, the absorption of the dimer into the channels of a nanoporous alumina "form," followed by conversion to ceramic and dissolution of the alumina membrane, generates free-standing fibers of much smaller dimensions that those obtainable by spinning techniques. Figure 8.17 shows a scanning electron micrograph of boron carbide nanofibers generated by the filling of 250 nm pores of 60 μm alumina membranes with the dimer, thermolysis at 1025 °C and dissolution of the alumina in HF. The result is fibers of uniform diameter defined by the pore size, and length defined by the membrane thickness. The fibers are composed of crystalline boron carbide. The thin layer of boron carbide left on one end of the fiber "brush" serves to retain the alignment generated by the alumina "form."

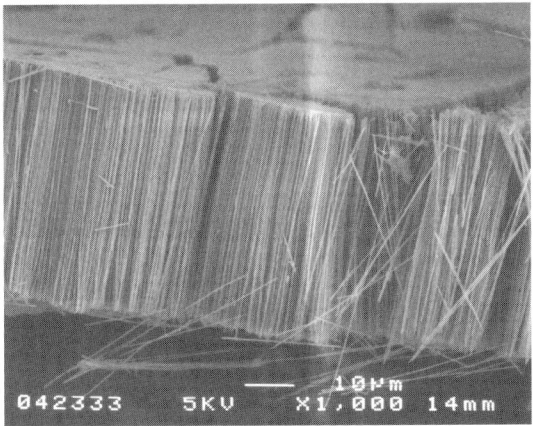

Figure 8.17

8.3.2 M–B clusters to amorphous metal-alloy films

Amorphous metal alloys formed from transition metals or rare-earth metals and main-group metalloids constitute an important class of materials valued for their chemical and magnetic properties. In contrast to crystalline substances, these materials exhibit short-range order but no long-range order. In contrast to polycrystalline materials, they lack grain boundaries that play a major role in determining the overall physical and chemical properties of the former, e.g., susceptibility to fracture when dislocations are trapped at grain boundaries. Single phase, chemically homogeneous, non-crystalline alloys are relatively recent additions to the array of materials available for technological application and are synthesized from the liquid state by a rapid quenching technique.

The local structure of some metallic glasses can be understood in terms of a disordered array of ligand-free main-group–transition-metal clusters. Thus, for example, amorphous iron borides have been viewed as containing cluster fragments similar to the repeating unit in crystalline intermetallic alloys, e.g., Fe_3B with trigonal-prismatic motif shown in Figure 8.18. Also illustrated in Figure 8.18 is a putative process of generating the same structural motif from B-capped triiron clusters in the manner of Figure 8.2. The appropriate ferraborane, $HFe_3(CO)_9BH_4$, is known and constitutes a potential low-temperature molecular precursor for amorphous Fe_3B that should crystallize at higher temperatures. This would permit deposition of a material with all of the properties of an amorphous metal alloy without destroying thermally sensitive components. By the same reasoning, $Fe_2(CO)_6B_2H_6$ and $HFe_4(CO)_{12}BH_2$ (Figure 8.19) are precursors for amorphous solids of compositions FeB and Fe_4B. The former would permit access to a composition not possible

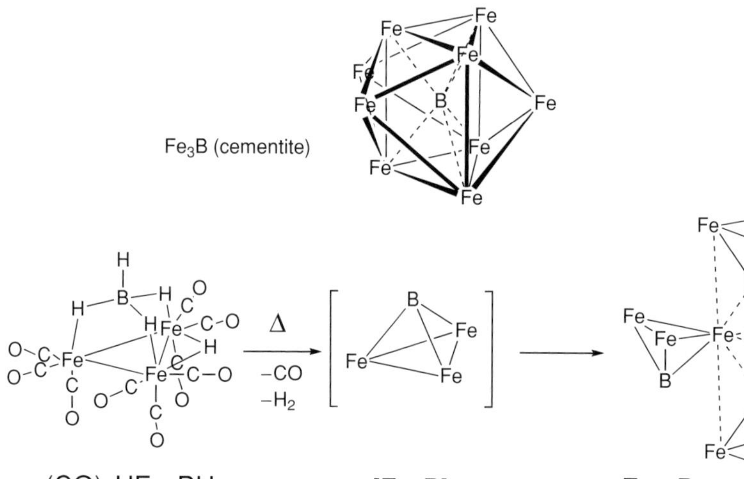

318   *Inter-conversion of clusters and solid-state materials*

Fe$_3$B (cementite)

(CO)$_9$HFe$_3$BH$_4$ [Fe$_3$B]   Fe$_{75}$B$_{25}$

Figure 8.18

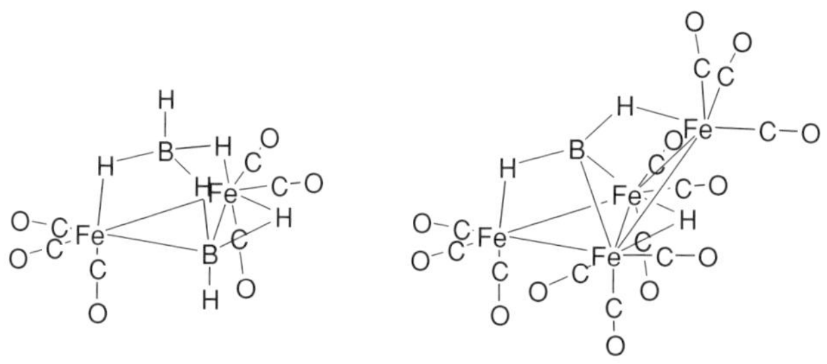

(CO)$_6$Fe$_2$B$_2$H$_6$ (CO)$_{12}$HFe$_4$BH$_2$

Figure 8.19

by conventional techniques as it lies outside the miscibility range of B and Fe in the liquid state.

Deposition of thin films of amorphous borides on SiO$_2$ at 200 °C generates iron boride films about 500 nm thick with good substrate adhesion and resistivities similar to liquid metals (Amini et al., 1990). Film composition is linearly dependent on precursor core composition. Films formed from HFe$_3$(CO)$_9$BH$_4$ crystallize at higher temperatures to known Fe$_3$B$_{1-x}$C$_x$ phases, i.e., C impurities occur by B atom replacement in the amorphous structure. Mössbauer spectroscopy shows similarities in local Fe environments to those of alloys prepared by other methods. Comparison of data on films from different ferraborane precursors shows that packing of the Fe and B atoms is not random and suggests, rather, that local structure results from

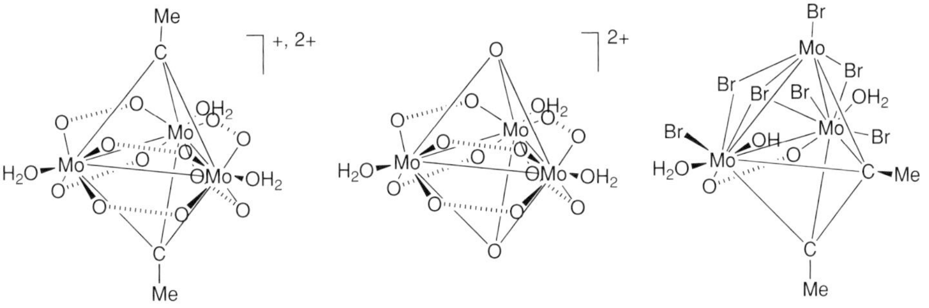

[Figure 8.20 structural diagrams]

I, II = [Mo$_3$(CCH$_3$)$_2$(OAc)$_6$(H$_2$O)$_3$]$^{+,2+}$ **III** = [Mo$_3$(O)$_2$(OAc)$_6$(H$_2$O)$_3$]$^{2+}$ **IV** = [Mo$_3$Br$_7$(OAc)(H$_2$O)$_2$(MeCCMe)]

Figure 8.20

random packing of Fe$_m$B$_n$ bare clusters. This is in accord with initial assumptions described above. Due to columnar growth, the magnetically ordered films have moments preferentially oriented normal to the film plane. Ribbons of similar compositions formed by quick-quench methods have magnetic vectors that lie within the plane of the ribbon making them less useful for some applications.

8.4 A final problem

The metal complexes in Figure 8.20 have been suggested as models for understanding the metal catalysis of alkylidyne chain lengthening and metathesis (Bino *et al.*, 2005). Consistent with this hypothesis, the redox reaction:

$$5\,\mathbf{II} + 2H_2O \rightarrow 4\mathbf{I} + \mathbf{III} + CH_3C \equiv CCH_3 + 2H^+,$$

results in the coupling of two [CH$_3$C]$^{3-}$ ligands. With this definition of carbyne ligand charge, the reaction constitutes oxidative coupling. Complex **IV** is an intermediate in the overall reaction as it was trapped from the reaction with aqueous HBr. In contrast to **I** and **II**, **IV** contains a C–C interaction. Its presence is taken as evidence of coupling of the ethylidyne fragments of **II** in the coordination sphere of the metal complex.

The appreciative reader of this text will immediately recognize that compounds **II** and **IV** and their inter-conversion have analogs in both carborane and metal-carbonyl cluster chemistry. Thus, B$_3$C$_2$ or M$_3$C$_2$ cores of six-sep *closo*-trigonal-bipyramidal clusters can have 1,2- or 1,5-C atom positions with the latter clearly more stable in the carborane. For the Mo compounds in Figure 8.20, the 1,2-isomer (**IV**, C adjacent) must form from the 1,5-isomer (**II**, C apart) if it is an intermediate. Another notable difference is that this Mo reaction system takes place in air in water solution.

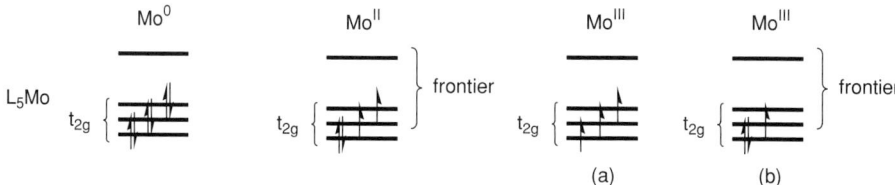

Figure 8.21

The authors of this work did not consider **I–IV** as cluster compounds nor did they view **II** and **IV** as cluster isomers. Let us see if we can relate the observed chemistry to cluster isomerization albeit one in which the forward (C apart to C adjacent) and reverse (C adjacent to C apart) have similar barriers. That is, an effective alkyne metathesis catalyst cannot have any species in the catalytic cycle of high stability (Bunz, 2005).

The Mo fragments contain electron-rich π-donor ancillary ligands so that we can expect the "t_{2g}" metal set to be involved in cluster bonding. The total charge on the Mo_3 fragment is 8+ and each Mo center is coordinated to five ancillary ligands. The frontier orbitals of a L_5Mo fragment derived from an octahedral L_6M complex are shown in Figure 8.21 for three different oxidation states. If the CH_3C fragments are considered neutral as appropriate for cluster chemistry, **IV** contains one Mo^{II} and two Mo^{III} centers (these are not the oxidation states given in the original paper as the C fragments were considered anionic). The Mo^{II} fragment is easily assigned as a three-orbital–two-electron fragment but there are two possibilities, labeled (a) and (b), for the Mo^{III} fragment. Possibility (a) again gives a three-orbital–two-electron fragment and, if we choose this scenario, **IV** with two three-orbital–three-electron CR fragments and three three-orbital–two-electron metal fragments is simply a six-sep cluster analog of 1,2-$C_2B_3H_5$! If so, the Mo^{III} centers will be paramagnetic and possibly weakly coupled by the carboxylate bridges. In fact **IV** is paramagnetic (see supplementary material of the original work). A similar analysis works for **II** thereby making it an analog of *closo*-1,5-$C_2B_3H_5$. The overall reaction, the conversion of the C-apart isomer to free alkyne, is driven by the oxidative addition of water to Mo to yield **III**.

Which approach is better? Wrong question! The cluster view simply adds perspective to the problem by changing the emphasis from the "oxidation" of two $[CH_3C]^{3-}$ ligands to a cluster isomerization driven by the charge-distribution requirements of the cluster framework. Note that these Mo "clusters" belong in the "rule-breaker" category. Like others we have commented upon earlier, "rule-breaker" clusters seem to possess a useful versatility in bonding. But keep in mind that this versatility only becomes understandable in the light of the concepts established by the "law-abiding" clusters.

8.5 Conclusion

Both cluster and solid-state chemistries have strong ties to materials chemistry. Materials chemistry is the broadest area and includes topics outside the corpus of inorganic chemistry. In addition, there is a distinct and appropriate flavor of practicality associated with the materials-science literature. Large clusters and nanoscale particles are of importance in many kinds of materials work. For example, nanoparticle/polymer composites and quantum-size semiconductor particle arrays are examples of materials-science systems with cluster roots. Other topics, such as sol-gel processing, syntheses of thin films and ceramics and development of electronic materials, are closely allied to solid-state chemistry. An understanding of cluster and solid-state chemistries, then, is fundamental to the understanding of a substantial fraction of materials science. There will always be an Edisonian component to a search for a material to solve a pressing practical problem, but the greater the understanding of the underlying chemistry, the more focused and efficient the search.

Appendix

Fundamental concepts: a concise review

A chemist's approach to understanding matter is conveniently divided into three stages. First, fixed stoichiometric relationships between the atomic constituents of matter exist in molecular compounds and in many compounds with extended structures. Hence, the composition of a pure substance provides initial definition to a new substance. Given the atomic nature of substances, it provides direct information on structure. In a historical sense, other early means of characterization were physical ones, e.g., melting point or dipole moment, and sensual, e.g., color or taste. As a consequence, they are less directly related to structure. In the second stage, geometric relationships between the constituent atoms as well as spectroscopic and theoretical information on electronic structure add dimensions and shape to the composition data. Finally, reaction chemistry, i.e., reaction stoichiometries, rates and mechanisms, provides connections between compound types as well as generating new substances for which the whole process begins again.

Stable compounds are the most thoroughly characterized and provide the corpus of chemistry. However, "stability" is one of the definable, but casually used, terms of chemistry. Stability is associated with energy, and energy and structure are inextricably combined. A chemist, well educated in a given area, has a good understanding of both energy and structure. In a practical sense, stable for the inorganic chemist is often defined empirically by isolation and storage at room temperature. Most stable is often implied by the term, but it is not always clear that the most stable (thermodynamic) products have been characterized in a given reaction. Noteworthy is the fact that unstable species are also valuable even if only partially characterized. For example, intermediates in reactions are intrinsically unstable under the reaction conditions yet provide considerable insight into mechanism.

Although the research chemist may well move on to new challenges when the rules governing compound stoichiometry, geometric/electronic structure and reaction chemistry are defined empirically and theoretically for a given system, these detailed results still must be fit into the "big picture" – the incomplete puzzle we call chemistry. There are two senarios: the new puzzle piece either fits with the existing picture or not. In the former case the new information, when combined and summarized for a given class of compound, reinforces existing chemical concepts while broadening their purview. In the latter case, the new information forces a reevaluation of existing chemical concepts and often provides impetus for the development of new chemistry.

Our approach to cluster chemistry addresses the question: how can compound stoichiometry, geometric and electronic structure, and reactivity for a structure type that ranges over many elements be integrated into existing chemistry? Specifically, we wish to make firm connections to contemporary understanding of molecular and solid-state chemistries. As with all chemical discussions, we attempt to show how element (atom) properties combined with the fundamental principles by which atoms interact, give rise to cluster structures. Without an understanding of how atomic properties are expressed in simple molecules, an understanding of the more complex bonding combinations encountered with clusters is not possible. Hence, a brief review of the basic concepts follows. Many standard texts are available for reference and a few are listed at the end of this Appendix.

A1.1 Elements

The atomic model of matter is a firmly established platform upon which our understanding of complex substances is based. Chemical characteristics of atoms are reflected in the structures and physical properties of the elements under normal Earth conditions. The majority of the elements are solids and metallic by nature (extended structures with high coordination numbers leading to ductility, weakly bound valence electrons permitting electrical conductivity with negative carriers and interaction of visible light with mobile electron density generating metallic luster). The last property serves as an obvious identifier of metals. The elements in the upper right and corner are non-metallic (often found as molecular species with weak intermolecular interactions and insulating behavior in the solid state). Although stoichiometry plays no role for the pure elements, the concepts of geometric/electronic structure and reactivity are certainly pertinent to a consideration of the bulk elements. However, before considering the complexities associated with an interacting assembly of atoms, we first review the properties of gaseous atoms. As atoms possess a single nucleus, geometric structure may not be considered an issue; however, atoms do exhibit size albeit only recognized and measured in relationship to other atoms either of identical or different type. Electronic structure and reactivity are also important factors associated with observed atom properties. The two are intimately related as the deficiencies of the former, if any, can be remedied by chemical reaction. Let's see how.

A1.2 Atomic properties

Three essential properties of isolated atoms are size, the energy required to remove the most weakly bound electrons (ionization energy) from the nucleus and other electrons, and the energy gained by adding an electron to a neutral atom (electron affinity). A common theme in all of chemistry is that energy and dimensions (size and shape for atoms, bond distances and angles for compounds) are connected. Indeed, one might collapse the purpose of chemistry into a search for the function that connects energy to structure. However, such a statement alone is no more useful than stating that the appropriate form of the Schrödinger equation is all we need to answer the question. Still, sometimes inorganic chemists, as they happily wallow in a sea of X-ray crystal-structure solutions, do neglect the energy parameter and the existence, even participation, of higher-energy structures in chemistry.

For atoms, the binding energy of the most weakly bound electrons and the atomic size reflect the strength of attraction of the nucleus plus core electrons for the valence electrons. According to Mulliken, the mean of the ionization energy and electron affinity constitutes a measure of the electronegativity of an atom. It is one of several measures, the most popular of which is based on bond energetics and developed by Pauling. Indeed, publications of new, improved electronegativity scales continue up to the present day as

A1.2 Atomic properties

						H 2.20											He
Li 0.98	Be 1.57											B 2.04	C 2.55	N 3.04	O 3.44	F 3.98	Ne
Na 0.93	Mg 1.31											Al 1.61	Si 1.90	P 2.19	S 2.58	Cl 3.16	Ar
K 0.82	Ca 1.00	Sc 1.36	Ti 1.54	V 1.63	Cr 1.66	Mn 1.55	Fe 1.83	Co 1.88	Ni 1.91	Cu 1.90	Zn 1.65	Ga 1.81	Ge 2.01	As 2.18	Se 2.55	Br 2.96	Kr
Rb 0.82	Sr 0.95	Y 1.22	Zr 2.20	Nb 1.60	Mo 2.20	Tc 1.90	Ru 2.20	Rh 2.28	Pd 2.20	Ag 1.93	Cd 1.69	In 1.78	Sn 1.88	Sb 2.05	Te 2.10	I 2.66	Xe
Cs 0.79	Ba 0.97	La 1.10	Hf 1.30	Ta 1.50	W 2.36	Re 1.90	Os 2.20	Ir 2.20	Pt 2.28	Au 2.54	Hg 2.00	Tl 1.8	Pb 2.1	Bi 2.02	Po	At	Rn

Figure A1.1

the Mulliken and Pauling scales reflect the broad sweep of atom characteristics, but do not measure finer variations caused by the different chemical contexts in which an atom may be encountered. Electronegativities are numbers constructed from empirical data such that the difference of any two measures the tendency and direction of electron displacement when the two atoms are allowed to interact. The periodic table in Figure A1.1 displays the Pauling electronegativities as these are the parameters universally used in teaching chemistry. The elements exhibiting metallic properties have distinctly lower values than the non-metals.

The numerical values of, for example, ionization energies, when listed in order of atomic number, display a periodicity reflected in the common form of the periodic table. However, by themselves they do not provide a rationale of this periodicity. The quantum-chemical model of the electronic structure of H, when extrapolated to the heavier atoms, not only provides a supporting rationale, but also the conceptual building blocks for describing bulk-element structure and reactivity. That is, the atomic model provides the basis for a molecular model as well as one for complex extended structures. For atoms it is a story taught in all beginners-chemistry classes and one that appears in some form in most chemistry texts. We simply repeat the essentials.

The single proton, single electron two-body problem can be solved explicitly to provide a quantum chemical description of the H atom. A set of solutions (the Legendre polynomials) are found, each with a different energy and wave function(s). For the lowest energy solution, the ground state, the wave function is spherically symmetric with a value that decreases exponentially as a function of distance from the nucleus. Polar coordinates are used by convention, and to accommodate our human inadequacies in the visualization of complex functions we separate the radial dependence (one coordinate) from the angular dependence (two coordinates) thereby producing representations such as that illustrated in Figure A1.2.

How do we connect the model to the three essential properties of an atom? An observable, such as size, depends on the square of the absolute magnitude of the wave function. Size, then, must be associated with the coefficient of the exponent of the radial function, but it is hard to say a priori when its value is small enough to be neglected. Hence, numerical values are obtained empirically under conditions where the atom of interest interacts with another atom. Rich sources of such information are solid-state structure determinations, but interaction with an atom of a tip of an STM constitutes another source. Consequently, different measures of size exist and reflect different types of interaction: van der Waals radii (non-bonding), covalent radii (shared-electron bonding) and ionic radii (electrostatic bonding resulting from complete electron transfer) are the common ones.

326 Appendix

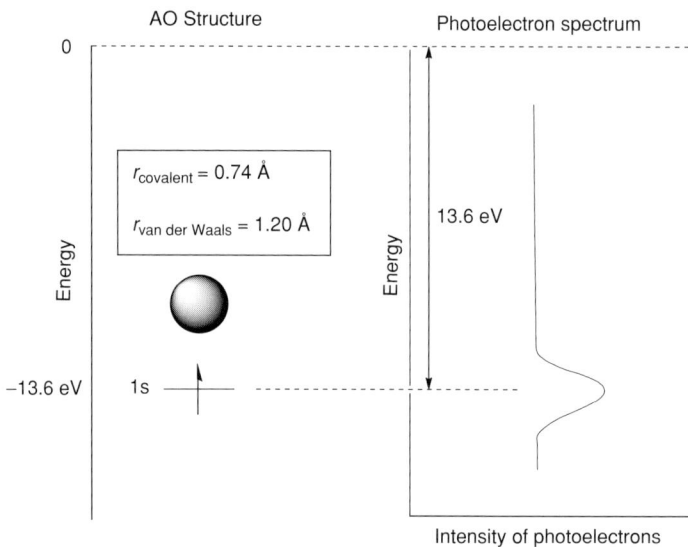

Figure A1.2

Ionization energy can be calculated if one spends the effort necessary to compute good energies for the atom and the ion. Those found in texts are experimental values and for H it is the energy necessary to completely remove the electron from the proton as shown in Figure A1.2. Two accurate ways of measuring the energy associated with this process are: photoionization, where the lowest frequency photon capable of generating the ion is measured (frequency $\upsilon = E/h$, where h is Planck's constant) and photoelectron spectroscopy, where it is the kinetic energy of the electron ejected by high energy photons that is measured ($IE = h\upsilon - KE$: the electron, being 10^3 lighter than the proton, carries off $> 99.9\,\%$ of the excess energy). The electron affinity can be measured in the same fashion by measuring the ionization energy of the negative H atomic ion instead of the neutral atom. As neither the neutral atom nor anion can be purchased from Aldrich it is necessary to prepare them from an appropriate precursor before photoionization. An interesting problem in itself!

All well and good – the H atom can be said to be understood. How does this help us with the other atoms as the n-body problem cannot be solved explicitly? Here is where the other solutions of the H atom problem come in handy. They provide the model the chemist uses every day as a fundamental part of his or her chemical language. To review, each solution is denoted by a primary quantum number n ($n = 1$ for the ground state, $E = -13.6\, Z^2/n^2 = -13.6\,\text{eV} = -IE$ and $Z =$ nuclear charge $= 1$ for H) but the number of wave functions for $n > 1$ is also larger than 1, i.e., these solutions are said to be degenerate. The higher-energy solutions have picked up a curious nomenclature derived from spectroscopic experiments predating the model. For $n = 2$, the "s" and "p" and for $n = 3$, the "s", "p" and "d" functions are degenerate. The degeneracy lies in the non-spherical symmetry of the angular parts of the "p" and "d" wave functions: "s" (one function spherically symmetric and no nodal planes), "p" (three functions, dipolar and directional with one nodal plane each), "d" (five independent functions, quadrupolar and directional with two nodal planes – by convention the d_{z^2} function is the linear combination of $d_{x^2-z^2}$ and $d_{y^2-z^2}$). However, the radial functions also change with n.

A1.2 Atomic properties

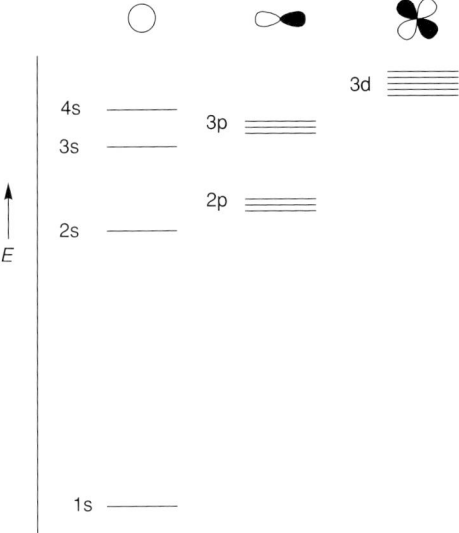

Figure A1.3

Specifically, the number of nodal surfaces in the radial function depends on the value of n. For $n = 1$, the total number of nodal surfaces is 0, for $n = 2$ the total number is 1, for $n = 3$, the total number is 2 and so on. Hence, the radial function for a 3s orbital has two nodal surfaces and the angular function has none whereas the radial function for a 3d orbital has none but the angular functions have two. A rule of thumb that also applies to molecules is that the greater the total number of nodal surfaces, the higher the orbital energy.

How does this help us with He, Li, Be, B...? Suppose an individual electron in one of these atoms experiences a positive potential that arises from some effective charge of the nucleus and all the other electrons taken together. Then we have a version of the H-atom problem yielding the same solutions, i.e., one well-defined particle (an electron) interacting with one ill-defined composite "particle" (the nucleus and the rest of the electrons). This "one-electron" approach differs in the sense that the nuclear charge experienced is not the actual nuclear charge but something smaller. In addition, each electron need not experience the same effective nuclear charge, as the composite "particle" can differ in each case – the ubiquitous crowd of electrons shields the nucleus from the electron one is specifically dealing with. Rules are developed, e.g., Slater's rules of shielding, based on measured atom properties and we find, for example, that 3s, 3p and 3d functions no longer have the same energy (Figure A1.3). Given such a general energy-level diagram, we can construct a model for heavier atoms provided we have a rule to guide the addition of electrons to these spatial orbitals. How many electrons will fit in a 3p orbital anyway? Important rules that are not violated are called principles and the Pauli principle states two electrons maximum for a single spatial orbital and then only when the electrons are of opposite spin. The ground-state configuration is the one in which the lowest energy H-like orbitals are filled. So now we have our Model T of atomic theory – a clunker that continues to chug away every day for the thousands of chemists who use it in describing exceedingly complex chemical systems. Well let's give the crank a turn and see how it drives.

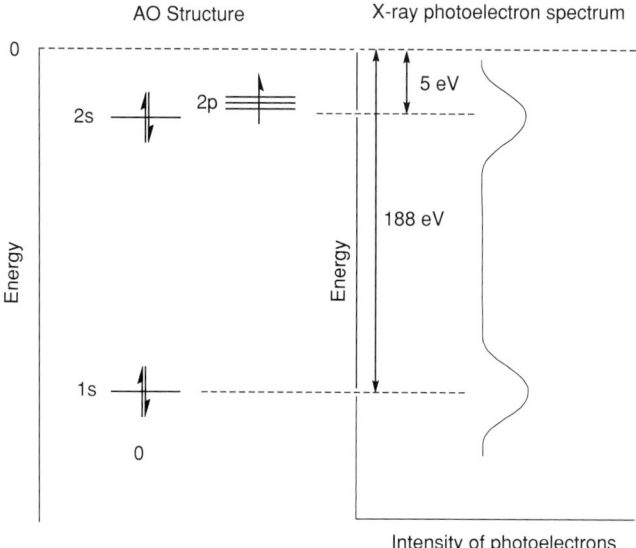

Figure A1.4

Consider B and Fe as representative examples. ^{11}B consists of five electrons and a nucleus containing five protons and six neutrons. What is its electronic structure? In the quantum-chemical approach it is described by an energy and wave function. With our H-atom model the latter is easy. It is: $(1s)^2(2s)^2(2p)$ which represents the product of individual atomic functions (Figure A1.4). The distribution of the five electrons in space (electronic structure) is given by the square of the absolute magnitude of the wave function. 1s and 2s functions are spherically symmetric whereas a single 2p function is not; hence, the shape will not be spherically symmetric. The ground-state energy, a single number, must be related to the energies of the individual H-like orbitals that are occupied; however, the simple sum of AO energies is a poor representation as the electron–electron interactions that we blithely ignore in this model are not small. On the other hand, the functional reality of this model gains considerable utility from the fact that the multiple atomic levels in this model do reflect observed multiple ionization energies each corresponding to the binding energies of electrons in the different filled AOs. Thus, in tables of photoelectron spectroscopic data one finds ionizations listed at 188 and 5 eV for B corresponding to removing an electron from the $n = 1$ and $n = 2$ shells (these are actually excitation energies from the ground state of the atom to the various excited states of the monocation which are $1s_{1/2}$ and $2s_{1/2}$, $2p_{1/2}$ and $2p_{3/2}$ where the $n = 2$ states are not individually distinguished experimentally as X-rays are used and line widths are >1 eV). The former is known as a core energy level and, for example, is used in surface analysis by X-ray photoelectron spectroscopy (XPS) to identify the element B. Atomic size depends on radial function and is inversely related to stability. Hence, the 1s electrons describe electron density close to the nucleus whereas the 2s and 2p valence electrons occupy a much larger volume and determine the atomic size as well as valence properties.

Now take a look at Fe with 26 electrons. The ground-state wave function is $(1s)^2(2s)^2(2p)^6(3s)^2(3p)^6(4s)^2(3d)^6$. The 4s shell fills before 3d as the former experiences a larger effective nuclear charge than the latter. The XPS book lists characteristic ionizations for Fe at 7114 ($1s_{1/2}$), 846, 723, 710 ($2s_{1/2}$, $2p_{1/2}$, $2p_{3/2}$), 95 ($3s_{1/2}$), 56 ($3p_{1/2}$, $3p_{3/2}$) and 6

A1.2 Atomic properties

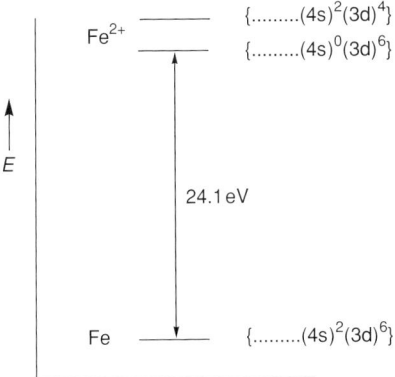

Figure A1.5

$(3d_{3/2}, 3d_{5/2})$ eV. Note that no $(4s_{1/2})$ ionization is listed! We must deal with the 4s vs. 3d energetics further below. Despite the similarities in outermost electron ionization energies with those of B, the Fe atom is significantly larger as the $n = 3$ level is now populated and the corresponding radial function yields an electron-distribution function with a maximum at larger values of r. What about shape? A half-filled (one electron in each space function with all spins parallel – Hund's rule) or filled p or d level is spherically symmetric, hence the shape will be determined by the $(3d)^6$ functions. Here Hund's rule of maximum multiplicity (spin) is the guide and four of the d functions contain one electron and the other two. Hence, the atom will not have a spherically symmetric electron distribution.

This is a great little model and it serves so well that sometimes the crude approximations made in generating it are forgotten and quantitative lapses become topics of discussion simply because the model is presented initially as some sort of fundamental truth rather than the approximate picture it is. For example, go back to the 4s vs. 3d filling problem above. The electron configuration of $[Fe]^{2+}$ is $(1s)^2(2s)^2(2p)^6(3s)^2(3p)^6(3d)^6$ not $(1s)^2(2s)^2(2p)^6(3s)^2(3p)^6(4s)^2(3d)^4$, i.e., the total energy of the former is lower than that of the latter configuration. There is one energy and one wave function for the ground state and one energy and wave function for the ion state (Figure A1.5). The crude one-electron model ignores electron–electron interactions that can tip the balance in favor of a non-intuitive configuration, e.g., the Fe ion has two fewer electrons and less electron–electron repulsion. The take-home message with these simple models is "keep it qualitative" and don't confuse the model with the facts. Models change. Carefully measured properties do not.

The periodicity in size and ionization energies observed empirically can now be rationalized in terms of recurring electron configurations. Indeed even the little jogs when sub-levels are half filled are understandable. If you need to you can review these points in any inorganic text. The model really works very well when used in a relative sense. More importantly we can now begin to explain why all elements don't exist as monatomic gases as do the elements of group 18 that occur at the turning points where a given n level is fully filled and the atom possesses a spherically symmetric closed-shell electron configuration. Turning this statement around, we associate a closed-shell electron configuration with stability; hence, the 8- and 18-electron rules discussed in more detail in Chapter 1. Simple ideas of valence arose from the fact that B with valence configuration $(2s)^2(2p)$ needs five more electrons to achieve a closed-shell configuration and Fe with valence configuration $(4s)^2(3d)^6$ needs ten more electrons. How each of these elements

330 Appendix

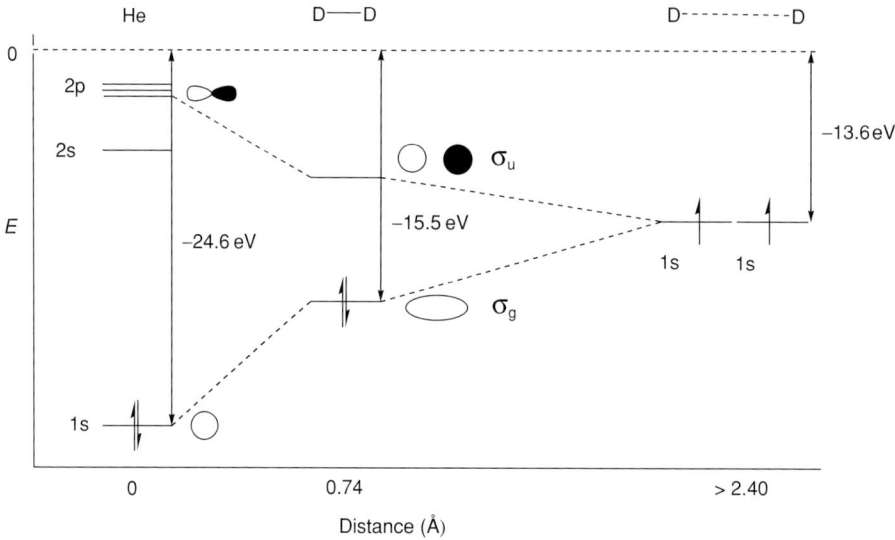

Figure A1.6

achieves electronic happiness by distributing the highest energy electrons over a complex geometric arrangement of nuclei is a principal topic of this text to be dealt with in due course. However, we can review how the process occurs for atoms that achieve electronic happiness in a much simpler fashion. The elements in the upper right-hand corner of the periodic table occur naturally as small gaseous molecules which are both easy to picture in terms of geometric structure and easy to explain in terms of electron distribution.

A1.3 Homoatomic substances

We begin with a review of the naturally occurring forms of the elements in terms of structure with emphasis on how nuclearity reflects Nature's energetically favorable solution to the demands of atomic structure.

A1.3.1 Diatomic species

Hydrogen again provides the paradigm albeit H_2 rather than H. H, with an electronic structure $(1s)^1$ is spherically symmetric but not closed shell. It needs another electron. Combination with another H atom does the trick giving a system with two protons and two electrons (Figure A1.6). Note that if we switch to D from H then we have a system with two protons, two neutrons and two electrons as also found in He. An early model for correlating the electronic structure of atoms with that of molecules is the united-atom model of Mulliken. That is, the distribution of electron density in the He atom as defined by its AOs should correlate with whatever function describes the electron distribution for D_2. The difference lies in the spatial locations of protons and neutrons as well as the thankfully huge barrier separating He and D_2 (think H bomb). Conceptually the united-atom model suggests that there should be orbitals that encompass both nuclei of D_2 that correlate with solutions of the H-atom problem when applied to He. These orbitals, called molecular orbitals (MOs) have the same fundamental meaning as AOs and can be

A1.3 Homoatomic substances

generated from the AOs (basis functions) of the two D atoms. The lowest energy MO of D_2 (σ_g) correlates with the 1s orbital of He whereas the other MO (σ_u) correlates with the $2p_z$ orbital of He. Two comments: following custom we let the z axis be the molecular axis as shown in Figure A1.6; however, the results are independent of the coordinate system. We use the traditional symmetry labels to identify the two MOs; however, for our purposes here they are just names (Greek symbols with German subscripts tend to keep the riffraff out of the area). Later we will point out how symmetry is a simplifying concept in these considerations.

We can make the energy axis quantitative by looking up the experimental ionization energies, 24.6, 155, and 13.6 eV for He, D_2 and D, respectively, as these provide a measure of the negative of the orbital energies (Koopmans' theorem). Empirically one sees that separating the protons of the He nucleus by 0.74 Å destabilizes the "1s" orbital and distorts it. This diagram suggests that the distortion is an uphill process energy-wise. Viewing the process from the opposite limit, that of the separated atoms, suggests a downhill process and the associated energy is known as the bond energy. This has been experimentally measured – two D atoms sharing their lone valence electron is an energetically more favorable situation than two separate D atoms by 437 kJ mol^{-1} (4.53 eV). Energy and structure are connected so it should be no surprise that the H–H distance is less that the sum of the van der Waals radii (atom–atom touching distance) and its value (0.74 Å) is taken to be characteristic of an H–H bond. One half of this distance is the covalent radius of H (0.37 Å). The bond is described as a single bond or a two-center–two-electron bond.

How is the dimeric molecular nature of elemental H reflected in its physical properties? The boiling point is low (20 K) such that hydrogen is a gas at room temperature and the associated heat of vaporization (0.8 kJ mole^{-1}) is very low in comparison to the H–H bond dissociation energy. That is, there are strong interactions between pairs of H atoms but very weak interactions between the dimers themselves. H_2 is just He (bp = 4 K, ΔH_{vap} = 0.08 kJ mole^{-1} with "atoms" containing twin nuclei! It has molecular orbitals that are modified atomic orbitals and can be understood in the same fashion.

Exercise A1.1. The difference between the energy of the separated atoms and the D_2 molecule in Figure A1.6 is not 4.5 eV, the energy of a D–D bond. Why not?

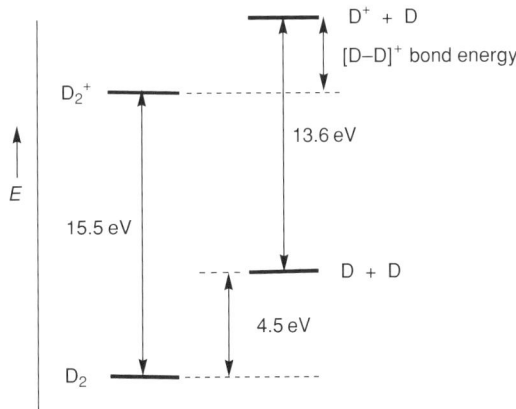

Exercise A.1

Answer. The appropriate energy-level diagram is shown below. The two ionization energies applied to Figure A1.6 to define the relative energies of the MO and AOs are energy differences between the neutral and cationic states. Hence, as shown below, they provide a measure of the [D–D]$^+$ bond of 2.6 eV which may be compared to a reported value of 2.65 eV.

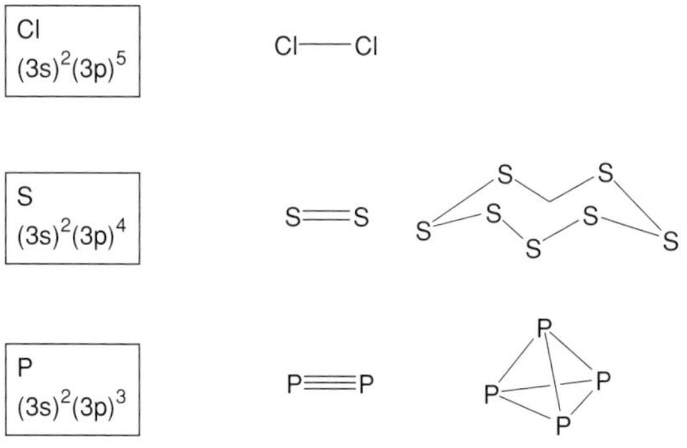

Figure A1.7

A1.3.2 Polyatomic species

The bond between the H atoms in the H$_2$ molecule serves as a model for a two-center–two-electron bond. It is a concept easily applied to heavier atoms. Let's apply it to the elements beginning in the upper right-hand corner of the periodic table. Group-18 atoms with valence configurations of $(ns)^2(np)^6$ possess a spherically symmetric closed shell and are stable as monotonic gases. The eight-electron configuration is thus set up as an electronic "goal" for main-group atoms with an $(ns)^x(np)^y$ valence shell. Thus, a group-17 atom with valence electronic structure $(ns)^2(np)^5$ can achieve an eight valence-electron closed-shell configuration by forming a dimer X$_2$ with one electron-pair bond just as found for H$_2$ (Figure A1.7). One might think of H or Cl in terms of a tinker-toy connector with a single hole – all one can make is dimers.

If so, a group-16 atom with configuration $(ns)^2(np)^4$ morphs into a connector with two holes. Hence, it should form rings. Experimental chemistry tells us (Figure A1.7) that, yes, S forms rings (S$_8$ plus many more sizes) but O is found as diatomic molecules (O$_2$). Apparently, both "holes" are capable of connecting to one adjacent atom giving a "doubly" bonded dimer (two-center–four-electron bond). Well, OK, but there is something funny here as O$_2$ is paramagnetic, i.e., not closed shell. We will have to look more carefully at this molecule as a simple double-bond model with paired electrons doesn't work. We need a better model to explain what is going on and will present it in Chapter 1.

On to group 15 with $(ns)^2(np)^3$ and a three-hole connector which suggests either a "triply" bonded dimer (N$_2$) or a three-dimensional cage (tetrahedral P$_4$) as shown in Figure A1.7. Here is a good place to mention coordination number. The number of bonded nearest neighbors to any given P atom is three; hence, we say it has a coordination number of three. Each atom also has three two-center–two-electron bond connections. But one shouldn't get the idea that coordination number specifies the number of two-center–two-electron bonds. It depends on the system. The focus of this book, clusters, is an area where the coordination number often does not specify the number of two-center–two-electron bonds.

A1.3 Homoatomic substances

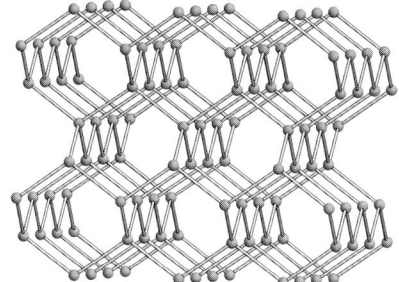

Figure A1.8

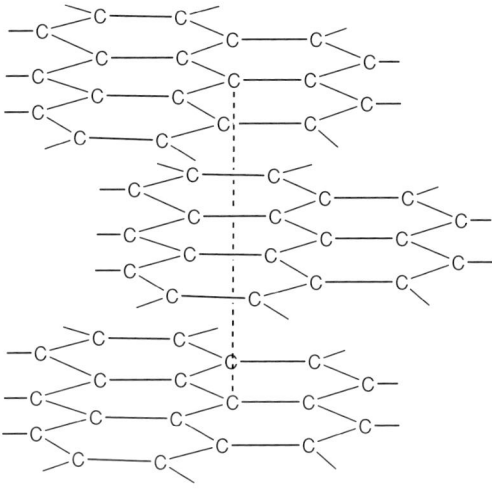

Figure A1.9

Dare we push on with the same simple idea to group 14? Well, the chemist's approach is to push an idea until it fails. Then one tries to figure out why it does so and see what modifications to the model are needed. An electron configuration of $(ns)^2(np)^2$ implies a four-hole connector and four two-center–two-electron bonds. A quadruple bond comes to mind, but for a main-group species this is not observed. The tinker-toy theory fails. However, quadruple bonds are possible – just not for main-group atoms. As will be seen in Chapter 1, examples are observed in dinuclear transition-metal systems.

With four holes in our connector we are forced into some kind of extended or never-ending structure. If we use Nature as a guide, we see in Figure A1.8 that the bonding requirements of C are satisfied by the diamond structure (four single bonds per C atom with a $d_{CC} = 1.54$ Å) and (Figure A1.9) the hexagonal graphite sheet structure (two single and one double bond per C atom within a sheet and weak interactions between the sheets). An additional possibility is a one-dimensional chain of alternating single and triple bonds (carbyne allotrope, see Chapter 6). An unexpected solution from Nature was revealed on the characterization of the C_{60} cluster (Chapter 2 and Chapter 7) made up of pentagonal and hexagonal fused rings in the form of a soccer ball. Note that the graphite structure generates a problem for our localized bonding model. The

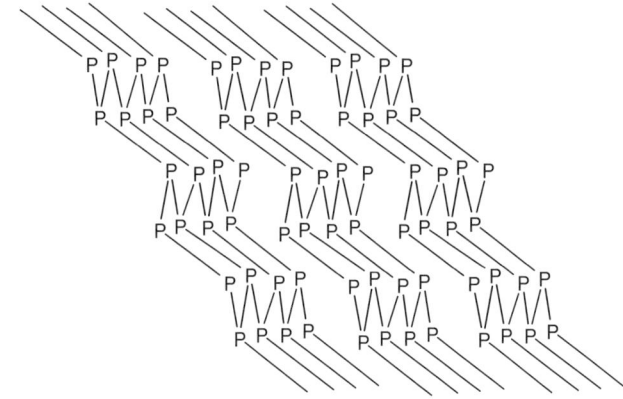

Figure A1.10

inter-sheet spacing is 3.35 Å consistent with weak inter-layer interactions. That's fine but the intra-sheet structure exhibits a single $d_{CC} = 1.42$ Å rather than alternating double and single bond distances. This discrepancy can be patched up by introducing "resonance" but we will describe a better way in Chapter 1.

With the exception of C_{60} these structures place us outside of molecule chemistry as the dimensions are only limited by the size of the crystallite. How is the infinite, extended nature of the structure of the diamond allotrope of C reflected in properties? All the C–C distances are less that the sum of the van der Waal's radii and characteristic of a C–C single bond. The melting point is high such that C is a solid up to about $2500\,^\circ$C. The associated heat of sublimation to atomic C (714 kJ mol^{-1}) is very high and approximately double the C–C bond dissociation energy (349 kJ mol^{-1}) as expected if four half C–C bonds can be associated with each C atom in diamond. Perhaps it is unnecessary to point it out, but, seeing as some introductory text titles proclaim chemistry as "the molecular science," it is worth while stating that there is no such entity that can be called a molecule in the diamond form of elemental C. One purpose of this text is to emphasize that molecules hold down one side of chemistry but the solid-state side is conceptually distinct and should not be marginalized. The thesis of this text is that clusters provide an intellectual bridge between the two.

Stare at the diamond structure of C and turn each C into Si. Then imagine putting a proton and a neutron into each nucleus generating $(P^+)_n$. If we neutralize this weird compound by adding n electrons to n P–P bonds, they become lone pairs. This particular P–P distance lengthens and becomes non-bonding. If we place the lone pairs symmetrically such that the diamond structure is cleaved into sheets, the puckered hexagonal sheet structure of rhombohedral black phosphorus is formed (Figure A1.10). But there are many other less symmetrical ways of distributing the lone pairs and bond pairs so it should come as no surprise that three crystalline forms and one amorphous (no long-range order) form of black phosphorus are known. Molecular P_4 is white phosphorus. In both extended forms and cluster form each P atom is associated with three single P–P bonds and one lone pair. Now you can let your imagination run wild. Take P_4 tetrahedra and break one edge to form an open diamond with two active connectors. A chain can be formed. Or just make a chain with alternating single and double bonds. Many allotropes of P are possible and lots (15!) are known albeit not always structurally characterized, e.g., amorphous red phosphorus.

A1.3 Homoatomic substances

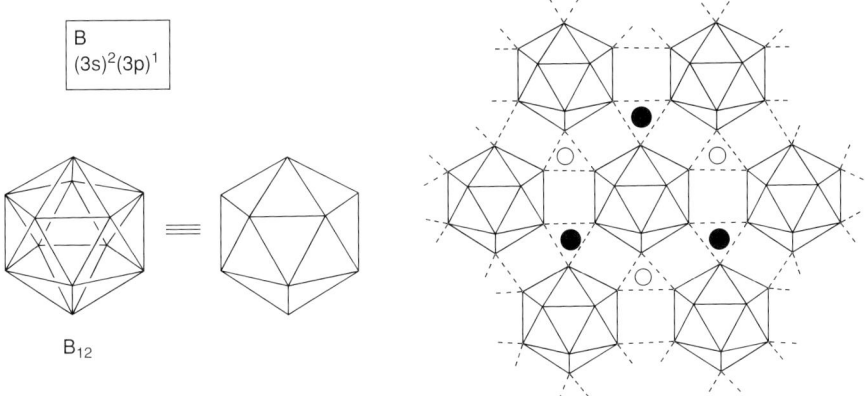

Figure A1.11

Backing up one more step to group 16 you can now imagine many more allotropes of S including larger rings as well as long chains – fond recollections of converting powdery yellow sulfur (S_8) into a rubbery material in the kitchen (we won't mention the clouds of SO_2 from the occasional disaster!). For groups 16 and 15, extended structures in one and two dimensions, respectively, are alternatives to the molecular forms.

Exercise A1.2. Discuss the energetic implications of the facts that elemental P exists as tetrameric molecules or as sheet-like extended structures under normal conditions but N is found as diatomic molecules. Focus your discussion on the relative energetics of E–E single vs. E≡E triple bond energies for E = N, P.

Answer. Consider the conversion of two E≡E molecules into one E_4 tetrahedron.

$$E_4 = 2E_2$$
$$\Delta H = 4D(\text{E–E}) - 2D(\text{E}\equiv\text{E})$$

where D refers to the bond energy. The observations suggest that for E = N, $\Delta H < 0$, whereas for E = P, $\Delta H > 0$. There is no evidence for the N_4 molecule; however, by heating white phosphorus, P_4, one can generate P_2 molecules, consistent with an endothermic process (based on the reported heats of formation of P_2 and P_4 the heat of reaction is -388 kJ mol^{-1} for formation of the tetramer from the dimer). In absolute terms one can attribute the difference to weaker than expected N–N single bonds or P≡P triple bonds or stronger than expected P–P single bonds or N≡N triple bonds or a combination of both. Reported values for E–E are: 252, 323 and those for E≡E are: 945, 487 kJ mol^{-1} for E = N, P, respectively albeit one finds a huge variation in the suggested average bond energy for P–P from different sources. The suggestion is that the difference in common molecular states for E = N and P is due to weaker N–N bonds (lone pair–lone pair repulsion) and weaker P≡P triple bonds (poor p π overlap). The key factor is the short N–N bonding distance relative to that for P–P bonding.

In the flush of success one proceeds to group 13 with configuration $(ns)^2(np)^1$. Somehow the atom must now become associated with five electrons. What does Nature tell us? A representation of α-rhombohedral B is shown in Figure A1.11 (the more complicated structure of the β-rhombohedral B allotrope will be avoided). It consists of layers of icosahedral B_{12} units (clusters) one of which is shown in Figure A1.11. There are B–B

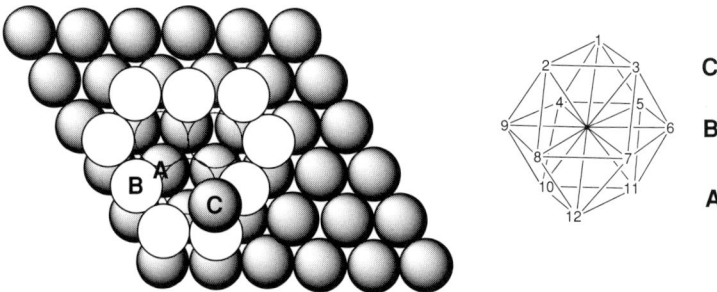

Figure A1.12

bonds between the icosahedra in this layer (dashed lines) but any of these six B atoms is found equidistant from two others. Thus, the six B atoms shown are each bound to two other B atoms. There are identical layers above and below the one shown. The one above has icosahedra centered over the black dots whereas the one below has icosahedra centered under the white dots. As a result, each icosahedron shown forms three connections to the upper layer and three to the lower layer.

Count the average number of connections for a single B atom in one icosahedron: six of the B atoms have five internal and two external bonding connections. The other six have five internal and one external bonding connection. On average, then, each B atom makes 6 1/2 bonding connections so if there were 6 1/2 two-center–two-electron bonds each atom would be associated with 13 electrons! But each B atom only has three electrons so ten electrons are missing in this absurd scenario. Play with it all you want but you will find no way you can construct a satisfactory representation with our simplest bonding idea.

In Chapter 2 (Section 2.9) we see how the cluster bonding requirements for the icosahedron, plus two-center and three-center inter-cluster bonds perfectly uses the three available valence electrons and four available valence orbitals in a covalently bonded cluster network. Once one has these advanced bonding models in hand, then the explanation of the B network structure is no more difficult than that of the C diamond structure. One purpose of this text is to provide these advanced models, but for now the solution to the problem remains hidden. Hey, a little suspense always helps the story line. At this empirical stage of the presentation you have learned that the nature of bonding (distribution of electrons) is expressed in geometry. The tricky bit is to interpret the empirical nuclear position in terms of a useful (simplest one that answers the question asked) model for the distribution of valence electrons.

Does Al (group 13, second full row) have the same bizarre structure as B? No. We have crossed the metal–non-metal divide and its structure (Figure A1.12) is typical of that possessed by those elements we recognize as metals. A quick review of the most common ways of representing these so-called close-packed structures on a two-dimensional sheet of paper using the example of elemental Al is in order. Represent each Al atom as a hard sphere of diameter 2.86 Å and form a two-dimensional layer in which the atoms are most closely packed (minimum free space). The result is shown in Figure A1.12. Simple geometry requires any given atom to be surrounded by six nearest neighbors. The three-dimensional solid is formed of identical sheets stacked to minimize free space. Place another sheet on the first and note that only half the gaps in the first sheet are covered by atoms in the second. Place the third sheet on top of the second. Now there is a choice to be made. The third sheet can be in registry with the first (stacking ABAB . . . designated hexagonal close-packing, hcp) or not (stacking ABCABC . . . designated cubic close-packed, ccp, or face-centered cubic, fcc). The latter is a representation of the structure of elemental Al, i.e., Al is fcc with $d_{AlAl} = 2.86$ Å. In the drawing at the side, a small chunk of Al is represented as a lattice of points connected by lines of identical length = 2.86 Å.

What about bonding? If we pick any Al atom then we find it surrounded by six nearest neighbors in its sheet, three in the sheet above and three in the sheet below to give a coordination number of 12. As with B two-center–two-electron bonding is out of the question. Again we need a different model but we can't use the same one that works for B. The dramatic difference in properties between B and Al tells us this. Like C and Si in the diamond structure, B is a semiconductor (conductivity is small and increases with increasing temperature) whereas Al is a metal (conductivity is large and decreases with increasing temperature). For some fundamental reason the nuclei of the former exert control over the least-tightly bound electrons whereas the nuclei of a metal, e.g., Al, allow them to move freely between and beyond the boundaries of the crystallite on application of an electric field. The two-center–two-electron bond that localizes the valence electrons between the two nuclei is consistent with the properties of the non-metals but clearly we need a different approach to understand what is going on with a metal. In fact, no localized model is going to work. A delocalized model will be provided in Chapter 6.

Let's go back to B again for a moment (Figure A1.11). In your mind convert each of the B_{12} icosahedra into a single spherical atom. Voilà, we have a fcc structure! So B, the element jammed into the corner created by the column of group-14 elements and the diagonal metal–non-metal boundary exhibits a crazy mixture of network structure with overtones of closed-packed features. In the diamond structure of C the lines nicely represented strong covalent bonds similar to that found in H_2. In Al metal the lines represent geometric relationships only. And in B we have a situation where some of the lines represent localized bonds (between the icosahedra) whereas others represent only geometry (within the icosahedra). Thus, later in this text we need to develop a delocalized model to provide a valid description of the bonding within the icosahedra as well as a modified localized model for the inter-icosahedral cluster bonding. The delocalized model is connected to that required for description of a metal, whereas the localized model is connected to that used for simple molecules. The hybrid nature of the bonding found for a borderline element structure is pleasing. Nature utilizes all of the features of its atomic tool box and refuses to be constrained by our limited intellectual capacity – a humbling thought!

A1.4 Heteroatomic substances

The reader might well wonder what the authors are smoking because this heading covers all of chemistry! True indeed, so all we can possibly present in this short section is a review of the ways in which the fundamental structural forms represented by the elements are modified by heteroatoms and how the introduction of heteroatoms can generate new types of structure. Again, in the simplest molecular forms the application of the two-center–two-electron bond model plus 8/18-electron rules will be our focus; however, places where this guide fails will be noted.

A1.4.1 Molecular stoichiometry

Stoichiometry is easy. Again consider the atoms as connectors with the number of prongs governed by electron configuration – a kindergarten approach but there is no need to make it any more complicated yet. Treated as such we can put together any given set of connectors and make up compounds and check the results against Nature. As we no longer are restricted to a single atom type we can match the requirements of a foundation

338   Appendix

atom (central atom) to those of the attached atoms in order to satisfy valence requirements. As illustrated by element structures, lone pairs, multiple bonds and extended structures of various types can be anticipated. Certain common atom combinations appear frequently so they are conveniently grouped together and called substituents or ligands with characteristic valence properties. Such a device permits a bootstrap approach to impressively complex compounds. It is one that is useful on paper as well as in the laboratory. An exercise will suffice to review the point.

Exercise A1.3. Make up a set of compounds containing P and H, i.e., define m and n in P_mH_n, and draw their two-center–two-electron structures using lines to represent bonds.

Answer. A systematic approach is best and there is more than one. Let's do it in order of number of P atoms. This yields: PH_3, P_2H_4, P_2H_2, (P_2), P_3H_5, P_3H_3, etc. Using a PH_2 group as a stand in for H, one generates P_3H_5 from PH_3 as $HP(PH_2)_2$. In fact any one-electron group, e.g., CH_3 can be used to replace an H on any of the compounds listed. The possibilities are enormous. Note that P is isoelectronic with CH so that these phosphorous hydrides are analogs of hydrocarbons. Mainly on account of their sensitivity and toxicity not many have been characterized, but, in principle at least, a large number of these so-called catenated species are possible. The fact that the nitrogen hydrides are more limited is consistent with the N_2 vs. P_4 problem mentioned above.

A1.4.2 Geometric structure

After stoichiometry comes structure – the arrangement of atoms in space. How do we rationalize and predict the geometric structures of simple main-group compounds like the phosphorus hydrides in the exercise above? Real prediction nowadays comes from quantitative calculations even for systems containing a large number of atoms. But "back of the envelope" predictions are useful if only to develop a feeling for the major factors that control, or appear to control, compound shape. The valence-shell electron-pair repulsion model (VSEPR) is the guide most often adopted even though it is of limited help in situations where some valence electrons "are not stereochemically active," i.e., when the rule doesn't work. The model uses the two-center–two-electron bond model and emphasizes the mutual repulsion of lone pairs and bonding pairs. Again an exercise is sufficient for review.

Exercise A1.4. Suggest three-dimensional structures for PH_3 and P_2H_4.

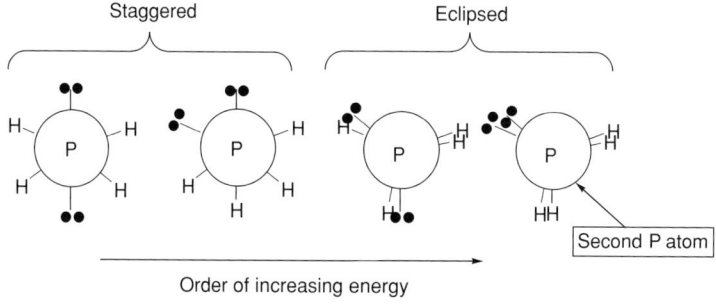

Exercise A.4

A1.4 Heteroatomic substances

Answer. PH$_3$ possesses three P–H bond pairs and one lone pair. Consider three negative charges on the surface of a sphere. Mutual repulsion tells us that the most stable disposition will be at the vertices of a tetrahedron. Hence, PH$_3$ will have a tetrahedral array of electron pairs and a trigonal-pyramidal arrangement of the four nuclei. There is only one type of P atom in P$_2$H$_4$ with two P–H, one P–P bond pair and one lone pair, i.e., four pairs each with a tetrahedral geometry at each P center. Take another look at the structural representations viewed down the P–P bond axis and note that the adjacent lone pairs can have a cisoid or transoid relationship in an eclipsed or staggered arrangement. If there are no P–P bonding interactions that require an eclipsed arrangement, mutual H–H and lone-pair–lone-pair repulsion will make the staggered arrangements of lower energy. There will also be a difference in energy between cisoid and transoid configurations if the repulsion between electron-pair types is not equal. If lone-pair–lone-pair repulsion is larger, as is usually the case, then the transoid arrangement will be favored and may dominate if the barrier to rotation around the P–P bond is much larger than kT. However, such subtleties are best left for the aficionados of VSEPR. This is not because conformational properties are unimportant, as they are. But experiment, interpreted in the light of quantitative calculations, is the best method for pinning down the lowest energy configuration. It also defines the number of minima in the energy surface describing the rotation as well as barrier magnitudes and the electronic factors ultimately responsible. Crude theory based on a single contributing factor is not helpful for low-barrier problems where electronic origins are often complex.

A1.4.3 Isomers

The mere presence of heteroatoms introduces interesting additions and changes to the picture presented by the elements. As long as the bonding requirements of each atom are satisfied, atom types can be distributed differently over the possible positions generating isomers. Linkage, geometric and optical isomers are the common types encountered and are illustrated in Figure A1.13. For some compositions and geometries the enumeration of possible isomers is an interesting challenge – a kind of a puzzle that one loves to set up for students to solve. A systematic approach is advisable for complex situations and many clear discussions are available in standard organic and inorganic texts. We will return to isomer count in later chapters when discussing cluster isomer possibilities.

The two-center–two-electron bond is also modified by the presence of heteroatoms. Two important effects concern how electron density is shared between the two atom centers of a two-center bond. First, each bond is now polarized due to the fact that the electronegativities of the two atoms differ. This leads to molecules with net electric dipole

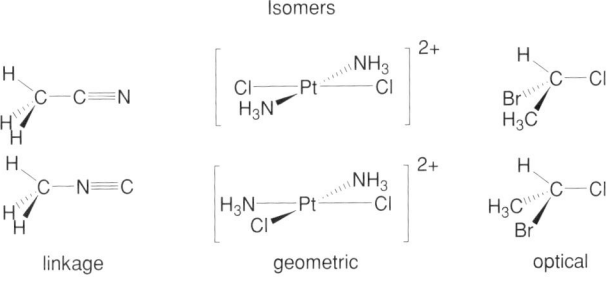

Figure A1.13

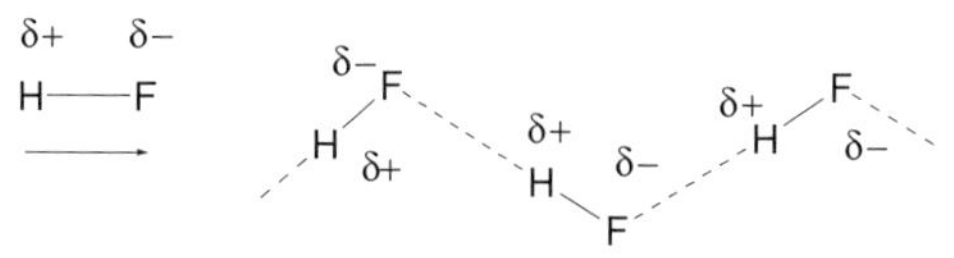

Figure A1.14

moments (each polar bond possesses a dipole moment and if the vector sum of all the bond moments is non-zero the dipole moment of the entire molecule will be non-zero). It also introduces sites of nucleophilic (electron-rich atom seeks positive nuclei) and electrophilic (electron-poor atom seeks negative electron) activity. Second, both electrons in the bond can arise from a single partner giving rise to a coordinate covalent bond (donor–acceptor bond). This spawns the important area of Lewis acid–base chemistry, a huge subsection of which is coordination chemistry. Let us review each in turn.

A1.4.4 Polar bonds

Take the hydrides of the first complete row of the periodic table. If we follow the rules we generate: FH, OH_2, NH_3, CH_4, BH_5, BeH_6, LiH_7. The last three probably make you smile as they don't exist. But better to say they haven't been made yet. For example, calculations suggest BH_5 as a BH_3 adduct of H_2 and, in fact, $[CH_5]^+$ is a prominent ion in methane chemical-ionization mass-spectrometric sources. Be that as it may, the compositions determined are: BH_3, BeH_2 and LiH none of which correspond to a molecular formula as they are observed as dimer, amorphous solid and high melting point crystalline solid, respectively. So the eight-electron rule isn't going to be satisfied our easy way for elements to the left of group 14. The way in which Nature solves the problem for BH_3 (dimerization to give B_2H_6) will serve in Chapter 1 to introduce the concept of a three-center–two-electron bond – a concept that loosens the constraint of the two-center bond that keeps our bonding pair between two nuclei. On the other hand we can play with charges and consider ions, i.e., $[NH_4]^+$ as found in the salt $[NH_4][Cl]$ and $[BH_4]^-$ as found in the salt $[Na][BH_4]$. More about ionic substances can be found in any inorganic text but for our purposes let's stick with neutral molecules and work our way through the series from FH to CH_4.

In a homonuclear diatomic the bond pair must be shared equally between the two nuclei. However, for FH the electronegativities are now very different which is another way of saying a potential difference between the two atom sites exists and is cancelled only by allowing some electron density to flow from H to F. A polar molecule is generated with the positive pole at H and the negative pole at F (Figure A1.14). The dipole moment generated depends on the charge separation – distance and magnitude. This charge imbalance has serious consequences. Both physical and reaction properties are strongly affected. No longer are the inter-molecular forces tiny. Indeed, there is evidence for FH existing in the gas phase as an equilibrium mixture of monomers and cyclic hexamers and the solid-state structure (Figure A1.14) consists of zig-zag chains. For the elements tucked into the upper right-hand corner of the periodic table these inter-molecular forces are large enough to be given a special name – hydrogen bonding. Hydrogen bonding is responsible for water being a liquid at room temperature rather than a gas like its heavier congener H_2S.

The polarity of NH_3 requires a little more thought. Each N–H bond will have a dipole with H positive and H–N–H angles of 107°. The vector sum gives a net moment along the three-fold axis. Should one also consider a contribution from the lone pair in the same

A1.4 Heteroatomic substances

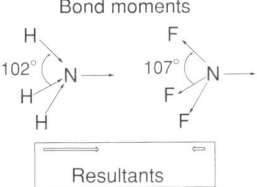

Figure A1.15

direction along the three-fold axis? The N–F bond is also strongly polar but of the opposite sense to the N–H bond. Hence, one might expect the dipole moment of NF_3 to be as large as or larger than that of NH_3 if only bond moments need be considered. It would be oriented in the opposite direction of course. Experimentally the dipole moment of NF_3 is only 10 % of that of NH_3 thereby suggesting the N–X bond moments alone don't tell the whole story. One explanation is that the lone pair does make a contribution and it adds to the moment generated by the N–H bonds but subtracts from the moment generated by the N–F bonds (Figure A1.15). Even an apparently simple idea like dipole moment requires careful thought, e.g., the small dipole moment of CO with C slightly negative.

Polarity is strongly involved in reactivity. Because of its symmetry the dipole moment of CH_4 would be zero even if the C–H bond moment were substantial. But because the electronegativities of C and H are similar, the bond moment is in fact small. Although CH_4 burns rather nicely, it is not so easy to convert it to functionalized derivatives under mild conditions. This is another way of saying the C–H bond is rather unreactive in the hands of the chemist. On the other hand, consider the interaction of HF with OH_2. Both molecules have polarized E–H bonds, E = F, O, with H positive and E negative; however, the electronegativity of F is considerably larger than that of O. Addition of HF to OH_2 permits the isolation of the compound $H_2O \cdot HF$ with melting point $-36\ °C$. Structural studies show the compound to be a salt $[H_3O][F]$ with strong hydrogen bonding. The H of HF has been transferred to O. Proton transfer chemistry, that is, Brönsted acid–base chemistry, is widespread and constitutes an important aspect of chemical reactivity.

Exercise A1.5. Combine the ideas of stoichiometry, geometric structure and polar bonds and estimate the dipole moments (large, medium, small, zero) of: (a) the simplest chloro derivative of Si, (b) a permethyl derivative of P, (c) a perfluoroderivative of H_2NNH_2.

Answer. (a) $SiCl_4$; four bond pairs, tetrahedral and zero dipole moment by symmetry even though the electronegativity difference is large (1.90 vs. 3.16).

(b) $(CH_3)_3P$; The P center has three bond pairs and one lone pair, pyramidal, non-zero dipole moment possible. Likewise the C center of each methyl group has three C–H bond pairs and one C–P bond pair, tetrahedral and a net group dipole moment is possible. However, similar C, P and H electronegativities (2.55, 2.19, 2.20) suggest the C–H and C–P bond moments are small. The net dipole moment arises mainly from the P lone pair and, based on the discussion of Figure A1.15, the moment should be somewhat smaller than that for NH_3. (The dipole moment of NH_3 is about three times that of PH_3.)

(c) Each NF_2 fragment has two N–F bond pairs, one N–N bond pair, and one lone pair, pyramidal and a non-zero dipole moment possible. A number of configurations of $F_2N–NF_2$ are possible (phosphine exercise above). The staggered, *trans* configuration which will have zero dipole moment by symmetry even though the N–F and lone-pair moments are expected to be large (electronegativities 3.04, 3.98).

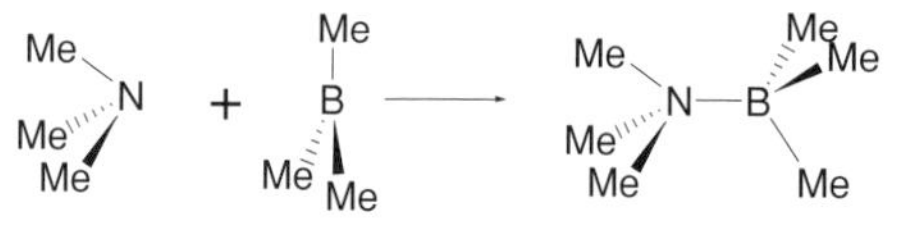

Figure A1.16

A1.4.5 Donor–acceptor complexes

Consider once more the EH_n compounds. Those of group 14, EH_4, are saturated in the sense that there are neither extra valence electrons nor orbitals unused. Those of group 15, EH_3, possess "extra" electrons as a lone pair, whereas those of group 13, when formulated with the simple valence rules employed so far as EH_3, possess an "extra" orbital. The species have complementary electronic properties and can form an adduct with an N–B two-center–two-electron bond in which the two electrons originate from N as shown in Figure A1.16 for the permethyl derivatives. The bond formed is analogous to a C–C bond, albeit not as strong, and the compound is isoelectronic with hexamethyl ethane. As a result of adduct formation the B center now satisfies the eight-electron rule. The amine is an example of a Lewis base and the borane a Lewis acid.

The formation and decomposition of donor–acceptor complexes constitute a large chunk of reaction chemistry. Based on the discussion of polar bonds above, one might jump to the conclusion that the N atom in $(CH_3)_3N–B(CH_3)_3$ will be positively charged and the B atom negatively charged. However, the charge redistribution is non-intuitive and electron density moves from the methyl groups of N to the methyl groups of B such that the overall charges on N and B remain close to those found in the monomeric species. Another interesting aspect is that the base–B donor–acceptor bond strength depends strongly on the nature of the Lewis base and vice versa. Hard–soft acid–base theory provides a rationale for understanding the factors that affect the strength of the interaction. There are important practical consequences. For example, a species like $THF.BH_3$ with a very weak O–B bond acts as a surrogate for free BH_3. It is packaged as a 1M solution and sold as a reagent whereas free borane itself has a strong tendency to dimerize. "Lightly stabilized" complexes of this type are important synthons in both main-group and transition-metal chemistry.

Exercise A1.6. Consider the following compounds as Lewis acid–base adducts and identify the Lewis acid(s) and the Lewis base(s): CH_3CNBF_3, $[SiF_6]^{2-}$, $[I_3]^-$.

Answer. The lone pair on the N atom of CH_3CN permits it to act as a base and the formally empty B 2p orbital on BF_3 constitutes a site of Lewis acidity. According to the octet rule, SiF_4 should be the stoichiometry of the binary compound. $[SiF_6]^{2-}$ would (and is) formed by reaction with two F^-. Clearly F^- can only function as a Lewis base so that leaves us with SiF_4 as the Lewis acid. How does one explain the bonding? See Problem 2 at the end of the Chapter 1 which deals with SF_6. Again, the simple ideas of bonding would give us I_2 as the stable homonuclear compound. $[I_3]^-$ must be formed from it by reaction with I^- which clearly is a Lewis base. Hence, I_2 must be the Lewis acid.

A1.4.6 Transition metals

At last we permit a discussion of transition metals in molecular compounds. Why the delay? Well, look at the electron configuration of the valence shell of a typical metal, Fe. If we blithely take the main-group perspective then its $(4s)^2$ valence configuration should

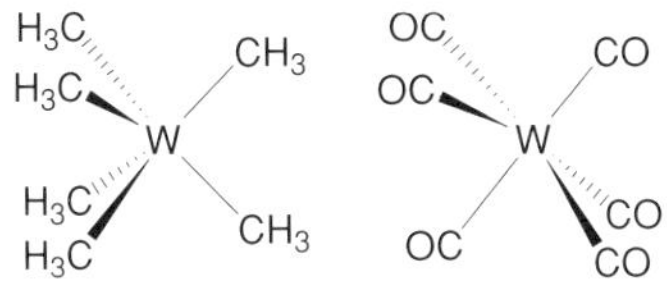

A1.4 Heteroatomic substances

Figure A1.17

make it behave like Be. Whoa, hold on, the 3d shell is at higher energy than the 4s. Therein lies both the problem and the fascination of the transition metals. Moving to Zn with a stable, filled, spherically symmetric d shell, one finds a loose analogy between its chemistry and that of group-2 metals. But in the middle of the transition series, compounds like $W(CH_3)_6$ have been isolated. These are vaguely reminiscent of group-14 species in the sense that six metal orbitals and six metal electrons generate six W–C single bonds in a distorted trigonal-prismatic complex (Figure A1.17). The same metal, W, employs six empty orbitals to accept six electron pairs from six CO ligands to generate $W(CO)_6$. The six metal-based electrons now fill three metal orbitals as "lone pairs" which are not, however, stereochemically active (the structure is octahedral and based on six bond pairs if one takes the VSEPR-model perspective). Apparently the tungsten alkyl is a 12-electron complex whereas the tungsten hexacarbonyl is an 18-electron complex. What gives?

$W(CO)_6$ is not the problem as it obeys a rule (18-electron rule) analogous to the eight-electron rule. In fact, the stoichiometry of numerous compounds in organometallic chemistry establishes the usefulness of the 18-electron rule suggesting that filled d, s and p shells lead to stable compounds. But life is even more complicated. In addition to $W(CH_3)_6$ and $W(CO)_6$, we can find octahedral complexes like $WBr_4(MeCN)_2$ which are also formulated as coordination compounds. The metal atom is considered $[W]^{4+}$, i.e., oxidation state W^{IV}, the Br atoms are considered Br^- which serve as two-electron Lewis bases as do the MeCN ligands. Hence, the W center is associated with 14 electrons not 12 or 18. Perhaps the high metal charge stabilizes the d, s, p valence set such that not all orbitals need be filled.

Well, OK, but what then is $W(CH_3)_6$ – a W^{VI} donor–acceptor complex with six $[CH_3]^-$ ligands or an ordinary covalent compound like CH_4 with six W–C two-center–two-electron bonds? No matter how you count it still has 12 valence electrons. You might as well know that it's a bit like a religion now, counting electrons. People get pretty passionate about partitioning electrons between metal and ligands even when they give lip service to the fundamental tenets of quantum chemistry that require electronic structure (geometric information on where the electrons are) to be understood in non-classical terms quite in contrast to geometric structure (nuclear positions) which can be discussed in comfortable classical terms. Consequently, neither method of partition tells the true story, i.e., the real charge of W probably lies somewhere between 0 and +6. On the other hand, there are large areas of metal chemistry where each view can be useful as a guide. The 18-electron rule rules in organometallic chemistry, but does not in classical coordination chemistry where the first step in analyzing the bonding in a compound is definition of the oxidation state of the metal. Beware of organometallic compounds with electronegative ligands as they can display aspects of Werner chemistry (paramagnetism). Complicated, yes, but the seemingly unlimited variations are a feature of transition-metal chemistry that experienced chemists find extremely exciting. We can anticipate that transition metals will complicate cluster chemistry in the same ways they complicate the chemistry of mononuclear compounds. In the following we outline some of the ways in which transition-metal compounds differ from those of main-group compounds.

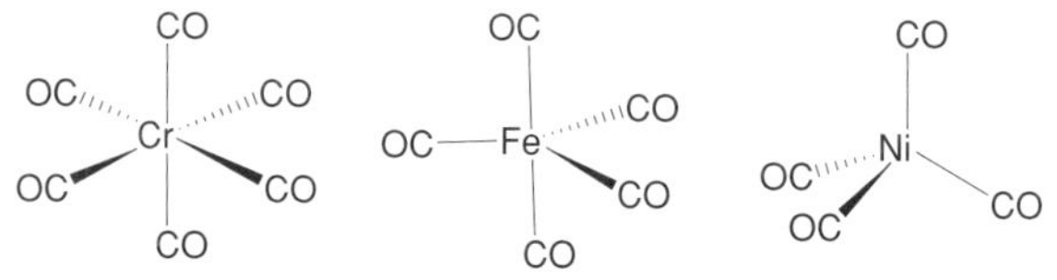

Figure A1.18

A1.4.7 Organometallic chemistry

Large books and multivolume series have been written on the varied aspects of organometallic chemistry for metals from the s-, p- and d-blocks. This brief treatment will be focused on compounds containing the types of d-block metal fragments often found in cluster compounds. An organometallic text is an excellent companion if additional information is needed by an individual reader.

Although the first metal compounds containing M–C bonds were prepared in the 1800s, it was only about 50 years ago that the understanding of metal complexes containing unsaturated organic ligands coordinated side-on to a metal center sparked an explosion of new chemistry. Ferrocene is the iconic compound of the area with all ten C atoms bound to a single Fe center but with a dynamic mobility characterized by the appropriate appellation "ring whizzer." But let's begin with Hieber's metal carbonyls that arrived on the scene in the first half of the twentieth century and then work our way back into π complexes. Fragments derived from both types of compounds are important in metal-cluster chemistry.

For metal carbonyls, the 18-electron rule is an effective guide to compound stoichiometry. As CO is a two-electron donor when bound via C alone, neutral mononuclear compounds are found only with metals with even numbers of electrons, i.e., groups 6, 8, 10 generate octahedral $M(CO)_6$, trigonal bipyramidal $M(CO)_5$ and tetrahedral $M(CO)_4$ complexes, respectively (Figure A1.18). Note that we are not counting bond pairs and lone pairs in the discussion of these structures. Metal carbonyls from groups 5, 7 and 9 need to be charged, e.g., octahedral $[V(CO)_6]^-$ and $[Mn(CO)_6]^+$, dimeric with a shared M–M two-electron bond, e.g., $(CO)_5Mn–Mn(CO)_5$, or associated with a one-electron ligand like H, e.g., $(CO)_5MnH$. Of course there is always the oddball, e.g., $V(CO)_6$ with 17 electrons, to keep life interesting.

An important aspect of the CO ligand, as well as other related organic ligands, is that it functions as a Lewis base and a Lewis acid at the same time. That is, the primary donor orbital has σ symmetry relative to the M–C bond axis but there are low-lying empty acceptor orbitals of π symmetry available as well. The former complements an empty σ metal orbital whereas the latter complements filled π metal orbitals. Hence, the CO ligand can form a strong bond with the metal via the double donor–acceptor interaction and electronic charge build up on the metal is minimized. In a sense the π orbitals of the CO ligand function as the methyl groups on B in the donor–acceptor adduct $(CH_3)_3NB(CH_3)_3$ discussed above. If one tries to make $W(NH_3)_6$, for example, there is a problem as the amine ligand is a good σ donor but a very poor π acceptor. A more detailed molecular-orbital model for a M–CO interaction will be reviewed below. The CO ligand acts as a two-electron donor when it is bound as a terminal ligand (one M–C interaction) or as a bridging ligand (μ- with two M–C interactions and μ_3- with three M–C interactions). So the two isomeric forms of cobalt carbonyl, $(CO)_8Co_2$ and $(CO)_8(\mu-CO)_2Co_2$ count exactly the same way.

The cyclopentadienyl ligand, $Cp^- = [C_5H_5]^-$ also has a significant presence in cluster chemistry. It is to organometallic chemistry what a polydentate ligand is to classical

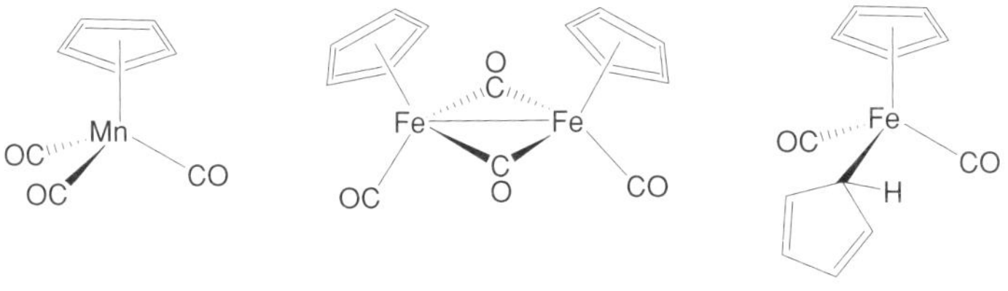

A1.4 Heteroatomic substances

Figure A1.19

coordination chemistry. Formally it occupies three coordination positions and donates three pairs of electrons if considered a negative ligand and five electrons if it is considered to be neutral. It is not as good an acceptor as CO and some of its metal derivatives fall in the crack between 18-electron organometallic complexes and Werner complexes, e.g., diamagnetic 18-electron Cp_2Fe vs. paramagnetic 16-electron Cp_2Cr with two unpaired electrons.

If one considers Cp a five-electron ligand then $CpM(CO)_n$ will be mononuclear for groups 15, 17 and 19, e.g., $CpMn(CO)_3$, but dimeric, charged, or contain hydrides for 16, 18 and 20, e.g., $[CpFe(CO)_2]_2$, $[CpFe(CO)_2]^-$ or $CpFe(CO)_2H$ (Figure A1.19). It is also possible for Cp to act as a three- or one-electron ligand; hence, it is necessary to specify the hapticity (number of metal-bonded C atoms) in a complex, e.g., $(\eta^5\text{-}Cp)Fe(CO)_2(\eta^1\text{-}Cp)$ contains one Cp ligand bound through five C atoms (η^5-five-electron donor) and one Cp ligand bound through one carbon atom (η^1-one-electron donor). Eight electrons from Fe and four from the two CO ligands give a total of 18. Alternatively, if the Cp ligand is viewed as an anion, then the count is $6 + 2 + 6 + 4 = 18$, the difference being Fe^{II} with six electrons. It's all a question of how one wants to partition the electrons. *Caution*: don't switch horses in midstream!

For the group-10 metals and the heavier group-9 metals, 16-electron complexes are often observed. One of the valence orbitals on the metal is unoccupied and the ligands adopt a square-planar arrangement around the metal center. Thus, two very famous complexes, Wilkinson's complex $(PPh_3)_3RhCl$ and Vaska's complex *trans*-$(PPh_3)_2(CO)IrCl$ are 16-electron complexes with four-coordinate, square-planar metal centers. However, the addition of small molecules concomitant with bond rupture readily takes place, e.g., the oxidative addition of H_2 to Wilkinson's complex yields an 18-electron dihydride *cis*-$H_2(PPh_3)_3RhCl$.

Exercise A1.7. Hieber found that CO forms three compounds with Fe in which the ratios of CO to Fe are 5, 4.5 and 4. On the basis of the 18-electron rule predict the molecular formulae of these three compounds.

Answer. $Fe(CO)_5$ ($8 + 5 \times 2 = 18$); $Fe_2(CO)_9$ ($2 \times 8 + 9 \times 2 = 34$ or 17 per Fe to which one is added for the shared formal Fe–Fe bond); $Fe_3(CO)_{12}$ ($3 \times 8 + 12 \times 2 = 48$ or 16 per Fe to which two is added for two shared formal Fe–Fe bonds if the structure is a cyclic trimer).

In summary, the rare-gas rule provides a basis for understanding the stoichiometries of a large fraction of organometallic compounds and offers a guideline for predicting likely compound types for a given set of ligand/metal combinations. Organometallic chemistry can be considered a perturbation of either organic or inorganic chemistry depending on the point of view of the practitioner. The action of the metal fragment on the organic ligand or the organic ligand on metal properties generated a revolution in synthetic organic and inorganic chemistry and reached its apex in the rational development of

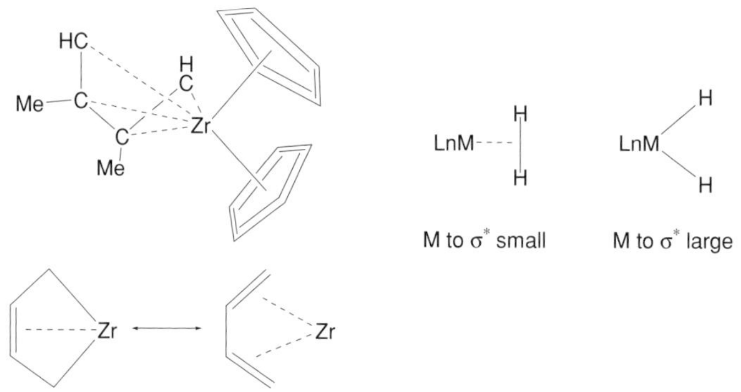

346   *Appendix*

Figure A1.20

homogeneous catalysts. Clearly it is desirable to understand the electronic origins of these perturbations and to do so one must go beyond compound composition and geometric structure. That is, one must go beyond the 18-electron rule.

In trying to do so it quickly becomes clear that the two-center–two-electron bond model needs to be patched up so often that it almost seems easier to just discard it and try for something better. Look at a couple of compounds and see what the nature of the problem is. Consider the dimer [CpFe(CO)$_2$]$_2$. In counting electrons for each metal center we have (considering all the ligands neutral) five from Cp, eight from Fe, four from two CO to give 17. An Fe–Fe single bond is assumed giving one electron to each metal center to make up 18. However, the structure (Figure A1.18) exhibits two bridging CO ligands and two terminal CO ligands. The former eliminate the necessity of a direct Fe–Fe bond in order to explain its diamagnetic character. Quite a few journal pages have been consumed dealing with the question: is there a direct Fe–Fe bond or are the metal atoms electronically coupled via the CO bridge interactions? It turns out there is no substantive Fe–Fe bond and that the metal–metal interaction is indeed mediated by the bridging ligand. Thus, we are faced with a situation not explained by a simple two-center–two-electron bond. However, qualitative molecular-orbital methods can easily deal with it. It stands to reason then that when we must deal with a transition-metal cluster with bridging CO ligands, the manner in which we satisfactorily count the electrons contributed to the bonding network need not imply any understanding of the actual bonding taking place between the metal atoms the CO spans.

Consider one more example where the facts show that a better model is needed to discuss the nature of the ligand–metal interaction. 1,3-butadiene forms a stable complex with the Cp$_2$Zr fragment and exhibits the geometric structure shown in Figure A1.20. One can now compare the C–C distances in the free ligand with those in the complex. The C=C distances increase on coordination (1.36 to 1.45 Å), whereas that of the central C–C bond decreases (1.45 to 1.40 Å). In addition the Zr–C distances are not equal with those to the terminal C atoms being distinctly shorter than those to the internal C atoms (2.30 vs. 2.59 Å). An explanation in terms of resonance with contributions from the two resonance forms shown in Figure A1.20 can be proffered; however, as already mentioned, resonance is only introduced to patch up deficiencies in the two-center–two-electron model. This butadiene binding problem is a more complex variation of the problem of how to describe the olefin–metal interaction in (CO)$_4$Fe(η^2-C$_2$H$_4$). Two limiting models exist: a metallacyclopropane with one C–C and two Fe–C single bonds and tetrahedral C atoms and a π complex with one C=C donor–Fe acceptor bond and trigonal C atoms. Based on

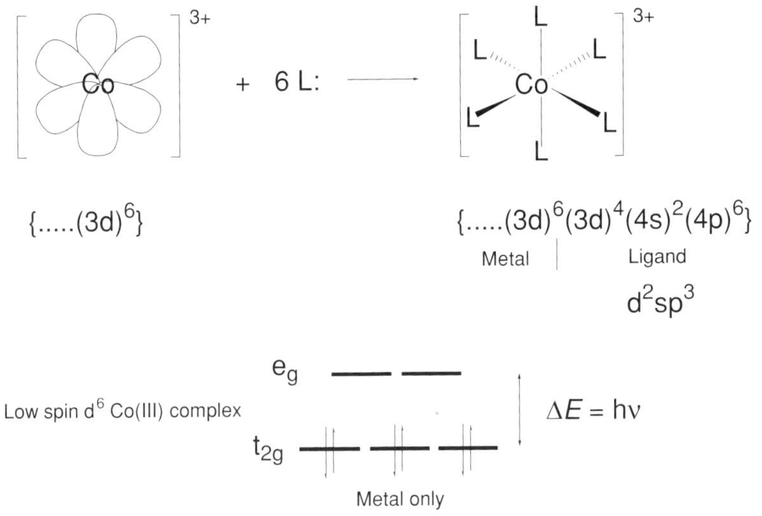

A1.4 Heteroatomic substances

$$[Co]^{3+} + 6\,L: \longrightarrow [CoL_6]^{3+}$$

$\{\ldots(3d)^6\}$ $\{\ldots(3d)^6(3d)^4(4s)^2(4p)^6\}$

           Metal | Ligand

            d^2sp^3

Low spin d^6 Co(III) complex e_g — — $\Delta E = h\nu$

          t_{2g} ⥮ ⥮ ⥮

          Metal only

Figure A1.21

measured geometric parameters (C–C distance and pyramidalization at C), known compounds span the range between the limiting models. A single model accommodating this variation and relating it to metal, olefin substituents and metal ancillary ligands is necessary to advance beyond a purely phenomenological understanding. The Dewar model based on ligand to metal σ donation and metal to ligand π (back) donation, just over half a century old now, does just that.

The same qualitative model describes the binding of H_2 to transition-metal centers in which the extent of metal to H_2 antibonding orbital (back) donation describes the continuous variation of H–H distance from that of coordinated H_2 to that of two individual hydride ligands (Figure A1.20). Although the Dewar model was there in the literature, it took the more recent experimental work of Kubas and those who followed him to demonstrate the realities of these so-called σ complexes of H_2.

A1.4.8 Werner complexes

Our understanding of classical metal-coordination chemistry developed well before organometallic chemistry blossomed in the 50s and 60s. It constitutes part of the core of any first course in inorganic chemistry and is usually introduced in first year general chemistry courses as well. Metal clusters are commonly derived from and related to mononuclear organometallic complexes, not Werner complexes. On the other hand, there is no clear dividing line between Werner and organometallic complexes and some consider the latter as a subset of the former. Consequently, typical characteristics of classical coordination chemistry, e.g., paramagnetism and multiple oxidation states, creep into cluster chemistry, e.g., the cubane clusters that will be discussed later in Chapter 5. Thus, the reader must be reminded of the principles that underpin classical coordination chemistry and a short summary follows.

Perhaps the best way to proceed is to simply list the principal properties of a typical Werner complex that needed to be explained: (a) "salts" that contained labile and non-labile anions and solvent; (b) coordination numbers and geometries not four and not tetrahedral; (c) absorption of visible light that depends on metal and ligand type; (d) paramagnetic compounds and variable metal oxidation state and (e) ligand lability that depends strongly on metal and oxidation state. Property (a) was explained by viewing metal ions as poly-Lewis acids and the anions or solvent molecules as mono- or polydentate Lewis bases (Figure A1.21). Property (b) was first addressed by selective utilization of the valence orbitals of the metal center (hybridization with two-center–two-electron donor–acceptor bonds as also shown in Figure A1.21). This model, in which the valence electrons are partitioned artificially as shown, allowed the paramagnetism of some compounds to be rationalized albeit with pragmatic assumptions that better theories showed unnecessary.

Property (c) and (d) demanded a better theory of bonding than the contemporary one being used for C chemistry. The first edition, crystal-field theory, is a concept transferred from a successful approach to a solid-state problem – one of several such inter-area "technology transfers" that molecular chemists do not always acknowledge. For a known geometry the characteristic splitting associated with the metal-ion d levels in a field defined by the set of negative ligands permitted visible colors, as well as the number of unpaired electrons and trends in selected energetics to be rationalized. For example, in the lower symmetry of the O_h point group, the 3d orbital set of Co^{3+} splits into a two over three pattern. Provided the splitting is large relative to the pairing energy, then Hund's rule no longer applies and the low spin configuration shown in Figure A1.21 holds. Geometry, metal identity, metal oxidation state and ligand character were thereby neatly tied together. Ligand effects were phenomenologically explained in terms of the spectrochemical series and therein another problem arose. Relative positions of ligands, such as, e.g., $[F]^-$ and $[CN]^-$, were not easily rationalized on the basis of an ionic model.

The structural strengths of the hybridization model were combined with the electronic strengths of the crystal-field model in a molecular-orbital model albeit with the loss of the simplicity of the earlier models. The essential aspects of this MO model will be discussed in Chapter 1. The key point here is that, if one wishes to understand the electronic structure of metal-coordination compounds, one need go beyond the Lewis model of two-center–two-electron bonds. It should be obvious, then, that this is also a requirement for organometallic complexes, metal clusters and extended solid-state systems containing metal atoms.

Additional reading
Section A1.1

Cotton, F. A. and Wilkinson, G. (1972). *Advanced Inorganic Chemistry*, 3rd edn. New York: John Wiley.
Porterfield, W. W. (1984). *Inorganic Chemistry*. Reading, MA: Addison-Wesley.
Greenwood, N. N. and Earnshaw, A. (1984). *Chemistry of the Elements*. Oxford: Pergamon Press.

Section A1.4

Elschenbroich, C. and Salzer, A. (1989). *Organometallics*. New York: VCH.
Mingos, D. M. P. (2001). *J. Organomet. Chem.*, **635**, 1.

Problem answers

Chapter 1

1. (a) Approach: Cr possesses six valence electrons, thus ligands must be chosen to supply 12 more to satisfy the 18-electron rule, e.g., six CO ligands; Mn possesses seven valence electrons, thus ligands need to supply 11, e.g., 5 CO + 1 H. (b) This problem requires recognition of the types of bonds in the molecule and which ones, if any, are unusual. The answer to part (a) suggests the Cr–CO bonds can be adequately described as two-center donor–acceptor bonds. The Cr–H–Cr with a two-coordinate H atom is clearly the unusual situation; hence, a fragmentation into two $Cr(CO)_5$ fragments and an H^- anion is appropriate. A $Cr(CO)_5$ fragment is a 16-electron species with an empty orbital available to accept an electron pair. The H^- anion possesses one filled orbital; hence, the three orbitals (two from the metal fragments and one from H) can be used to form one bonding, one non-bonding and one antibonding three-center orbital with the bonding combination containing the two available electrons.

2. S contributes four AOs and the six H atoms contribute six AOs for a total of ten leading to ten MOs. The central atom has functions of symmetry a_{1g} (3s) and t_{1u} (3p), whereas the six ligand functions have symmetry-adapted combinations of a_{1g} (3s) and t_{1u}, and e_g just like an octahedral transition-metal complex. As shown below, interaction between the central atom and the ligands generates four bonding MOs plus their antibonding partners and two non-bonding orbitals having ligand character only for a total of ten MOs.

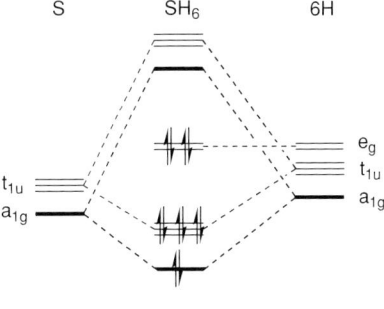

Problem 1.2

3. (a) The MO diagram for this ten-electron diatomic molecule is related to that of isoelectronic N_2 and CO except that the difference in electronegativities of the two elements is even larger than that for CO.

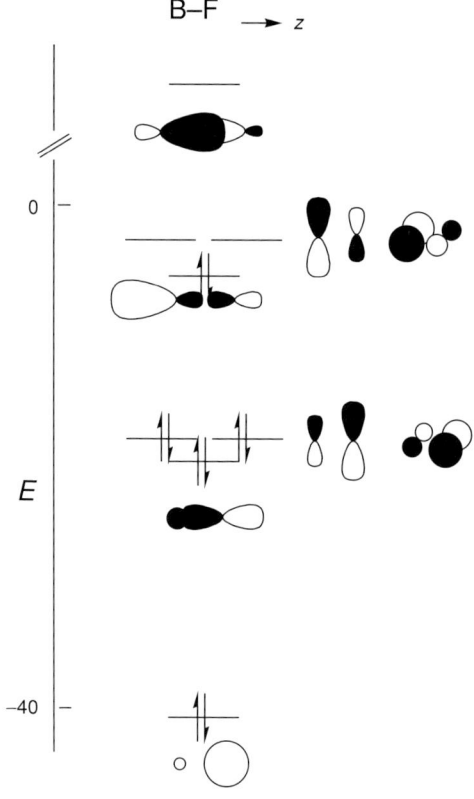

Problem 1.3

(b) The HOMO is of σ symmetry relative to an M–L bond axis and orbital amplitude lies predominantly on the B atoms; hence, as a ligand one would expect it to bind to the metal through B. The LUMO is doubly degenerate and of π symmetry relative to an M–L bond axis and again the orbital amplitude is largely on B suitable for accepting electrons from a filled metal t_{tg} set. The analogy with CO is clear.
(c) The charge distribution suggests a dipole with the negative end at F.
(d) The σ interaction yields a total overlap population of $0.08 + 0.19 = 0.27$ for the s and p interactions, plus the two π interactions add 0.22 each. Formally the bond order is three; however, without additional information the strength of the bond cannot be judged. A similar style calculation on CO yields 0.68, 0.47, 0.47 suggesting BF has a weaker covalent interaction.
4. The series Cl_2, SCl_2, PCl_3, $SiCl_4$ might suggest $NaCl_7$ or $[NaCl_4]^{3-}$ with eight electrons associated with the central atom. However, this is not observed. For the NaCl molecule, the MO diagram shown below obtains. With a total of eight valence electrons, it can be compared with the Si_2 homonuclear diatomic molecule. One obvious difference is the disparity of the AO energies for the two bonding partners. In

Si$_2$ they are a perfect match, but in NaCl the AO energies of the electropositive atom are much higher than those of the electronegative atom thereby making the 3p functions of Na effectively inaccessible.

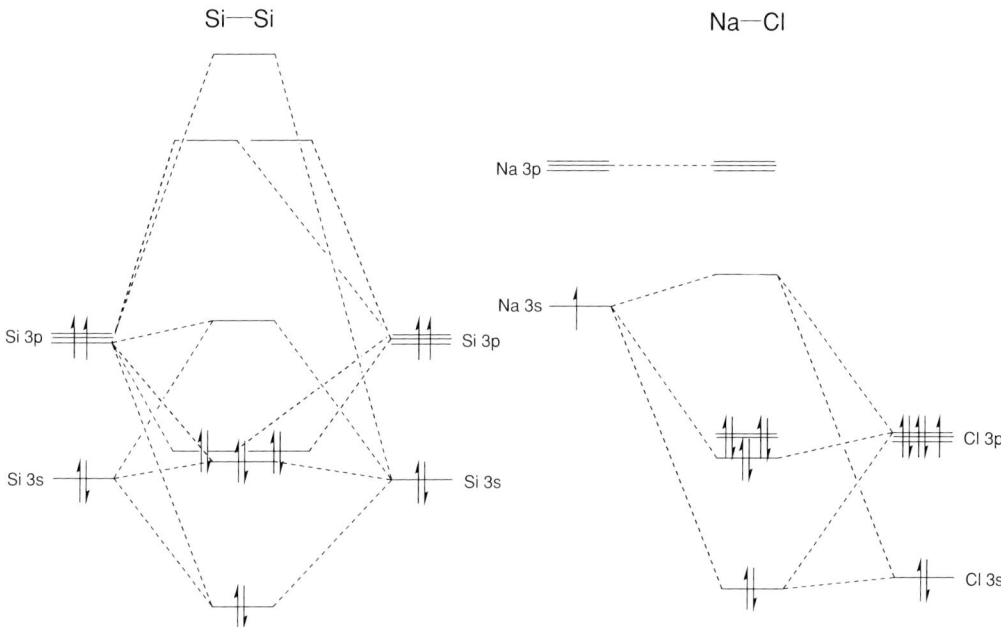

Problem 1.4

5. (a) The HOMO and LUMO of H$_2$ are shown in Figure 1.1.
   (b) According to the frontier-orbital concept, the LUMO of the acid must interact with the HOMO of H$_2$. A side-on structure should maximize overlap.
   (c) Removal of one CO from Cr(CO)$_6$ generates an empty orbital; hence, Cr(CO)$_5$ is primarily a Lewis acid like BH$_3$ and H$_2$ should bind in a similar fashion, i.e., side-on. The LUMO of H$_2$ has π symmetry relative to the Cr–H$_2$ axis, hence it has the proper symmetry to interact with the filled t$_{2g}$ set of the metal center. As with CO, the double acid–base interaction strengthens the M–H$_2$ bond such that metal complexes can be isolated whereas BH$_5$ and isoelectronic [CH$_5$]$^+$ are species with short lifetimes.
6. The four π MOs of free butadiene are shown below with the C$_4$H$_8$ framework in the plane of the paper. The HOMO is bonding between the outer pairs of C atoms and antibonding between the central C atoms whereas the LUMO is bonding between the central C atoms and antibonding between the outer pairs of C atoms. Effective transfer of electrons from the former to the latter will decrease the net bonding between the outer pairs of C atoms and increase it between the central C atoms.

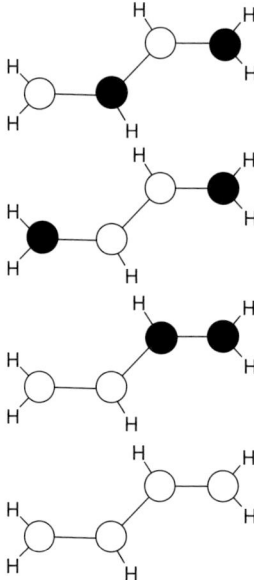

Problem 1.6

Chapter 2

1. The molecular formula for a hydrocarbon with the "butterfly" structure of tetraborane is C_4H_6. The structure with two double bonds is 1,3-butadiene shown alongside. Is it a surprise then that the structure shown below was once proposed for B_4H_{10}, i.e., two protonated double bonds in a *trans* arrangement?

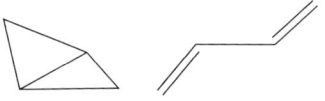

Problem 2.1

2. The appropriate fragment analysis is n SnR fragments each of which possess three orbitals and three electrons perfectly set up for forming three Sn–Sn bonds. Hence, likely shapes are the three-connect clusters of Figure 2.3, i.e.,

Problem 2.2

3. 1,2-$C_2B_3H_5$, 2 CH + 3 BH = $(2 \times 3 + 3 \times 2)/2 = 6$ sep appropriate for a trigonal-bipyramidal shape with the CH fragments in adjacent axial and equatorial

positions (non-zero dipole); 1,6-$C_2B_4H_6$, 2 CH + 4 BH = $(2 \times 3 + 4 \times 2)/2 = 7$ sep appropriate for an octahedral shape with the CH fragments in *trans* positions (zero dipole); 2-Cl-1,6-$C_2B_4H_5$, 2 CH + 3 BH + 1 BCl = $(2 \times 3 + 3 \times 2 + 1 \times 2)/2 = 7$ sep appropriate for an octahedral shape with the CH fragments in *trans* positions (the Cl substitution in the 2-position produces a non-zero dipole); 2,4-$C_2B_5H_7$, 2 CH + 5 BH = $(2 \times 3 + 5 \times 2)/2 = 8$ sep appropriate for a pentagonal-bipyramidal shape with the CH fragments in non-adjacent equatorial positions (zero dipole); 1,2-Me_2-1,2-$Si_2B_{10}H_{10}$, 2 SiMe + 10 BH = $(2 \times 3 + 10 \times 2)/2 = 13$ sep appropriate for an icosahedral shape with the SiMe fragments in adjacent positions (non-zero dipole); $SB_{11}H_{11}$, 1 S + 11 BH = $(1 \times 4 + 11 \times 2)/2 = 13$ sep appropriate for an icosahedral shape (non-zero dipole); $[AsB_{11}H_{11}]^-$, 1 As + 11 BH + (−) = $(1 \times 3 + 11 \times 2 + 1)/2 = 13$ sep appropriate for an icosahedral shape (non-zero dipole).

4. (a) 2,3-$C_2B_4H_8$, 2 CH + 4 BH + 2 H = $2 \times 3 + 4 \times 2 + 2 = (6 + 8 + 2)/2 = 8$ sep appropriate for a pentagonal-bipyramidal shape with one unoccupied axial vertex, adjacent equatorial CH fragments and two BHB bridging H on the open pentagonal face; 1,2-$C_2B_4H_6$, see Problem 3; 1,5-$C_2B_3H_5$, see Problem 3; B_4H_{10}, 4 BH + 6 H = $(4 \times 2 + 6)/2 = 7$ sep appropriate for an octahedral shape with two adjacent unoccupied vertices ("butterfly" shape), four BHB bridging H and two extra terminal BH at the "butterfly" wing tips; $C_3B_3H_7$, 3 CH + 3 BH + 1H = $(3 \times 3 + 3 \times 2 + 1) = (9 + 6 + 1)/2 = 8$ sep appropriate for a pentagonal-bipyramidal shape with one unoccupied axial vertex, BH in the axial position, adjacent BH fragments in equatorial positions and one BHB bridging H.
   (b) The pentagonal pyramid requires eight sep and, for a *nido*-structure, six fragments to generate this electron count. With a maximum of four BH fragments, adding SiH and P plus two BHB gives $(4 \times 2 + 3 + 2 + 2)/2 = 8$ sep. Place one BH in the apical position and two BHB bridging H on the open pentagonal face.

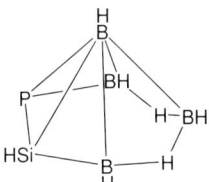

Problem 2.4

5. The sep $= 9 = n + 1$: $n = 8$. Removing two adjacent five-connect vertices from the parent deltahedron (four four-connect and four five-connect vertices) yields the framework shown (as well as flattened out for convenience at the bottom left) which has C_2 symmetry (pairwise equivalence of Bs and Hs) and therefore consistent with the observed spectroscopic data.

   To analyze the bonding first remove the four BHB bridging H as protons to give $[B_6H_8]^{4-}$. This species has a total of 32 AOs and 30 valence electrons. If we remove 16 orbitals and 16 electrons to form the eight two-center–two-electron B–H bonds, we are left with 16 orbitals and 14 electrons. Hence, two three-center–two-electron bonds (six orbitals and four electrons) plus five two-center–two-electron bonds (ten orbitals and ten electrons) nicely utilizes them all and is consistent with the fact that there are two four-connect vertices in the framework.

6. The boundary conditions are: one donor fragment per acceptor fragment, acceptors and donors at right angles and a sep count of $n + 1$. Adjacent functional fragments on

354 Problem answers

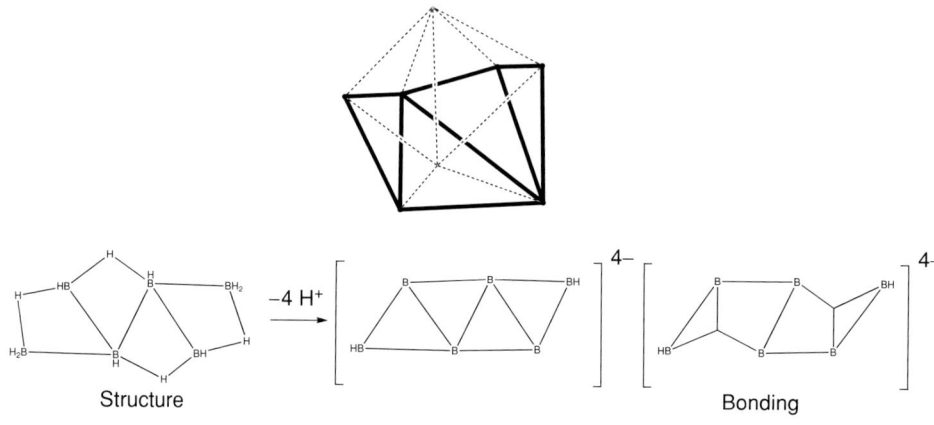

Problem 2.5

an octahedral cluster framework meet the criteria. A protected acid site is a three-electron B–THF fragment whereas a base site is a three-electron P atom with a lone pair. One could choose to place one each on a single cluster or make two different clusters each with two acid or two base sites. Hence, 1-P-2-(BTHF)B_4H_4 with seven sep and adjacent heteroatom fragments constitutes an appropriate target.

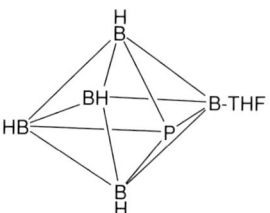

Problem 2.6

7. (a) Two six-connect and six four-connect vs. four five-connect and four four-connect.
   (b) A hexagon may be constructed from equilateral triangles; hence, such a bicapped-hexagonal shape would have the apical atoms superimposed.
   (c) The greater the uniformity of the vertex connectivities the more spherical the shape, i.e., the dodecahedron is more spherical.

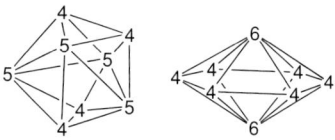

Problem 2.7

8. Of the five symmetry-adapted combinations of the out of plane π orbitals of $[B_5H_5]^{4-}$, the three lowest-lying ones (four electrons) have the proper symmetry to interact with the σ and π orbitals of the B–H fragment (two electrons) to generate a

six-center–six-electron bond (five empty orbitals over three filled orbitals). The five two-electron B–B bonds plus these three ring-cap bonding pairs gives a total of eight framework bonding pairs equal to the eight sep required by the electron-counting rules.

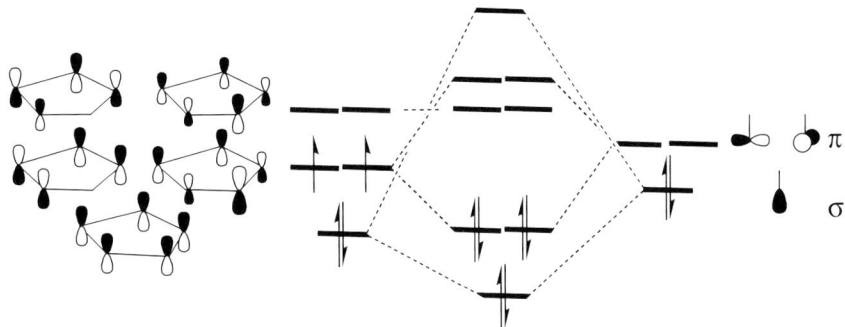

Problem 2.8

9. The anion $[2,3\text{-}C_2B_4H_6]^{2-}$ can be formally generated from $closo\text{-}2,3\text{-}C_2B_5H_7$ by removal of an axial $[BH]^{2+}$ fragment. Working Problem 8 in reverse, but removing $[BH]^{2+}$, generates five orbitals on the open five-membered face containing a total of six electrons.

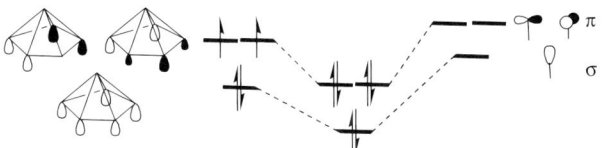

Problem 2.9

10. The structure can be described as a pentagonal bipyramid with two bridging GaR groups which requires eight sep or 30 cve. The mononegative cluster contains four two-electron RGa fragments and three one-electron (possibly three-electron) Ga atoms plus two two-electron RGa bridges. Hence, the sep $(4 \times 2 + 3 \times 1 + 2 \times 2 + 1)/2 = 16/2 = 8$ appropriate for its shape. The cve count comes from seven Ga, four R, one $-$ charge, and two RGa bridges $= 21 + 4 + 1 + 4 = 30$. The "unusual" feature is that the Ga centers in the RGa bridges are six- rather than eight-electron centers. If this cluster were treated according to the general cluster fusion rules of Chapter 3, Section 3.3.3, eight-electron centers would be assumed and the observed count of 34 (9 Ga + 6 R + 1) would be four electrons less than the required count of 38 (30-cve cluster fused to two 18-cve triangles with elimination of two 14-electron dimers).
11. As is a three-orbital–three-electron fragment and AsR or $[As]^-$ are three-orbital–four-electron fragments *or* two-orbital–two-electron fragments. The latter implies some two-connect vertices. Try three-connect or lower structures (Figure 2.3) and you will find the following are consistent with the 1:3:3 ratio of atom environments. Alternatively, the cve = 38, whereas for a three-connect structure one expects $5n = 35$; hence, three of the As units must be two-connect to generate three additional lone pairs to accommodate the extra electrons.
12. (a) 8 Sn + 4 R give $32 + 4 = 36$ valence orbitals and the same number of electrons. However, for an edge-bonded cube and either external lone pairs or bonds, we need

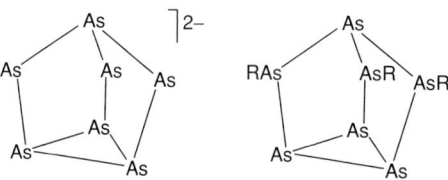

Problem 2.11

$5n/2 = (5 \times 8)/2 = 20$ pairs of electrons. A localized model will not be adequate. (b) Treat the 12 Sn–Sn edges and the four R–Sn bonds as two-center–two-electron bonds (16 total) thereby utilizing 32 orbitals and 32 electrons of the 36 orbitals and 36 electrons available. The remaining four orbitals and four electrons can then be used in four-center bonding of the rectangle formed by the two Sn_2 fragments. The two net bonding combinations of the set of four shown below are filled and the two net antibonding combinations are empty. Note that all four of these MOs have "lone pair" character.

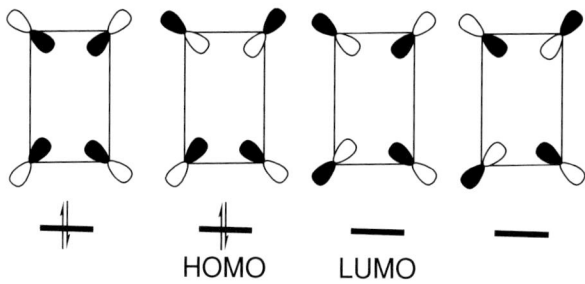

Problem 2.12

13. The molecular formula gives: 5 MeC + Si + (+) = $(5 \times 3 + 2 - 1)/2 = 16/2 = 8$ sep (assuming a lone pair on the Si atom). Hence, the structure should be based on a seven-vertex deltahedron with one vertex unoccupied. The observed structure, a pentagonal pyramid with Si occupying the axial position, is one of the possible *nido*-isomers generated. The implication of this model is that three Si orbitals are engaged in bonding to the C ring as discussed in the ring-cap model of Section 2.9.1.
14. A 13-vertex deltahedron, the geometry exhibited requires $4n + 2 = 54$ cve if the rules for fully ligated clusters are followed. $[Ga_{13}R_6]^-$ has only 46 cve, however. If one applies the approach based on capping a polyhedrane skeleton (Section 2.12.5) we find: five triangles, three rectangles and one pentagon in the skeleton. For the five triangles and four capped faces larger than triangular, we have $(5 \times 2 + 3 \times 6 + 1 \times 6) = 34$ to which we add $(6 \times 2) = 12$ for the external pairs associated with the six R groups for a total of 46 electrons.
15. The number of cluster valence electrons is 66. Based on the examples of metalloid clusters in the text, we don't expect the borane paradigm to work for this closed cluster and it does not ($4n + 2 = 58$ cve). It can be viewed as a hexacapped three-connect (highly distorted) cube giving, via the cluster fusion rule: $5 \times 8 + 6 \times 24 - 6 \times 24$ [or $- 6 \times 22$] = 40 [or 52] cve. This doesn't work and besides, four of the cube edges have very long Sn–Sn contacts. Finally, it can be viewed with the modified Schleyer approach (no lone pairs on Sn atoms) or the Schleyer approach

(lone pairs on Sn atoms). For the first we have: six squares capped by six SnR giving $6 \times 6 + 6 \times 2 = 48$. For the second we have $6 \times 6 + 6 \times 2 + 8 \times 2 = 64$ cve which is within two electrons of the actual count. Although the precise count is not achieved, we have seen (and will see in subsequent chapters) that there are several examples where the actual and model-based counts differ by two electrons. Most such discrepancies arise from idiosyncrasies of the cluster symmetry; hence, the next step here is to look in actual MO calculations for the origin of the difference.

Chapter 3

1. $[HFe_7B(CO)_{20}]^{2-}$: 102 cve. View as a capped trigonal prism $90 + 12 = 102$ or, alternatively, as a square pyramid fused to a rectangular face of a trigonal prism; $90 + 74 - 62 = 102$.

   $Os_7(CO)_{21}$: 98 cve. View as face-fused octahedron and tetrahedron (or capped octahedron) giving $86 + 60 - 48 = 98$.

   $[Ni_9(CO)_{18}]^{2-}$: 128 cve. View as triangular face-fused octahedron and trigonal prism giving $86 + 90 - 48 = 128$.

   $[Fe_4Pt_6(CO)_{22}]^{2-}$: 138 cve. View as tetrahedron edge fused to two five-atom triangular rafts. Each five-atom triangular raft may be obtained by edge fusion of two triangles to another triangle giving $48 + 2(48 - 34) = 76$. Thus, we have $60 + 2(76 - 34) = 144$ which is six higher than observed. The discrepancy can be attributed to the presence of Pt in the cluster thereby generating additional unavailable cluster orbitals.

2. The cve count is 90. Table 3.1 gives one possibility: a trigonal prism. The raft in Figure 3.12 provides another possibility. Note that the variation in the raft structure shown below also has 90 cve.

  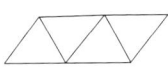

Problem 3.2

3. The cve count is 110 which is too low for either a single deltahedral cluster ($14n + 2 = 114$ cve) or a three-connect cluster ($15n = 120$ cve); hence a capped or fused-cluster system is likely. A bicapped octahedron ($14n + 2 + 2(12) = 110$ cve) fits the bill. Alternatively, we have eight $Re(CO)_3$ fragments $+ C + 2- = (8 \times 1 + 4 + 2)/2 = 7$ sep suggesting a octahedron as the cluster core of a capped system. Two isomers are possible. One has C_{2v} symmetry and four types of metal environments as shown below whereas the other has only two.

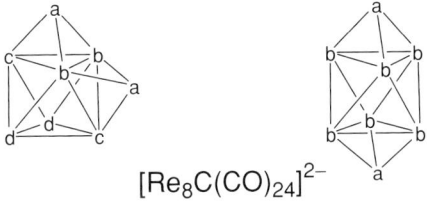

$[Re_8C(CO)_{24}]^{2-}$

Problem 3.3

4. To satisfy the 18-electron rule for each metal center and let each line be a two-center–two-electron bond, the trigonal prism can be constructed out of six 15-electron metal fragments (three orbitals and three electrons) whereas the raft can be constructed out of two 16-electron fragments (at the two-connect vertices, two orbitals and two electrons), two 15-electron fragments (at the three-connect vertices, three orbitals and three electrons) and two 14-electron fragments (at the four-connect vertices, four orbitals and four electrons). Note that the last is not possible for the equivalent borane (30 cve *arachno*-B_6H_{12}) and we saw in Problem 5, Chapter 2, that we needed two three-center–two-electron bonds to provide a localized description of the main-group system.

5. The "stretched" cluster with 14 CO ligands has 76 cve whereas the other has 72. Table 3.1 shows that the latter is the normal count for a trigonal-bipyramidal metal cluster. Hence, we are asked to explain an increase in cve without change in qualitative shape, i.e., we cannot invoke 16-electron Pt as usually this leads to a decrease in cve. The experimental result suggests that the LUMO of the 72-cve trigonal-bipyramidal metal cluster is doubly degenerate, with apical–basal triangle antibonding character and low enough in energy to be occupied, e.g., it becomes the HOMO of the 76-cve cluster. How is this possible? The trigonal bipyramidal cluster has been the topic of considerable discussion (see Burdett and Eisenstein, 1995). One limiting model is bonding via six two-center–two-electron bonds as shown below which would leave no B–B bonding in the equatorial plane. If we use the HOMO and LUMO of $[B_5H_5]^{2-}$ as a guide then the MO version of the crude localized model shows that the LUMO pair contains about 50 % character of in-plane Walsh-type orbitals also shown below. These have equatorial B–B bonding character and would contribute to the bonding of the B triangle if occupied. Why is the LUMO accessible for the metal clusters and not the boranes? We have already seen that the intrinsically weaker M–M vs. E–E bonding makes the HOMO–LUMO gap smaller for the metal systems and consequently the likelihood of counting-rule violations increases. As to the observed structural distortion in going from 72 to 76 cve, Lauher has shown that the energy of the LUMO of a metal-cluster version of $[B_5H_5]^{2-}$ falls as the apical–equatorial plane distance increases at constant equatorial triangle size thereby demonstrating its apical–equatorial antibonding character.

Walsh Orbitals

Problem 3.5

6. The 44 M cluster was successfully counted using the limiting model of a fully radially and surface-bonded internal octahedral cluster surrounded by 38 metal fragments bonded to the octahedral core by radial bonding only (metal-fragment ligands to the central cluster). As this problem involves the same cluster from which 20 metal-fragment "ligands" have been stripped away, a reasonable first approach is the same but with 18 ligands rather than 38. The cve count from the composition is 302 and that from the model is $86 + 12(18) = 302$.

7. 86-cve $Rh_2Os_4(CO)_{18}$ + 18-cve $Fe(CO)_5$ generates a 102 cve intermediate **1** on the way to the 98-cve capped octahedral product. A reasonable structure is the "spiked" octahedron which can be formed by fusing an 86-cve octahedron with a 34-cve dimer losing an 18-cve monomer (86 + 16 = 102). Loss of another CO leads to 100-cve intermediate **2** for which the edged-bridged intermediate is reasonable (86-cve octahedron + 48-cve triangle − 34-cve edge = 100). Loss of another CO and closing the edge bridge to form a cap is the final step.

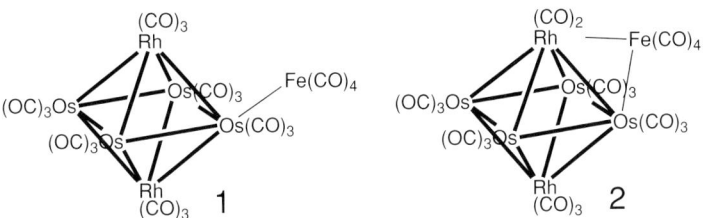

Problem 3.7

8. The cluster stoichiometry gives 6 Re + 8 Se + 6 I + 4− = 6 × 7 + 8 × 4 + 6 × 1 + 4 = 84 cve. Hence, it can be described with a $[Re_6]^{18+}$ core with 24 valence electrons sufficient for 12 two-center–two-electron M–M bonds. Alternatively the MO diagram of Figure 3.17 can be used to describe its qualitative electronic structure.

9. The cluster stoichiometry gives 11 Ru + H + 27 CO + 3− = 146 cve. As a deltahedral cluster the cve count would be $14n + 2 = 156$. Alternatively, the cluster geometry can be generated by fusing a pair of octahedra and a pair of trigonal bipyramids as shown below eliminating a triangle and two butterflies yielding 86 + 86 − 48 + 2(72) − 2(62) = 144 cve. Finally, an octahedral cluster surrounded by five metal "ligands" gives 86 + 12(5) = 146 cve which agrees with the observed composition.

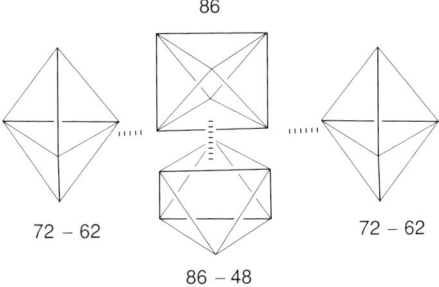

Problem 3.9

10. The tetrahedral intermediate proposed is an attractive one as rupture of one M–M bonding interaction (and some ligand rearrangement) takes one to one isomer or the other depending on the M–M bond chosen. Observe that the cluster $HPtOs_3(CO)_{10}(dppm)\{Si(OMe)_3\}$ and its isomer have planar butterfly structures and electron counts of 60. The typical butterfly structure is not planar; however, bulky ligands can readily flatten the butterfly (see discussion of $[Re_4(CO)_{16}]^{2-}$ in Section 3.2.5). The typical cve count for a butterfly structure is 62; however, with Pt clusters,

cve counts are often two or four electrons lower. Now cve = 60 is the usual count for a tetrahedral metal cluster; hence, even though the planar butterfly is the observed lowest-energy structure, a higher-energy structural alternative is the tetrahedral cluster. Apparently it is accessible via photon absorption and then decays thermally into isomer **2**; hence, isomer **2** is of lower stability. The intermediate for the thermal conversion of **2** to **1** is likely tetrahedral as well but probably has a different ligand arrangement than that proposed for the forward reaction.

Chapter 4

1. SiH is a three-orbital–three-electron fragment and CpRh is isolobal with BH, a three-orbital–two-electron fragment. The *nido*-square-pyramidal cluster requires seven sep; hence, without using any other fragments or an overall cluster charge, only the composition $(SiH)_4(CpRh)$ with $(4 \times 3 + 2)/2 = 7$ sep fits. Two isomers are possible with the metal fragment in the 1- and 2- positions, respectively. The former may also be viewed as an analog of a metal η^4-cyclobutadiene complex.

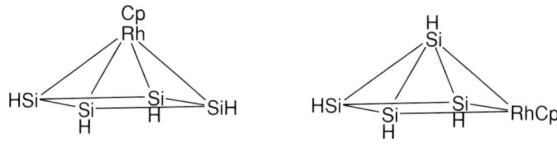

Problem 4.1

2. The $Fe(CO)_4$ fragment is capable of acting both as a fragment with a single acceptor orbital isolobal to BH_3 and as a fragment with two frontier orbitals containing two electrons isolobal with CH_2. BH_3, with a more limited valence set, can act in the former manner but not the latter.

3. The $Fe(CO)_4$ fragment can be isolobal with the CH_2 fragment, hence, the first compound is analogous to cyclopropane. The 15-electron $CpCr(CO)_2$ fragment can act as a one-orbital–minus one-electron, two-orbital–one-electron or three-orbital–three-electron (see Figure 4.8) fragment. Treating the second compound as a cluster we have: $2\ Se + 2\ CpCr(CO)_2 = (2 \times 4 + 2 \times 3)/2 = 7$ sep suggesting a structure based on an octahedron with two occupied vertices. If we choose adjacent vertices the observed structure is generated, i.e., it is analogous to B_4H_{10}. The 16-electron $CpMn(CO)_2$ can act as a one-orbital–zero-electron or two-orbital–two-electron fragment (Figure 4.7). The As atom can be viewed as CH so, considering the metal fragment isolobal with CH_2, we have an analog of an allyl radical (C_3H_5) dimer joined through the central C atoms.

4. The As center in the dichromium compound has an out of plane empty p orbital of the right symmetry to interact with filled d orbitals on the metal center. This three-center–four-electron interaction produces a set of three orbitals: Cr–As–Cr bonding, Cr–As–Cr non-bonding and Cr–As–Cr antibonding, the first two of which are filled. Assume the absorption energy corresponds to the HOMO–LUMO gap in each molecule. In comparing the two compounds, the HOMO energies will be about the same (t_{2g} metal-orbital energy vs. Cr–As–Cr non-bonding), whereas the LUMO in the first case (the e_g^* on the metal caused by two-center ligand–metal donor–acceptor interaction) will be at higher energy than the LUMO on the second because the energy of the Cr–As–Cr antibonding MO will be lower.

5. Assuming each metal center has three filled "t_{2g}" cluster non-bonding orbitals, 4 Cr + 2 CpCr + CO = $(4 \times 3 + 2 \times (-1) + 2)/2 = 6$ sep which is two sep short of the eight required for a *nido*-pentagonal-pyramidal cluster. Delocalized bonding will be favored by metal and main-group fragments of similar electronegativities. i.e., in this case the Ph_4C_4 fragment is best viewed as a complex bridging ligand donating two electrons to the "basal" Cr and four electrons to the "apical" Cr which, with a triple bond, gives 18 electrons at each metal center.
6. First consider the Fe center as an 18-electron metal; hence, the B atom coordinated to it must provide a pair of electrons. As shown in the drawing below the cluster is [C_5Me_5B:], where the external lone pair on B is emphasized in the formula. Now, using the sep formalism we have 5 CR + 1 B: = $(15 + 1)/2 = 8$ sep consistent with its *nido*-pentagonal-pyramidal shape.

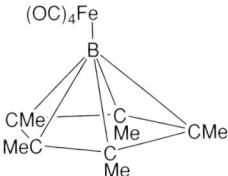

Problem 4.6

7. In the first compound, viewing everything as neutral we have (W + Ru + 3 CO + Cp + dicarbolide + CR + B–H–Ru) = $6 + 8 + 6 + 5 + 4 + 3 + 2 = 34$. At the W center we have W + 2 CO + CR + dicarbolide + W–Ru = $6 + 4 + 3 + 4 + 1 = 18$ and at the Ru center Ru + CO + Cp + B–H–Ru + W–Ru = $8 + 2 + 5 + 2 + 1 = 18$ in agreement with the bonding denoted. In the second compound the total electron count is unchanged; however, the B–Ru bond has replaced the B–H–Ru interaction thereby requiring a C–Ru bond to retain the 18-electron count at Ru. W now has a W=C interaction and picks up a formal negative charge.
8. In the first compound, viewing everything as neutral, we have (W + Mo + 3 CO + Cp + dicarbolide + μ-CR + B–H–Mo) = $6 + 6 + 6 + 5 + 4 + 3 + 2 = 32$. In the second compound the B–H–Mo interaction is simply replaced by a R_3P–Mo two-electron donor leaving the count unchanged.
9. Count: molecular formula = 9 B, 7 H, 2 C, 2 Me, W, 2 Fe, 8 CO, 1 C, 1 Me, 1 − charge = 88 cve; theory = icosahedron + square pyramid + tetrahedron = $170 + 74 + 60 − 2 \times 48 = 208$ for an all metal system − 12×10 for replacing 12 metal atoms with main-group atoms = 88 cve.
10. Considered as a *closo*-pentagonal bipyramid, the compound should possess eight sep; however, with $Mo(CO)_3$ + 5 Pb + 4− = $(4 \times 0 + 5 \times 2 + 4)/2 = 7$ sep assuming each Pb has an external lone pair and the metal fragments have filled "t_{2g}" orbitals. Considered as a triple-decker complex it has $6 + 6 + 0 + 6 + 6 + 4 = 28$ valence electrons and one would expect a Mo–Mo bonding interaction based on the arsenic model mentioned. However, the Mo–Mo distance is clearly non-bonding. In Chapter 2 we saw that the removal of bare-atom lone pairs from the cluster bonding problem is problematical because these orbitals lie at the filled–unfilled orbital interface and can mix with other orbitals of similar symmetry thereby generating unexpected unavailable unfilled orbitals. How this plays out in this case can be found in the literature reference where density functional calculations show strong interactions of

both in plane as well as out of plane Pb_5 fragment interactions with frontier orbitals of the two metal centers.

Chapter 5

1. $Os_6(CO)_{17}S_2$: $6 \times 8 + 17 \times 2 + 2 \times 6 = 94$ cve. Consider it a pentagonal bipyramid fused to a triangle with an all-metal count of $100 + 48 - 34 = 114$ cve. Replacing two metals with main-group atoms yields cve = 94. $Fe_3(CO)_9Sn_2\{CpFe(CO)_2\}_2$: Sn-(CpFe(CO)_2) = Sn–R. 3 Fe(CO)_3 + 2 Sn–R = $(3 \times 2 + 2 \times 3)/2 = 6$ sep which is appropriate for a trigonal bipyramid. $Fe_3(CO)_9(NH)_2$: 3 Fe(CO)_3 + 2 NH = $(3 \times 2 + 2 \times 4)/2 = 7$ sep which is appropriate for an octahedron with one vacant vertex. $(CpCo)_3(BPh)(PPh)$: 3 CpCo + BR + PR = $(3 \times 2 + 2 + 4)/2 = 6$ sep which is appropriate for a trigonal bipyramid. $Mn(CO)_3B_9H_{12}THF$: Mn(CO)_3 + 8 BH + B–THF + 4 H = $(1 + 16 + 3 + 4)/2 = 12$ sep which is appropriate for an 11-vertex deltahedron (Figure 2.7) with the vertex of connectivity six vacant. $Os_6(CO)_{18}P(AuPPh_3)$: Considering the AuPPh_3 fragment as isolobal with H, we have $6 \times 8 + 18 \times 2 + 5 + 1 = 90$ cve which is appropriate for a trigonal prism.

2. $[Rh_9(CO)_{21}P]^{2-}$, monocapped square antiprism with interstitial P; $Fe_3(CO)_9C_2BH_3$, octahedral; $Co_3(CO)_9Bi$, tetrahedral; $HRu_3Fe(CO)_{12}N$, butterfly with interstitial N or trigonal bipyramid depending on the butterfly deltahedral angle; $Fe_3(CO)_{12}(CH)As$, trigonal bipyramid.

3. The Ge–Co(CO)_4 is equivalent to a Ge–R three-orbital–two-electron fragment; hence the cluster has eight sep/68 cve instead of the expected seven sep/66 cve. Hence, the electronic problem is the one dealt with in Section 5.2.1 where octahedral $Fe_4(CO)_{12}(\mu_4\text{-PR})_2$ was considered.

4. $Cp_4Fe_4(CO)_4$: a 15-electron CpFe(CO) fragment is isolobal with CH giving sep = 6. The cve count as a four-metal cluster is 4 Fe + 4 Cp + 4 CO = $32 + 20 + 8 = 60$ cve. Both match the fully bonded metal tetrahedron observed. As a cubane, it has $60 - 48 = 12$ electrons in the metal orbitals thereby just filling the six M–M bonding orbitals (Figure 5.26). $Cp_4Fe_4(C_2H_2)_2$ As a cubane, the four CH fragments are three-electron ligands and 16 electrons occupy the metal orbitals filling two Fe–Fe antibonding MOs as well as the six bonding orbitals. This is consistent with four Fe–Fe bonding distances as observed. However, the C–C distances are also bonding and the alkyne apparently acts as a six-electron ligand. Considered as an eight-atom cluster the reduced cluster has eight sep (4 CpFe + 4 CH = $(4 + 12)/2$) or 72 cve (4 Fe + 4 Cp + 4 C + 4 H = $32 + 20 + 16 + 4$), two less than prescribed by the counting rules. Note that $Cp_4Fe_4(CH)_4$ is isoelectronic to rule-breaking $Cp_4Co_4(BH)_4$ (Chapter 4, Figure 4.11), which suggests both are better viewed as variants of metal cubane clusters.

5. Each vertex may be considered as an octahedral ML_3S_3 metal fragment so that if one ignores the differences in metals and ligands the diagram in Figure 5.26 applies. The total valence electron count is 66 and using 48 for M–L bonding leaves 18 metal electrons to fill the metal orbitals of the cubane. Hence, six M–M bonding and three M–M antibonding orbitals will be filled predicting three M–M bonds in agreement with observations.

6. The cluster has 12 sep (7 Sb + 3 Ni(CO) + 3– = $(21 + 3(0) + 3)/2 = 12$) and ten occupied vertices so would be classified *nido* based on electron count. Its geometry, however, differs from *nido*-geometry which is derived from a 11-vertex deltahedron (Figure 2.13) by removing the vertex of connectivity six. Because it has two square faces it cannot be derived from a deltahedron by the removal of a single vertex. Hence,

Chapter 6 363

it is described in the literature reference as "a new structural type for *nido*-ten-vertex polyhedral clusters."

7. Re(CO)$_5$ is isolobal with CH$_3$ and the MeIn fragment is a three-orbital–two-electron cluster fragment. Re(CO)$_3$ can be viewed as a three-orbital–one-electron fragment giving a total of $(4 \times 2 + 4 \times 1)/2 = 6$ sep appropriate for the tetrahedral cluster upon which the structure is based.

Chapter 6

1. We need to find the COs for the special points in k space (kx, ky); $\Gamma(0, 0)$, M$(\pi/d, \pi/d)$, and X$(0, \pi/d)$. You know that 0 means no change in sign of the function along the coordinate specified and π/d means an alternating sign. Hence, $\Gamma(0, 0)$ will have all B 2p functions in phase along x and y; M$(\pi/d, \pi/d)$ will have a change in sign for every lattice point along x and y; and X $(0, \pi/d)$ will have no change in sign along x and a change at every lattice point along y. The differing consequences for the p_x, p_y and p_z functions are shown in the drawing below. To understand the energy positioning of the various COs label each row–row interaction in terms of bonding or antibonding. Sketch the band structure. Notice that the p_x and p_y functions are not degenerate at X but are at Γ and M.

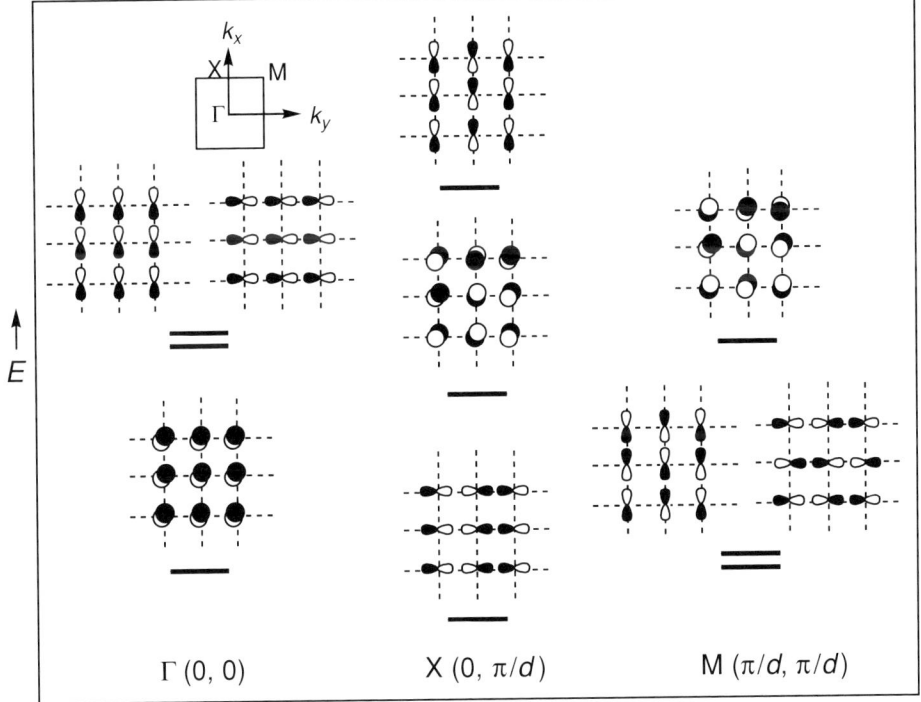

Problem 6.1

2. The π MOs of N$_2$ will interact in the same manner as the p_x and p_y functions in Problem 1. The van der Waals radius of N is 1.55 Å; hence, the intramolecular N$_2$ interactions will be weak and the π bands will be narrow.

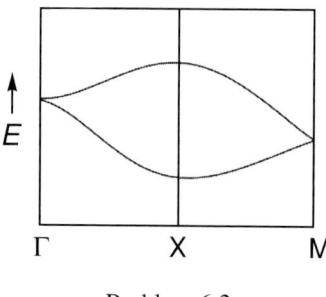

Problem 6.2

3. Exercise 6.8 gives the solutions at the special points for each array separately, i.e., this diagram will be the same for both acid and base albeit at different energies. The band width for both will be small as the distances between acids and bases are large; hence, the DOS will be narrow. The diagram expressing the result is reminiscent of a simple MO diagram where the splitting between the two DOS on interaction is a measure of the strength of the donor–acceptor bond.

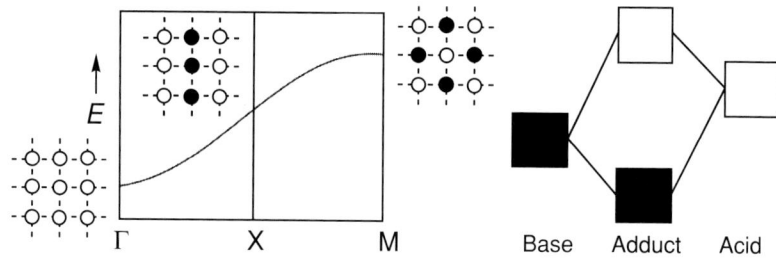

Problem 6.3

4. In a ccp lattice there are two tetrahedral holes per lattice point; hence, the stoichiometry is 1:1.
5. This problem contains elements of Exercises 6.4 and 6.5 that treat infinite CH and BN chains, respectively. Use the same approach. Instead of two π-band pairs there is only one for planar poly-$\{-BHNH-\}$ and, as for poly-BN, there will be a band gap. With two π electrons per unit cell, the valence band is just filled. Hence, it would be a semiconductor and not subject to a Peierls distortion.
6. a) The repeating unit is C_6H_6 with HOMO and LUMO shown below. Stacking the benzene rings with C aligned generates a band from each. The bands will run as the $2p_z$ band of linear C_n shown in Figure 6.12. As d will be longer than a chemically bonding distance, the bands will be flat with a substantial gap between the bands. The lower band will be completely filled; hence, the stacked material will be a semiconductor. Since all the inter-molecular bonding and antibonding states of the valence band are occupied, there is no significant bonding between the benzene units. (b) Reduction of the stack will partially populate the conduction band leading to metallic conductivity along the stacking axis. As described in Section 6.3.4, graphite consists of stacked sheets of fused benzene rings. The inter-layer distance is long and the sheet–sheet interaction weak. Neither the hexagonal nor rhombohedral forms of graphite have the six-membered rings aligned as described for the stacked benzene molecules in this problem. However, the stacked benzene model reasonably explains the metallic properties of $([C^{-0.125}])_n$ in a graphitic geometry. The counterions would reside between the C layers.

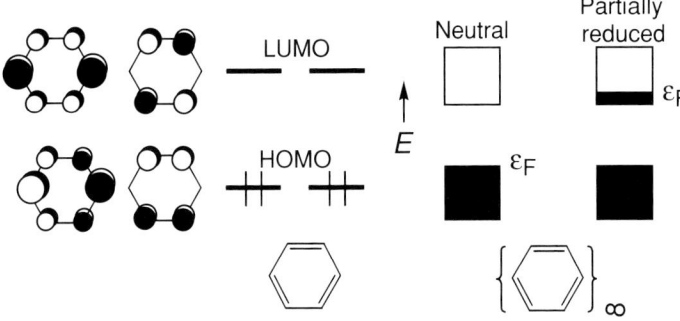

Problem 6.6

7. (a) The π band generated by the TCNQ⁻ SOMO is half occupied, thus "dimerization" is expected (right side of the diagram below). However, the π–π overlap (interaction) is small, thus the bandwidth is small. In other words, the energy gained by the "dimerized" phase upon bond alternation is not large relative to the electron pairing energy of a high-spin non-distorted phase (left side of the diagram below). We are in a situation often encountered with transition-metal complexes where a distorted low-spin and an undistorted high-spin species are close in energy. A small perturbation of, e.g., temperature or pressure, can induce transformation of one phase to the other. The "dimerized" phase is a low-spin semiconductor, whereas the regular phase is a high-spin semiconductor.

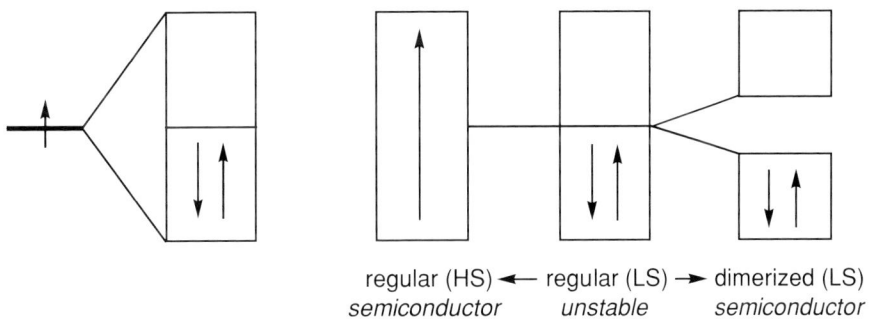

Problem 6.7–a

(b) In (NEt$_4$)(TCNQ)$_2$ we know that the cation is tetramethylammonium NEt$_4^+$. Cation/anion charge balance results in TCNQ$^{-1/2}$. Don't be afraid, this is an average charge attributed to the repeat motif of an extended one-dimensional structure. This means that the HOMO of TCNQ$^{-1/2}$ is 1/4 occupied. In the extended stack, it generates a band which is 1/4 occupied. This induces Peierls "tetramerization." In Rb$_2$(TCNQ)$_3$, the cation/anion charge balance leads to TCNQ$^{-2/3}$, i.e. its HOMO is 1/3 occupied. In the extended stack, it generates a band which is 1/3 occupied. This induces Peierls "trimerization."

(c) Given the properties of each component, (TTF)(TCNQ) can be reformulated (TTF$^+$, TCNQ$^-$). One might expect co-existence of dimerized stacks of TTF$^+$ and of TCNQ$^-$ since both bands are half occupied. However, the Fermi levels of both subsystems are not equal. That of the anionic stacks has higher energy than that of

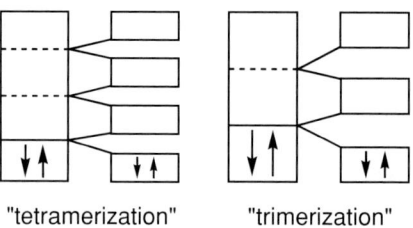

"tetramerization"   "trimerization"

Problem 6.7–b

the cationic stacks. Two different Fermi levels cannot coexist in the same crystal, i.e., the two stacks are in "electric contact." They equalize by electron transfer analogously to water-level equalization and (TTF)(TCNQ) should be reformulated $(TTF^{\delta+}, TCNQ^{\delta-})$ ($\delta < 1$). The result is unique Fermi level and TTF and TCNQ bands which are no longer $1/2$ occupied. As a consequence, there is no Peierls distortion and the compound is a good isotropic conductor.

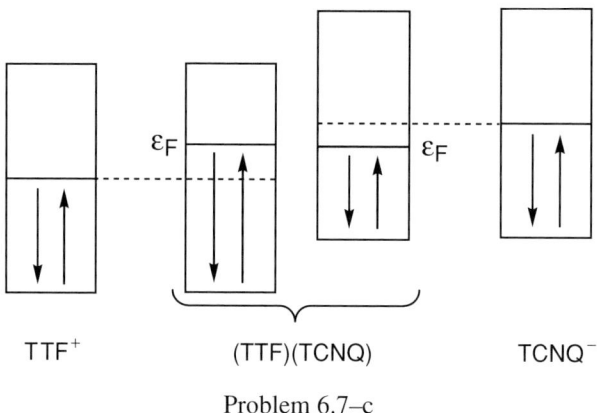

TTF$^+$ (TTF)(TCNQ)   TCNQ$^-$

Problem 6.7–c

Chapter 7

Problems

1. Gd strongly prefers a $+3$ oxidation state leading to $[Gd^{3+}]_{10}[Cl^-]_{18}[B^{3-}]_4$. Hence, we have $[B_2]^{6-}$ dumb-bells in the octahedral cavities which are 12-electron species analogous to C_2H_4 or $[C_2]^{4-}$ encountered in the related solid $Gd_2Cl_2C_2$. The B–B distance should be less than the sum of the single-bond covalent radii (1.74 Å). With a half-filled π^* band, the hypothetical compound should be an electrical conductor. We hope you were not fooled by the representation of the geometry of the system to consider the system as fused octahedral clusters.
2. Apply the Zintl–Klemm paradigm with the cations Mg^{2+} and Na^+. This leads to counts of 18 and 16 electrons for $[NiH_4]^{4-}$ and $[PtH_4]^{2-}$, respectively. The former should be tetrahedral (T_d symmetry) whereas a square planar arrangement (D_{4h} symmetry) is appropriate for the latter. Both are confirmed experimentally. The pertinent MO diagrams can be found in Figures 1.10 and 5.28.

3. (a) If x is the oxidation state of the metal, $6x - 14$ (Cl$^-$) $= -2$, so $x = 2$ (Mo^{2+} d^4). Each Mo^{2+} atom is bound to four Mo congeners and five Cl anions. The 18-electron count around each metal is achieved by adding to its four electrons, one electron from each Mo congener, and ten electrons from five Cl$^-$ ligands. The total number of metallic electrons is 24 (4 (Mo^{2+}) × 6). Assuming two-electron–two-center Mo–Mo bonds, all these electrons are metal–metal bonding. None is non-bonding. The metal atom uses all its nine AOs to form nine bonds (5 Mo–Cl bonds + 4 Mo–Mo bonds). None is left for housing non-bonding electrons.
   (b) The MO diagram shows 12 Mo–Mo bonding orbitals largely separated from 12 Mo–Mo antibonding orbitals.
   (c) The charge balance requires [Mo$_6$]$^{16+}$, i.e., a metallic electron count of 20 for the Mo$_6$ core. This implies that the e$_g$ MO, HOMO for [Mo$_6$(μ_3-Cl)$_8$Cl$_6$]$^{2-}$, becomes the LUMO for Mo$_6$(μ_3-S)$_8$(PEt$_3$)$_6$. This e$_g$ set lies in the middle of an energy gap. It can therefore be fully or partially occupied or empty. Counts intermediate between 20 and 24 should be possible. Indeed, compounds with these intermediate counts have been characterized. See Exercise 3.10 for an alternative way of treating the cluster bonding in Mo$_6$(μ_3-S)$_8$(PEt$_3$)$_6$.
4. (a) The computed DOS of PbMo$_6$S$_8$ is given below (the MO diagram of an isolated 22-electron cluster, similar to that in Problem 3, is recalled on the right). With 22 electrons, the band derived from the e$_g$ MO is half filled and the compound is an electrical conductor. It is even a superconductor at low temperature.

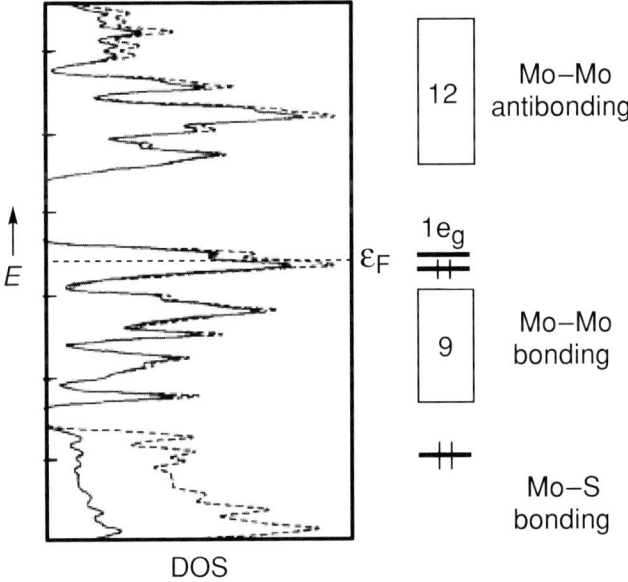

Problem 7.4

(b) The charge balance for Mo$_2$Re$_4$S$_8$ gives 24 metallic electrons per Mo$_2$Re$_4$ core. Assuming a DOS comparable to that of PbMo$_6$S$_8$, the e$_g$ band will be filled and the compound will be semiconducting. It is.

References

Adams, R. D. and Captain, B. (2005). *Angew. Chem. Int. Ed.*, **44**, 2531.
Amini, M. M., Fehlner, T. P., Long, G. J. and Politowski, M. (1990). *Chem. Mater.*, **2**, 432.
Adams, R. D., Cortopassi, J. E., Aust, J. and Myrick, M. (1993). *J. Am. Chem. Soc.*, **115**, 8877.
Beauvais, L. G., Shores, M. P. and Long, J. R. (2000). *J. Am. Chem. Soc.*, **122**, 2763.
Beswick, M. A., Choi, N., Harmer, C. N., Hopkins, A. D., McPartlin, M. and Wright, D. S. (1998). *Science*, **281**, 1500.
Bino, A., Ardon, M. and Shirman, E. (2005). *Science*, **308**, 234.
Brynda, M., Herber, R., Hitchcock, P. B., Lappert, M. F., Nowik, I., Power, P. P., Protchenko, A. V., Růžička, A. and Steiner, J. (2006). *Angew. Chem. Int. Ed.*, **45**, 4333.
Bunz, U. H. F. (2005). *Science*, **308**, 216.
Burdett, J. and Eisenstein, O. (1995). *J. Am. Chem. Soc.*, **117**, 11939.
Charles, S., Eichhorn, B. W. and Bott, S. G. (1993). *J. Am. Chem. Soc.*, **115**, 5837.
Cleaver, W. M., Späth, M., Hnyk, D., McMurdo, G., Power, M. B., Stuke, M., Rankin, D. W. H. and Barron, A. R. (1995). *Organometallics*, **14**, 690.
Corbett, J. D. (1985). *Chem. Rev.*, **85**, 383.
Crawford, N. R. M., Hee, A. G. and Long, J. R. (2002). *J. Am. Chem. Soc.*, **124**, 14842.
Eichhorn, B. W., Haushalter, R. C. and Pennington, W. T. (1988). *J. Am. Chem. Soc.*, **110**, 8704.
Eichler, B. E. and Power, P. P. (2001). *Angew. Chem. Int. Ed.*, **40**, 796.
Fassler, J. F. (2001). *Coord. Chem. Rev.*, **215**, 347.
Fumagalli, A., Pergola, R. D., Bonacina, F., Garlaschelli, L., Moret, M. and Sironi, A. (1989). *J. Am. Chem. Soc.*, **111**, 165.
Gaines, D. and Schaeffer, R. (1964). *Inorg. Chem.*, **3**, 438.
Grimes, R. N. (1992). *Chem. Rev.*, **92**, 251.
Jutzi, P., Mix, A., Rummel, B., Schoeller, W. W., Neumann, B. and Stammler, H.-G. (2004). *Science*, **305**, 849.
Kanatzidis, M. G., Salifoglou, A. and Coucouvanis, D. (1986). *Inorg. Chem.*, **25**, 2460.
Kehrwald, M., Köstler, W., Rodig, A., Linti, G., Blank, T. and Wiber, N. (2001). *Organometallics*, **20**, 860.
Kubas, G. J. (2001). *J. Organomet. Chem.*, **635**, 37.
MacInnes, A. N., Power, M. B. and Barron, A. R. (1993). *Chem. Mater.*, **5**, 1344.
Masumori, T., Seino, H., Mizobe, Y. and Hidai, M. (2000). *Inorg. Chem.*, **39**, 5002.

Mingos, D. M. P. and Wales, D. J. (1990). *J. Am. Chem. Soc.*, **112**, 930.
Okazaki, M., Ohtani, T., Inomata, S., Tagaki, N. and Ogino, H. (1998). *J. Am. Chem. Soc.*, **120**, 9135.
Rogel, F. and Corbett, J. D. (1990). *J. Am. Chem. Soc.*, **112**, 8198.
Schmettow, W. and Schnering, H. G. v. (1977). *Angew. Chem. Int. Ed.*, **16**, 857.
Schnepf, A. and Schnöckel, H. (2002). *Angew. Chem. Int. Ed.*, **41**, 3533.
Shores, M. P., Beauvais, L. G. and Long, J. R. (1999). *J. Am. Chem. Soc.*, **121**, 775.
Sneddon, L. G., Pender, M. J., Forsthoefel, K. M., Kusari, U. and Wei, X. (2005). *J. Eur. Ceram. Soc.*, **25**, 91.
Tian, Y. and Hughbanks, T. (1995). *Inorg. Chem.*, **34**, 6250.
Ugrinov, A. and Sevov, S. C. (2002). *J. Am. Chem. Soc.*, **124**, 2442.
Ugrinov, A. and Sevov, S. C. (2003). *Inorg. Chem.*, **42**, 5789.
Xu, L. and Sevov, S. C. (1999). *J. Am. Chem. Soc.*, **121**, 9245.
Yong, L, Hoffmann, S. D., Fässler, T. F., Riedel, S. and Kaupp, M. (2005). *Angew. Chem. Int. Ed.*, **44**, 2092.
Zhou, H., Day, C. S. and Lachgar, A. (2004). *Chem. Mater.*, **16**, 4870.

Index

absorption properties 15
agostic H 154
alkyne cluster [Ta(t-Bu$_3$SiO)$_3$]$_2$C$_2$ 264
alkyne complex
   [Mn(CO)$_5$]$_2$C$_2$ 263, 265
   [RuCp(PPh$_3$)$_2$]$_2$(C$_4$) 269
   [ScCp*]$_2$C$_2$ 263
alkyne metathesis 320
amorphous boride films 318
amorphous metal alloy films 317
atomic orbitals
   core energy level 328
   electron–electron interactions 328
   H atom 325
   Hund's rule 329
   one-electron approximation 327
   Pauli principle 327
   Slater's rules 327
atomic properties
   binding forces 257
   electron affinity 324
   electronegativity 324
   ionization energy 324
   model 324
   size 324

band theory
   band folding 219
   band structure 214
   band structure cubic H 238
   band structure, multidimensional 236
   band structure, poly-BHNH 253
   band structure, stacked benzene 253
   band width 214
   Bloch function 212, 217
   Bloch function, Al metal 243
   Bloch functions, p-block element 226
   Bloch functions, transition metal 229
   Brillouin zone 213, 217, 225, 236
   conduction band 224
   COOP d-block element chain 231
   crystal orbital overlap population 216
   crystal orbital vs. molecular orbital parameters 235

   density of states 215
   dispersion 214
   electron count and instability 224
   Fermi level 214
   Fermi level, d-block element chain 231
   H cubic lattice 237
   H square array 238
   insulator 223
   LiH chain 232
   linear H$_N$ 212
   Peierls instability 222
   semiconductor 337
   semimetal 246
   valence band 224
bond parameters
   B–B distances 50
   B–B–B 57, 66
   B–H distances 50
   B–H–B 20, 21, 54, 64
   B–H–B bridge 19
   C–H–M 20
   double M=C 154
   double M=Ge 155
   double M=Si 155
   Ga–Ga–Ga 37
   Ge–Ge–Ge 68
   M–H–M bond 20
   P–H–P 39
bond properties
   coordinate covalent 340
   covalent radius 325, 331
   donor–acceptor 340
   hypercoordinate 27
   hypervalent 27
   hypovalent 27
   ionic radius 325
   multicentered 39
   three-center 18, 27, 28, 53, 56
   three-center–four-electron bond 21, 28
   three-center–two-electron bond 18, 19, 37, 86, 108, 336
   two-center–two-electron bond 8, 28, 29, 34, 35, 85, 108

borane anions
    $[B_{10}H_{10}]^{2-}$ 44, 64
    $[B_{12}H_{12}]^{2-}$ 41, 42
    $[B_{14}H_{14}]^{2-}$ 60
    $[B_{20}H_{18}]^{2-}$ 64
    $[B_2H_4]^{2-}$ 20
    $[B_5H_{10}]^-$ 49
    $[B_6H_6]^{2-}$ 54, 57, 93, 116, 141, 166, 168, 180, 251, 283
    $[B_9H_9]^{2-}$ 42, 70
    $[B_nH_n]^{2-}$ 43
borane clusters
    $[B_5H_9]_2$ 60, 283
    $B_{10}H_{14}$ 50, 316
    $B_{10}H_8(N_2)_2$ 44
    $B_{12}H_{10}(CO)_2$ 45
    $B_{18}H_{22}$ 63
    $B_{20}H_{16}$ 60, 62
    B_2H_6 27, 28, 41, 340
    B_4Cl_4 37
    B_4H_{10} 79, 170
    B_5H_{11} 48, 170
    B_5H_9 54, 56, 57, 97, 147, 150, 168
    $B_5H_9(PMe_3)_2$ 53
    B_6H_{10} 47, 169
    B_6H_{12} 79
    B_9Cl_9 70
boride
    $Gd_5Si_2B_8$ 286
    B_6 lattice 283
    CaB_6 283
    $Gd_5Si_2B_8$ 286
    GdB_4 285
    LaB_6 284
    Li_2B_6 285
    Li_3B_{14} 285
    LiB 275
    Na_2B_{29} 285
    Na_3B_{20} 285
boron carbide 315
    B_4C 315
    LaBC 274
    Sc_2BC_2 271, 273
    ThBC 276
    YBC 276
boron wheel
    $[B_8]^{2-}$ 71
    $[B_9]^-$ 71
bridging H 47
Brönsted acid 341

C_{60} 73, 121, 245, 278, 282, 333, 334
    Rb_2CsC_{60} 281
    electronic structure 279
    K_3C_{60} 279
C_8K 253
carbide
    Ag_2C_2 264
    CaC_2 267
    $Gd_{10}Cl_8C_4$ 264
    $Gd_{10}Cl_8C_4$ 264
    $Gd_2Cl_2C_2$ 264
    LaC_2 268
    NbC 260
    RuC 261
    Sc_3C_4 270
    $Ti_8(C_2)_6$ 266
    WC 260
    YC_2 268
    ZrC 261
carbon nanotubes 245
carborane complex
    $[Co(C_2B_9H_{11})_2]^-$ 157
    $CpCo(C_2B_9H_{11})$ 158
    $Si(C_2B_9H_{11})_2$ 156
ceramic fibers 316
ceramic polymer precursor 315
Chevrel phases 291
close packing 336
    cubic close packed (ccp) 336
    face-centered cubic (fcc) 336
    hard sphere model 208
    octahedral hole 208
    tetrahedral hole 208
cluster
    bicapped square antiprism 43
    borane paradigm 166
    building blocks 308
    closed (*closo*) 43
    debor process 51
    definition 33, 34
    diamond–square–diamond rearrangement 71, 183
    electron precise 35
    expanded solids 305
    face-capped deltahedron 91
    four-connect 39, 90
    fusion 60, 209
    icosahedral 40
    ion beams 120
    isomers 45, 48, 50, 63, 206
    linked 60
    naked 64, 73, 120
    networks 283
    non-deltahedral shape 60
    octahedron 40, 89
    open (*arachno*) clusters 48, 89
    open (*nido*) clusters 45
    oxidative coupling 64
    polar deltahedra 40
    polyfunctionalized 307
    precursors 303
    ring-cap model 54, 74, 95
    shape relationships 51–53
    three-connect 34, 35
    trigonal bipyramid 40
    trigonal prismatic 89
    unsaturated 174
    vertex connectivity 39
conductivity
    metallic 224
    semiconductor 224

Index 373

Coulomb blockade 131
crystal field theory 348
crystal orbitals 212
cubane
   $Cp_4Fe_4(C_2H_2)_2$ 202
   $(CpM)_4E_4H_4$ 152
   $(NO)_4Fe_4S_4$ 193
   $[(RS)_4Fe_4S_4]^{2-}$ 194
   $[BuGaS]_4$ 305
   $[Cp_4Fe_4S_4]^{n+}$ 189, 192
   cluster 189, 305
   $Cp^*_3Ru_3Co(CO)_2(BH)_3(CO)$ 192
   $Cp_4Co_4B_4H_4$ 192
   $Cp_4Co_4S_4$ 189, 192
   $Cp_4Cr_4S_4$ 195
   $Cp_4Fe_4(CO)_4$ 192, 202
   $Cp_4Ni_4B_4H_4$ 192
   $Cp_4Ti_4S_4$ 195
   $Mo_2Ir_2S_4$ core 202
cubic clusters 198
   $(CO)_8Ni_8(PPh)_6$ 198
   $[I_8Fe_8S_6]^{3-}$ 198
CVD deposition 305

deltahedra 39
Dewar model 266, 347
diamond 58
diborane 18, 20, 27, 28
dipole moment 5, 339, 340
DOS 260
   Al (111) surface 250
   Al metal 243
   Al metal COOP 243
   Al_{77} nanoparticle 251
   C non-metal 245
   C_{60} 279
   CaB_6 284
   Co_9S_8 297
   cubic H 238
   d-block element chain 229
   d-block metal 242
   diamond 245
   Fermi level 262, 273, 277, 294
   GdB_4 285
   graphite 245
   $K_2Mo_9S_{11}$ 294
   metallic nanoparticle 251
   Ni metal 244
   p-block element chain 227
   p-block metal 242
   Sc_2BC_2 density of states 273
   Sc_2BC_2 273
   ThBC 277
   thin film 250
   TiO_2 290
   YBC 277

elementary bonding
   $[HP_4]^+$ 39
   $[I_3]^-$ 342
   $[SiF_6]^{2-}$ 342
   CH_4 341
   BMe_3 26
   HF 340
   Hund's rule 348
   N_2F_4 341
   N_2H_4 341
   NF_3 341
   NH_3 340, 341
   P_2H_4 338, 339
   P_3H_5 338
   PH_3 338, 341
   phosphorus hydrides 338
   PMe_3 341
   rotational barrier 339
   $SiCl_4$ 341
elementary bonding rules
   16-electron complex 345
   8- and 18-electron rules 22, 24, 28, 29, 34, 39, 140, 287, 329
electron count
   cluster-fusion rule 62
   cve count for open metal clusters 94
   count deltahedral clusters $4n + 2$ cve rule 41, 42, 143
   deltahedral clusters $n + 1$ rule 41, 42
   hypercloso 182
   isocloso, isonido, isoarachno 182
   metal cluster capped cve count 100
   metal clusters $14n + 2$ rule 90, 93, 143
   three-fold axes 70
   three-connect clusters $3/2n$ sep rule 36, 87
   three-connect clusters $5n$, $15n$ cve rule 36, 88, 143
   unusual count, unusual shape 182
   variation count, constant shape 179
electron counting rule 22, 35, 42
   $6m + 2n$ 75
   cluster valence electrons (CVE) 36
   mno rule 61
   skeletal electron pairs (sep) 36
   styx rule 57
electron deficient 26, 41
electron precise 48, 50
electrophilic 340
element properties 324
   Al 336
   B, 58
   B rhombohedral 335
   C, diamond 333
   C, allotropes 258
   C, C_2 263
   C, C_{60} 278, 282
   C, carbyne 333
   C, chains 269, 270
   C, graphite 333
   C, single-walled nanotubes 281
   P, P_4 143
   P, rhombohedral 333, 334
   solids cubic close packed Al 206
elementoid clusters 73, 128
endo-H 49

films 309
fragments
  BH 147
  Ru(CO)$_3$ 150
  CpFe(CO)$_2$ 149, 155
  CpML$_n$ 146
  CpMo(CO)$_3$ 149
  CpW(CO)$_2$ 150
  Fe(CO)$_4$ 145
  Ni(PR$_3$)$_2$ 146
fragment concepts
  building blocks 197
  frontier orbitals 140
  isolobal EH$_n$ 140
  isolobal ML$_m$ 142, 143
  non-bonding electrons 88
  utilization of t$_{2g}$ set 148
fullerenes 278

heteroboranes 43
  [C$_2$B$_4$H$_6$]$^{2-}$ 80
  [C$_2$B$_9$H$_{11}$]$^{2-}$ 46, 56
  [CB$_{11}$H$_{12}$]$^-$ 44
  C$_2$B$_{10}$H$_{12}$ 44
  C$_2$B$_3$H$_5$ 79
  C$_2$B$_3$H$_5$ 151
  C$_2$B$_3$H$_7$ 79
  C$_2$B$_4$H$_6$ 45, 47, 79
  C$_2$B$_4$H$_8$ 48, 79
  C$_2$B$_8$H$_{10}$ 44
  C$_n$B$_{6-n}$H$_{1-n}$ 166
  NB$_{11}$H$_{12}$ 44
  role of electronegativity 48
hexagonal close packed (hcp) 336
hydrides
  [(Pr$_3$P)$_6$Rh$_6$H$_{12}$]$^{2+}$ 113
  H$_2$Os$_6$(CO)$_{18}$ 101, 107
  H$_4$Re$_4$(CO)$_{12}$ 86
  Mg$_2$CoH$_5$ 288
  Mg$_2$FeH$_6$ 287
H bonding 340

isolobal analogy 139, 265
  ancillary ligand geometry 145
  ancillary ligands 145
  C$_1$ analog complexes 154
isomers 339

LCAO-MO model 1
Lewis acid–base complexes
  BH$_3$NH$_3$ 342
  BMe$_3$NMe$_3$ 342, 344
  CH$_3$CNBF$_3$ 342
Lewis acids, bases 9, 35, 159, 342
ligands
  [C$_5$H$_5$]$^-$ (Cp) 344
  [C$_5$R$_5$]$^-$ (Cp*) 48
  C$_5$H$_6$ 50
  Cp analog [C$_2$B$_9$H$_{11}$]$^{2-}$ 155, 156
  Cp analog [C$_2$B$_3$H$_7$]$^{2-}$ 156
  Cp analog [P$_5$]$^-$ 156
  Cp analog [SC$_2$B$_2$H$_4$]$^-$ 156
  hapticity 345
  R$_4$C$_4$ 38
  TCNQ 253
  TTF 254

macropolyhedral clusters 62
magnetically oriented films 319
melting point
  NbC 262
  NbN 262
metal 337
metal catalysis 319
metal cluster
  alkali-metal clusters 119
  alkoxide ligands 109
  carbonyl 93
  central cluster with metal ligands 124
  early metals 107
  face capping 99
  fragment cone angle 106
  fusion 103, 126
  gold clusters 117
  group-10 metals 114, 115
  group-11 metals 117
  high nuclearity 122
  interstitial atoms 96, 123
  interstitial cluster with bonded outer cluster 125
  isomers 95, 101
  large metal clusters 123
  late transition metals 114, 115
  multiple redox states 114
  naked clusters 120
  sep count 95
  steric effects 105
  π-acceptor ligands 93, 108
  π-donor ligands 93, 107, 108
metal cluster compounds
  [(Au$_{13}$)$_{13}$]$_n$ metals 131
  [Au$_{13}$Cl$_2$(PR$_3$)$_{10}$]$^{3+}$ 117
  [Au$_6$(PR$_3$)$_6$]$^{2+}$ 119
  [Au$_6$C(PR$_3$)$_6$]$^{2+}$ 120
  [Au$_7$(PR$_3$)$_7$]$^+$ 119
  [Au$_9$(PR$_3$)$_8$]$^{3+}$ 118
  [Co$_3$(μ-dppm)(CO)$_7$]$_2$(C$_{26}$) 269
  [Co$_6$N(CO)$_{15}$]$^-$ 98
  [Co$_6$Ni$_2$(C$_2$)(CO)$_{16}$]$^{2-}$ 263, 265, 266
  [Co$_6$P(CO)$_{16}$]$^{2-}$ 259
  [Co$_8$S$_2$(SPh)$_8$]$^{4-/-}$ 2, 298
  [Cp*$_4$Rh$_4$H$_4$]$^{2+}$ 87
  [Fe$_4$(CO)$_{12}$CC(O)Me]$^-$ 171
  [Fe$_4$(CO)$_{13}$]$^{2-}$ 107
  [Fe$_4$C(CO)$_{12}$]$^{2-}$ 98
  [Fe$_4$RhC(CO)$_{14}$]$^-$ 98
  [Fe$_6$C(CO)$_{16}$]$^{2-}$ 97
  [HOs$_8$(CO)$_{22}$]$^-$ 104
  [Mo$_6$Cl$_8$L$_6$]$^{4+}$ 107, 114, 115
  [Nb$_6$Cl$_{12}$(CN)$_6$]$^{4-}$ 307
  [Ni$_{12}$Ge(CO)$_{22}$]$^{2-}$ 97, 259
  [Ni$_{38}$Pt$_6$(CO)$_{48}$H$_{(6-n)}$]$^{n-}$ 123, 124
  [Ni$_8$C(CO)$_{16}$]$^{2-}$ 259

Index 375

$[Ni_n]^+$ 121
$[Os_{10}C(CO)_{24}]^{2-}$ 101, 125, 207
$[Os_6(CO)_{18}]^{2-}$ 100
$[Os_8(CO)_{22}]^{2-}$ 102
$[Pb_5\{Mo(CO)_3\}_2]^{4-}$ 163
$[Pt_{24}(CO)_{30}]^{2-}$ 133
$[Pt_{38}(CO)_{44}H_m]^{2-}$ 124
$[PtIr_4(CO)_{12}]^{2-}$ 133
$[PtIr_4(CO)_{14}]^{2-}$ 133
$[Re_4(CO)_{16}]^{2-}$ 94
$[Re_4Se_8I_6]^{4-}$ 134
$[Re_6Te_8(CN)_6]^{4-}$ 305
$[Re_8C(CO)_{24}]^{2-}$ 133
$[Rh_{14}(CO)_{26}]^{2-}$ 105
$[Rh_{24}(CO)_{26}]^{2-}$ 126
$[Rh_2Fe_4(CO)_{16}B]^-$ 172
$[Rh_2Fe_4B(CO)_{16}]^-$ 98
$[Rh_6C(CO)_{13}]^{2-}$ 98, 259
$[Ru_{11}H(CO)_{27}]^{2-}$ 134
$[Ru_5N(CO)_{14}]^-$ 98
$[Ru_6(CO)_{18}]^{2-}$ 92, 95, 97, 107, 166, 168
$[SFe_3(CO)_9]^{2-}$ 175
$[Ta_6Cl_{12}L_6]^{2+}$ 107, 114, 115
$[W_6CCl_{18}]^{2-}$ 113, 262
$[Zr_6(B)Cl_{18}]^{5-}$ 313
$[Zr_6CCl_{18}]^{4-}$ 261
$[Zr_6Cl_{18}B]^{5-}$ 110
$\{Re(CO)_3\}_2Pt_3(PBu_3)_3$ 116
$\{Re(CO)_3\}_2Pt_3(PBu_3)_3H_3$ 116
$Au_{55}(PR_3)_{12}Cl_6$ 130
$Ba_2Zr_6Cl_{17}X$ 112
$CFe_5(CO)_{15}$ 168
$Co_4(CO)_{12}$ 86, 106, 143, 144
Co_9S_8 296
$Co_9Se_{11}(PPh_3)_6$ 295
$Cp_3Rh_3(CO)C_2R_2$ 151
$Cp_3Rh_3C_2R_2$ 151
$CRu_5H_2(CO)_{15}$ 168
$Cs_2Mo_{12}Se_{14}$ 293
$CsKZr_6Cl_{15}B$ 111
$Cu_{26}Se_{13}(PEt_3)_{14}$ 314
$Fe_3(CO)_{12}$ 142
$Fe_3(CO)_9(C_2R_2)$ 152
$Fe_3(CO)_9N_2R_2$ 176
$Fe_4(CO)_{11}(PR)_2$ 95, 180
$Fe_4(CO)_{12}(PR)_2$ 180
$Fe_4C(CO)_{13}$ 171
$Fe_5C(CO)_{15}$ 97, 258
$HFe_4(CO)_{12}EH_n$ 171
$HPtOs_3(CO)_{10}(dppm)\{Si(OMe)_3\}$ 134
$Ir_4(CO)_{12}$ 85, 107
$K_2Mo_9S_{11}$ 294
$K_3Zr_6Cl_{15}Be$ 111
$KZr_6Cl_{15}C$ 111
$La_2Mo_{16}O_{28}$ 295
$Li_2Zr_6Cl_{15}Mn$ 112
$LiZr_6Cl_{15}Fe$ 112
$M_4S_4(PR_3)_4$ 314
$M_{55}L_{12}Cl_x$ 129
$M_8S_8(PR_3)_6$ 314
$Nb_2Mo_{16}O_{28}$ 295

$Os_3(CO)_{12}$ 88
$Os_4(CO)_{13}S$ 94
$Os_5(CO)_{16}$ 96
$Os_5C(CO)_{16}$ 166, 168
$Os_5C(CO)_{16}$ 168
$Os_5S(CO)_{15}$ 99
$Os_6(CO)_{21}$ 132
$Os_7(CO)_{21}$ 99
$PbMo_6S_8$ 292
$PCo_3(CO)_9$ 143
$Pd_{59}(CO)_{32}(PMe_3)_{21}$ 127
$Pt_{309}phen*_{36}O_{30}$ 132
$Rb_2Mo_{12}Se_{14}$ 293
$Rb_3Mo_{15}Se_{17}$ 293
$Rb_5 Zr_6Cl_{18}B$ 110
$RCCo_3(CO)_9$ 144
$Rh_{13}(CO)_{24}H_5$ 125
$Ru_5(C_2)(CO)_{11}(PPh_2)(SMe)_2$ 263
$Ru_5(CO)_{15}(PPh)$ 94, 151
$Ru_6C(CO)_{17}$ 97, 259
$W_6S_8(PR_3)_6$ 114
$Zr_6(Z)Cl_{12}\cdot(EtNH_2)_6$ 313
$Zr_6Cl_{15}Co$ 112
$Zr_6Cl_{15}N$ 111
metal complexes
  $(CO)_4FeBCp$ 162
  $(CO)_4FeH_2$ 161
  $(CO)_4FePMe_3$ 161
  $[(CO)_4Mo(BH_4)]^-$ 154
  $[BH_4]^-$ 154
  $[Co(NH_3)_6]^{3+}$ 15
  $[Fe(C_2B_9H_{11})_2]^{2-}$ 156
  $[FeCp(CO)_2]^-$ 345
  $[FeCp(CO)_2]_2$ 149, 345, 346
  $[Mn(CO)_5]_2$ 344
  $[Mn(CO)_5]_2C_2$ 263, 265
  $[Mn(CO)_6]^+$ 344
  $[Ni(C_2B_9H_{11})_2]^{2-}$ 157
  $[ReCp*(PPh_3)(NO)]_2(C_{20})$ 269
  $[RuCp(PPh_3)_2]_2(C_4)$ 269
  $[ScCp*]_2C_2$ 263
  $[Ta(t-Bu_3SiO)_3]_2C_2$ 264
  $[V(CO)_6]^-$ 344
  $\{CpMo(CO)_2\}_2$ 149
  $C_2H_4Fe(CO)_4$ 142, 145
  $CH_2Fe_2(CO)_8$ 142
  $CH_3Mn(CO)_5$ 142
  $Co_2(CO)_8$ 344
  $Co_2(CO)_8(C_2R_2)$ 144
  $Cp*(CO)_2CrHSiR_3$ 154
  $Cp*_2Cr_2(CO)C_4Ph_4$ 173
  $CpCoC_4H_4$ 150
  $CpMn(CO)_3$ 149
  $CpTi(CO)_2$ 316
  $Fe(CO)_4(C_2H_4)$ 346
  $Fe_2(CO)_6S_2R_2$ 176
  $FeCp(CO)_2H$ 345
  ferrocene 344
  $Ir(PPh_3)_2(CO)Cl$ 345
  $Mn(CO)_5H$ 344
  $Mn_2(CO)_{10}$ 142

metal complexes (cont.)
  Ni(C$_2$B$_9$H$_{11}$)$_2$ 157
  Rh(PPh$_3$)$_3$Cl 345
  Rh(PPh$_3$)$_3$Cl(H)$_2$ 345
  V(CO)$_6$ 344
  W(CO)$_6$ 343
  WBr$_4$(MeCN)$_2$ 343
  WMe$_6$ 343
metal complex types
  carbyne 154
  multidecker 158
  spectrochemical series 15
  tripledecker 158
metallaboranes 166
  HFe$_3$(CO)$_9$BH$_4$ 317
  (C$_6$H$_6$)RuB$_9$H$_9$ 184
  (Cp*Re)$_2$B$_5$H$_2$Cl$_5$ 185
  (Cp*Re)$_2$B$_6$H$_4$Cl$_2$ 185
  (Cp*Re)$_2$B$_7$H$_{11}$ 186
  (Cp*Re)$_2$B$_8$H$_8$ 184
  (Cp*Ru)$_2$(C$_6$H$_6$)RuB$_7$H$_7$ 184
  [(C$_6$H$_6$)RuB$_9$H$_9$]$^{2-}$ 184
  {Cp*Co}$_2$B$_3$H$_7$ 150
  Cp*$_2$Cr$_2$B$_4$H$_8$ 173, 183
  Cp*$_2$Ir$_2$B$_2$H$_8$ 171, 172
  Cp*$_2$Ir$_2$H$_2$B$_4$H$_8$ 171
  Cp*$_2$M$_2$B$_4$H$_8$ 172
  Cp*$_2$Re$_2$B$_4$H$_8$ 172, 185
  Cp*$_2$Re$_2$B$_m$H$_m$ 183
  Cp*$_2$Ru$_2$(CO)$_2$B$_3$H$_7$ 170
  Cp*$_2$Ru$_2$B$_4$H$_8$ 172
  Cp*CoB$_4$H$_{10}$ 170
  Cp*CoB$_4$H$_8$ 170
  Cp*Ir(H)$_2$B$_3$H$_7$ 170
  Cp*IrB$_4$H$_{10}$ 170
  Cp$_2$Co$_2$B$_8$H$_{12}$ 176
  Cp$_3$Co$_3$(CO)B$_3$H$_3$ 176
  Cp$_4$Co$_4$B$_2$H$_2$PPh 176
  CpCoB$_4$H$_8$ 150
  Fe$_2$(CO)$_6$B$_2$H$_6$ 317
  HFe$_4$(CO)$_{12}$BH$_2$ 171
  HFe$_4$(CO)$_{12}$BH$_2$ 317
  Os$_5$(CO)$_{16}$BH 166, 168
metallaborane types
  closo 166
  metal variation 172
  arachno 170
  nido 168
metallocarbohedranes
  (met-cars) 266
  Ti$_8$(C$_2$)$_6$ 266
metalloid clusters 73
main-group clusters
  (CR)$_6$(AlH)$_6$(AlNMe$_3$)$_2$ 78
  (H)(CR)$_5$(AlMe)$_8$ 78
  [(R$_2$C$_2$B$_4$H$_4$)$_2$Ga]$^-$ 62
  [Al$_{12}$R$_{12}$]$^{2-}$ 45
  [Al$_{14}$R$_6$I$_6$]$^{2-}$ 72, 76
  [Al$_{69}$R$_{18}$]$^{3-}$ 76, 123, 128, 206, 207, 242, 313
  [Al$_{77}$R$_{20}$]$^{2-}$ 76, 129, 200, 247, 251
  [Al$_7$R$_6$]$^-$ 314

[Ga$_9$R$_6$]$^-$ 80
[Ga$_{13}$R$_6$]$^-$ 81
[Ga$_{19}$R$_6$]$^-$ 74, 76, 128, 201
[Ga$_{26}$R$_8$]$^{2-}$ 75
[Ge$_{10}$R$_6$I]$^+$ 67, 74
Al$_4$R$_4$ 37
As$_7$(SiMe$_3$)$_3$ 80
Ga$_{10}$R$_6$ 69
Ga$_4$R$_4$ 37, 87, 88
Ga$_9$R$_9$ 70
Ge$_6$R$_2$ 67, 76
In$_8$R$_6$ 101
P$_2$ 335
P$_4$ 38, 332
SiAl$_8$(AlCp*)$_6$ 77
Sn$_{15}$R$_6$ 81
Sn$_8$R$_4$ 76, 80
Sn$_n$R$_n$ 79
[Cp*Si]$^+$ 81
MO descriptions
  [B$_6$H$_6$]$^{2-}$ to [B$_6$]$^{2-}$ 283
  [H$_3$]$^+$ 21, 27
  [HeH]$^+$ 5
  [I$_3$]$^-$ 13
  [Re$_2$Cl$_8$]$^{2-}$ 16
  [Ru$_6$C(CO)$_{18}$]$^{2+}$ 260
  Al$_6$ 251
  B$_2$ 5
  B$_2$H$_6$ 18, 20
  BF$_3$ 26
  bond energy 331
  C$_2$ 16
  CO 8, 9
  Cr(CO)$_6$ 13, 15
  H$_2$ 330
  H$_2$ model 2, 210, 214
  H$_3$BNH$_3$ 12
  H$_n$ oligomers 211
  I$_2$ 13
  LiH 4
  N$_2$ 8, 332
  O$_2$ 8, 332
  P$_3$H$_5$ 11
  R$_4$E$_4$ 37
  S$_8$
  SF$_4$ 27
  SF$_6$ 28
  XeF$_2$ 27, 28
MO concepts 2
  basis functions 5
  bonding–antibonding 2
  closed shell principle 22
  coordination compounds 13
  Dewar model 16
  diatomic molecules 330
  donor–acceptor complex 12
  Fe–CO bond 10
  fragment analysis 264
  frontier orbitals 9
  ground state 3
  HOMO 9

HOMO–LUMO gap 15, 23, 24, 26
ionization 3
Jahn–Teller first order instability 23
Jahn–Teller second order instability 23
Jahn–Teller stability 22, 221
Koopmans' theorem 3, 331
LUMO 9
MO vs. CO parameters 235
Mulliken populations 5, 9
one-electron model 7
polyatomic molecules 10, 332
protonated double-bond model 20
quadruple bond 16
symmetry 15
united atom model 330
Walsh correlation 10, 221
π complex 16
σ aromaticity 72, 75

nanofibers 316
nanoparticles 129, 206, 252
  water soluble 131
nanowheels 71, 72
networks
  $(Me_4N)_2[Mn(salen)]_2[Nb_6Cl_{12}(CN)_6]$ 308
  $Fe_4[Fe(CN)_6]_3 \cdot xH_2O$ 305
nitrides
  NbN 262
nucleophilic 340

oxides
  $La_2Mo_{16}O_{28}$ 295
  $Nb_2Mo_{16}O_{28}$ 295
  TiO_2 289
  TiO_2 structure 289
  TiO_2 electronic structure 290
  VO_2 291

p-block–d-block clusters 165
  $[Br_6Fe_6S_6]^{3-}$ 196
  $[Ni_{12}As_{21}]^{3-}$ 188
  $[Rh_9(CO)_{21}P]^{2-}$ 201
  bare 177
  cluster internal atoms 187
  $Co_3(CO)_9Bi$ 201
  $Co_3(CO)_9GeFe(CO)_2Cp$ 176
  $Co_3(CO)_9SiCo(CO)_4$ 175
  $Fe_3(CO)_{12}(CH)As$ 201
  $Fe_3(CO)_9C_2BH_3$ 201
  $HRu_3Fe(CO)_{12}N$ 201
  M_4E_2 175
  M_4E_4 189
  M_5E 175
  $Re_8In_4(CO)_{32}$ 203
Peierls distortion 275, 276
periodic table 325
photodiodes 309
photoelectron spectra 3, 10
poly(hexenyldecaborane) 316
polyacetylene 224

polyhedranes 74
polyketone 277
porous materials 305
Prussian blue 197, 305

single-walled nanotubes (SWNT) 281
solvents
  $(ImCl)/AlCl_3$ 313
  molten salt 313
solid-state compounds
  $[Sb_7Li_3 \cdot HNMe_3]$ 308
  Co_9S_8 296
  $Cs_2Mo_{12}Se_{14}$ 293
  GaS 304
  GdB_4 structure 285
  graphene 246, 281
  III/VI semiconductor 303
  $K_2Mo_9S_{11}$ 294
  metal properties 324
  $M^I_x[Zr_6(Z)Cl_{12}]Cl_n$ 312
  NbC arrangement 260
  $PbMo_6S_8$ electron count 292
  $PbMo_6S_8$ structure 292
  $Rb_2Mo_{12}Se_{14}$ 293
  $Rb_3Mo_{15}Se_{17}$ 293
  $Tl_2Mo_6Se_8$ 293
solid-state systems
  BN chain 233
  Bravais lattice 242
  C chain 225
  carbyne 228
  glide plane 225
  heteroatomic chains B_5C_5 274, 276
  heteroatomic chains B_5C_8 275
  heteroatomic chains BC, infinite linear 275
  heteroatomic chains BC, infinite zigzag 275
  linear $(H_2)_N$ 217
  metal chains 229
  Miller indices 250
  Ni chain 229
  periodicity 210
  polymorphism 241
  reciprocal unit cell 236
  repeat unit size 219
  screw axis 225
  semiconductor 223
  sphalerite structure 253
  steps to band structure 235
  translational vector 224
  unit cell 239
  unit cell cubic close packed 240
  unit cell hexagonal close packed 240
stability 323
superconductors
  K_3C_{60} 280
  Rb_2CsC_{60} 281
supraicosahedral clusters 59
surface properties
  array of Lewis bases on Lewis acidic surface 253
  band structure, square B-atom array 253
  d-block metal surface states 248

surface properties (*cont.*)
  dangling bonds 40, 248
  defects 250
  polarization 249
  reconstruction 206
  square N_2 surface array 253
  states 247
surface vs. bulk atoms 205
surface vs. bulk properties 205

thin films 250
tripledecker complexes 153, 183
  [Cp*$_2$Co$_2$(Et$_2$C$_2$B$_3$H$_3$)] 159
  CpM(C$_3$B$_2$H$_5$)M'Cp 158

united atom model 2, 11

valence shell electron pair repulsion (VSEPR) model 338
van der Waals radii 325

Werner complexes 343, 347

X-ray photoelectron spectroscopy 328

Zintl ion clusters 66, 177, 180, 303, 308, 312
  [(CO)$_3$MSn$_9$]$^{4-}$ 178
  [(Ph)(Ph$_2$Sb)-Ge$_9$]$^{2-}$ 310
  [(Ph$_2$Bi)$_2$Ge$_9$]$^{2-}$ 310
  [(Ph$_2$Sb)$_2$Ge$_9$]$^{2-}$ 310
  [(PhSb)$_2$-(Ge$_9$)$_2$]$_2$ 310
  [As$_7$]$^{3-}$ 80
  [B$_5$H$_8$]$_2$ 310
  [Bi$_9$]$^{5+}$ 70
  [E$_7$]$^{3-}$ 177
  [Ge$_4$]$^{4-}$ 66
  [Ge$_4$]$^{4-}$ 310
  [Ge$_5$]$^{2-}$ 67
  [Ge$_9$]$^{4-}$ 67
  [Ge$_9$]$^{4-}$ 310
  [Ge$_9$=Ge$_9$=Ge$_9$]$^{6-}$ 311
  [Ge$_9$=Ge$_9$=Ge$_9$]$^{8-}$ 311
  [Ge$_9$-Ge$_9$]$^{6-}$ 310, 312
  [In$_{11}$]$^{7-}$ 71
  [Sb$_7$]$^{3-}$ 308
  [Sb$_7$Ni$_3$(CO)$_3$]$^{3-}$ 203
  [Sn$_9$]$^{4-}$ 70
  [TlSn$_9$]$^{3-}$ 67
Zintl–Klemm concept 263